WORSHIP AND THEOLOGY
IN ENGLAND

WORSHIP AND THEOLOGY IN ENGLAND

WORSHIP
AND THEOLOGY IN
ENGLAND

FROM CRANMER TO HOOKER

1534-1603

BY HORTON DAVIES

PRINCETON, NEW JERSEY
PRINCETON UNIVERSITY PRESS
1970

Publication of this book has been aided by
the Whitney Darrow Publication Reserve Fund
of Princeton University Press.

HORTON DAVIES is currently the Henry
W. Putnam Professor of the History of Christianity at Princeton University. He was born
in Wales, became an honors graduate in both
Arts and Divinity at Edinburgh University,
and obtained his Doctorate of Philosophy at
Oxford University. Mr. Davies has served successively as minister of an English Congregational Church in London, Dean of the Faculty
of Divinity at Rhodes University in the Union
of South Africa, and Head of the Department
of Church History at Mansfield and Regent's
Park Colleges at Oxford University. In 1970
he was awarded the Doctorate of Letters of
Oxford University for the first three volumes
of this series to be published.

This book was composed in Linotype Monticello

Printed in the United States of America

by Princeton University Press, Princeton, New Jersey

ACKNOWLEDGMENTS

A scholar's literary indebtedness is properly acknowledged in the Bibliography. The other debts that he has incurred should not be recorded so impersonally. It is a great pleasure to thank Dr. James Thorpe, the Director, and the Trustees of the Huntington Library and Museum for the award of a grant-in-aid which enabled me to spend the spring of 1968 in the most congenial setting of San Marino, California. I am also grateful to Mrs. Noelle Jackson of the same city for the care with which she typed my manuscript. It is also my privilege to thank the members of the Princeton University Research Committee on the Humanities and Social Sciences for two summer travel grants that made it possible for me to research in the British Museum and the Bodleian Library.

It is with pride that I thank my son, Hugh Marlais Davies of the class of 1970 at Princeton University, for his help in drawing up the Bibliography and the three Indices, and for the understanding of art and architecture that went into chapter X of this volume. I cannot begin to thank my wife, Brenda, for all her concern and encouragement.

H. D.
Christmas, 1969

CONTENTS

    1. Pope or Prince: which was to be the Church's earthly
ruler? 2. Scripture as the Primary Category for Doctrinal
Authority. 3. The Primitive Church as an Important Sec-
ondary Norm. 4. Justification by Faith Alone, and its
Consequences. 5. The Redefinition of the Church. 6.
Church Order and the Sacraments. 7. The Liturgical
Consequences of Theological Change.

    1. The Puritans. 2. The Strategies of Puritanism. 3.
Anglican and Puritan Concepts of Scripture. 4. Theories
of Human Nature. 5. Predestination: God's Rule and the
Triumph of the Elect. 6. Understandings of the Church.
7. Views of the Sacraments. 8. Concepts of Ethics and
Eschatology. 9. Two Christian Styles. 10. Theology's
Consequences for Worship.

    1. Four Eucharistic Theories. 2. The Catholic Case:
Fisher, Tunstall, and Gardiner. 3. The Teaching of the
Early English Reformers. 4. The Sacramental Theology
of Cranmer's Advisors. 5. Cranmer's Eucharistic Theology. 6. Eucharistic Doctrine after Cranmer.

PART TWO: THE LITURGICAL ALTERNATIVES

CONTENTS

PART THREE: LITURGICAL ARTS AND AIDS

ILLUSTRATIONS

Preceding page 349

1. THOMAS CRANMER, 1489-1556, ARCHBISHOP OF CANTERBURY, ANGLICAN MARTYR

   The portrait is by Gerlach Flicke in 1546. Reproduced by permission of the National Portrait Gallery, London.

2. THOMAS MORE, 1478-1535, LORD CHANCELLOR OF ENGLAND, CATHOLIC SAINT

   Portrait by Holbein, 1527. More was canonised by Pius XI in 1935. Reproduced by permission of the Frick Collection, New York City.

3. JOHN JEWEL, 1522-1571, BISHOP OF SALISBURY, ANGLICAN APOLOGIST

   This panel painting is by an unknown artist. Reproduced by permission of the National Portrait Gallery, London.

4. WILLIAM PERKINS, 1558-1602, PURITAN DIVINE

   This portrait of the distinguished Cambridge tutor, preacher, and theologian is from Henry Holland's *Herωologia Anglica* (1620), p. 218. Reproduced, as are plates 7-11, by permission of the Henry E. Huntington Library and Art Gallery, San Marino, California.

5. EDMUND CAMPION, 1540-1581, JESUIT MARTYR

   An important member of the Jesuit mission to England, Campion was racked and executed at Tyburn on December 1, 1581. He was beatified by Leo XIII in 1886.

6. RICHARD HOOKER, 1554-1600, ANGLICAN APOLOGIST

   An engraving by Faithorne of Hooker's monument in Bishopsbourne Church, where Hooker was rector.

7. THE PROTESTANT SIGNIFICANCE OF THE REIGN OF EDWARD VI

   An intriguing symbolic illustration from the folio edition of John Foxe's *Acts and Monuments* (1573), vol. 2, p. 1,294. Features of the Protestantism of Edward's reign are the primary authority of the Bible over against tradition, the impor-

xi

tance of the two Dominical sacraments of Baptism and the Lord's Supper, the eager listening to the preached word of God, the preference for Renaissance to Gothic architecture, and the removal of the "idolatrous gear" of Catholicism.

8. THE MARTYRDOMS OF LATIMER AND RIDLEY IN OXFORD

This event occurred October 16, 1556. The reproduction also shows Cranmer observing the martyrdoms of the English bishops from his prison (upper right). The insensitive aptness of the preacher's text should be noted. The illustration is taken from the folio edition of Foxe's *Acts and Monuments*, vol. 2, p. 1,770.

9. THE MARTYRDOM OF ARCHBISHOP THOMAS CRANMER, MARCH 21, 1556

This illustration from Foxe's *Acts and Monuments*, vol. 2, p. 1,888, depicts the archbishop thrusting into the fire the hand which, previously through fear, had subscribed Catholic articles of belief.

10. INFANT BAPTISM IN THE ELIZABETHAN CHURCH

An illustration from John Daye's highly popular *A Booke of Christian Prayers* (1578 edition), a collection for the use of households and private individuals.

11. HOLY COMMUNION IN THE ELIZABETHAN CHURCH

This illustration from Daye's *Booke of Christian Prayers* shows the shape of the Protestant Communion cup, intended to communicate many parishioners, as contrasted with the smaller Catholic chalice, which is included with such impedimenta of "idolatry" as the pax, a monstrance, a crucifix, a rosary, and a bishop's crozier, shown at the upper left. The bread, it should be noted, is a small loaf, not in the least reminiscent of a Catholic wafer.

WORSHIP AND THEOLOGY
IN ENGLAND

INTRODUCTION

THE AIM of this series of five volumes, *Worship and Theology in England*, is to give an account of two aspects of the Christian life, worship and the theology that undergirds it, from the Reformation to the present. Worship, however, is understood in no narrow or rubrical sense, but as the corporate offering of thought, emotion, and decision-making as a response of the Christian churches to the divine saga of Christ and His followers throughout history. This series, then, is a study of the art of Christian adoration as expressed through the ritual and ceremonial of the major Christian denominations in England in the past 400 years. It is an adoration expressed in prayers and preaching, in sermons as in sacraments, in religious architecture and sacred music, in devotion and in duty. I have concentrated particularly on three traditions: the oldest, the Roman Catholic; the new Anglican; and the newest, the Puritan, the two latter originating in the sixteenth century.

This volume bears the subtitle *From Cranmer to Hooker, 1534-1603*, thereby indicating that the lion's share of the interest is devoted to the origin and early development of worship of the Church of England, which remarkably combines the characters of Catholic and Protestant worship. The first name in the subtitle is that of the virtual founder of the Church of England, Thomas Cranmer, the first non-Catholic Archbishop of Canterbury. He is— like Calvin—a scholar who was reluctantly pulled into the maelstrom of a religio-political revolution. Though not so great a theologian, his liturgical work, the Book of Common Prayer, has endured longer and become the chief channel of the doctrine and devotion of the Anglican Communion for four centuries. The second name in the title is that of Richard Hooker, notable defender of the Church of England against the attacks of the Puritans, as his mentor, Bishop Jewel, had been of the same church against the attacks of the Roman Catholic controversialists. Hooker gloried in a church freed of the superstition of Rome and the scrupulosity and intensity of Geneva, affirming that Anglicanism's first loyalty was to Scripture, its second to the pure traditions of the primitive and undivided church, and its third to reason. The opening date of this volume is that of the breaking away of the English church from the international Roman Catholic church affirming allegiance to the Pope, into an autonomous national religious community. The closing date, three years after the death of Richard Hooker, is that of Queen Elizabeth's death, and is intended to recall that she, the "Supreme Governor" of the Church of England had been the stabilizer of both church and nation during her long rule. Without her there could have been no "Elizabethan Settlement."

The organisation of this volume reflects in the first part, which is concerned with the history and theology of the period, the impassioned partisanship of the period. This was a time when a man gave no quarter in upholding his particular form of Christian allegiance, unless, of course, he tamely submitted to the principle of *cuius regio eius religio* and adopted the religious tenets of his prince. But with the rapid shifts of religion dependent on each successive change of sovereign in England, not only a time-serving parson but even a moderately cautious minister had to become a veritable Vicar of Bray. In Tudor England conscience could cost a man his consciousness.

Thus the first three chapters demonstrate that it was a continuing struggle to affirm Anglican doctrine against the contentions of the Catholics on the right and those of the Puritans on the left.

Chapter I shows the struggle between the Catholic theologians who affirmed the adamantine strength of a tradition of faith that was a thousand years old in England against a puny upstart babe, while they accused the *nouveaux pauvres* Anglicans of heresy in doctrine and schism in organization. On the other side, the Anglican theologians, under the leadership of Bishop John Jewel of Salisbury, retorted that it was not Rome but Canterbury which had refused to depart from the norms of the Bible and of the Primitive church; hence Catholics were the horrid innovators.

Chapter II reviews the various facets of the Puritan argument that the Church of England was a miserable compromise, "a leaden mean," a mere halfway house from Rome, in retaining so many vestiges of an idolatrous church in its ritual and ceremonial, as well as in its discipline and polity, and in demanding that Canterbury become as Protestant in its devotions as it was in its doctrine. The Puritan apologetic is traced from the Vestiarian to the Admonitions Controversies, with its unwearying demand for the purgation of impure practices.

Chapter III, on the Eucharistic Controversy closes Part One. I attempt to show that the apparent triviality of the acrimonious discussion of the mode of Christ's presence in the sacrament of Holy Communion did in fact lead to a significant watershed of opinion, not only between Catholic and Protestant, but among Protestants. One is indeed tempted to agree with the position of A. L. Rowse, that "No one who does not know the literature of the time would believe the whole libraries written for and against Transubstantiation, for and against the presence of Christ in the sacrament of the Eucharist, the precise nature of the presence, whether a sacrifice, or merely commemorative, or not."[1] Yet however much one may lament the bitterness of the disputants, he cannot dismiss it as a superficial and unremunerative waste of intellectual energy. In the nature of things it makes a vast difference to one's faith and practice whether he believes Transubstantiation to be the explanation of an amazing miracle or the crassest of crude superstitions masked by scholasticism. It will also have many practical consequences, including the relative strengths of faith and reason, the importance of Communion compared with the Sermon, and the relative significance of the priestly compared with the prophetic role of the ministry. Nor should it be forgotten, as Luther's *Table Talk* so often reminds us, that Luther believed that the Gospel he had rediscovered had overthrown the Papacy, Monasticism, and the Mass in the Reformation.[2] It is only the humanist or the sceptic who dismisses the Eucharistic Controversy in the sixteenth century as a semantic squabble, much as Gibbon derided the debate in the Council of Nicea as an argument over a diphthong! We are concerned in this chapter to show that philosophical differences underlay eucharistic theories, that there were four intelligible theories of the presence, and that these theories led to differences of spiritual formation and outlook.

Part Two describes, from the original documents and wherever possible in the original spelling, the four alternative types of Christian worship available in Tudor England, under the heading "Liturgical Alternatives."

In Chapter IV I review Roman Catholic worship from two perspectives. The first considers the weaknesses of Catholic worship at the end of the Middle Ages and of attempts made to rectify and standardise it during the sessions of the Council of Trent, and rescues from undeserved oblivion the contribution made by the Theatine Bishop of Asaph, Thomas Goldwell. The second perspective gives a view of the desperate difficulties of the Recusants who

---

[1] *The England of Elizabeth*, p. 434.
[2] Luther, *Works, Table Talk*, ed. and tr. Theodore G. Tappert, p. 134.

after the Bull of 1570 were forced to choose between Catholicism and patriotism and between celebrating clandestine masses in remote country houses in England in ever present fear of the agents of the king, or go into exile. There are fascinating glimpses of the stratagems of a spiritual resistance movement.

Chapters V and VI review the development, respectively, of Anglican worship and preaching, although Anglican sacramental thought and practice receives attention in Chapter III. Chapter IV concentrates on the novelty and the character of the successive revisions of the English Book of Common Prayer, especially those in 1549 and 1552. Considerable attention is devoted to the Prayer Book of 1549 to determine whether its theology is crypto-Catholic, Lutheran, Calvinist, or Zwinglian in tinge; an assessment is made of the strengths of the worship of the "bridge-Church," which has had so profound an impact on the worship of English-speaking countries and even beyond the Anglican Communion. This vernacular, participatory, responsive, Biblical, and sacramental liturgy, has also given a significant place to the proclamation of the oracles of God in preaching. This is assayed in Chapter VI, which deals chiefly with a comparison of the pulpit contributions of the racy, humorous, and popular Latimer with those of the meditative and oratorical Hooker.

Chapters VII and VIII are the complement of the Anglican chapters. They are devoted, respectively, to the worship and preaching of the Puritans. In Chapter VII I discuss the Puritan critique of the Book of Common Prayer, the long controversy over whether the most sincere and Spirit-led prayer uses or rejects prayer-books, and considers the major Puritan ordinances of worship. Chapter VIII elucidates the distinctive Puritan development of the sermon according to the structure of doctrine, reason, and use, and discusses the characteristic Puritan concern for developing a psychology of spirituality.

Chapter IX presents the most radical, iconoclastic, and charismatic alternative of worship, that of the various groups of Separatists, whether Barrowist, Brownist, or Anabaptist, the forebears of the later Free church tradition in worship in England.

Chapters IV-IX attempt to provide a full and objective descriptive and evaluative account of four of the major liturgical options open to Englishmen in the sixteenth century.

Chapters X-XII form Part Three, "Liturgical Arts and Aides." Chapter X concentrates on religious architecture and art, showing the Anglican preference for Gothic as illustrating its intention to give a sense of continuity from the old Catholic to the new Anglican Church in England, while demonstrating significant adaptations of its inherited fanes for a worship that used the eye less and the ear more than Catholic worship had, which was yet more intellectual and more congregational in character. Nonetheless it seems and is symbolically more bare and austere than Catholic worship. An important question propounded in this and the following chapter is: how far has Anglican church art moved in a secular direction? The elaboration of the tombs of the small gentry in the parish churches as status symbols and the reduced role of the choirs in Anglican parish churches might suggest that secularism had moved far. But then, as Chapter XI on religious music reminds us, this was the great age of English composers, when Tallis and Byrd were at their zenith, and also when metrical psalmody ran through the masses like a musical measles, so contagious was its popularity.

Chapter XII is a kind of summary and climax of this book. It takes the shape of a comparative analysis of the forms of spirituality approved and used by Catholicism, Anglicanism, and Puritan-Protestantism. It aims to show how the different theological emphases and modes of the formation of the spiritual life led to varied types of spirituality which, however much they had in common, were distinctive and authentic. I argue that the Catholic type of spirituality was essentially a spirituality for the "religious," the professionals, whereas Protestant spirituality was characteristically that of the laity in their "callings" in the secular world. It is also argued that Catholic spirituality concentrates on imitating the imitators of God—the saints—while Protestant spirituality concentrated less on the interior reproduction of the historic "Christ-event" in the individual than on the responsive love of neighbour for Christ's sake. In a word, it is the devotional as ethical. Unquestionably each type of spirituality, however different its rationale (and sixteenth-century Anglicanism is closer to Protestant than to Catholic piety, although seventeenth-century Anglican piety in Lancelot Andrewes, Laud, and Cosin approximates Catholicism), produces saints remarkably alike, whether we think of Sir Thomas More or of a Protestant martyr like John Bradford, both of whom left behind them noble prayers on their journey to God.

PART ONE
HISTORICAL AND THEOLOGICAL

CHAPTER I

CATHOLICS AND PROTESTANTS
IN CONTROVERSY (1534-1568)

IN THE COURSE of the sixteenth-century Christians in England experienced the confusion of five different changes of the established religion, all (to be sure) versions of the Christian faith and life, but each dissimilar and each demanding an absolute obedience to the royal change of faith.[1] In the early years of Henry VIII's reign Henry was an orthodox Catholic, proudly rejoicing in the title of *Fidei Defensor* bestowed on him by a grateful Pope for his defense of the seven Catholic sacraments[2] against Luther's iconoclastic reduction of them to three (subsequently to two). England at the time was as Catholic as France or Spain.

Once Henry found the Pope unwilling to declare his marriage with Catherine of Aragon null and void so that he might produce a male heir from another wife, and declared himself to be "Supreme Head" of the Church in England, England's faith changed. This non-Papal and nationalistic Catholicism veered from a humanistic and liberal (even Erasmian) form to the rigid conservative mode of the religion of "The Six Articles." From the Roman Catholic standpoint England was in schism. But apart from the denial of the Papal supremacy there was no trace of heresy, and the doctrine and ceremonies of England were much nearer to Catholicism than to the detested Lutheranism.

The third change came with the accession of the adolescent Edward VI who was surrounded by Protestant protectors, Somerset and then Northumberland. The religious thrust of the reign was increasingly Protestant. This can be seen in the more radical character of the 1552 Book of Common Prayer compared with the first version of 1549, in the profusion of Scripture translations, in the virtualist or memorialist doctrines of the Holy Communion current in this period, in the official encouragement of the proclamation of Protestant doctrine from the pulpits and in the Articles, and

---

[1] Sir Maurice Powicke writes: "The one definite thing which can be said about the Reformation in England is that it was an act of State." *The Reformation in England*, p. 1.

[2] *Assertio Septem Sacramentorum Adversus Martin. Lutherum.* Henry was probably assisted by Fisher, More, and Lee. Cf. E. G. Rupp, *Studies in the Making of the English Protestant Tradition*, p. 90.

in the permissions for ministers to marry. Never had England been more Protestant, whether in ceremonies, its admiration and delight in preaching, or its Communion doctrine. Certainly no one had ever scrupled the pomp and circumstance of episcopal dress more than John Hooper, bishop-designate of Gloucester who was sent to prison to cool off because he protested that he would not make his oath of allegiance in episcopal vestments. No greater exemplar of Protestant preaching on the bench of bishops than Latimer could be found, and it was Nicholas Ridley, Bishop of London, who insisted that in his diocese all altars had to be replaced with tables, because for him the Communion was the commemoration of a sacrifice, not its re-presentation in the Mass.

The fourth "about turn" came with the accession of Catherine of Aragon's daughter, the sanguinary Mary, who tried to turn the clock back to pre-Reformation times with the full restoration of orthodox Papal Catholicism.

The fifth and most enduring change came with the accession of Elizabeth in 1558, and brought an attempt to make the Church of England as comprehensive as possible. This form of Protestantism, combined Calvinistic doctrine with a modified Catholicism in worship and in church order. It was quite deliberately a *via media*, hoping to gain the allegiance of Englishmen, neither stiffly Papistical nor incorrigibly Puritan.

These five changes in the established religion of England offered, on one hand, either a Catholicism of the conservative (Marian) or liberal (Henrician) kind, or, on the other, Protestantism of the conservative (Elizabethan) or liberal (Edwardian) kind.

In the sixteenth century there were two major religious debates and controversies. The first, to which this chapter is devoted, was between the Catholics and the Protestants. The second, to which the energies of men in Elizabeth's reign were devoted, was between Anglicans and Puritans. Later still, there will be brief consideration of the debate between the patient Puritans and the impatient Puritans (or Separatists). In the present chapter the concern is exclusively with the grand debate between the English Catholics and Protestants as they developed their apologetic and their distinctive theologies, searching desperately for decisive authorities in doctrine and ethics. It was these theologies in controversy that would ultimately be the determinants of their differing forms of worship, which, in turn, through capturing the affections and

4

aspirations of the people would determine their religious ideals and their moral actions.

There is one great obstacle to defining the differences between Catholic and Protestant theologies in England, as distinguished from attempting the same task on the European Continent. It is that the earliest shapers of English Protestantism were surreptitious proclaimers of the truths they affirmed in closed gatherings of friends and supporters or through the medium of illicit books; for, with the exception of Edward VI, the Tudor sovereigns were most reluctant Protestants. Indeed, it has been wittily said that in England the most distinctively Protestant doctrine was that of the divine right of kings! Certainly there were no great theological reformers in England of the stature of Luther or Calvin or Zwingli, even though in Edward's time Bucer ornamented the University of Cambridge and Peter Martyr Vermigli the University of Oxford; but both were imports, while both Luther and Zwingli were German and Swiss, respectively. Protestantism, except for the brief reign of Edward VI, was a *religio illicita* in England until 1558, as Christianity was for so many centuries in the Roman empire. It has had its translators of the Bible, notably Tyndale and Coverdale, and its martyrs for the faith to whom Foxe's *Book of Martyrs* bore valiant testimony for many generations. But its theologians were rather apologists for the religion of an establishment than the constructors of theological systems, which, like Luther's, were hammered out through courageous experience, or like Calvin's, through a reconstruction of the testimonies of the Bible with an admirably clear and cogent architectonic. John Jewel was a good and Hooker an excellent apologist, but neither was an original or creative theologian, though I will make considerable use of the writings of each. Erastianism discourages originality, particularly so ferocious and dangerous a form of it as Henry's.[3]

1. *Pope or Prince:*
Which Was to be the Church's earthly ruler?

The very first issue to be raised was a largely political and practical one: was the Pope, as faithful Catholics maintained, or the Christian prince, as the Protestants affirmed, to be the head of the church? If the church was thought of as an international

[3] See A. G. Dickens, *The English Reformation*, p. 184: "If this curious situation tended to prohibit constructive thought, it did not prevent the promulgation of a series of commonsense, practical reforms."

body, transcending the continents and the centuries, then unquestionably in the medieval West the Pope was the rightful head. If there was, with growing nationalism, a desire to conceive of national churches in the plural, then ancient precedents might be found in the Israelite people of God ruled by their godly sovereigns or in the autocephalous and national orthodox churches of the East. The separation of the English church from the spiritual dominion of the Papacy was essentially a political act of the sovereign; there was hardly a less religiously motivated Reformation than Henry VIII's English one, even if the German Reformation was also partly motivated by politics as Luther's shrewd tract, *An den christlichen Adel deutscher Nation* of 1520, demonstrates.

Luther, with insight, had seen in this "Appeal to the Christian Nobility of the German Nation" that the Papal tyranny was buttressed by three Papal claims: that spiritual power was superior to temporal power; that the Pope alone could interpret Scripture; and that he alone had the right to convene a general council of the church. These were the three walls that effectively prevented reformation of the Church of God. The first claim negated the right of the Christian magistracy to effect reformation either in his country or by means of calling a general council, as Constantine had done in convening the great Christological Council of Nicaea. The second prevented any appeal to the Scripture, so that all claims that the church of the present might be renovated by a return to the apostolic and primitive pattern, unless seconded by the Pope himself, were immediately declared to be heresy. Luther might well have preferred to control the rate of the Reformation in Germany through a council of bishops, but so few of them turned to the Reformed faith that the only alternative was to turn to the lay political leaders, the dukes. This appeal Luther made to the magistrates in virtue of the priesthood of all believers. It was the only way to break the tyranny of the Papacy and the clericalism of the Roman Catholic church, even if there was always the possibility that the temporal sovereign might be a worse tyrant than the Pope.

It would be entirely wrong to suggest that the Reformation in England was, as Hilaire Belloc has termed it, the 'English Accident.' For apart from the king's interest in securing his second-generation throne by male issue and his rapacity for the property of the church as was evident in the dissolution of the monasteries, there were two forces moving towards reformation, or at least spiritual renovation, in England. One was the secret brotherhood of Lollards[4] who longed to see the "dominion of grace" overwhelming the legal, institutional, and all too wordly external carapace of the church, of whom, since they were covert companies, we know very little. The other force was the band of Cambridge scholars who gathered from time to time in the White Horse Inn and who must have included Erasmians as well as Lutherans; from them the greatest Protestant episcopal leadership was to come in King Edward's reign. The outright supporters of Reformation were forced, like Tyndale, Coverdale, Joye and Roy, to live on the Continent and to propagate their views in writing, publishing them often on secret presses and having them smuggled into England from about 1520 to 1535.[5] If the Protestants espoused their doctrines in England more openly or more carelessly than that, they were burned at the stake.[6] Unquestionably, however, there were Protestants in England committed to a national Reformation while the king was still a "defender of the faith" of Rome, and for whom his later turning from Rome was not quick or thoroughgoing enough.

Clearly the first task on the Protestant side was to establish the right of the Christian prince to act in the Pope's stead. Tyndale, the great translator, produced *The Obedience of a Christian Man* in 1527. There was in the back of his mind the common Catholic accusation that one recent effect of the preaching of the gospel was the sheer anarchy of the Peasants' Revolt in Germany in 1525. It was therefore a rather stiff dose of political authoritarianism that he prescribed. "The King is in the room of God," he wrote, "and his law is God's law, and nothing but the law of nature and natural equity which God graved in the hearts of men."[7] The King's power was so absolute, according to Tyndale, that governors of nations "are even the gift of God, whether they be good or bad. And whatsoever is done to us by them, that doth God, be it good or bad."[8]

---

[4] A. G. Dickens, *Lollards and Protestants in the Diocese of York, 1509-1558*, pp. 243-46; and J. E. Oxley, *The Reformation in Essex to the death of Mary*, pp. 3-15.

[5] William A. Clebsch, *England's Earliest Protestants*, pp. 10f., writes of 1520-1535 as, "that initial and most difficult time," when "the fountain of faith was a banned Bible, and when the gospel rediscovered by Luther rallied Englishmen to martyrdom."

[6] Rupp, *English Protestant Tradition*, pp. 196f.

[7] *Doctrinal Treatises*, p. 240.

[8] Such a doctrine hardly served the needs of the Protestants under Catholic Queen Mary; Ponet's treatise *Of Politick Power* recognized the possibility of temporal as well as spiritual tyranny and its right to be overthrown. See Winthrop S. Hudson, *John Ponet (1516?-1556), advocate of limited Monarchy*.

In the early days there was hardly time to formulate (even if the king had allowed it) a doctrine of the powers and duties of the Christian prince. There was, instead, an assertion of the king's right to wield ecclesiastical power than any statement of his duty to lead the church toward a spiritual reformation. In the expression of the divine right of the sovereign there was nothing to choose between the "Henricians," whether they were Foxe and Cranmer on one side or Tunstal and Gardiner on the other. Indeed, all would have subscribed to the strong defense of the royal supremacy given in Gardiner's classic, *De Vera Obedientia* (1536). Gardiner saw the king as the head of a Christian nation and inevitably and naturally as the head of the Christian church within that kingdom: "Seinge the churche of Englande consisteth of the same sortes of people at this daye, that are comprised in this worde realme of whom the kinge his [*sic*] called the headde, shall he not being called the headde of the realme of Englande be also the headde of the same men when they are named the churche of Englande. . . . They [*sic*] kinge (saye they [the Catholics]) is the headde of the realme but not of the churche: whereas notwithstonding ye churche of Englāde is nothing elles but ye cōgregatiō of men and women of the clergie and of the laytie in Christe's profession, that is to say, it is justly to be called the churche because it is a communion of Christen people."[9] The finest expression of the interdependence of ruler and people in a Christian commonwealth is to be found in the opening paragraph of the *Homily of Obedience*, prescribed for the Edwardian Church in 1547 and republished in Elizabeth's reign: "Almighty God hath created and appointed all things in heaven and earth and water in a most excellent and perfect order. In heaven he hath appointed distinct and several orders of archangels and angels. In earth he hath assigned kings, princes, and other governors, under them in all good and necessary order . . . every degree of people in their vocation, calling and office . . . [and] hath appointed unto them their duty and order: some are in high degree, some in low, some kings and princes, some inferiors and subjects, priests and laymen, masters and servants, fathers and children, husbands and wives, rich and poor: and every one hath need of other: so that in all things is to be lauded and praised the

[9] Sigs. D verso—Dii recto et verso. Bishop Gardiner was a great supporter of the King's spiritual rights in Henry VIII's reign, but when the political tables were turned under Elizabeth, his chaplain, Thomas Harding, was bitterly critical of a temporal prince's usurpation of the spiritual role of Pope or bishop. See *A Confutation of a Booke intituled An Apologie of the Church of England*, pp. 317 recto, 320 recto, 299 verso.

goodly order of God without which no house, no city, no common-wealth can endure or last."[10] From this excerpt it can be seen that the whole interdependence of society was thought to depend on the subordination of the people to the prince, both politically and reli-giously. Moreover, in a society still predominantly Christian (unlike the secular world of the present day), it could be assumed that the king would act upon Christian motives (or, at least, motives mixed with Christianity) and according to Christian doc-trinal, liturgical, and ethical standards. The sovereign was seen as the protector of the social order, a dike against the chaos that would ensue if "degree" and "dependency" were overthrown.

Two complementary concepts controlled the political thinking of the Erastian Henricians; both were due to the Erastian tradition and to the renewed interest in legal studies that had "brought Con-stantine and Byzantine notions of the powers of Christian princes and the claim of civil law"[11] to their attention. The first concept was that the Christian prince has sacred obligations, since he rules by the providence and under the law of God, with responsibility for God's people. The counterpart of this concept was that the Christian prince has Christian subjects, pledged to honor the authority of God in their prince. Thus a claim of law becomes an interior claim of conscience, because the "Powers that be are ordained of God."[12]

Assuming that there was a general agreement among those who had broken with the Papacy, that a Christian prince was the appro-priate instrument to undertake the reformation of the church, there was an equally strong negative conviction that the Popes had so far departed from their evangelical office as not to be in any true sense the successors of Peter. Such was the continuing jeremiad of the Protestant pamphleteers. This contention in the controversy with Rome was effectively if crudely propagated with woodcuts or etchings contrasting the meek Christ riding on a lowly ass into Jerusalem with the proud and pompous Pope donning his triple crown as he was carried in state in the gestatorial chair borne by Italian noblemen.

This view is set forth at length, but in a much more dignified fashion, by John Jewel, Elizabethan bishop of Salisbury and author of *An Apology or Answer in Defence of the Church of England* (1564). Here is Jewel in his most ironical vein, asking:

---

10 Certain Sermons . . . [Homilies], ed. Corrie, p. 104.
11 P. M. Dawley, *John Whitgift and the English Reformation*, p. 18.
12 Rom. 13:1.

What one thing (tell me) had Peter ever like unto the Pope, or the Pope like unto Peter? Except peradventure they will say thus: that Peter, when he was at Rome never taught the Gospel, never fed the flock, took away the keys of the kingdom of heaven, hid the treasures of his Lord, sat him down only in his castle in St. John Lateran and pointed out with his finger all the places of purgatory and kinds of punishments, committing some poor souls to be tormented and other some again suddenly releasing thence at his own pleasure, taking money for so doing; or that he gave order to say private Masses in every corner; or that he mumbled up the holy service with a low voice and in an unknown language; or that he hanged up the sacrament in every temple and on every altar and carried the same about before him, whithersoever he went, upon an ambling jennet, with lights and bells; or that, he consecrated with his holy breath oil, wax, wool, bells, chalices, churches, and altars; or that he sold jubilees, graces, liberties, advowsons, preventions, first fruits, palls, the wearing of palls, bulls, indulgences, and pardons; or that he calleth himself by the name of the head of the church, the highest bishop, bishop of bishops, alone most holy; or that by usurping he took upon himself the right and authority over other folk's churches; or that he exempted himself from the power of any civil government; or that he maintained wars, set princes together at variance, or that he, sitting in his chair, with his triple crown, full of labels, with sumptuous and Persian-like gorgeousness, with his royal scepter, with his diadem of gold, and glittering with stones, was carried about, not upon palfrey, but upon the shoulders of noblemen.[13]

Then Jewel argues that the true Petrine succession would be seen if the Pope went from country to country and from house to house preaching the Gospel, if he proved to be "the watchman of the house of Israel," and if he "doth not feed his own self, but the flock." Furthermore, Jewel wanted to see Peter's authority as ministerial, not judicial, and his acts on an equality with those of all other bishops instead of lording it over them. But the onus of the accusation was the departure from the simplicity and spirituality of a preeminent servant of God by one who, amid the panoply and

[13] Jewel, *An Apology of the Church of England*, ed. John E. Booty, pp. 130-31. See also Booty's *John Jewel as Apologist of the Church of England*; and W. M. Southgate, *John Jewel and the Problem of Doctrinal Authority*.

10

sheer worldliness of Renaissance Rome, could call himself *servus servorum Dei* only in the bitterest irony. Rather was he *princeps principorum*, as if he had claimed the very throne of Christ as *rex regum*, who was supposed to be only his vicar.

2. *Scripture as the Primary Category for Doctrinal Authority*

It was settled, then, that the Christian prince, presumably advised by his theologians, was the true instrument for the reformation and disciplining of the national church. But a larger question remained: Which criteria should the reformers use as their patterns for the renovation of the church? Were they to follow Scripture, or tradition, or reason, or the guidance of the Holy Spirit? Were they to follow one or more of these criteria, and which, in case several were to be used, was to have priority? Difficult as the question of criteria was, it would be even more difficult if one questioned these weasel words. Since "Scripture" contained two testaments and could be found in several, often tendentious, translations, and according to some Christians included the Apocrypha as a secondary authority; since there were different levels of authority in Scripture; and further, since theologians might and did use a different controlling concept or congeries of concepts in interpreting Scripture, it was hard to judge what was meant by accepting the primary authority of Scripture. "Tradition" was also a very elastic concept, because there were believed to be unwritten traditions that Christ had passed on secretly to his apostles between his resurrection and ascension, and Western Roman Catholics and Eastern Orthodox, while agreeing substantially on the authority of the Ecumenical Councils, accepted other councils or refused to accept them. The Anglicans appealed to the undivided and primitive church, and specifically to the *traditio quinquesecularis*, that is, to the tradition of the first five centuries of the church. In the eyes of the Roman Catholics, as the Council of Trent formulated the dogma, Scripture and tradition were to have coordinate authority; but for the Anglicans, Scripture had absolute primacy. When it came to the Puritans, Scripture was the almost exclusive criterion, not only of faith but of worship, church government, and ethics; the only meaning they could give to tradition, apart from apostasy, was the pure and primitive tradition of the apostolic church as reflected in the pages of the New Testament. Reason, which played a large part in the Anglican apologetics of Hooker and his successors, was to have little importance for Puritanism.

11

The Puritans were to give an absolute and almost exclusive role to the Scriptures, but there was a subordinate stress on the role of the Holy Spirit as providing the *testimonium internum*, or inner witness, to the truth of Christ. With the Society of Friends the Spirit was to become primary and the Scriptures merely a check on the integrity and authenticity of Christian experience. In moving from the settled Catholic recognition of the coordinate authority of Scripture and tradition, and the placement of ultimate power in the hands of the Pope in disputed issues, Protestants were opening a Pandora's box.

In the early decades of the sixteenth century, however, it was clear that opinion was that only the authority of the word of God in the Holy Scriptures was sufficient for the reform and rebuilding of the church. There was a constant appeal to the word of God as the only standard authoritative enough to purge the corruptions of the church. This had been inherent in the appeal of Erasmus to the newly translated New Testament. It was not until 1535 that Coverdale published the first complete translation of the Scriptures in English, which was published not in England but on the Continent. Moreover, the translation appeared 19 long years after Erasmus had finished work on his translation in England. The adoption of the Bible as the instrument and criterion of the Reformation in England was slow. Even though the Convocation petitioned Henry VIII in 1534 that the entire Bible be translated into English, it was only in 1537 that a revised English translation, known as Matthew's Bible, appeared; a generally approved translation, heavily dependent on Coverdale, though not mentioning his name, appeared in 1539 and, with a preface by Thomas Cranmer, in 1540. The latter was "the Bible with the largest volume in English," which the Injunctions of 1538 had ordered to be set up in every church in England.

Cranmer's typically vivid, clear, and modest preface revealed his deep appreciation for the Bible "in the vulgar tongue." He warned those who would refuse it (presumably the conservative Catholics) and those who would abuse it by disputatiousness (presumably the argumentative, radical Protestants). How new a vernacular translation was in England can be seen by the curiously medieval comparison Cranmer used to indicate the surpassing value of the Bible, for he described it as "the worde of God, the moost preciouse Juell and moost holy relyque that remayneth upon earth."[14] He

14 Preface to *The Great Bible* (1540), p. 4.

further claimed that "yt is as necessary for the lyfe of mans soule, as for the body to breath."[15] Cranmer was constantly aware that the purpose of studying the Bible was not for men to gain a reputation for learning and become masters of theology, but for God to master them. He therefore concluded that every man who comes to read the Bible should bring with him the fear of God, "and then next a fyrme and stable purpose to refourme his owne selfe," and to be a good example to others by his life, "whych is sure the moost lyvelye and effecteous fourme and maner of teachynge."[16]

It would have been impossible to have attacked the international power and authority of the Roman Catholic church, with its immense prestige and immemorial traditions, unless it was believed that the Scriptures were the very word and expression of the will of God. It should not be forgotten that the Reformation in every land was a Reformation, "according to the pure Word of God" to give its full title as its rationale; the "purity" of the Word of God was always contrasted with the impurity of the traditions of men. Cranmer would not have dared to affirm the doctrines of the reformed faith on his own authority as a scholar. The ultimate authority was his own careful reflection on Scripture, as he indicates in his controversy with Stephen Gardiner on the mode of Christ's presence in the Holy Communion: "I, having exercised myself in the study of Scripture and divinity from my youth (whereof I give most hearty lauds and thanks to God), have learned now to go alone. . . ."[17] Archbishop Cranmer was almost certainly the author of the preface to the first prayer-book of Edward VI (1549), which declared that the lections were so arranged, "that all the whole Bible (or the greatest parte thereof) should be read over once in the year, intendyng thereby, that the Cleargie, and specially suche as were Ministers of the congregacion, should (by often readyng and meditacion of Gods worde) be stirred up to godlines themselfes, and be more able also to exhorte other by wholsome doctrine, and to confute them that were adversaries to the trueth."[18] There was another aim in this order, namely, "that the people (by daily hearyng of holy scripture read in the Churche) should continuallye profite more and more in the

15 *Ibid.*, p. 5.
16 *Ibid.*, pp. 6-7.
17 *Works of Thomas Cranmer*, 2 vols., I, 224.
18 Cranmer, *The First and Second Prayer Books of Edward VI*, ed. E.C.S. Gibson, p. 3.

knowledge of God, and bee the more inflamed with the love of his true religion."[19]

Bishop Jewel, the first Elizabethan apologist, placed great stress on the primacy of God's Word in Scripture as the basic criterion for doctrine and life. The Scriptures of the Old and New Testaments were for him, "the heavenly voices whereby God hath opened unto us his will"; they alone quieted the otherwise restless heart of man, and they contained all that was necessary for salvation. However, their supreme value for him as an apologist of the Church of England was "that they be the foundations of the prophets and apostles whereon is built the church of God; that they be the very sure and infallible rule whereby may be tried whether the church doth stagger and err and whereunto all ecclesiastical doctrines ought to be called to account; and that against these Scriptures neither law, nor ordinance, nor any custom ought to be heard; no, though Paul himself, nor an angel from heaven, should come and teach the contrary."[20] Jewel had sounded the authentic Protestant note of the absolute primacy of Scripture.

It is only when this is understood that we can sympathise with the vast endeavours in the Protestant fold to train ministers to expound the Scriptures in the original languages of Hebrew and Greek, with the assistance of the learned commentaries both patristic and contemporary, and, above all, the unwearying insistence from Latimer onwards that the soul of man would perish without the bread of the word of God delivered in preaching—for, in Cranmer's words, "in the Scryptures be the fatte pastures of the soule."[21] Even the very sacraments were meaningless unless the people learned what promises God had attached to these signs of the Gospel.[22]

Latimer, the greatest popular preacher of the age, declared, "this office of preaching is the only ordinary way that God hath appointed to save us all thereby."[23] Moreover, his memorable sermon "Of the Plough" was a brilliant application of the Parable of the Sower, with its theme of God's Word as the seed, the congregation the field, and the preacher the ploughman: "He hath first a busy work to bring his parishioners to be a right faith . . . and then to confirm them in the same faith; now casting them down with the law and with threatenings of God for sin; now ridging

---

[19] Ibid., p. 3.   [20] Jewel, Apology, p. 30.
[21] Preface to The Great Bible, p. 3.   [22] See Tyndale, Works, I, pp. 274f.
[23] Latimer, Works, I, p. 306. The sermon was preached in London on January 18, 1548.

them up again with the Gospel and with the promises of God's favour; now weeding them by telling them their faults and making them forsake sin; now clotting them by breaking their stony hearts and by making them supple-hearted, and making them to have hearts of flesh, that is, soft hearts and apt for doctrine to enter in; now teaching . . . now exhorting . . . so that they have a continual work to do."[24]

Edmund Grindal, Parker's successor as Elizabeth's Archbishop of Canterbury, was planted firmly in the same tradition. He faced the wrath of Elizabeth and subsequent sequestration rather than fulfill her command to prohibit "prophesyings" or gatherings at which local ministers could improve the exposition and application of Scripture by conference. In his defense of the refusal he told the queen that homilies were no substitute for preaching: "The godly preacher is termed in the Gospel *fidelis servus et prudens, qui novit famulitio Domini cibum demensum dare in tempore*, who can apply his speech according to the diversity of the times, places and hearers, which cannot be done in homilies; exhortations, reprehensions, and persuasions are uttered with more affection, to the moving of the hearers, in sermons than in homilies."[25]

In the task of reforming the Church of England it was agreed that it was the task of the Christian prince to undertake the reforming, and that the standard should be according to the Scriptures, the Pure Word of God which alone had sufficient authority, as the expression of the divine will, to destroy the vain traditions of men.

3. The Primitive Church as an Important Secondary Norm

But were all the traditions of the early church during the period when it was most pure and primitive to be jettisoned and abandoned? By no means, was the answer of the Anglican divines, as from Cranmer on there emerged a singularly fine tradition of patristic scholarship, itself fed by the conviction that the Tudor church had much to learn from the first six centuries of Christian history before the establishment of the Roman primacy of the medieval type. There seem to have been two basic reasons for the acceptance of the early church's guidance as an authority second only to Scripture. Since a private man might read his idiosyncrasies into Scripture and use a text as a pretext, there was a need for a rule where the Scripture was obscure or did not speak with unam-

[24] *Ibid.*, pp. 59ff.
[25] Edmund Grindal, *Remains*, ed. W. Nicholson, pp. 382f. The letter is dated December 20, 1576.

biguous voice. This very point was at the heart of the Catholic polemic urging the dangerous subjectivity to which the church was prone by the publication of the Scriptures in the vernacular. John Jewel expresses this argument for the Catholics as he reports it. The Catholics had called the Scriptures "a bare letter, uncertain, unprofitable, dumb, killing, and dead"; they added "a similitude not very agreeable, how the Scriptures be like to a nose of wax or a shipman's hose; how they may be fashioned and plied all manner of ways and serve all men's turns."[26] The second ground for seeking the approval of the apostles and fathers of the early church was the desire to return beyond corruption to the first five centuries of Christian history, where the foundation of the primitive church was in the Scriptures, and its explication in the four general councils and the writings of the early Church Fathers.

This linking of Scripture and the primitive church was central to the establishment of religion in England. The chime rang as clearly in the apologists as it did in the official theological formulas and liturgies of the Anglican church. It is reechoed in Jewel's *Apology* at the outset, where he observes what his strategy in controversy will be: "Further, if we do show it plain that God's Holy Gospel, the ancient bishops, and the primitive church do make on our side, and that we have not without just cause left these men [the Roman Catholics], and rather have returned to the apostles and old catholic fathers."[27] When he reached the conclusion, Jewel argued that the Church of England was no innovator, but a renovator, "and we are come, as near as we possibly could, to the church of the apostles and of the old catholic bishops and fathers, which church we know hath hitherto been sound and perfect and, Tertullian termeth it, a pure virgin, spotted as yet with no idolatry nor with any foul or shameful fault; and have directed according to their customs and ordinances not only our doctrine but also the sacraments and the form of common prayer."[28] Similarly the preface to the first Book of Common Prayer appeals to the "auncient fathers" whose order it is attempting to restore; even when Cranmer seems most to be moving away from tradition, as in the publication of the Great Bible in English, he is careful to insist that the Saxon forefathers of the English had their Saxon translations, and his very motives and considerations for reading the vernacular Scriptures are drawn from John Chrysostom and Gregory Nazian-

26 Jewel, *Apology*, p. 77.   27 *Ibid.*, p. 17.
28 *Ibid.*, pp. 120-21.

zus. Not only was there useful guidance to be obtained from the undivided church of the first five or six centuries, but the more attentive to its teaching the Church of England was, the less could it be accused of what was the most original form of original sin for the traditionalists, namely, the sin and scandal of innovation. The study of the Fathers was a proof that her firm intention was *renovation*.

Cranmer, Jewel, and Whitgift exemplify the great patristic erudition of the leaders of the Anglican church in Tudor times.[29] Even so, it was neither a blind nor servile deference to the authority of the Fathers. Latimer, for instance, while respecting the judgment of the Fathers, refused to enslave himself to them: "These doctors, we have great cause to thank God for them, but I would not have them always to be allowed. They have handled many points of our faith very gladly, and we have a great stay in them in many things; we might not well lack them; but yet I would not have men sworn to them, and so addict as to take hand over head whatsoever they say."[30] The vigorous and independent Thomas Becon, Cranmer's chaplain, was of a similar opinion. If the Fathers swerved from the doctrine of Christ, he did not in the least care how venerable, ancient, or saintly they might be. "When Christ saith in the gospel, 'I am the truth,' He said not, 'I am the custom.' "[31] In short, the Anglican theologians were true to the Fathers as long as the latter followed Scripture as their primary authority. Through their patristic scholarship Anglicans claimed to show that what Rome regarded as Anglican heresies were in reality primitive orthodoxies.

4. *Justification by Faith Alone, and its Consequences*

It was clear that the Church of England, in separating from the great and historic Roman Catholic church, would have to articulate its own ecclesiology and provide its own definitions of the scope and authority of the ministry and the nature of the sacraments. These will be considered shortly. But it was in their understanding of soteriology and, in particular, the mode of the appropriation of salvation that Anglicans differed the most from Catholicism. It was the doctrine of Justification by Faith, especially in

29 Cranmer, *Remains*, pp. 77-78; and H. F. Woodhouse, *The Doctrine of the Church in Anglican Theology (1547-1603)*, p. 22.
30 Latimer, *Sermons*, 2 vols., I, 218.
31 T. Becon, *Prayers and Other Pieces*, p. 390.

its Lutheran form of *sola fide* (by faith alone) that challenged the entire Catholic system.

This was the primary doctrine the great German Reformer had learned from the Epistles of St. Paul to the Romans and the Galatians, a doctrine that was essential to his rediscovery of the Gospel. The Henrician formularies[32] show a slow change from a combination of justification by faith combined with the predispositions of contrition and charity[33] toward a closer approximation to Luther's doctrine in the fourth part of *The Institution of a Christian Man*.[34] The most unqualified expressions of the doctrine, however, are to be found in those writers who had renewed the experience of Luther, who knew both the imponderable gravity of sin and the extreme costliness of grace, and who were appalled by the "cheap grace"[35] which the Roman Catholic church was then offering through such indulgence hucksters as Tetzel, or through the much more respectable austerities by which it was thought possible to purchase at least part of one's own salvation on the installment plan. The word of liberation, as it came to Luther and those like him, was that salvation had already been won by Christ; it had only to be accepted in faith, it had no longer to be achieved. Such a Protestant saw, as Gordon Rupp has aptly summed it up, "that the relation between God and Man in Jesus Christ, of which the divine side is Grace and the human Faith, was the ground of Christian experience from beginning to end."[36] With such a conviction he was delivered from the bondage of guilt and anxiety, condemnation and fear, and translated from the kingdom of darkness into the light, love, and liberty of the sons of God. Rome thought such a salvation was too easily and conveniently obtained, and suspected, quite understandably, that the doctrine of justification by faith alone masked an invitation to Anti-nomianism. The Roman church also saw its sacramental system, especially the sacrament of penance; its doctrines about the future life, especially

[32] Conveniently collected in Charles Lloyd, ed., *Formularies of Faith* (Oxford, 1825).

[33] *Ibid.*, pp. xvi-xxvii, 16.

[34] *Ibid.*, pp. 209f.: "Item, that sinners attain this justification by contrition and faith, joined with charity, after such sort and manner as is before mentioned and declared in the sacrament of penance. Not as though our contrition or faith, or any works proceeding thereof, can worthily merit or deserve to attain the said justification. For the only mercy and grace of the Father, promised freely unto us for his Son's sake Jesu Christ, and the merits of his blood and passion, be the only sufficient and worthy causes thereof."

[35] A modern phrase of Dietrich Bonhoeffer's for a phenomenon as old as man.

[36] Rupp, *English Protestant Tradition*, p. 164.

18

purgatory; its carefully articulated canon law; its judicious balance of the threatenings of law with the promises of the Gospel; and its whole system of spirituality in the monasteries and its casuistical ethics threatened by the rude hands of iconoclasts who would tear down in a few weeks the work of dedicated men through a millennium. Justification by faith was indeed an article for the standing or falling of the church.[37]

It is not always easy to recognise this fact because of the apparent similarity of the terminology of the disputants on each side. It is essential, however, to recognise that the Continental Reformers and the English Protestants after them made a fundamental distinction between Justification as God's forgiving acquittal of sinful man, for which they reserved the term, and Sanctification which is its consequence, by which they intended the inner regeneration of the Holy Spirit; for the latter the Catholics reserved the term Justification. The Protestant doctrine of Justification "by faith alone" depends on the distinction that it makes between Justification and Sanctification in order to safeguard the fact that its Gospel is all the operation of God's free and unearned grace. Otherwise, it was felt that man will attempt to argue that his own works or predispositions are preconditions for the reception of grace. One of the earliest English Protestants and martyrs, Barnes, affirms both objectively, that justification is the work of God in Christ on the cross, and subjectively, that it is appropriated by faith: "The Scripture doth say that Faith alone justifieth because that it is that alonely whereby I do hang of Christ, and by faith alonely am I partaker of the merits and mercy purchased by Christ's blood."[38] There was no Anti-nomianism in Barnes' exposition of the doctrine, for he saw its true consequence as, "finally of a fleshly beast it maketh me a spiritual man: of a damnable child it maketh me a heavenly son; of a servant of the devil it maketh me a free man of God."[39]

One of the clearest expressions of the doctrine is found in Cranmer's *Homily of Salvation*, in which he describes Justification as a triple work: "Upon God's part, his great mercy and grace:

[37] The Catholic teaching is aptly stated by Edmund Campion (see [A. Nowell], *A true report of the Disputation or rather private Conference had in the Tower of London, with Ed. Campion Iesuite, the last of August, 1581 . . . Whereunto is ioyned also a true report of the other three dayes conference had there with the same Iesuit . . .* published 1583, sig. F.f.1.): "Fayth onely as it is a good worke, ioyned with hope and charitie, doeth iustifie."

[38] Barnes, *Works*, ed. Foxe, p. 241, cited in Rupp, *English Protestant Tradition*, p. 167.

[39] Barnes, *Works*, p. 235.

upon Christ's part, justice, that is the satisfaction of God's justice or price of our redemption, by the offering of his body and shedding of his blood in the fulfilling of the law perfectly and thoroughly: and upon our part, true and lively faith in the merits of Jesus Christ, which is not yet ours, but God's working in us."[40] Cranmer had clearly appropriated Luther's distinction between faith as an intellectual assent to doctrine, and as *fiducia*, or staking one's whole life in trust. This is what he meant by "lively faith," which he defined as "a sure trust and confidence of the mercy of God through our Lord Jesus Christ and a steadfast hope of all good things to be received at God's hand."[41]

While a man's salvation owed everything to Christ, yet he was afterwards expected to live as a sanctified being. It was not that good works justified a man before God, but rather that the justified man did good works because he was justified. Tyndale insisted, "Faith only maketh a man safe, good and righteous and the friend of God, yea and the son and heir of God and of all his goodness, and possesseth us with the Spirit of God."[42]

One of the most radical conclusions derived from the acceptance of the sole mediatorship of Christ the Justifier and Redeemer was the great reduction in the status and role of the Virgin Mary and the saints. The diminution of the invocation of the saints in Henry VIII's reign can be charted with almost mathematical precision. The traditional list of saints' names in the Litany of the Sarum use included 58 saints. When Bishop Hilsey published his *Manual of Prayers or the Prymer in English and Laten* in 1539 the number was reduced to 38, while Cranmer's English Litany[43] of May 27, 1544 abandoned the long catalogue of saints formerly invoked, replacing it with a comprehensive petition addressed to patriarchs, prophets, apostles, martyrs, confessors, virgins, and all the company of heaven.

Becon poked fun at the numerous mediators of the Catholic church who virtually reduced the humble laymen to living as practicing polytheists. He imagined them as saying: "If we fast the

---

[40] Cranmer, *Works*, II, p. 129.
[41] Cranmer, *Remains*, p. 135. See also Tyndale's distinction between a "historical" and a "lively" faith in *The Exposition of the Fyrste Epistle of Seynt Jhon* (1531), cap. II, verse 2, in *English Reformers*, Library of Christian Classics, ed. T.H.L. Parker, p. 113.
[42] *Wicked Mammon*, from *Doctrinal Treatises*, p. 50. See also Hooper, *Early Works*, p. 50.
[43] Its full title is *An exhortation unto prayer, thoughte mete by the Kynges maiestie and his clergy, to be read to the people in every church afore processyons. Also a Letanie with suffrages to be said or song in tyme of the said processyons.*

blessed saints' evens and worship them with a *Paternoster*, *Ave* and Creed, they will do for us whatever we ask. St. George will defend us in battle against our enemies. St. Barbara will keep us from thundering and lightning. St. Agasse [Agatha] will save our house from burning. St. Anthony will keep our swine. St. Luke will save our ox. St. Job will defend us from the pox. St. Gertrude will keep our house from mice and rats. St. Nicholas will preserve us from drowning. St. Loye will cure our horse. St. Dorothy will save our herbs and flowers. St. Sith [Osyth] will bring again whatever we lose. St. Apolline will heal the pain of our teeth. St. Sweetlad and St. Agnes will send us maids good husbands. St. Peter will let us in at heaven-gates—with a thousand such-like."[44]

Unquestionably the banning of the invocation of the saints in Reformed worship and the inevitable iconoclasm of requiring the images and shrines of the Virgin and the saints to be removed from all the churches must have left many vacant spaces in the affections of the simple believers, as if they had lost their friends, for God was the distant and awesome Creator and Christ was the Judge of Doomsday. Some echo of this loss undoubtedly lay behind various Catholic uprisings, from the Pilgrimage of Grace to the revolt of the Western counties. An echo of it remains even in the Homily prepared to discredit this lingering affection, that entitled *Against Peril of Idolatry*, which contrasts the pure Anglican worship and the more exciting Roman Catholic worship and reports: ". . . a women said to her neighbour, 'Alas, gossip, what shall we now do at church, since all the saints are taken away, since all the goodly sights we were wont to have are gone, since we cannot hear the like piping, singing, chanting, and playing upon the organs, that we could before?' "[45]

All this wealth of piety was condemned as superstition; the rich

---

[44] Becon, *Works*, 3 vols., II, 536. With this may be compared the following words of Jewel in his *Apology*, ed. Booty: "Neither have we any other mediator or intercessor, by whom we may have access to God the Father, than Jesus Christ, in whose name all things are obtained at his Father's hand. But it is a shameful part, and full of infidelity, that we see everywhere used in the churches of our adversaries, not only in that they will have innumerable sorts of mediators . . . so that, as Jeremiah saith, the saints be now 'as many in number, or rather above the number of the cities'; and poor men cannot tell to which saint it were best to turn them first; and though there be so many as cannot be told, yet every one of them hath his peculiar duty and office assigned unto him of these folks, what thing they ought to ask, and what to give, and what to bring to pass. But besides this also, in that they do not only wickedly but also shamelessly, call upon the Blessed Virgin, Christ's mother, to have her remember that she is a mother and to command her Son and to use a mother's authority over him." p. 38.

[45] Gardiner, *Certain Sermons*, p. 351.

art associated with it had to go. The reason was not that the Protestants despised art but that they loved religion more. The gravamen of their objection to the prayers and devotions addressed to the Virgin Mary and the saints was that these derogated from the sole mediatorship of Christ.

There were, of course, strong Catholic counterarguments. Catholics insisted that there was no worship of the saints, only the paying of homage to the saint and not to the image. The Protestant reply was that the saints would refuse to accept honour which was defrauding God of his due. If the Catholics argued that tradition gave the approval to bedecking the images of the saints as eminent servants of God, the Anglican reply was that the true honouring of the saints was to live in charity, giving generously and living in simplicity as they did. "No service of God, or service acceptable to him can be in honouring of dead images, but in succouring of the poor, the lively images of God."[46]

If the first consequence of the emphasis on Justification by faith alone in Christ reduced the status and importance of the Virgin and the saints, a second equally radical consequence was the downgrading of the importance of good works based on an emphasis that they played no part in the gaining of salvation. If Pelagianism was the danger into which Catholicism could easily fall with its emphasis on good works, the peril confronting Protestantism was Antinomianism. The former attributed too much to man and too little to God; the latter gave God all the glory in the initiation and accomplishment of salvation at the risk of making man too passive a recipient of God's gifts. The extraordinary paradox at the heart of Protestantism was that predestinarians should have been so ethically active. It was because no Protestant was to presume upon his election, and also because with the Reformation there came a new sense of vocation which Max Weber aptly called an intra-mundane asceticism. Previously the "religious" in the great medieval Catholic church were those who tried to obey the counsels of perfection and by withdrawal from the world into monasteries became the athletes of spirituality seeking perfection through chastity, poverty, and perfection.[47] Those who lived in the workaday world were

[46] Ibid., Homily "Of Alms Deeds," p. 269. Is any significance to be read into the fact that, on the dissolution of the monasteries which had provided many alms freely for the poor and with the new doctrine of justification by faith and not by good works, it was necessary to provide a three-part homily, "Of Alms Deeds," extending to 104 large pages in a modern edition?

[47] See the Anglican critique in the third part of the Homily, "Of Good Works," Gardiner, Certain Sermons, pp. 56-58.

doomed to live a second-class Christian life. With the onset of the Reformation, however, Luther had insisted that God could be served through one's secular vocation as farmer, teacher, house-wife, mother.[48] In fact, Luther had even insisted that the Christian's sense of indebtedness could never be paid directly to God, but could be paid only by regarding the neighbour as Christ. It was this double safeguard of living to serve God in the world through one's daily calling and by treating one's neighbour as a Christ, that prevented the main body of Protestants from falling into any pits of Anti-nomianism or passivity.[49]

The Church of England was careful in defining the relationship of faith to good works in order to safeguard the primacy of the former without minimizing the latter. The 11th of the 39 Articles insisted that the meritorious cause of Justification is the atoning work of Christ: "We are accounted righteous before God only for the merits of our Lord and Saviour Jesus Christ . . . and not for our own works." The instrumental cause of Justification was asserted in the same article to be faith: "We are accounted righteous . . . by faith. . . . Wherefore that we are justified by faith alone is a most wholesome doctrine, and very full of comfort."[50] It is significant that Article 11, "Of Good Works," was one of the four new articles added by Parker and that there was nothing corresponding to it in earlier versions of Anglican Articles. It states: "Albeit that good works, which are the fruits of faith, and follow after justification, cannot put away our sins, and endure the severity of God's judgment: yet are they pleasing and acceptable to God in Christ, and do spring out necessarily of a true and lively faith, in so much that by them, a lively faith may be as evidently known, as a tree discerned by the fruit." It is but another example of the balance that characterised the Elizabethan *via media*.

This radical doctrine of Justification by faith alone, the central theological tenet of the Reformation, discovered (or, as he believed, rediscovered in St. Paul) by Luther, was recovered by the English Protestants. It changed, as I have hinted, the very style of religious life in England in the most profound ways—theologically, liturgically, ethically, and aesthetically. It removed much anxious

[48] C. S. Lewis once remarked that the Reformation lowered the honors standard of the degree in spirituality and raised the passing level.

[49] See Luther's *Von der Freiheit eines Christenmenschen* (1520); and Karl Holl, *The Cultural Significance of the Reformation* (New York, 1959).

[50] The New Testament sources for the doctrine of justification are: Romans 3:28-30, 4:2-5, and Galatians 2:16, 3:5ff. But the Epistle of James 2:14-26 denies that man is justified by faith alone without good works.

23

scrupulosity, banishing, as it were, all thermometers for fevered hypochondriacs. Although a sinner, man was assured of his salvation by faith; *simul justus et peccator*. Out with *angst*. Man did not have to calculate his salvation and wonder how many more good deeds might ease his way through purgatory or would tip the scales of divine justice in his favour on the Day of Judgment. So the demand for more masses, more pilgrimages, more scourgings, more fastings, more penances, and more indulgences came to an abrupt end. The concentration on the victory of salvation achieved on the cross by Christ as the sole mediator between God and man reduced the role of all the other mediators from the Blessed Virgin to the great company of the saints, and banished most of them and their shrines from the churches and ultimately from the consciousness of the worshippers. Here too there would be a sense of relief, for how could one be sure in the older dispensation if one's prayers had been addressed to the correct celestial postal box number? Now it was enough to trust the overruling providence of God in the assurance that Christ, who had worn the vesture of humanity, thoroughly understood the human plight and was the incarnation of the love of God. (There would also be a deep sense of loss, for some of the local or national saints seemed closer to the devout, less awesome than the greater dignitaries of heaven.[51] There would certainly be a grievous aesthetic loss as the rood-screens with their representations of the crucified Savior and the watching Mother and St. John were pulled down together with all the images in stone and carved wood, or the representations of divine persons in the stained-glass windows.) The respect for the "second-milers"— the ascetical and mystical athletes—the monks and nuns and the celibate priesthood, and, indeed, for the entire hierarchical conception of the church, would be greatly reduced; and a married clergy would bring the parson and people closer in their understanding of family problems. The sacraments would be reduced in number (from seven to two) and in significance, and the pulpit would come into greater prominence, as would the role of teaching the common people the meaning of faith, requiring more schools and

[51] *The Bishops' Book*, which is "Henrician Catholic" in character, still preserves the sense of the saints all assisting the humble believer in his pilgrimage, as the following passage shows: "And I believe, that I being united and corporated as a living member into this catholic church . . . not only Christ himself, being head of this body, and the infinite treasure of all goodness, and all the holy saints and members of the same body, do and shall necessarily help me, love me, pray for me, care for me, weigh on my side, comfort me, and assist me in all my necessities here in this world." Charles Lloyd, ed., *Formularies of the Faith*, p. 58.

24

schoolmasters, and forcing the parish minister to catechise every Sunday afternoon. The liturgy would be radically revised so that nothing forbidden in the Bible would find a place in the worship; moreover, the biblical material would oust many of the saints' legends. As we shall see, even the shapes of the chalices would change. But the biggest change of all would come with the sense that the laity were not second-class citizens, spectators at the Mass, but participants in worship who by their daily conversation and callings were God's servants in the world; the godly father now became at family prayers in his home a father-in-God. This Protestant faith was simpler and more austere than Catholicism, but it was exciting for those who desired more responsibility in religion.

5. *The Redefinition of the Church*

England's severing of the bonds that linked it to the international Roman Catholic church required a radical revision of the doctrine of the church, as also of the dependent doctrines of the ministry and the sacraments. The Roman Catholic church defined itself in terms of institutional continuity, to distinguish itself from heretical and schismatical groups. St. Augustine, for example, stressed the institutional continuity and geographical spread of the church, as two of its marks, its apostolicity and catholicity, to distinguish Catholic Christians from the Donatist dissidents in North Africa. In doing so he was moving away from his earlier stress on the church as constituted by grace and faith. Now the Reformers returned to the earlier Augustine who had defined the church as the *coetus electorum*, the congregation or community of the elect, who had saving faith in Christ.[52] This was a mystical rather than an institutional doctrine of the church, stressing that divine initiation in salvation which was the primary concern of reformers all too conscious of the weaknesses where faith was too anthropocentric.

The great strength of the Catholic church had been its insistence on unity, for it was a unity achieved by charity that was the unmistakable sign of the authenticity of the church. The Protestants, however, believed that the Catholic church in the West had forfeited both its apostolicity and its holiness and that it had been invaded by worldliness. The Catholic church might, indeed, claim that it was the largest church and possibly the historically longest church, but the Protestants replied to the charge that they were schismatics and breakers of the unity of the church by stressing

52 Woodhouse, *Doctrine*, p. 57.

25

that this was the only way to return the church (or their part of the church) to holiness and to apostolic ways, that truth mattered more than tradition, and that charity could dwindle into indifference. Thus of the four traditional marks of the church—unity, holiness, apostolicity, and catholicity—it might be said that the Catholics stressed the first and fourth and the Protestants the second and third. The Catholic church denied that the Church of England was a true church, but the Anglicans admitted that the Catholic church was a true church.[53]

The Bishops' Book (1537) begins the interpretation of the ninth article of the Apostles' Creed by asserting that there is among the saints in heaven and the faithful people of God on earth "a company of the elect" which is the church and "mystical body of Christ" who is its "only head and governoor." The source of its holiness is the Holy Spirit, though it is a mixed community consisting of true wheat mingled with chaff. It affirms a belief that "particular churches, in what place of the world soever they be congregated, be the very parts, portions, or members of this catholic and universal church."[54] In obvious reference to the Roman church it is denied that "any one of them is head or sovereign over the other," while it is insisted that "the church of Rome is not, nor can worthily be called, the catholic church, but only a particular member thereof. . . ."[55] The "King's Book" (the popular name for the Necessary Doctrine and Erudition for any Christian Man) of 1543, defined "catholic" as "not being limited to any place or region of the world," because "God of his goodness calleth people, as afore, without exception of persons or privilege or place."[56] It affirms that all these differently governed churches are together "one holy catholic church." The unity of this church "is not conserved by the Bishop of Rome's authority or doctrine; but the unity of the catholic church, which all Christian men in this article do profess, is conserved and kept by the help and assistance of the Holy Spirit of God, in retaining and maintaining of such doctrine and profession of Christian faith, and true observance of the same, as is taught by the scripture and the doctrine apostolic."[57] The over-

---

[53] Richard Hooker affirmed in his sermon on Justification: "The best learned in our profession are of this judgment, that all the corruptions of the Church of Rome do not prove her to deny the Foundation directly; if they did, they should grant her simply to be no Christian Church." B. Hanbury, ed., The Ecclesiastical Polity and other Works of Richard Hooker, III, 409.

[54] Lloyd, Formularies, p. 55.   [55] Ibid.

[56] Ibid., p. 246.   [57] Ibid., p. 247.

riding concern expressed in this ecclesiology is to affirm that there is a spiritual unity of all Christians and that "this unity of the holy church of Christ is not divided by distance of place nor by diversity of traditions, diversely observed in divers churches, for good order of the same."[58] Clearly the Reformation had shattered the external framework of unity; all that could be gathered from the ruins was the assertion of a spiritual unity between national churches.

Following the examples of Wittenberg, Zurich, and Geneva, the more Protestant *Forty-Two Articles* had a functional definition of the Church: "The visible Churche of Christ is a congregation of faiethfull Menne, in the whiche the pure worde of God is preached, and the sacraments be duelie ministred, according to Christes ordinaunce, in all those thinges that of necessitie are requisite to the same."[59] This article remained unchanged in the Elizabethan *Thirty-Nine Articles*, probably because it was borrowed from the corresponding article in the Augsburg Confession.[60] The ensuing Elizabethan article—that is, the 20th—declares that the church has power "to decree Rites or Ceremonies, and authority in controversies of faith"[61] yet this is qualified by the flat declaration that the church may not "ordain anything that is contrary to God's word written," nor should it to enforce any unscriptural doctrines "to be believed for necessity of salvation." In 1563 a significant clause was added to Article 34, "Of the Traditions of the Church," which read: "Every particular or national Church hath authority to ordain, change, and abolish ceremonies or rites of the Church ordained only by man's authority, so that all things be done to edifying."

It was left to Hooker to develop the most impressive apologia for the Elizabethan Settlement. He argued that the powers of church and state were coordinate. This was stated in the famous definition: "We hold, that seeing there is not any man of the Church of England but the same man is also a member of the commonwealth; nor any man a member of the commonwealth, which is not also of the Church of England . . . so with us, that no person appertaining to the one can be denied to be also of the

---

[58] *Ibid.*, p. 246.

[59] Article xx; the *Forty-Two Articles* of 1553 are reprinted in Latin and English on pp. 70-89 of E.C.S. Gibson, *The Thirty-Nine Articles of the Church of England* (1898).

[60] This reads: "Est autem ecclesia congregatio sanctorum, in qua evangelium recte docetur, et recte administrantur sacramenta."

[61] This clause was added in 1563.

other."[62] These are not independent but interdependent societies ruled by one head, the Christian ruler of the realm. Indeed, Hooker maintained that they were two aspects of *one* society, "being termed a Commonweal as it liveth under whatsoever Form of secular Law and Regiment, a Church as it liveth under the spiritual Law of Christ."[63] In theory this is admirable; as we saw earlier, the insistence on the divine right of the king made for a great sense of national security under the Christian prince. But the king who controlled the church could also crush tender consciences and all but the most courageous prophets. What was civilly advantageous might be spiritually weakening. Thus it was that during Elizabeth's reign many who cared most about religion, the Catholic Recusants and the Puritans, were almost equally abhorrent to the equable and compliant Erastianism of the middle way.

In addition to the discussion of the marks of the church, which occupied so much of the time of the controversialists and which inevitably concentrated on the legal, institutional, and exterior fabric of the church, there was a recognition of its deeper aspects, namely, its personal and interpersonal dimensions. Anglican teaching on the church's nature neglected neither the Biblical nor the patristic understanding of the relationship of the church to Christ and of Christians to each other. Woodhouse has characterized this aspect of Anglican ecclesiology as "summed up in four words—incorporation, dependence, obedience, and representation."[64] Christ is the head and the church is his body; by this Pauline metaphor it was understood that the direction of the Christians is under the rule of Christ, and that the unity of the members of the church inheres in him. The members of the church further depend wholly for their lives on him, a teaching that is expressed in the Johannine metaphor of Christ as the vine and the members as the branches. The obedience required of Christians is that of servants to their divine master, or of soldiers to the captain of their salvation, the "obedience of faith." The church also consists of Christ's representatives, those who through their commitment are his witnesses or ambassadors, to use two other familiar New Testament metaphors. Enough has been said by now to show that the Anglican understanding of the church had devotional depth which prevented it from being merely a rationalisation of the political domination of a spiritual society. For the church was seen most profoundly as

[62] *Ecclesiastical Polity*, Book VIII, i, 2.
[63] *Ibid.*   [64] *Ibid.*, p. 31.

that organism through which Christ, by means of the Holy Spirit, perpetuated his work in England.

The claims of the nascent Church of England did not go unchallenged. The Roman Catholic apologists hammered away at what they considered to be the four weaknesses in the Anglican position: schism, heresy, divisiveness, and immorality.[65] Anglicans had fractured the unity of Christendom and lacked the charity that is the essential bond of Christians. Thomas Harding, the able Recusant controversialist, challenged Jewel with these words: "Ye have divided the Church of God, ye have rent our Lordes nette, ye have cut his wholewoven cote, which the wicked souldiers that crucified him, could not finde in their hartes to do."[66]

Even worse, from the Catholic standpoint the Anglicans had fallen away from the foundation of apostolic and Catholic doctrine, and were teaching heresy, such as the denial of the primacy of the Pope and of the doctrine of transubstantiation. The Jesuit Robert Persons gives as the first reason why orthodox Catholics might not attend Anglican worship: "because I perswading my selfe their doctrine to be false DOCTRINE, and consequently to be venomous doctrine, I may not venture my soule to bee infected with the same."[67]

The Catholic propagandists skillfully used the charge that Protestantism was a destructive and revolutionary force that divided kingdoms, freed men of their oaths of loyalty, and was followed by all who were lax and lewd. Persons, in his subtle Epistle Dedicatory to Queen Elizabeth, while affirming Catholic loyalty, contrived to show that she should be suspicious of the loyalty of her Protestant subjects, for it was Luther's teachings that had ignited the Peasants' Revolt of 1525; he glossed Calvin (in the *Institutes* III:9), as exempting the consciences of the faithful from the power of all men by reason of their liberty in Christ.[68]

The fourth charge was that Protestantism was an easy, convenient faith that attracted the immoral. All the pretence to return to

[65] See Booty, *Jewel*, p. xvii.

[66] *A Confutation of a Booke intituled An Apologie of the Churche of England*, p. 262 recto. See also Robert Persons, *A brief discourse containing certain reasons Why Catholikes refuse to goe to Church* (Douai, 1601), sigs. C8 verso, C11 recto.

[67] *Ibid.*, sig. B12 recto. See also Harding, *A Confutation of a Booke intituled An Apologie of the Church of England*. Harding compares the fountain of the water of life of Catholicism with the Anglican "leaking pittes, which holde no pure and holesome water, but myre and puddles, with the corruption whereof ye have poysoned many soules." See also pp. 7, 24 verso, 25 verso, 132, 133 verso.

[68] *Ibid.*, sigs. A8 recto and verso.

Biblical simplicity was only an excuse of libidinous libertarians to throw off the healthful discipline of the church, to neglect the grace of the sacraments, and to throw off those Christian standards which had been maintained for a millennium, to serve their own rapacity and lust. Harding posed the alternative ethical demands of Anglicanism and Protestantism: "Let any man of reason iudge, whether he that maketh his belly his God would not holde with that gospell which mainteineth only faith to iustifie, the keping of the commaundementes to be impossible, confession of sins not to be necessarie, stedfast trust in Christes passion to be the only sufficient waye to save al men, lyve they never so loosly and disorderly: rather then with the gospell, which preacheth no salvation to be without keping the whole lawe (so farre as man may kepe it), or rising again from synne by penaunce, that only faith to iustifie which worketh by charitie, that all mortal synnes under paine of damnation must be confessed to a priest if occasion suffer, that Christes passion is applied by meanes of the sacraments unto us, and not by confidence of our own phansies. . . ."[69]

The Anglicans were no less confident of their own superiority to the Roman Catholic church; their general charges against Rome were as vitriolic as Rome's against Anglicanism. They accused Catholics of diminishing the gravity of sin and therefore depreciating the glory of God and the generosity of divine grace, as well as overestimating the power of man's free will by their stress on good works. Their teaching of the Mass, which they regarded as a renewal of the sacrifice of the cross, was a denial of the original sufficiency of Christ's sacrifice, and their doctrine of Transubstantiation was a fable contrary to Scripture and common sense alike. The Church of Rome, from the Pope and Cardinals down, sought its own glory. Its doctrinal faults were as nothing compared to its glaring moral evils, which only proved that far from being the only true church of Christ, it was disobedient to the voice of Christ.[70]

6. *Church Order and the Sacraments*

In church order there seemed to be a close approximation between Rome and Canterbury, for the Church of England retained the threefold historic ministry of bishop, priest, and deacon, and the leaders of the church even inherited some of their former

[69] *Ibid.*, pp. 301 recto and verso. The charge is repeated on pp. 8, 34, 253 verso.
[70] Woodhouse, *Doctrine*, pp. 137-38.

political duties as the ecclesiastical agents of royalty. The same traditional character of the ministry was to be seen also in the retention of a liturgy, instead of giving the ministers the right to frame their own prayers within a set of general directions, as was the case in the Reformed churches of Switzerland.

Nonetheless, the difference between the respective understandings of the status and the functions of the sacred ministry was great. The Anglican ministers of the word and sacraments were not as hierarchically raised above the laity as their Roman Catholic counterparts, nor, of course, were they required to be celibates. Permission for ministers to marry meant that they shared to a considerable extent the same problems as the laity and that their rectories and vicarages became centers of hospitality and understanding. Furthermore, faced (at first in Elizabethan times) with an intransigent Presbyterian claim that the Genevan Church Order alone was the true Biblical pattern of church order, the Anglicans modestly affirmed that episcopacy was of the *bene esse* but not of the *esse* of the church. While they could have made strong historic claims for episcopacy, they did not presume to unchurch other Protestants because they lacked this form of the ministry. Nor did the Church of England in our period assert that the validity of the sacraments depended on the Apostolical succession of bishops. Norman Sykes has rightly observed that "the absence of any statement concerning the doctrinal significance of episcopacy is noteworthy."[71] The concept of Apostolicity in early Anglican theology was no "pipe-line" theory of the transmission of episcopal power and authority, but rather the concern to safeguard the fidelity of teaching in line with that of the apostles as recorded in the New Testament. The bishops were indeed regarded as the successors of the Apostles, but it was never affirmed that they held the powers of the apostles or that the church was as dependent on the bishops in the way it was dependent on the Apostles. Anglicanism in this age was content to stress the fact that Christ had given authority to the church and that the importance of the bishops was in maintaining order, charity, and true doctrine in the church, as well as officiating as Christ's representatives at ordinations and confirmations. Not until the last years of Elizabeth's reign was there an insistence that new ordinations must be at the hands of a bishop of the Church of England, and not until 1662 were all ministers

71 *The Church of England and Non-Episcopal Churches in the Sixteenth and Seventeenth Centuries*, p. 7.

of the Church of England required to be episcopally ordained.[72] In short, Apostolic faith alone was essential, but a particular type of church government was secondary to the faith; indeed, it existed to safeguard the faith. Quite typical in its alegalistic attitude was Hooker's affirmation, in relation to the two sacraments of baptism and the Lord's Supper, that Christ gave gifts to his church, but that what he did was far more important than how he did it.[73]

The primary tasks of the ministers of the Church of England were to preach the word and administer the sacraments.[74] As the new primary authority of the Anglican church was the Scriptures; as its concern was to reform both faith and morals by the Word of God, the preacher of that word had a very important role. While it would be simplistic to state that the Protestant word replaced the Catholic sacraments in the Church of England, because its genius was to combine Reformed and Catholic elements, there was a new interest in and concern for the fuller and more frequent proclamation of the word. Latimer was no exception in affirming, "the true ladder to bring a man to heaven is the knowledge and following of the scripture," rather than the Mass.[75] The high emphasis given to sermons by Anglicans and Puritans was, according to Christopher Hill, part of the Protestant intellectual concern, "to elevate teaching, discussion, the rational element in religion, generally, against the sacramental and memorial aspects."[76] Since all were required to attend the sermon and the service of worship, preaching became a major educational instrument from the time of Elizabeth on. Louis B. Wright has emphasized that this was a means of entertainment as well as an intellectual exercise. It was understood that the master of every household would catechise his children and servants about the sermon; several would be seen taking notes during the sermon. The most noteworthy sayings of the minister would be dutifully recorded in commonplace books for further consideration.[77] Unquestionably the spiritual formation of the Elizabethan people depended more on preaching and teaching than on the sacraments.

The number of the sacraments was reduced from seven to two,

[72] Woodhouse, *Doctrine*, p. 101. Cf. Hooker, *Ecclesiastical Polity*, III, ii, 2.
[73] *Ibid.*, V, lxvii, 3.
[74] Themes which will be treated fully in succeeding chapters, but which are briefly outlined here.
[75] Hugh Latimer, *Sermons*, p. 97.
[76] Christopher Hill, *Society and Puritanism*, p. 55.
[77] *Middle-Class Culture in Elizabethan England*, pp. 277-78. This also accounts for the many family and private devotional books published in this period.

though what the Catholics termed sacraments were often retained in modified form as rites, services, or ordinances. For example, Holy Orders was a Catholic sacrament, but Ordination was an important ceremony for the setting apart and authorising of the Anglican ministry. Confirmation was an important service, but was not defined as a sacrament; the same was true of Holy Matrimony. Even Extreme Unction was retained in the form of Communion of the Sick. For Penance alone there is no equivalent, except the shadow of it in the general Absolution in the prayer-book given after the General Confession of sins by the minister in Christ's name. The two major Gospel sacraments were retained—Baptism, the sacrament of initiation, and Holy Communion, or the Lord's Supper, the sacrament of spiritual nourishment.

Anglicans differed greatly from Catholics in their interpretation of the meaning of the Eucharist, and in the manner in which they celebrated it. The Anglican objections to the Roman Mass were comprehensively and tersely listed by John Jewel in his notable "Challenge Sermon," preached at St. Paul's Cross November 26, 1559 and again on March 31, 1560, which had been repeated at Court exactly a fortnight earlier. Jewel criticised: using Latin and not the vernacular, Communion in one kind, the teaching in the Canon on sacrifice, the adoration of the Sacrament, and private celebration.[78] Anglicans were eager to assert the "Real Presence" of Christ in the Eucharist, which conveyed the grace of God that it represented, but the *mode* of the presence is not contended for as being both too common a matter of disputation and too high a mystery for human intelligence.[79]

[78] *Works*, I, 9-16. For more scurrilous criticism see Bridges on transubstantiation in *A Sermon preached at Paules Crosse*, in which he affirms that the Catholics "turned Chryst out of his owne likenesse, and made him looke lyke a rounde cake, nothyng lyke to Iesus Christe, no more than an apple is lyke an oyster, nor so mutche, for there appereth neyther armes nor handes, feete nor legges, back nor belly, heade nor body of Chryst: but all is visoured and disguysed under the fourme of a wafer, as lyghte as a feather, as thinne as a paper, as whyte as a kerchiefe, as round as a trenchour, as flat as a pancake, as smal as a shilling, as tender as the Priestes lemman that made it, as muche taste as a stycke, and as deade as a dore nayle to looke upon. O blessed God, dare they thus disfigure our Lord and Saviour Iesus Christ?" (1571), p. 125.

For a fuller treatment of Catholic and Anglican Eucharistic doctrine see Chap. III.

[79] See Hooker, *Ecclesiastical Polity*, v, lxvii, 6: ". . . why do we vainly trouble ourselves with so fierce contentions, whether by consubstantiation or else by transubstantiation the sacrament be first possessed with Christ or no? — a thing which no way can either further or hinder us however it stand, because our participation of Christ in this sacrament dependeth on the co-operation of His omnipotent power which maketh it His body and blood to us, whether with change or without alteration of the elements such as they imagine, we need not greatly to care or inquire."

Only Baptism and Holy Communion are appointed by Christ as sacraments of his Gospel, with the appropriate sign of water for the cleansing of Baptism, and the appropriate signs of bread and wine for the body and blood of Christ in the Lord's Supper. Bradford declared: "There are two Sacraments in Christ's Church: one of initiation, that is, wherewith we be enrolled, as it were, into the household and family of God, which sacrament we call Baptism; the other wherewith we be conserved, fed, kept, and nourished to continue in the same family, which is called the Lord's Supper, or the body and blood of our Saviour Jesus Christ, broken for our sins and shed for our transgressions."[80] These are, however, in no sense bare signs, but communicating and "effectual signs of grace and God's will towards us, by the which he doth work invisibly in us," as the 25th of the 39 Articles said. As noted by Philip Edgcumbe Hughes, the efficacy has a double reference, both to the promise of the Gospel and to the faith of the believers who are the recipients of the sacrament.[81] Cranmer affirmed that the sacraments were not *signa nuda*, empty signs: "As the washing outwardly in water is not a vain token, but teacheth such a washing as God worketh inwardly in them that duly receive the same, so likewise is not the bread a vain token, but showeth and teacheth the godly receiver what God worketh in him by His almighty power secretly and invisibly. And therefore as the bread is outwardly eaten indeed in the Lord's Supper, so is the very body of Christ inwardly by faith eaten of all them that come thereto in such sort as they ought to do, which eating nourisheth them unto everlasting life."[82]

What seemed the greatest novelty was the mode of the celebration of Holy Communion. The service was in English, and the communicants received consecrated bread and wine. Former spectators at a mystery were conscious of participating in the Lord's Banquet. Some of the mystery may have been dissipated, along with the gain of intelligibility, simplicity, and sharing for the people.

7. The Liturgical Consequences of Theological Change

There remains only to be considered in outline the consequences of theological change on the nature of Anglican worship, taking

[80] John Bradford, *Works*, 2 vols., I, 82.
[81] See Hughes, *The Theology of the English Reformers*, an admirably systematic doctrinal anthology with careful critical introductions, pp. 158f.
[82] *Works*, I, 17.

the First Prayer Book of Edward VI as the norm.[83] There is little evidence that the king had replaced the Pope as the earthly head of the Church in England, except for the requirement that the second in the series of collects at Communion is one of two alternatives for the king, and that he is briefly mentioned in the Suffrages for matins and evensong. The most prominent anti-Papal utterance was the round condemnation in the Litany: "From all sedicion and privye conspiracie, from the tyrannye of the bishop of Rome, and his detestable enormities . . . Good lorde, deliver us."[84] The Erastianism of the prayer-book is also possibly reflected in the collect for peace which emphasizes twice a day that God is "author of peace and lover of concord."

There is, however, abundant evidence of the seriousness with which the primacy of the Scriptures was reflected in Anglican worship. It is most prominent in the elimination of non-Biblical lections and in the arrangement for the reading of the Old Testament through once a year as the first lesson at both matins and evensong, while the New Testament is appointed for the second lesson at matins and evensong, "and shal be red over orderly every yere thrise, besides the Epistles and Gospelles: except the Apocalips, out of the whiche there be onely certain Lessons appoynted upon diverse proper feastes." The Psalms were, of course, a very important feature of Anglican worship. The other place where the Scripture's importance was plainly evident was in the requirement at every Communion service for a sermon to be preached or a homily read. Since the people possessed copies of the prayer-books in which they could follow the services and anticipate each lesson, prayer, and ceremony, the very unpredictability of a sermon, as contrasted with a homily, must have made this the most exciting part of worship.

Justification by faith found its place among the homilies and in the Pauline epistles, all of which were included three times a year as lessons. The influence of the doctrine was, however, clear in the modifications introduced into Roman collects to make them Reformed in character. The standard ending for a collect was,

[83] While the First Prayer Book of Edward VI (1549) lasted for only three years, it yet had a profound influence on subsequent Anglicanism. The Second Prayer Book of Edward VI (1552) lasted for only a year, and was extremely Protestant in character.

[84] This and all subsequent citations in this chapter will be to the Edward Whitechurche (March 1549) edition in the Huntington Library copy. References will also be given, for convenience, to the Everyman edition, introduced by Bishop E.C.S. Gibson. The reference here is to p. 232.

"Through Jesus Christ Our Lord," thus stressing the mediator-ship of Christ. Clearly a collect for a saint's day was composed to suggest that he be imitated in his dependence on God's grace or in the following of Christ. In no case was there an invocation of the saint. The denial of the merits of the saints and the affirmation of faith in the sole merit and mediatorship of Christ can be seen plainly in the collect for Saint Mary Magdalene's Day: "MERCY-FUL father, geve us grace, that we never presume to synne thorough the example of any creature; but if it shall chaunce us at any time to offende thy divine maiestie; that then we may truely repent, and lament the same, after the example of Mary Magda-lene, and by lyvely faith obtaine remission of all our sinnes; through the onely merites of thy sonne our saviour Christ."[85] Moreover, there was a considerable reduction in the number of saints' days; almost all of those retained had some connection with the earthly life or Gospel of Christ. The emphasis was more on the Christological than the sanctoral cycle in the Anglican prayer-book.

The same desire to instruct the people in the Christian faith and give the Gospel of Christ priority, is certainly responsible for the prayer-book's being prepared in English, as was the notable Angli-can invention of providing in a single volume a book that priest and people could share. The Preface, almost certainly the work of Cranmer, shows how deliberately these features were considered; so that, "is ordeyned nothyng to be read, but the very pure worde of God, the holy scriptures, or that whiche is evidently grounded upon the same;[86] and that in suche a language and ordre, as is most easy and plain for the understandyng, bothe of the readers and hearers."[87]

This concern for edification (a Pauline term for building up in the faith),[88] combined with giving a greater responsibility to the laity because they were recognised as being an important part of the people of God and not merely those who were not the "reli-gious," must also account for the expectation that the laity would model their devotions on the prayer-book and that heads of house-

[85] P. 198.

[86] This evidently refers to the Biblically-based doctrines propounded in the various exhortations in the prayer-book, as, for example, in the office for holy communion.

[87] Particular exception is taken to the fact that, "the service in this Churche of England (these many yeares) hath been read in Latin to the people, whiche they understood not; so that they heard with theyr eares onely; and their hartes, spirite, and minde, have not been edified thereby." Cf. Article XXIV.

[88] The main references are: I Corinthians 14:5, 12, 26 and Ephesians 4:12, 16, 19.

holds would lead the devotions in their own homes and test their children and servants on how well they had understood and remembered the sermons preached in church. All this may have reflected a growing sense of "the priesthood of all believers"; it certainly reflected an understanding of the church that was less hierarchical than the Catholic and that made much less rigid the difference between the clergy and the laity. It should also be noted that in 1549, when Cranmer was replying to the West Country rebels who wanted among other things the restoration of the Latin rite, he wrote that "worship should be the act of the people and pertain to the people, as well as to the priest: And it standeth with reason that the priest should speak for you and in your name, and you answer him again in your own person, and yet you understand never a word, neither what he saith nor what you say yourselves."[89]

The spirit of simplicity, of reverence for God and respect for men, orderliness, and dignity so typical of Anglicanism, is finely expressed in two homilies. The first, "Of the Right Use of the Church," states that: ". . . the material church or temple is a place appointed . . . for the people of God to resort together unto; there to hear God's holy word, to call upon his holy name, to give thanks for his innumerable benefits bestowed upon us, and duly and truly to celebrate his holy Sacraments. . . ."[90] The second, "Of the Time and Place of Prayer," stresses the Anglican ethical emphasis in worship: ". . . Churches were made . . . to resort thither, and to serve God truly; there to learn his blessed will; there to call upon his mighty name; there to use the holy Sacraments; there to travail how to be in charity with thy neighbour; there to have thy poor and needy neighbour in remembrance; from thence to depart better and more godly than thou camest thither."[91]

The veil is removed from the sensitive and reserved spirit of Anglicanism only once and that is in the Prayer of Humble Access that cries, "down peacock feathers" to all human pretensions, and, characteristically combines a Protestant beginning with a Catholic ending. The rubric introducing it goes: *Then shall the priest turning him to gods boorde, knele down, and say in the name of all them that receyve the Communion, this prayer following.* "We do not presume to come to this thy table (o mercifull lord) trusting

---

[89] *Works*, III, 169f. See also A. L. Rowse, *Tudor Cornwall*, pp. 262ff.; and *Troubles connected with the Prayer Book of 1549*, ed. Nicholas Pocock, n.s., 37, pp. 148f.

[90] *Certain Sermons*, p. 156.   [91] *Ibid.*, p. 350.

in our owne righteousness, but in thy manifold and great mercies: we be not woorthie so much as to gather up the cromes under thy table: but thou art the same lorde whose propertie is always to have mercie: Graunt us therefore (gracious lorde) so to eate the fleshe of thy dere sonne Jesus Christ, and to drynke his bloud in these holy Misteries, that we may continuallye dwell in hym, and he in us, that our synfull bodyes may bee made cleane by his body, and our soules washed through hys most precious bloud. Amen."[92]

Perhaps the single most impressive characteristic of the Anglican revision of worship was its deep simplicity. Instead of many diverse "Uses" of various cathedrals in the realm of England, there was now one uniform order of *Common* Prayer. Instead of four principal service books,[93] and three supplementary ones[94] being necessary for the conduct of divine worship, as had been the case for Roman Catholic celebrants of worship, the single Book of Common Prayer served alike for bishops, the lesser clergy and the part of the laity. Instead of an unintelligible service in Latin there was now an easily understood service in English. Instead of many burdensome and mystifying as well as elaborate ceremonies, as in the Roman rites and ceremonies, these had been reduced to a few simple and significant signs. Instead of the lustrations, prostrations, and frequent crossings, the deeply significant signs were those given in the Gospel itself, such as the affusion of water in Baptism, and the fraction of the bread and libation of wine in the Holy Communion, and a few simple and time-honoured customs such as bowing in prayer and at the reception of Holy Communion, signing with the cross in Baptism, and the use of the ring in marriage. The ornaments, too, were greatly simplified, whether reference was made to the surplice as the usual vestment for matins and evensong, with the alb, tunicle, and cope for the celebration of Communion, or to the decoration of the churches, which were to lose the holy figures on the Roodlofts, or the stone or glass images of Christ, the Virgin, or the saints. Similarly the subtle polyphonic Gregorian chanting was to be substituted for the

[92] P. 225 of the Everyman edition of *The First and Second Prayer Books of King Edward VI.*

[93] The four principal service books were: the Missal or Mass book, the Breviary or Book containing the seven daily Offices, the Manual with the occasional rites as Baptism or Burial, and the Pontifical containing such rites as Ordination which required the presence of a bishop.

[94] The supplementary books consisted of: the Lectionary (with the lessons needed at Mass), the Gradual with the musical portions of the Mass, and the Antiphoner with the musical portions of the Breviary Office.

English mode of chanting (one syllable to a note). In short, simplicity, order, intelligibility, and fidelity to the Bible were to replace the riches of tradition and an impressive mystery and pageantry. Spectators became participants and ethics were preferred to aesthetics. It was felt, rightly or wrongly, to be a return to the simplicity of Christ and to the purity of the primitive church.

CHAPTER II

ANGLICANS AND PURITANS
IN CONTROVERSY (1564-1603)

CANTERBURY fought with Rome in the days of Henry VIII and Edward VI and with Geneva in the days of Elizabeth. The first Anglican apologetic in the writings of Jewel produced a polemic against Catholicism. The second Anglican apologetic in the hands of Whitgift, Bancroft and especially Hooker, produced an armoury against the slings and arrows of Puritanism. What made the second battle so acrimonious was that it was between foes in the same Protestant household.

As early as 1529 there were signs of a rift in Protestant solidarity, the harshest evidence of which was in the inability of the Saxon and Swiss Reformers at the Colloquy of Marburg to produce a unifying interpretation of the Eucharist that would satisfy Luther's demand for the 'real presence' and Zwingli's allegorical and memorial requirements. Even more ominous for the future unity of Protestants in England was the war between the "Coxians" and the "Knoxians" among the Marian exiles in Frankfurt which foreshadowed the controversy over liturgy between Anglicans and Puritans as it would break out with the "Admonitioners" in the early 1570s.[1] Significantly the "Coxians" desired to maintain their loyalty to the second prayer-book of Edward VI of 1552, for the defence of the doctrines of which Cranmer, Ridley, and Latimer had been burned at Oxford. The "Knoxians" fought for a form of worship modelled on Calvin's Genevan Order, which, in their judgment, was more in accordance with the model of Scripture and the custom of the earliest church, the very criteria the Puritans were to apply in their criticisms of the Book of Common Prayer in Elizabeth's reign. Even the beginning of the controversy could be seen

---

[1] For the importance of the influence of the exiles see Christina H. Garrett, *The Marian Exiles* (Cambridge, 1938). Miss Garrett shows that in the eight communities of the exile (Emden, Aarau, Wesel, Zurich, Strassburg, Frankfort, Basel, and Geneva), there were about 788 exiles, among whom were 67 clergymen and 117 theological students. Of the 788, two became Privy Councillors of Queen Elizabeth, 16 became bishops (and three archbishops), and three were printers and publishers who became "ardent servants of the Elizabethan Reformation." For the genesis and development of the liturgical squabble see my *Worship of the English Puritans*, pp. 27-34. For a relatively accessible edition of the source see Edward Arber's 1908 edition of *A Brief Discourse of the Troubles begun at Frankfort in the year 1554 about the Book of Common Prayer and Ceremonies.*

in advance in Edward VI's reign with Hooper's great reluctance to wear episcopal robes which seemed to him to symbolize the Lord Bishop rather than the Father in God or chief pastor of the diocese. For it was the "Vestiarian Controversy"[2] in 1566 that began to divide Elizabethan Protestants into two camps.

1. *The Puritans*

The term "Puritan" came into use during the Vestiarian Controversy late in 1567 according to Stow, who, however, says that it was used by Anabaptists to refer to themselves as being Puritans or "unspotted Lambs of the Lord."[3] It is much likelier that the term was used by less iconoclastic Protestants who wanted to "reform the reformation" or by those who criticised the pretensions of such purists or precisionists. Unquestionably the entire movement was characterised by a concern that the Reformation should proceed according to the norm of the "pure" word of God further than it had already done in England; the impurity of the Church of England was seen in its retention of the vestments of Catholicism along with some Catholic ceremonies. The movement was also notable for an admiring adherence to Geneva (or Zurich) for its theology, its Presbyterian church order of ministers, teachers, elders, and deacons, its Genevan service book or liturgy, and its Genevan translation of the Bible. In the seventeenth century Puritanism included Calvinistic Anglicans, Presbyterians, and Independents, together with several smaller sectarian groupings in the days of the Commonwealth.

Perhaps the two most important components in the term were the implicit seriousness and intensity of this uncompromisingly Biblical faith and the way that it is supposed by the more compliant to be shot through with superiority and hypocrisy. Both shades of meaning are suggested by Shakespeare's taunt at the vain, cross-gartered Malvolio in *Twelfth Night*, where the Puritan is challenged: "Dost thou think because thou art virtuous there

---

[2] See J. H. Primus, *The Vestments Controversy, An Historical Study of the Earliest Tensions within the Church of England in the reigns of Edward VI and Elizabeth*, p. 71: "The Marian exiles' first-hand experience of the methods of 'the best reformed churches' on the continent was largely responsible for various religious tensions in the years of Elizabeth's reign, not least of all, for the tensions regarding the use of vestments." See also Powel Mills Dawley, *John Whitgift and the English Reformation* (New York, 1954), pp. 69f.

[3] See "Puritan" in the *New English Dictionary*. William Haller, *The Rise of Puritanism*, p. 8, insists that Puritanism was generated by the shock of disappointment that Queen "Elizabeth did not reform the church but only swept the rubbish behind the door." See also H. M. Knappen's *Tudor Puritanism*, Appendix 2: Terminology.

shall be no more cakes and ale?"[4] The intensity and strictness is suggested by naming one offensively Puritan character in Ben Jonson's play, *Bartholomew Fair*, "Mr. Zeal-of-the-land Busy." These are, however, more caricature than character sketch, and it will do no harm if we are reminded that Edmund Spenser and John Milton were proud to be designated Puritans, combining seriousness in religion with courtesy and learning.

"Puritan" was a handy smear term for the opponents of further reformation in Elizabeth's reign. Field and Wilcox, authors of the *First Admonition to Parliament*, complained that the term was applied to poor simple people who were serious about their religion and who were thus charged as heretics, "calling them Puritanes, worse than the Donatists."[5] In his reply Whitgift made matters worse by stating that the term aptly recalled the *Cathari*, not because they were purer than others but only because they claimed to be, and "separate themselves from other churches and congregations as spotted and defyled."[6] As Knappen points out,[7] it took the Puritan apologists 20 years to accept the designation for themselves, after vainly trying to show its inapplicability to themselves or its greater suitability for the Anglicans who, by their strictness over vestments and ceremonial which should be *adiaphora*, were creating new schisms. It was a term used several times by the critics of Anglicanism in the bitingly shrewd and scurrilous *Marprelate Tracts*.[8] In the end their successors were to speak warmly of the "good old Puritans," and Puritans themselves were to bear the designation as a mark of pride, as a scar in spiritual warfare. The Puritans, as the Quakers and Methodists later did, turned a sarcastic nomenclature invented by their enemies into a badge of courage and fidelity.

Because of the flexibility of the term "Puritan," it is best not to circumscribe it too narrowly. Certainly it cannot be equated with English Presbyterianism, for the Puritans included in their camp those who disagreed on church government. We shall find Anglican (that is, Episcopalian) Puritans, Presbyterian Puritans, and Independent Puritans. Nor can Puritanism be the name for English Calvinists, since that archenemy of Puritanism, Archbishop Whitgift, was a Calvinist in doctrine, and the sources of English Puri-

---

4 Act II, scene iii, line 123.
5 Field and Wilcox, *First Admonition*, sig. A 1 b.
6 *Answer to the Admonition*, p. 18.
7 *Tudor Puritanism*, p. 488.
8 The Marprelate Tracts, ed. Pierce, pp. 22, 242, 245, 257.

tanism, especially in the reign of Elizabeth, flowed almost as much from Zurich as from Calvin's Geneva, and Wittenberg and Strassburg were not without their mediating influence.

What is clear is that Puritanism was marked by the desire, as expressed by the admonitioners, for the "restitution of true religion and the reformation of the church of God"[9] according to the scriptural norm in all things, doctrine, worship, and ecclesiastical government. Perhaps the most objective account of the origin of the nickname, given by one of the sternest critics of Puritanism, appears under the caption "Anno Reg. 7" in Heylyn's *Ecclesia Restaurata*, or *History of the Reformation*: "This year the *Zuinglian*, or *Calvinian* Faction began to be first known by the name of *Puritans*, which name hath ever since been appropriated to them, because of their pretending to a greater Purity in the Service of God, than was *held forth* unto them (as they gave it out) in the *Common-Prayer* Book; and to a greater opposition to the Rites and Usages of the Church of Rome, than was agreeable to the Constitution of the Church of *England*."[10] The aptness of this description lies in its suggestion that the Swiss Reformers were the mentors of the Puritans, and that the Puritans wished to make worship and ceremonies conform more closely to the Biblical pattern and the norm of the Continental Reformed churches.

Is it possible to interpret Puritanism more strictly without losing the spirit and ethos of the movement? William Haller sees its essence as a vital form of English Calvinism centered on "the dynamic Pauline doctrine of faith, with its insistence on the overruling power of God, on the equality of men before God, and on the immanence of God in the individual soul"[11] which was lucidly and trenchantly formulated in Calvin's theology and strikingly exemplified in the city of God in Geneva. The steel in the Puritan soul was the conviction of the truth of the doctrine of predestination with all the ringing assurance of St. Paul's battle-cry: "If God be for us, who can be against us?"[12] H. M. Knappen follows G. M. Trevelyan in using the term "Puritanism" to mean "the religion of all those who wished either to 'purify' the usage of the established church from the taint of popery or to worship separately by forms

9 *The First Admonition to Parliament*, in *Puritan Manifestoes*, ed. W. H. Frere and C. E. Douglas, p. 8.

10 *Ibid.*, edition of 1661, p. 172.

11 *Ibid.*, p. 8.

12 Rom. 8:31. For the thrust of predestination in Paul's thought see Rom. 8:28-35.

so 'purified.' "[13] The dominating characteristics of the Puritans, in his view, were bibliolatry, a tendency toward individualistic interpretation of their authority, a seriousness that made them wary of gaiety, a passionate love for civic freedom, and a moral earnestness.[14]

My own view is that the flexibility and unity of Puritanism are best preserved if I describe their subdivisions as conforming or nonconforming Puritans, or even better, as patient and impatient Puritans. In the first category I include all who desired further reformation according to the Word of God and following the examples of the best Reformed churches, but who saw this happening in the context of the nation; these were the Anglican and Presbyterian Puritans. In the second category we see those Puritans who, like Robert Browne, wanted a "Reformation without tarrying for anie," and who instead adopted the "gathered church" concept of ecclesiology. The church was then to consist not of the "mixed multitude" but only of committed and covenanted Christians, living like island colonies in the midst of a sea of half-Christians or nominal Christians.[15] This second group, or series of groups, organised themselves independently of the national church. Imprecise as this inclusive use of the term may prove to be, it at least avoids the awkwardness of other alternatives such as "Separatist" and "Semi-separatist," which deny the fundamental unity in the desire to build a Biblical theocracy in England, which was shared by all the groups.

2. The Strategies of Puritanism

Protestantism seemed to be united in the first years of the reign of Elizabeth by a sense of the importance of maintaining a common front against the inflexible Catholicism they had so recently experienced under Mary. That Catholicism was steeling itself to win back the provinces lost to Protestantism by the Council of Trent and by the creation of Ignatius Loyola's new Order, the Society of Jesus, sworn to a "cadaveric" obedience in defense of the Pope and of the church, and with its priests superbly trained in will and imagination through the *Exercitia Spiritualia*. The patience of the established church and of the Puritans gave away after the defeat

[13] *Tudor Puritanism*, p. 488. Trevelyan's proposal was made in *England under the Stuarts*, pp. 60-71.

[14] Haller, *Rise of Puritanism*, p. 489.

[15] Browne, the founder of the Independent (or later, Congregational) ecclesiology, expounded his views in *A Treatise of Reformation without tarrying for anie* (1582).

of the armada in 1588 had ensured the relative safety of England from foreign and Papist attack, and there was little need to maintain a common Protestant front.

However much they might hope that the Elizabethan Settlement would rid itself increasingly of the vestiges of Catholic vestments and ceremonies, the Puritans at the beginning of Elizabeth's reign kept themselves in patience through hope. The advanced wing of Edwardian Reformers had gone into exile in Emden, Wesel, Zurich, Strassburg, Frankfurt, Basel, Geneva, and Aarau, where they had become familiar with the customs in worship and church government of the Reformed churches of Germany and Switzerland, as well as with their leading ministers and theologians. No less than 67 clergy and 117 theological students, with the backing of William Cecil and several merchants, were preparing to be the Protestant vanguard in England when the tide should turn against Marian Catholicism. They were sustained by the superb examples of the Marian martyrs such as Latimer, Ridley, and Cranmer who at the end of Edward VI's reign were preparing further steps to bring England closer to the Continental Reformation. Naturally, since 16 of the exiles became bishops and three archbishops, not to mention those who gained professorships, prebendaries, and canonries in the Church of England, there was every reason to hope that the exiles would take over the Church of England.

There were, however, three forces working against such a result. There was the queen herself, who desired that the externals in religion should remain much the same, because this might quiet her more conservative subjects belonging to the old faith and because it pleased her love of pageantry. She also disliked the egalitarianism of Presbyterianism, and even more, its spokesman John Knox, who had condemned women rulers in his tactless *The First Blast of the Trumpet against the Monstrous Regiment of Women*. It is true that this was directed against Elizabeth's sister, Mary, but might not a second blast be directed at her by this courageous and uncompromising Scot in whom the iron of the French galleys had entered? Elizabeth was determined to control both church and state, which she did with a firm hand as Archbishop Grindal, the Marian exile, was to know to his regret, through four years of suspension although he was Primate of All England. The second obstacle was the division between the conservative and the radical exiles, corresponding to the "Coxians," who followed the example of Cox, the avaricious Bishop of Ely, in making the best of both worlds, and the "Knoxians" like Humphrey, Sampson, and Whittingham who had to be content with deaneries which they forfeited rather than wear the "idolatrous gear" prescribed by Archbishop Parker's "Advertisements" in 1566. The third and saddest obstacle was the tendency of youthful reformers to become middle-aged conservatives, to lose their own fire as they basked in the rays of court preferment. The Elizabethan bishops, from Jewel to Parkhurst, from Aylmer to Horne, soon developed a fatty degeneration of the conscience as they became religious civil servants charged with creating a religiously compliant people for their sovereign. Only the stoutest hearts resisted the blandishments of the balmy breezes of flattery and the sunshine of promotion.

The Puritans' first strategy was developed only after their initial disappointment during the Vestiarian Controversy. Indeed, no Puritan protest was necessary at first because Puritan-minded ministers took matters much into their own hands, often in iconoclastic ways. The disorder in worship in the realm was complete.

Archbishop Parker was requested by the queen to provide diocesan certificates describing the "varieties and disorders" existing in the ceremonies of the church. The results showed that some ministers said the service in the chancel, others in the nave; some led worship from the pulpit, others from a chancel seat. Some kept the order of the Book of Common Prayer, others "intermeddled" metrical Psalms; some read the service in a surplice, others without one. The Communion table was variously placed, either "altar-like distant from the wall a yard" or facing north and south in the midst of the chancel, or even in the nave of the church. Sometimes it was a trestle with a table top, usually it was a table. Occasionally it had a carpet on it, often it bore none. Communion was administered by some clergy with a surplice and cope, by some with only a surplice, by others wearing neither vestment. Some clergy used a Catholic chalice, others an Anglican communion cup, and yet others merely a common cup. Some clergy used unleavened, and other leavened, bread. There was equal variety in the postures for the reception of Communion—some knelt, some sat, some stood. As for the Sacrament of Baptism some ministers administered it with surplice, some without; some used a font, others used a basin; and some used the signation with the cross, while other refused to use it.[16]

---

[16] Lansdowne MSS, viii, 7, cited vol. 1, pp. civ-cv of W.P.M. Kennedy, *Elizabethan Episcopal Administration*, 3 vols.

Clearly some order was desirable. It was Parker's unhappy duty to require all the clergy to accept his "Advertisements."[17] The first vestiarian troubles had broken out as early as 1563. It seems Archbishop Parker drew up the "Advertisements" in 1564 and submitted them, unsuccessfully, to Cecil for the royal signature of authority on two separate occasions. The queen presumably refused to bear the odium caused by this demand for uniformity. They were finally issued in 1566 under the signatures of the archbishop and bishops as commissioners in causes ecclesiastical. The principal minister in cathedral and collegiate celebrations of Holy Communion was to use a cope, while the celebrant in a parish church "shall wear a comely surplice with sleeves."[18] All communicants were to receive kneeling. Each man in holy orders was to promise to preach only if he had the license of the bishop, to read the service prescribed in the Book of Common Prayer, to wear the appointed apparel in church and the appropriate clerical garb when travelling, to accept no secular employments, to read a chapter in both Testaments daily, and to give several unimportant assurances in addition.[19]

The leaders of the Anglican church could urge that it was a church aiming at the greatest comprehensiveness in its demand for uniformity, and that it was avoiding the Scylla of Rome and the Charybdis of Geneva on its middle course; but the true friends of Reformation considered the Anglican establishment to be less a golden mean and more of a "leaden mediocrity."[20] And Elizabeth, in the words of John Knox, was "neather gude Protestant nor yit resolute Papist."[21] This did not satisfy the firm Calvinists who had been temporary sojourners in the city of Geneva during their exile, such as Thomas Sampson, Laurence Humphrey, William Whittingham, Anthony Gilby, and Thomas Lever.[22] Dean William Whittingham argued that there were four grounds on which something indifferent in the church (that is, neither approved nor prohibited by Scripture) might be used: it must contribute to God's glory, consent with His word, edify His church, and display Chris-

---

[17] H. Gee and W. J. Hardy, *Documents illustrative of English Church History*, pp. 467-75.

[18] *Ibid.*, p. 471. [19] *Ibid.*, p. 475.

[20] *Zurich Letters*, 2 vols., I, 23.

[21] Knox, *Works*, ed. David Laing, II, 174.

[22] It is noteworthy that only three of the exiles who later accepted bishoprics—Pilkington, Bentham, and Scory—had been to Geneva, and of these, Pilkington was tender to those clergymen in his diocese who had Vestiarian scruples. See J. H. Primus, *The Vestiarian Controversy*, p. 74.

tian liberty. When applied to the vestments, he found the first three principles absent by reason of Papal and superstitious origin and use, and the last broken by the demand for absolute conformity.[23] Alexander Nowell, also a Marian exile, and the dean of St. Paul's cathedral, produced an eirenicon urging disputants to accept the magisterial authority to prescribe vestments, while petitioning for their removal on the grounds that they were subject to abuse, as a public testimony to the rejection of superstitions, as a profession of Christian liberty, and a removal of a cause of dissension among brethren in the same household of faith.[24] The importance of the Vestiarian Controversy was that it polarised Puritan and anti-Puritan opinion and once and for all put the Puritans on notice that they could expect no further moves on the part of the queen in a more Reformed church direction.

Though the queen was an obstacle, Parliament might not be. The Puritans remembered happily their near victory in 1563. In the Convocation of that year there were both radical and moderate reform proposals put forward to ease consciences on the matters of the vestments and ceremonies. The radical proposals would have eliminated all the ancient vestments. The moderate proposals would have required the surplice alone for all services, with the provision that no minister should officiate except in "a comely garment or habit," a deliberately vague term to allow for the substitution of the black Genevan gown.[25] The proposals included the omission of crossing in Baptism, making kneeling at the reception of Holy Communion optional, with the abolition of organs and saints' days. This was the "moderate policy of the exilic party";[26] it was defeated by only one vote in the lower house.[27]

The result of this narrow defeat was to alert the leaders of Puritanism to the necessity for a protracted literary and political warfare. The Vestiarian Controversy had touched only the periphery of the zone of battle. Now the Puritans would concentrate on the centre. From this time on the Anglican Book of Common Prayer and the ecclesiastical government by bishops become the chief objects of Puritan attack, always with the presumption that the Genevan Service Book and the Presbyterian form of church

23 *Ibid.*, p. 88.
24 *Ibid.*, p. 93.
25 D. Wilkins, *Concilia Magnae Britanniae* (1737), 4 vols., IV, 237-40.
26 A phrase of W. H. Frere and C. E. Douglas, *Puritan Manifestoes*, p. ix.
27 J. Strype, *Annals of the Reformation*, 4 vols., II, 335; also H. Gee, *The Elizabethan Prayer Book*, p. 164.

order were to be preferred for their Biblical basis. The focal point of the Puritan attack changed from Convocation to Parliament.

The new strategy was quite successful at first. In 1566 a group of Puritans in the House of Commons initiated the first ecclesiastical bill of an uncontentious nature, in fact to give civil standing to the Articles of Religion. It had passed the three readings in the lower house and one reading in the House of Lords when it was stopped at the special command of the queen who disliked this attempt to interfere with her ecclesiastical prerogatives.[28]

Before the next Parliament met in 1571 the Puritans organised for the attack. Cartwright had already attacked the organisation of the English church and its worship on the pretext of lecturing in Cambridge University on the Acts of the Apostles. He was to be followed rapidly by the authors of *The First Admonition to Parliament*, whose aim was "to proffere to your godly considerations, a true platforme of a church reformed, to the end that it being layd before your eyes, to beholde the great unlikenes betwixt it & this our english church. . . ."[29] In their eyes the Book of Common Prayer was "an imperfect booke, culled & picked out of that popishe dunghill, the Masse book full of all abhominations."[30] *The Second Admonition to Parliament* has been attributed to Cartwright; but whether Cartwright or another Puritan was the author, he was entirely out of patience with the bishops: "What talke they of their being beyond the seas in quene mariee's days because of the persecution, when they in queene Elizabethes dayes are come home to raise a persecution?"[31] From this time on there was a growing body of dissident Puritan literature, both scholarly and scurrilous, from Cartwright and Travers to the anonymous author of the *Marprelate Tracts*, which, as their name implies, were merciless in their attacks on the bishops.

From then on the Puritans would use every means at their disposal to disseminate their views. They would publish tracts, pamphlets, and broadsides from underground presses in England and overseas. They would use the great medium of the pulpit. They would found a college like Emmanuel in Cambridge especially to train a Puritan ministry and godly laity through the munificence of Sir Walter Mildmay, who had previously given generously to

---

28 Parker, *Correspondence*, ep. ccxxiv, p. 291, ccxxv, p. 293.
29 *Puritan Manifestoes*, p. 8.
30 *Ibid.*, p. 21.
31 *Puritan Manifestoes*, p. 112. For the question of authorship see A. F. Scott Pearson, *Thomas Cartwright and Elizabethan Puritanism*, p. 74.

another Puritan centre in the same university, Christ's College.[32] They would endow scholarships to maintain the supply of trained Puritan ministers from the universities. They would endow lectureships in the greater towns to ensure that the country people flocking to the markets there would hear sound preaching leading to the practise of piety. They would arrange to supplement theological training with conferences, or "prophesyings,"[33] gatherings in which ministers would discuss the import of Scripture passages, translating them from the original languages and applying them to the needs of the contemporary situation—and incidentally providing a superb channel for Puritan propaganda. Beginning in 1564 these became spread all over England within the next decade and were only halted temporarily by the queen's suspension of Grindal, her Archbishop of Canterbury, because he countenanced such means of more rapidly educating ministers and responsible laity in theology.[34] Puritans published primers and books of devotion[35] for families and private persons through such printers as John Day, which supplemented the Puritan interpretation of the Bible as found in the marginal annotations to the Geneva Bible, or the Puritan interpretation of history as found in the superb *aide-mémoire*, Foxe's *Acts and Monuments of matters happening in the Church*.[36] The Puritans set up a complete Presbyterian system of church organisation in the Channel Islands (an English possession a few miles from the coast of France), and a presbytery

[32] Fuller, in *The Worthies of England*, ed. J. Freeman, p. 177, tells of Mildmay, Elizabeth's Chancellor of the Exchequer, "coming to court after he had founded his College, the Queen told him, 'Sir Walter, I hear you have erected a puritan foundation.' 'No, Madam,' saith he, 'far be it from me to countenance anything contrary to your established laws; but I have set an acorn, which when it becomes an oak, God alone knows what will be the fruit thereof.' "

[33] See Leonard J. Trinterud, "The Origins of Puritanism," in *Church History*, XX, i, 46, who shows that these seminars for Biblical exposition were frequently organised by the Swiss and Rhineland cities where they would have become familiar to the Marian exiles. For a contemporary English account see *Harrison's Description of England in Shakespeare's Youth*, ed. F. J. Furnivall, 2 vols., I, 18-19.

[34] Archbishop Grindal sent his famous letter in defence of "prophesying" to Queen Elizabeth on December 20, 1576. It is to be found in Strype's *The History of the Life and Acts of Edmund Grindal*, pp. 578ff. On May 7, 1577 the queen, ignoring her primate, sent letters to all the bishops commanding the suppression of "prophesyings" in their dioceses; Grindal was sequestered by the royal order and confined to his palace in Lambeth for six months. Since Grindal remained impenitent, his suspension was never fully removed in the ensuing six years until his death in 1583, despite appeals made to the queen by bishops and the Convocation.

[35] Puritan, Anglican, and Catholic devotional manuals will be considered in the final chapter.

[36] Commonly known as "Foxe's Book of Martyrs," it appeared in Latin in Strassburg in 1544 and was published in an enlarged English edition in 1563.

in Wandsworth in 1572, the very year the Admonitioners were pressing their propaganda on Parliament.[37] When all else failed, the persistent Puritans would establish their colonies in exile in Holland and eventually found theocracies in the New World in the Plimouth Plantation and the Bible Commonwealth of Massachusetts. Such was the assiduity of these predestinarians, and such was the variety of their strategies for creating the kingdom of God on earth.

We now turn to the theological differences between official Anglicanism and the Puritans during the reign of Elizabeth, and to a consideration of the liturgical consequences of these differences.

3. Anglican and Puritan Concepts of Scripture

The Puritans believed that the Church of England had accepted the primacy of the Scriptures in its doctrinal statements (certainly the Thirty-Nine Articles of 1571 were as Calvinistic a summary of Pauline doctrines as could be wished), but that it had not followed the Scriptures either in its form of church order or in its worship, both of which retained too much of the false traditions and accretions of Roman Catholicism. Their very position indicated that the Puritans believed the Scriptures to be an authority for more than doctrine and ethics, whereas the Anglicans restricted the authority of Scripture to belief and behavior. In brief, the Anglicans, following the example of the conservative Luther, argued that whatever traditions of the ancient church were not forbidden by Scripture might be retained by a Protestant national church; whereas the Puritans argued that the Scriptures were a pattern and model for worship and the form of church government very much in the spirit of John Calvin. For Luther and the Anglicans the Scriptures were a *Trostbuch*, a book eliciting faith. For Calvin and the Puritans, however, the Bible was *la saincte parole et loi de Dieu*; its teaching was to cover all aspects of life and especially to direct life within the church, both liturgical and administrative. To ignore its teaching was merely to prove that one was not yet freed from the absurd pride and interior darkness which was the legacy of original sin and of unredeemed nature.

Cartwright insisted that "the word of God contains the direction of all things pertaining to the church, yea, of whatsoever things

37 Knappen, *Tudor Puritanism*, p. 304; Pearson, *Cartwright*, pp. 74ff., 157-66; Richard Bancroft, *Daungerous Positions and Proceedings* (1593), p. 43.

can fall into any part of man's life."[38] Should any questions of ceremonies still remain obscure, he had four infallible Pauline tests which he would put to them. First, do they offend any, especially the Church of God? His authority here was I Corinthians 10:32. Second, is all done in order and comeliness? Here his authority was I Corinthians 14:40. Third, is all done for edifying, building up in the faith? The authority now is I Corinthians 14:26. Fourth, is all done to the glory of God? This authority, Romans 14:6-7.

It was left to a later Puritan theologian to express with greater precision the Puritan concept of Biblical authority. This was the learned William Ames, whose *Medulla theologica*[39] was translated as the *Marrow of Sacred Divinity* and had a great success as a classic of Reformed theology and piety. Ames affirms that "All things necessary to salvation are contained in the Scriptures and also those things necessary for the instruction and edification of the church," and draws the conclusion: "Therefore, Scripture is not a partial but a perfect rule of faith and morals. And no observance can be continually and everywhere necessary in the church of God, on the basis of any tradition or other authority, unless it is contained in the Scriptures."[40]

On the Anglican side, Whitgift answers Cartwright's assertion of the comprehensive applicability of the authority of Scripture by agreeing that "nothing ought to be tolerated in the Church as necessary unto salvation, or as an article of faith, except it be expressly contained in the Word of God, or may manifestly thereof be gathered";[41] but he also draw the corollary, "Yet do I deny that the Scriptures do express particularly everything that is to be done in the church."[42] He goes on to show that while the Scriptures urge the necessity of Baptism they do not prescribe its mode; whether "in fonts, basins, or rivers, openly or privately, at home or in the church, every day in the week or the sabbath day only" was left to the church to determine.[43]

The best expositor of the Anglican understanding of the author-

---

[38] Whitgift, *Works*, 3 vols., I, 190.

[39] The *Medulla theologica*, published in Amsterdam in 1623, was the fruit of Ames's lectures to the sons of Leyden merchants. It was followed by 12 Latin printings; three printings of an English translation appeared between 1638 and 1643. The best modern English translation with historical introduction is by John D. Eusden, *The Marrow of Theology, William Ames, 1576-1633*.

[40] Ames, *Marrow*, p. 189.   [41] Whitgift, *Works*, I, 180.
[42] *Ibid.*, p. 191.   [43] *Ibid.*, p. 201.

ity of Scripture was Hooker, the greatest apologist of the Elizabethan age. For him the Bible was the revelation of the nature of God and His purposes for mankind, but God had not intended his Word to prescribe the detailed ordering of worship, the method of church government, or the details of conduct. All these he left to the discretion of men. Such matters were to be determined by the use of reason; reason would take into account propriety, decency, proportion, according to the circumstances of the case. The Bible gives man such information only as he cannot obtain by his reason. Otherwise, if the Bible were the only and exclusive criterion, the law of reason and the law of nature, which both reflect their Creator, would be wholly abrogated by the Scriptures. Hooker spoke directly to the contentions of the Puritans in the following passage: "Let them with whom we have hitherto disputed consider well, how it can stand with reason to make the bare mandate of sacred Scripture the only rule of all good and evil in the actions of mortal men. The testimonies of God are true, the testimonies of God are perfect, the testimonies of God are all sufficient unto that end for which they were given. Therefore accordingly we do receive them, we do not think that in them God hath omitted anything needful unto his purpose, and left his intent to be accomplished by our devisings."[44] Thus for Hooker and Anglicanism the Bible was authoritative in doctrine, and prescribed the outstanding ordinances in worship, such as prayer, praise, preaching, and the two sacraments of Baptism and Holy Communion, but it was never intended to be authoritative in the determination of the details of worship. Times, circumstances, and ceremonies are properly to be decided by the church rulers in accordance with the guidance of antiquity and the imperatives of right reason.

The Puritan replied, in effect, that the Anglican accepted the authority of the Bible in theory but not in practise. It was inconsistent of the Anglican, he charged, to accept the biblical authority for doctrine but not for ecclesiastical government or worship. The Anglican seemed to set aside any parts of the Bible that seemed to go against the contemporary usages of the Anglican church by dismissing them as appropriate only for the times of their origins and therefore no part of the eternal law of the Christian religion. It was too convenient to be true.

By contrast, the Puritan held that the Bible was the revealed will of God from end to end, through and through. He believed it

44 *Ecclesiastical Polity*, II, viii, 5.

53

to be authoritative not only for doctrine and ethics, but also for every aspect of ecclesiastical and even human life. It was an absolute code. It was the expression of the Divine will in matters theological, moral, judicial, political, military, economic, ecclesiastical, and even sartorial. Therefore it was necessary to look to its pages not only for general laws or pointers but for detailed guidance.[45]

4. Theories of Human Nature

Both Anglicans and Puritans accepted the doctrine of original sin, but they estimated differently the seriousness of the effects of man's wound. Anglicans found man to be deficient in spiritual capacity; his other powers were weakened, but not desperately wounded and in need of redemptive blood transfusions, as the Puritans claimed. Man's reason was, for the Anglicans, unimpaired; it had a natural capacity to distinguish between good and evil in a moral order. Cranmer assumed, for example, that men could choose the good without the help of sanctifying grace.[46] Jewel affirmed that, "Natural reason holden within her bonds is not the enemy, but the daughter of God's truth."[47] Donne held that reason must be employed when the meaning of Scripture is unclear, but, "Though our supreme court . . . for the last appeal be Faith, yet Reason is her delegate."[48]

In contradistinction the Puritans interpreted the Fall, with Calvin, as resulting in a total interior perversion of man's perspective and powers. Some remnants of man's intelligence remain in matters not connected with his salvation, such as in liberal studies, mechanical arts, jurisprudence, and architecture. Had Calvin known Freud he would have agreed with his diagnosis of man's psyche as being given to both wishful thinking and deep rationalizations. "We are," he complained, "so utterly mastered under the power of sin that our whole mind, heart, and all our actions bend

[45] Peter Bayne, in *Puritan Documents*, p. 10, observes: "in considering that urgency of appeal to Scripture and Scripture alone, which throughout its whole history was made by English Puritanism, an appeal which with our modern prepossessions, may seem to us to be a wilful searing of the eye-balls of reason and conscience, it is essential to recollect that it was against the authority of Rome that Calvin and his followers asserted the supremacy of God's written Word."

[46] *Works of Thomas Cranmer*, 2 vols., II, 143.

[47] Jewel, *Works*, 4 vols., I, 501.

[48] Donne, *Essays in Divinity*, ed. E. M. Simpson, pp. 114-16, cited by John H. A. New, *Anglican and Puritan, The Basis of their Opposition, 1558-1640*, p. 8, to which this chapter owes much, even though I take issue with New on several points.

towards sin."[49] In brief, as St. Augustine had seen, man is hopelessly crippled, being *incurvatus in se*, twisted in on himself.

William Perkins, the popular Puritan theologian of Cambridge, claimed that "Original sin is nothing else but a disorder of evil disposition in all the faculties and inclinations of men, whereby they are all carried inordinately against the law of God." Its location is not a part, but rather the whole of man, body, and soul. In the first place, the very appetites are greatly corrupted, as are the outward senses. Even the very understanding of man, according to the Spirit of God, "is only evil continually: so we are not able of ourselves to think a good thought."[50] Perhaps the clearest statement of the difference between Anglican and Puritan estimates of human nature was that of John H. A. New, to the effect that the Puritans did not deny that man had a true knowledge of good and evil, but rather, "they asserted a total inability in man to desire and to choose rightly, and in that lay their alienation from Anglicanism."[51] It was further explained by the Puritans that man's inability to choose aright was because he was blinded by pride and disordered in his appetites. To trust in reason was to trust in a constant temptation to pride and therefore away from the humility and tractability that true repentance brings.

Though the Puritans depreciated the role of unredeemed reason in religion, it would be an egregious error to conclude that they were despisers of human learning, whether theological or humanistic. They shared with their opponents the rich cultural legacy of the Renaissance. Their leaders were almost all men of learning, many of whom had held high positions at Oxford and Cambridge. Cartwright was Lady Margaret Professor of Divinity in the University of Cambridge, which was a hive of Puritan activity.[52] In New England they were apostles of religion and learning. Perry Miller pays them this tribute: "The greatness of the Puritans is not so much that they conquered a wilderness, or that they carried a religion into it, but that they carried a religion which, narrow and starved though it may have been in some aspects, deficient

[49] Calvin's *Letters*, ed. Jules Bonnet, II, 189. This information is contained in a letter to the Protector Somerset, 22 October 1548. Other pertinent Calvin references are to the *Institutes*, II, ii, 13-14; II, iii, 1; IV, xv, 10; and the Commentary on Romans, at Romans 7:14.

[50] Perkins, *The Works of that Famous and Worthy Minister of Christ in the University of Cambridge*, 3 vols., I, 165.

[51] New, *Anglican and Puritan*, p. 10.

[52] For a lively and comprehensive account of Cambridge Puritanism see Harry C. Porter, *Reformation and Reaction in Tudor Cambridge*.

in sensuous richness or brilliant color, was nevertheless, indissolubly bound up with an ideal of culture and learning. In contrast to all other pioneers, they made no concessions to the forest, but in the midst of frontier conditions, in the very throes of clearing the land and erecting shelters, they maintained schools and a college, a standard of scholarship and competent writing, a class of men devoted entirely to the life of the mind and of the soul."[53] Harvard College was founded by the Puritans in 1636, only seven years after they had reached Boston and its environs. The Puritan was a true son of Jerusalem, but he paid frequent visits to Athens.

If the Anglican apologist claimed that all God reveals must be comprehended by human reason, because the author of both revelation and reason is God, the Puritan replied that this was to make reason a judge over revelation, and man an arbiter of God. Had the Anglican replied that there could be no discrepancy between what is rational and what is revealed, the Puritan would have retorted that human reason is unfitted for the task of comprehending the majesty and mystery of God's ways. The Puritan would have pointed to the perverseness of human nature, and to history's confirmation of the wickedness of men. You may give men (we may suppose him to argue) an education in the wisdom of antiquity and in modern skills, but they will continue to be mean and cruel, selfish and carnal. The galleon of human nature, richly laden with the treasures of antiquity, and steered by the natural reason of man, is bound to shipwreck on the rock of human perversity.

The twofold legacy of Calvin found secure lodging in the Puritan safes: the all-sufficiency of Scripture and a radical restatement of original sin to make clearer not only the utter inadequacy of man, but the creating, providing, and directing omnipotent adequacy of God.

5. Predestination: God's Rule and the Triumph of the Elect

Predestination is not an Anglican, Puritan, or even a Protestant doctrine; it is a Christian doctrine. But Calvinism (I refer to the later rather than the earlier Calvin) made much of it. It appears as the 17th of the Thirty-Nine Articles of the Church of England, formulated in 1563. The first paragraph reads: "Predestination to life, is the everlasting purpose of God, whereby (before the foundations of the world were laid) He hath constantly decreed by His

---

[53] Perry Miller and T. H. Johnson, *The Puritans*, pp. 11f.

counsel secret to us, to deliver from curse and damnation, those whom He hath chosen in Christ out of mankind, and to bring them by Christ to everlasting salvation, as vessels made to honour. Wherefore they which be endued with so excellent a benefit of God, be called according to God's purpose by His Spirit working in due season: they through grace obey the calling: they be justified freely: they be made sons of God by adoption: they be made like the image of His only-begotten Son Jesus Christ: they walk religiously in good works, and at length by God's mercy, they attain to everlasting felicity."[54]

Essentially the doctrine was an extension of justification by faith, a reasseveration that salvation is wholly the work of God, "before the foundations of the world," so that man cannot take any credit for his salvation. Even though it seems, especially in the explicit form of 'double predestination,' to accuse God of arbitrariness in making a few elect and a multitude reprobate, regardless of their response to His mercy, for those who have been chosen or elected it is a doctrine of great comfort, in that it assures them that however great temptation or suffering may be, they cannot lose their salvation because God has predestined them to salvation. It is a doctrine that offers a strong consolation where the faithful are conscious of being God's resistance movement, as, for example in Catholic France, where the Huguenots defined their Protestant community as an anvil which has worn out many hammers.

This sense of the strengthening or comfort[55] to be derived from the doctrine is to be found in the second paragraph of the same article: "As the godly consideration of predestination, and our election in Christ, is full of sweet, pleasant, and unspeakable comfort to godly persons, and such as feel in themselves the working of the Spirit of Christ, mortifying the works of the flesh, and their earthly members, and drawing up their mind to high and heavenly things, as well because it doth greatly establish and confirm their faith of eternal salvation to be enjoyed through Christ, as because it doth fervently kindle their love towards God: so, for curious and carnal persons, lacking the Spirit of Christ, to have continually before their eyes the sentence of God's predestination, is a most dangerous downfall, whereby the devil doth thrust them either into

---

[54] See E.G.S. Gibson, *The Thirty-Nine Articles of the Church of England*, pp. 459-92, for an account of the history of the interpretation of the 17th article.

[55] Modern usage has weakened the original force of "comfort" (derived from the Latin *fortis*), which in Elizabethan times meant to strengthen or invigorate rather than to console.

desperation, or into wretchedness of most unclean living, no less perilous than desperation."

Calvin recognized that this was a *decretum horribile* and cautioned that people should not delve into the hidden counsels of God precisely because of the dangers of despair or anti-nomianism.[56] Yet despite the weaknesses of the doctrine, either in the dark shadow that it throws on the grace of God or in its denial of human freedom and responsibility that are supposedly the prerequisites of moral life and accountability, it was a conviction that God could give man "new dignity for old degeneracy,"[57] an ineluctable assurance.

The Puritan recognized that his confidence came from the immutability and omnipotence of God, and from the firm grasp of Christ on the elect soul (rather than from its feeble and wavering hold on Christ). Cartwright had a succinct confirmation: "We learn that they that are truly reconciled and called shall abide for ever. This is a true doctrine, a saint once, a saint forever. It is impossible that they which believe should perish."[58] Predestination leads inevitably to the doctrine of the perseverance of the saints.

Both Anglicans and Puritans accepted predestination, but the Puritans held more vigorously to both consequences of the doctrine, positive and negative. Anglicans *implied* the negative consequence, namely, reprobation. Some Anglicans even went so far as to dilute the doctrine of predestination with the cold water of God's indiscriminate charity. The eternal damnation of the reprobate was qualified by the belief that the wicked, by their obstinate refusing of the repeatedly proffered grace, were to blame, rather than God; this was where the wedge of Arminianism slightly opened the closed door of reprobation. By the same token, the sovereignty of God was weakened as His mercy widened, and the mystery of God diminished as His will for all humanity appeared more rational, and the power of God was demeaned as man's freedom and power were elevated.[59] Officially, however, Anglicanism,

---

[56] There is good reason to believe that while Elizabethan Puritanism stressed the doctrine of predestination, as in the writings of William Perkins, yet his great pupil William Ames gave the doctrine less prominence in his theological system, and, by stressing the covenant aspect of the dealings of God with men, lessened their arbitrariness. See Ames, *Marrow of Theology*, pp. 27-28. See also Norman Pettit, *The Heart Prepared: Grace and Conversion in Puritan Spiritual Life*.

[57] New, *Anglican and Puritan*, p. 20.

[58] Thomas Cartwright, *A Commentary upon the Epistle of St. Paul written to the Colossians* (1612), expounding Colossians 1:23-29.

[59] Two Anglican Arminians in the last decade of the sixteenth century were

which had sent representatives to the Synod of Dort in 1618 to maintain predestination against the Arminian inroads of the Remonstrants, was orthodoxly predestinarian.

William Haller has seen a further strength of Puritanism in its full acceptance of the doctrine of predestination—that is, an implicit egalitarianism in the acceptance of the will of God who is no respecter of persons, and who, in casting down the mighty from their seats and raising the humble and meek, is the great revolutionary, as the Magnificat shows.[60] There is no more democratic doctrine than the universal depravity posited by the doctrine of original sin, and no more aristocratic doctrine than the doctrine of election, by which God chooses the godly select. The last in the world's estimation are to be first in the kingdom of God in the most radical transvaluation. The conviction that this was the Puritan's august destiny as a "saint" may also be the source of his ethical dynamism and ceaseless energy, which contrasted so vigorously with the compliance, not to say docility, inculcated so frequently by Erastian Anglicanism.

6. *Understandings of the Church*

Anglican and Presbyterian Puritan alike believed in the necessity of a Protestant national church in England. The Anglican theory was that church and state were coterminous, and that as one was born an Englishman so was he christened a member of the Church of Christ. "There is not," said Hooker, "any man of the Church of England but the same is also a member of the Commonwealth, nor any man a member of the Commonwealth, which is not also of the Church of England."[61]

To a considerable extent the same was true for Presbyterian Puritans, with the exception that they believed that, although each parish comprised a 'mixed multitude' of the elect wheat and reprobate tares or weeds which had to be allowed to grow up together until the harvest of the day of judgment, yet it was the duty of the parish minister, assisted by his godly churchwardens (when godly elders were not yet openly allowed in England), to discipline

Peter Baro and William Barrett, but they were rapped over the knuckles for their beliefs. In the 1630s and 1640s Richard Montagu, John Cosin, and William Laud belonged to the influential group of Anglican Arminians. See Charles and Katherine George, *The Protestant Mind of the English Reformation* (Princeton, 1961), pp. 67-68.

60 *Rise of Puritanism*, pp. 89ff. It is Peter who confesses (in Acts 10:34): "I perceive God is no respecter of persons."

61 *Ecclesiastical Polity*, VIII, i, 2.

evildoers and keep them away from the celebrations of Holy Communion until they amended their lives.

The third ecclesiology was that proposed by Robert Browne, the father of the 'gathered church' idea, and practiced by Separatist Puritans such as Barrowe, Penry, and Greenwood. They believed that the local congregation should consist only of committed Christians, or "visible saints"—a doctrine to gain its strongest following with the rise of the Independents to power and influence in the days of the Commonwealth and Protectorate.[62] The Anglican and Separatist positions were consistent; that of the Presbyterian or "Consistorial" Puritans was inconsistent in trying to combine a parochial congregation of "all comers," which was indiscriminate in character, with a disciplined nucleus (or "gathered church" group) within each parish.[63]

The most decisive differences between Anglican and Presbyterian Puritan concepts of the church concerned the pattern of church order or government, and the relation of the church to the state. In the matter of ecclesiastical government Anglicans kept the threefold ministry of bishop, priest, and deacon, hallowed by centuries of tradition. The Genevan order was much more to the liking of the Presbyterian Puritans, with its provision for pastors, doctors (teachers), elders, and deacons. The first two categories were ordained clergy, but the third consisted of ordained laymen who were charged with sharing the spiritual overseeing of the flock's faith and morals with the clergy, and who had the privilege of distributing the sacred elements to the members in the Lord's Supper. It was a remarkable attempt to express in the very structure of church order the priesthood of all believers. In this respect the followers of Geneva thought they were recovering the pattern of the Apostolic church. This they intended to introduce in their "Discipline." It was an attempt to maintain the purity of the members of the church of Christ, a matter of the greatest concern to Puritans as a group who were impatient with a merely nominal or superficial Christian witness.

The claim that the Presbyterian church order was restoring the pattern of the apostolic and primitive church was not conceded

[62] For the full development of the Independent ecclesiology see Geoffrey E. Nuttall, *Visible Saints: The Congregational Way, 1640-1660*. The writings of Barrowe, Greenwood, and Penry have been edited by Leland H. Carlson in the "Elizabethan Nonconformist Texts" series.

[63] "Consistorial" is a term applied to Cartwright, Travers, Field, Fenner, Gilby, and other leading Genevan Puritans by Richard Bancroft in *A Survey of the Pretended Holy Discipline* and *Daungerous Positions and Proceedings*, both of which appeared in 1593.

by the Anglican apologists. Whitgift, for example, said that the form of the church depends on time and place. Indeed, he insisted that the Apostolic Church cannot be taken in all respects as a model for later churches because there was a total absence of Christian magistrates. In Christianity's early days of persecution Whitgift granted there might well have been the need for seniors (or elders) to assist in setting up the organization of the congregations, as well as for deacons to look after the needs of the poor, but that in more enlightened Elizabethan days such offices were not necessary. The coming of the Christian state made any distinction between church and state a mere anachronism. Cartwright rebutted this argument by stating that there was greater need for elders in the present because ministers needed greater assistance than when there was a more plentiful effusion of the gifts of the spirit. But church government, Cartwright insisted, cannot be decided, as Whitgift had suggested, by considerations of utility or efficiency or convenience, but only by obedience to the pattern plainly shown in God's Word, and God "will plane the ways be they never so rough."[64]

On the issue of the proper relationship of church and state, both Anglicans and Presbyterian Puritans asserted the importance of joint responsibilities; yet the Presbyterians believed in a "two-kingdom" doctrine. The Anglican church is ultimately subsumed under the English state, and this is fittingly expressed in the acknowledgment of the queen as the 'supreme Governor' of the Church of England. The Presbyterian theory, however, refused to place the church under the state, for the princes or magistrates were subject to the sovereignty of God and needed the guidance of the Word of God as preached. Thomas Cartwright clearly envisioned the primacy and priority of the claim of the church: "As the hangings are made fit for the house, so the Commonwealth must be made to agree with the Church, and the government thereof with her government. For, as the house is before the hangings, and therefore the hangings which come after must be framed to the house which was before, so the Church before there was any Commonwealth and the Commonwealth coming after, must be fashioned and made suitable unto the Church."[65] The church must lead the state in the theocracy envisioned by Cart-

[64] The debate on ecclesiology between Whitgift and Cartwright is summarised in A. F. Scott Pearson, *Church & State: Political Aspects of Sixteenth Century Puritanism*, pp. 120f.
[65] Cited in Whitgift, *Works*, III, 189.

wright, which would turn England into a second and larger city of God, like Geneva. In short, the Anglicans with a one-kingdom theory, leaned toward secularising the church as the state's department of religious affairs, while the Puritan two-kingdom theory looked in the direction of spiritualising the state.

7. *Views of the Sacraments*

Anglicans and Puritans were agreed in much of their understanding of the nature of the sacraments. They denied Transubstantiation, Consubstantiation, and a naked Memorialism (the latter is often wrongly attributed to Zwingli). Both Anglicans and Puritans insisted that Communion was the commemoration (not the repetition) of Christ's sacrifice, and that it was a means of grace, as also that it must be received in faith.[66] The sacraments, in short, commemorate but also communicate the grace of God.[67] Anglicans and Puritans reduced the number of sacraments from the traditional seven of the Catholic church and the Orthodox churches to two—Baptism and Holy Communion—believing that these two were Sacraments of the Gospel. That is, they believed that the two sacraments represented the forgiveness and new life which were the chief gifts of Christ's death and resurrection, and were given in ordinances for which there were a Dominical warrant and example.

There were, of course, differences of emphasis in the understanding of Baptism and Communion. For Anglicans the sacraments were the chief means of grace, whereas for Puritans, in the tradition of Calvin the sacraments were seals of grace, confirmations of a prevenient grace already received, proofs that the divine promises of the Gospel are fulfilled. Cartwright defined a sacrament: "a signe & seale of the covenant of grace, or an action of the Church, wherein by outward things done according to the ordinance of God, inward things being betokened, Christ and his benefits are offered to all, and exhibited to the faithfull,

[66] Hooker declared: "The real presence of Christ's most blessed body and blood is not therefore to be sought for in the sacrament, but in the worthy receiver of the sacrament." *Ecclesiastical Polity*, v, vi, 67.

[67] New's *Anglican and Puritan* is accurate in stating (p. 62) that for both the Sacraments were "commemorations and participations in grace with a living Christ," but no Puritan would have agreed with the conclusion of the same sentence "by virtue of His real spiritual presence in the consecrated elements." The Puritans did not reserve the Sacrament, as would have been appropriate if they were conscious of a presence in the consecrated elements; rather, they insisted that Christ was present in the action in general, in fulfillment of His promise, and in the hearts of His faithful people.

for the strengthening of their faith in the eternall Covenant."[68] The Anglican did not feel it necessary to have a sermon preached at a Holy Communion service,[69] but the Puritan did, so that the sign would 'speak'—that is, that the meaning of the affusion of water in Baptism and of the bread broken and the wine poured out in the Lord's Supper should be proclaimed in the preaching.

The Anglicans were frequently told that the Eucharist was the sacrament of incorporation into the body of Christ, but there was little in the arrangement of the service to suggest this, for the participants received the consecrated elements singly and successively. On the other hand, the Puritans tried to recapture the banquet aspect of the Last Supper in their Lord's Supper, and were acutely conscious of being guests at Christ's table; they received the sacred elements as a group, not as individuals. Their emphasis on the importance of the Communion as a testimony or badge of Christian commitment also stressed the corporate aspect of the Communion service.

John New was probably right in seeing Anglicanism's appreciation of the sacraments as implying a more dynamic view than the more static view of the Puritans.[70] Surely Anglicanism's keeping of marriage as "holy matrimony," of confirmation and burial as *sacramentalia* or quasi-sacramental rites, pointed in this direction. Puritans, on the other hand, found no need for confirmation; they regarded marriage as a civil ordinance on which it was appropriate to seek the divine blessing, and burial as the termination of a life, at which prayer was appropriate. In no case did they think these occasions marked stages where divine grace was appropriately given for strengthening.

As I will have fuller opportunity to point out later,[71] there was a different order of priorities in the Christian life for Anglicans and Puritans. Anglicans give the primacy to sacraments, and Puritans to sermons, as channels of the knowledge and the grace of

<hr>

68 *A Treatise of Christian Religion or The Whole Bodie and substance of Divinitie* (1616), published posthumously, p. 212. William Ames wrote in *The Marrow of Theology*, p. 197, that "the special application of God's favour and grace which arises from true faith is very much confirmed and furthered by the sacraments."

69 However, it should be remembered that the only place where a sermon was called for in the Prayer Book was at Holy Communion. Presbyterians had infrequent Communions preceded by a preparatory sermon and followed by a thanksgiving sermon on each occasion, as is attested by Robert Baillie, *A Dissuasive from the Errours of the Time* (1645), p. 29.

70 *Ibid.*, p. 76.

71 In separate chapters on the Eucharist and on preaching I will discuss these differences in much greater detail.

God. The supreme moment for the Anglican in worship was when, on his knees before the altar, he received the sacramental symbols of the sacrifice of his Savior. The supreme moment for him was when the preacher, ascending like a Moses to his Sinai, revealed the oracles of God and the internal testimony of the Holy Spirit assured him inwardly of its truth and transforming power.

This differing evaluation of sacraments and sermons led to functional differences between the Anglican priest and the Puritan minister. The Anglican clergyman needed to spend less time in preparing sermons and so could concentrate on spiritual contemplation and on the tasks of administration and visitation. The Puritan minister, on the other hand, spent long hours in his study, preceded and succeeded by prayer, reading the Scriptures in their original languages and conning every reputable Scripture commentary he could lay his hands on. In other words, for the Anglican the sacramental was the primary mode of Christ's presence, but for the Puritan the primary mode of Christ's presence was kerygmatic.

8. Concepts of Ethics and Eschatology

The softening or weakening of the negative aspect of the doctrine of predestination in the interests of God's universal charity led to a weakening of the presentation of the two eschatological alternatives, heaven or hell, in Anglican thought. Less the sovereignty than the benevolence of God was stressed; the strong implication was that the truest service of God was the service of the neighbour. For the Puritans, however, living "in my Great Taskmaster's eye" (as Milton was to phrase it[72]), actions had abiding consequences; the sovereign God could never be interpreted as our Grandfather who art in heaven. God's sovereignty involved his severity, so although the elect would be saved they still had to undergo the severity of the day of judgment. In their diaries (which were their spiritual accounting books) Puritans often anticipated the great assize in their imaginations.

Both Anglicans and Puritans insisted that good works were incapable of contributing to a man's salvation, that they were the harvest of the indwelling of the Holy Spirit and a necessary proof that the redeemed man was also the sanctified man.

Anglican ethics tended to be more perfectionist and Puritan ethics more pragmatic, in keeping with the Anglican presupposi-

---

[72] "On his having arrived at the age of twenty-three."

tion inherited from the Middle Ages, that grace completes nature,[73] and in contrast with the Puritan disjunction between the natural and the spiritual man. Also, as we have seen, the Puritan took more seriously the disruptive force of original sin and the stain and corruption it spread in persons and institutions. This was a strong countervailing force against any tendency toward sentimental utopianism (which arose in Commonwealth Puritanism under strong eschatological pressures, coupled with an overwhelming conviction of the possession of the saints by the Holy Spirit; but the Presbyterians were never intoxicated by such theological drink). Furthermore, the ethics of Puritanism were forged in a world of everyday work and politics, and the attempt to subordinate politics and economics to the way of righteousness was a very disenchanting occupation.

By contrast, Anglican ethics (in the one-kingdom concept of the state-church entity) tended to support the status quo, to sanctify what *was* rather than what *should be*. There were remarkable examples of individual Anglican generosity to the poor that deserve to be called sacrificial generosity. The spirituality that fed the ethics of Anglicanism was too other-worldly to produce a transforming ethic. (We may suppose that for one George Herbert there were hundreds of Robert Herricks in the rectories of England.)

The Puritan's conception of "calling" leads us into the fullest appreciation of his understanding of human life and destiny. The reference is not to the general calling to faith and obedience in the Gospel, but to the particular vocation—which, in the thought of Perkins, can be subdivided into the familial obligations imposed by one's position in a household (father and husband and master of servants) and the obligations imposed by one's profession or trade and one's role as citizen.[74] In this calling a man had to have the necessary gifts from God and the approbation and confirmation of men (exactly, it should be noted, the requirements before a man can be ordained in the Puritan ministry).[75] Robert Bolton insisted a Christian's first duties were fruitful performance of holy tasks, but that his second duty was to decline idleness, "the very rust and canker of the soule," and "be diligent with conscience and faithfulnesse in some lawfull, honest particular Calling . . . not so much to gather gold . . . as for necessary and moderate provi-

---

[73] St. Thomas Aquinas wrote: "Grace completes rather than destroys nature."
[74] "A Treatise of Vocations," in *Works*, I.
[75] *Ibid.*, p. 760.

sion for family and posteritie: and in conscience and obedience to that common charge, laid upon all the sonnes and daughters of Adam to the worlds end."[76] But what was supremely important was that God was served in the secular calling: "whatsoever our callings be, we serve the Lord Jesus Christ in them," John Dod said, and he added the consoling thought for the underdog: "Though your worke be base, yet it is not a base thing to serve such a master in it."[77]

This then was an intramundane asceticism to which Puritans were called. Perhaps here we can see most clearly the difference between Anglican and Puritan ethics. In Anglicanism there was a sense that the true following of God is to be found in the self-deprivation and restraint of nature that were inherent elements of the monastic emphasis on evangelical poverty and chastity—this was completely lacking in Puritanism. But even in Anglicanism it was temporary, seasonal, Lenten. The Puritan ethic was totally integrated with general and secular callings. Anglican writers such as Hall and Sanderson were far from oblivious to instruction on callings, but they did not as strikingly insist that all callings are equal in God's sight and make no difference in terms of man's ultimate destiny. The Anglican was too clearly aware of the differences of degree in society and of the existing social hierarchy to be able to say with Perkins: "the action of a sheepheard in keeping sheep, performed as I have said, in his kind, is as good a worke before God, as is the action of a Judge, in giving sentence, or of a Magistrate in ruling, or a Minister in preaching."[78]

In comparison with Puritanism, Anglican ethics seem static. Puritanism was a spiritual dynamo, charged with the responsibility of creating New Jerusalems in England, after the Biblical pattern renewed in Geneva. Anglican thought was too paralysed by rational hesitations and compromises, by the smoothing away of the rough edges of severity in the interests of Divine charity, and by a greater tolerance. The result was seen in an imprecision and vagueness of attitude, all tending to elasticities of conscience in the half-light of uncertainty. The narrow but absolute certainties of the Puritan provided infinitely greater drive and impetus in ethics. The certainty of man's depravity, the absolute trustworthiness of God combined with his sovereign power to effect

[76] *Works*, 4 vols., IV, 66-67.
[77] John Dod and Robert Cleaver, *Ten Sermons . . . of the Lord's Supper* (1609), p. 82.
[78] *Works*, I, 758.

His will against all obstacles, the unmistakable obligation of His commandments, the certainty and clarity of the pattern of His purpose for all aspects of human life in the world of Scripture, all drove the Puritan toward an ethics of challenge, not stoic acceptance, toward fight rather than flight. Only 50 years after the period under study the meaning of human life would be expressed in sublime epitome at the Westminster Assembly of Divines, of both Presbyterian and Independent wings of Puritanism, as follows: "man's true end is to glorify God and to enjoy him for ever."[79]

9. *Two Christian Styles*

We have yet to try to discover the heart of Anglican and Puritan piety and where their distinctive styles of Christian devotion and life are to be found.

At the heart of Anglican piety there was the awed wonder at the condescension of the God-man, sheer adoring amazement at the humility of the Incarnation. The greatest Anglican sermons, such as those of Andrewes,[80] Donne,[81] or, for that matter, Newman[82] before 1845, took as their starting point the paradox of Bethlehem, where the Divine Son stoops to conquer and is most compelling in His weakness. They seemed to be happiest when they narrated the chief events of the Gospel from the Gospels, beginning with the Virgin Birth, where the second Eve recovers what the first Eve lost, and provides through her Son the Savior

[79] The Westminster Shorter Catechism.

[80] In Andrewes' *Ninety-Six Sermons*, 5 vols., there are 17 on the Nativity, three on the Passion, 18 on the Resurrection, 15 on Pentecost (Whitsunday), seven on Christ's Temptations, and 15 on the Lord's Prayer—a total of 60 "Christological" sermons.

[81] "Donne preached at St. Paul's his great series of Easter Sermons, which, together with his series of Christmas Sermons, may be held perhaps to represent his finest achievement as a preacher." (*The Sermons of John Donne*, ed. G. R. Potter and E. M. Simpson, 10 vols., IV, 6.) A sample of the paradoxes of the Incarnate Christ and the tenderness such contemplation inspires is the following excerpt from a sermon on the Crucifixion in Donne's *LXXX Sermons* (1640), p. 401: ". . . I see those hands stretched out, that stretched out the heavens, and those feet racked, to which they that racked them are foot-stooles; I heare him, from whom his nearest friends fled, pray for his enemies, and him whom his father forsooke, not forsake his brethren; I see him that cloathes this body with his creatures, or else it would wither, and cloathes this soule with his Righteousnesse, or else it would perish, hang naked upon the Crosse; And him that hath, him that is, *the Fountain of the Water of Life*, cry out, He thirsts. . . ."

[82] Newman's great Incarnational sermons, "Christ the Son of God made man" and "The Incarnate Son a Sufferer and a Sacrifice," are found in Vol. 5 of his *Parochial and Plain Sermons*, 8 vols. (1868); they were preached during his Anglican days. The finest of all on the same theme is, however, his Catholic sermon, "Omnipotence in Bonds," from *Sermons preached on Various Occasions*.

a second chance for humanity, on through the agony and bitter sweat of the Passion to the miracle of the Resurrection, where God tolls a bell for the beginning of a new eternal life for redeemed man. For the Anglican, then, the heart of piety was devout meditation and adoration evoked by contemplation on the mysteries of the Divine revelation concentrated in the saga of Christ's life. It became the Sacrament of the love of God, from the Nativity to the Passion to the Resurrection. This single, richly meaningful event was like a diamond presenting a different facet at every festival in the round of the Christian Year, inviting the Christian to imitation. In this meditation Christ was seen as the supreme revelation of the nature of God and of the possibilities open to man; the response evoked was always a gratitude of tenderness. It seems to be the continuation of the mysticism of medieval England.

English Puritanism, however, did not lose itself in the mystery of the Incarnation, which it regarded as a subject for theological explanation rather than meditation. It had little interest in the retrospective gaze and little incentive, since it had rejected the Christian Year because of the multitude of saints' days, which hid the solitary splendor of the king of saints, Christ. Its center of interest was not in the incidents of the Gospels, but in the Epistles of St. Paul and in the Acts of the Apostles, for it was absorbed in trying to live in the Spirit. Puritanism's point of departure was not the Nativity, the Passion, or the Resurrection, but the Ascension of Christ and the descent of the Holy Spirit. The Puritans were interested in the glorification of Christ at His ascension rather than in His humiliation at Bethlehem or Calvary. They had little interest in the historic drama of the past, only with the civil war of the soul in the present, in which Christ fought with Satan for possession. Theirs was not so much an imitation of Christ, as Anglican piety was, but a recapitulation in themselves of the story of Everyman Adam, from temptation and fall, through reconciliation, restoration, and renewal. They were interested in the stages of the redeemed soul's progress: election, vocation, justification, sanctification, glorification. "Here," said Haller, "was the perfect formula explaining what happened to every human soul born to be saved."[83]

The Puritan was not oblivious to the major events in the life of Christ, but the birth he cared for was the rebirth of the soul, his own regeneration. The crucifixion he bothered about was the crucifixion of the old Adam in him by the power of the new Adam,

[83] *Rise of Puritanism*, p. 90.

Christ. The Christian life was not for him the recapitulation of the main events in the life of Christ as it was for Anglican piety, so that in him the Biblical promise, "if we suffer with Him, we shall reign with Him"[84] is fulfilled. Rather, for the Puritan, life was a struggle against the forces of Satan under the orders of Christ the Captain of salvation, together with Christ's elect. It was a warfare from which there was no release, but the tired troops were encouraged by the assurance that the victory of Christ and His saints on earth was sure and would be the prelude to the triumphant return of Christ with all His saints to heaven. Ultimately the difference of style between the Anglican and the Puritan was the difference between life as a pilgrimage toward the shining towers of heaven glimpsed mystically as the clouds part, and life as a fighting the good fight of faith, with the courage of obedience, empowered by the sword of the Spirit.[85]

10. Theology's Consequences for Worship

The theological differences between Anglicanism and Puritanism inevitably led to differences in the practice of worship. The chief differences were the following: First, Anglicans were free to use the customs of the ancient church, provided Scripture did not veto them, whereas the Puritans demanded a positive warrant in Scripture for all their ordinances and even for the details of their organisation. Second, the chief means of grace for the Anglicans were the sacraments, especially Holy Communion, while for the Puritans it was unquestionably the lively oracles of God in preaching (though it should be recalled that there was an admirable tradition of preaching in the Church of England, especially in cathedral and university pulpits, and that there was a strong tradition of eager and reverent attendance at the Lord's Supper on the part of Puritans). This Anglican emphasis on the primary role of the sacraments was partly responsible for the ever-present dangers in the Church of England of sacerdotalism and the "sacristy mind," or of a play-acting in the form of ritualism. On the other hand, the Puritan mania for sermons often brought its revenge in a dreary and desiccated didacticism in devotions in the Puritan tradition of worship. The third consequence of the different theological outlooks was a deep loyalty to liturgical worship in Anglicanism and

84 2 Timothy 2:12.
85 For an excellent study of Puritan thought on the holy spirit see Geoffrey F. Nuttall, *The Holy Spirit in Puritan Faith and Experience.*

more than a little suspicion of its formality in the Puritan tradition. Puritanism, however, was not cut all of one cloth; the Presbyterians preferred a national liturgy with some variety and freedom of words in the pastoral prayer, while the Separatists and the Independents appreciated more impromptu praying by the minister. In the fourth place, Anglicans kept such ancient vestments as the surplice and the cope, and such time-honoured ceremonies as kneeling for the reception of Holy Communion, the signing of the Cross in Baptism, and the use of the ring in marriage. All these vestments and ceremonies were rejected by the Puritan iconoclasts as the remnants of Romish superstition. Fifth and finally, the Christian calendar, celebrating the chief events in the life of the Incarnate Son of God and commemorating the Virgin and the leading saints, was retained in streamlined form by the Church of England, but was discarded by the Puritans, although they had their own special days such as the weekly sabbath and special days of humiliation and thanksgiving, by which they marked the judgments and the providences of God as related to the nation or the family. These five major differences of interest and emphasis must be examined in greater detail.

The Puritans, as just mentioned, were fully committed to the acceptance of the Holy Scriptures as the supreme liturgical criterion and canon. This was made plain by William Bradshaw, a friend of William Ames, in his *English Puritanisme* (1605), who asserts at the outset: "IMPRIMIS they hould and maintaine that the word of God contained in the writings of the Prophets and Apostles, is of absolute perfection, given by Christ the Head of the Churche, to bee unto the same, the sole Canon and rule of all matters of Religion, and the worship and service of God whatsoever. And that whatsoever done in the same service and worship cannot bee iustified by the said word, is unlawfull."[86]

In trying to reproduce the worship of the Apostolic church of the New Testament the Puritans discovered six perpetually binding ordinances: Prayer, Praise, the proclamation of the Word, the administration of the sacraments of Baptism and the Lord's Supper, Catechising, and the exercise of Discipline. They sought for the types of prayer and the matter of suitable prayers in the New Testament, as well as for the varieties of approved praise. They even sought for the details of their ordinances in the Scriptures. They found the authority for holding two services on a Sunday

[86] *Ibid.*, p. 1.

from the requirement of a double burnt offering recorded in Numbers 28:9, and the sanction for introducing the Lord's Supper with a sermon in Acts 20:27.

Biblical fidelity could all too easily degenerate into Bibliomania in the more extreme forms of text-hunting and strained interpretation. Perhaps the most extreme example was arguing that God abhorred the responses in the Litany of the Book of Common Prayer from the prohibition of cuckoo's meat to the Jews in Leviticus.[87] Disregarding such special pleading and hair-splitting interpretation as mere eisegesis, it must still be recognized that the value of providing a Biblical warrant for all the ordinances of Puritan worship was that each was directly related to the divine will, and that this gave these ordinances an august authority for those who used them, as the Puritans did, in the obedience of faith. By contrast, the Anglican was free to use all decent, edifying, and comely ceremonies not prohibited by the Scriptures, and the church had the right, in his view, to arrange the details of worship as suited its convenience and traditions.

The excitement the Anglican kept for the sacrament, the Puritan exhibited over the sermon. The "sacramental principle," itself based on the supremely sacramental encounter of God with man in the Incarnation, enabled the Anglican in Platonist fashion to see in all the beauties of earth shadows of the more glorious realities in heaven by an analogy of ascent, for there was no fundamental dichotomy for him between nature and grace. He believed that the sacraments conveyed grace to fortify his soul; as bread and wine formed the staple diet of his earthly body, so the grace of the sacrament strengthened his spirit. The mystical approach to the sacrament elated Hooker with an almost Baroque ecstasy: "The very letter of the Word of Christ giveth plain security, that these mysteries do, as nails, fasten us to his very Cross, that by them we draw out, as touching efficacy, force, and virtue, even the blood of his gored side; in the wounds of our Redeemer we there dip our tongues, we are dyed red both within and without, our hunger is satisfied, and our thirst for ever quenched; they are things wonderful which he feeleth, great which he seeth, and unheard of which he uttereth, whose soul is possest of this Paschal Lamb, and made

[87] The author owes this example of overingenious biblical interpretation to a suggestion by the Rev. Dr. William McMillan, author of *The Worship of the Scottish Reformed Church 1550-1638*, in a book review of *The Worship of the English Puritans* in the *Dunfermline and West of Fife Advertiser*, issue of 24 July 1948. Unfortunately Dr. McMillan gave no Puritan bibliographical source.

joyful in the strength of this new Wine; this Bread hath in it more than the substance which our eyes behold; this Cup hallowed with solemn benediction availeth to the endless life and welfare both of soul and body, in that it serveth as well for a medicine to heal our infirmities and purge our sins, as for a sacrifice of thanksgiving; with touching it sanctifieth, it enlighteneth with belief, it truly conformeth us unto the image of Jesus Christ."[88]

The Puritan preacher made exactly the same claim for the proclamation of the word of God. The minister was, in the Pauline phrase, Christ's ambassador, and his Master's messages were delivered from the pulpit with the authority of a royal representative. It is surely a means of grace which could transform and convert the stony heart into a sensitive and charitable spirit, or supply new motives for living, or expel fears and anxieties, or create a divine dissatisfaction with the world as it is and a desire to rebuild it, or make it possible to bear suffering without bitterness—and these were all the triumphs of the Holy Spirit won from the pulpit. As that great Puritan author of *The Arte of Prophecying*, William Perkins, wrote: "Preaching of the word is Prophecying in the name and roome of Christ whereby men are called to the state of Grace, and conserved in it."[89] The sacrament had the advantage that its objectivity was guaranteed by the operation of God, although even it could be taken for granted and become too familiar. Preaching had the advantage of variety but the disadvantage of subjectivity.

It was only in the later development of Puritanism that there was a clear-cut opposition to all set forms of prayer and liturgies, as "stinting" the Spirit. Cartwright, for example, disliked the Book of Common Prayer, he was looking for another national formulary of prayer to replace it with. In general, he approved of liturgies: "First, for testifying the consent of all true Churches, in the things that concerne the worship and service of God, which may appeare by such bookes. Secondly, for direction of the Ministers to keepe in their administration (for substance) like soundnesse in doctrine and prayer. Thirdly, to helpe the weaker and ruder sort of people especially; and yet so, as the set forme make not men sluggish in stirring up the gift of prayer in themselves, according to divers

[88] *Ecclesiastical Polity*, v, 67.
[89] The full title is *The Arte of Prophecying, Or, a Treatise Concerning the Sacred and Onely True Manner and Methode of Preaching*, and is to be found in Vol. II of the three-volume edition of the *Works* of Perkins issued posthumously in 1613. It had first appeared in Latin in 1592. The quotation is II, 646.

occurents."[90] However, William Perkins, in his important *Cases of Conscience*, in answer to the question, Is a set form of prayer lawful? replied, "it is not a sinne," because the Psalms of David are in a form and most of them are prayers. Still he seemed to think prayer in the Holy Spirit was preferable since it required great gifts which many lack and therefore fall back on forms of prayer. His words were: "Secondly, to conceive a forme of prayer requires gifts of memorie, knowledge, utterance, and the gifts of grace. Now every child and servant of God, though he have an honest heart, yet hath he not all these gifts; and therfore in the want of them may lawfully use a set forme of prayer, as a man that hath a weake backe, or a lame legge, may lean upon a crutch."[91] The Anglicans had no alternative but to approve of the liturgy since it was one of the requisite modes for establishing an uniformity in religion. It is significant, however, that the Book of Common Prayer was defended, not only on the grounds of convenience and utility, but on the strongest historic grounds that all liturgies came from one common mould and that the churches of the past had always worshipped God in this way.[92]

This consideration of the Vestiarian Controversy has shown how unwilling the Puritan ministers were to accept the garments they considered hopelessly polluted by Roman Catholic use and association. I have not, however, shown with sufficient force how radical Anglican opinion detested the piling up of ceremony on ceremony and considered it as characteristically Roman Catholic. One of the most amusing castigations of excessive ceremonialism comes from the pen of the martyrologist, John Foxe: "What Democritus or Calphurnius could abstaine from laughter, beholding only the fashion of their masse, from the beginninge to the latter end, wyth suche turning, returning, halfe turning and hole turning, such kissinge, blessing, crowching, becking, crossing, knocking, ducking, wasshing, rinsing, lyfting, touching, fingring, whispering, stoping, dipping, bowinge, licking, wiping, sleping, shifting with

90 *Treatise*, p. 256.
91 *Works*, II, p. 67.
92 Hooker writes (*Ecclesiastical Polity*, v, 26): "No doubt from God it hath proceeded, and by us it must be acknowledged a work of his singular care and providence, that the Church hath evermore held a prescript Form of Common Prayer, although not in all things everywhere the same, yet for the most part retaining the same analogy. So that, if the Liturgies of all ancient Churches throughout the world be compared among themselves, it may be easily perceived that they had all one original mould, and that the public Prayers of the people of God in Churches throughly settled, did never use to be voluntary dictates proceeding from any man's extemporal wit."

a hundreth thinges mo. What wise man, I saye, seing such toyish gaudes can keepe from laughter?"[93] In utter contrast, the Puritans aimed at the simplest celebration of the Lord's Supper. The whole thrust of their worship, with the requirement of family preparation at home, and the need for children to recall the main points of the sermon to the satisfaction of their parents, was less on the external and legal requirements of worship than on the spiritual and interior demands it made. Significantly Perkins distinguished between outward and inward worship. The outward worship was the use of the ordinances, but only for the purpose of inward worship. Inward worship consisted of adoration of God and cleaving to Him. The adoration of God was expressed through the cultivation of four virtues: Fear, Obedience, Patience, and Thankfulness. Cleaving to God was possible through the cultivation of faith, hope, love and inward invocation.[94]

It is to be noted in the fifth and final place that the Anglicans kept the structure of the Christian calendar, while the Puritans utterly rejected it. The advantage of the structure of the Christian Year was that it led to that annual remembrance of the chief events of the Christian saga, so that it was another *itinerarium in mentis Deum*, and required its minister to preach on the Gospel or the epistle chosen for the day. Its other advantage was that it kept alive, through its reduced sanctoral cycle and the commemoration of certain saints, the sense of the communion of saints and of the unity of the church triumphant in heaven with the church militant on earth. Puritanism, in concentrating on the spiritual meaning of the Scriptures for the present, was to take a more instrumentalist view of Christ, leaving to Anglicanism a more contemplative view. Puritanism was afraid of being stranded on the shores of the Palestine, and of not making England the holy land and the English people the holy nation. But as a result, relevance was often found at the cost of highly tendentious and idiosyncratic exegesis, because the ark of the church, having cut the painter that bound it to history, was free to move with the winds of the Holy Spirit or as led on by the will-of-the-wisp of private fantasy. The great Anglican preachers found the historic Christian doctrines of the Incarnation, Reconciliation and Resurrection to be no more burdensome than sails are to a boat. Puritan freedom could very easily degenerate into license, as Anglican circumscription could easily

[93] *Acts and Monuments* (1563), sig. EEEeb. A similar list, with better rhythm and cadence, can be found in Thomas Becon's *The Jewel of Joy* (1560), folio xxx.
[94] *Cases of Conscience*, in *Works*, II, 62-63.

degenerate into a set of clichés. Both form and freedom had their rights and their representatives, but—what was not apparent in this controversy—they were complementary, not alternative needs. Form needs freedom to keep it fresh, and freedom needs form to prevent it from turning to irresponsible chaos.

In the last analysis there was a difference of spirit between the Anglican and Puritan ways of worship. On one side were the Anglicans whose historic and aesthetic bent found delight in the continuity of the church from Apostolic times, and rejoiced to read in the Fathers the testimony of a far-off witness to the same God and Father of Christ, and was encouraged by the examples of the saints. The Anglicans found joy in the symbolism that spoke to the eye as the sermon and organ music to the ear. They were happy that the architecture reflected the crucifixion in its very shape of nave and chancel crossed by transepts, and in the furnishings which praised God through the crafts of men in carved wood and stone, as in painted window and curiously wrought metal.

On the other side were the Puritans who are too readily dismissed as color-blind and tone-deaf Philistines in church. Yet for them all these Anglican delights were compromises with Roman Catholicism, debilitating distractions from the true service of God and man, outward shows as pretty and irrelevant as carousels, the otiose gilding of the lily and the varnishing of sunlight. The Puritans believed that worship was essentially the turning of the heart in reverent gratitude to God, and the slow shaping of the will by long discipline until it is the responsive instrument for the Spirit's bidding. When they came to build their own meetinghouses, the houses were scrubbed, white, bare, as austerely naked as the soul should be in the sight of God, stripped of all its disguises and pretensions.

The ultimate contrast, though never absolute, seemed to be between the Anglican's sense of the beauty of holiness combined with the holiness of beauty, sacramentally understood and perfectly mirrored in the poems of George Herbert, rector of Bemerton, and the Puritan's sense of the primacy of the Holy Spirit introspectively understood moving reluctant man through the stages of spiritual growth and disciplined training. The Anglican was a pilgrim and the Puritan a soldier; the former's piety was solitary or familial, the latter's was as a member of God's elect, one of the predestined and invincible ironsides of God. A pilgrim might go alone; but a soldier fights most effectively as a part of a disciplined regiment.

CHAPTER III

THE EUCHARISTIC CONTROVERSY[1]

THE EXAMPLES of the Christian humanists like Erasmus and Colet, the labours of Tyndale and Coverdale as translators of the Bible into English, and the conviction of all the Reformers, that the Gospel alone had the power to overthrow the old strongholds of corruption and superstition, combined to give a primary place to teaching and preaching from the pulpit in the sixteenth century. By contrast, nowhere was disagreement greater than in the interpretation of the meaning of the Catholic Mass, Lutheran or Anglican Holy Communion, or Reformed Lord's Supper. This disagreement was dramatically exposed at the Colloquy of Marburg in 1529. The Colloquy had been convened by Philip of Hesse to unite the forces of the Saxon and Swiss Reformers, but it only revealed the depths of the yawning chasm between Lutheran and Zwinglian interpretations of the Eucharist.[2] It is interesting that the English divines, both Catholic and Protestant, devoted a vast amount of attention to this theme, particularly from 1540 to 1560, when political changes forced them to take cognizance of religious change. The Anglican church, proud of its patristic heritage from the time of Cranmer, followed the early fathers in their devotion to Christ inspired by the chief sacrament, without depreciating the role of preaching.[3]

From the moment King Henry VIII entered the lists against Luther with his *Assertio Septem Sacramentorum* in 1521, up to the publication of the fifth book of Hooker's *Treatise of the Laws*

[1] The only modern scholarly histories of Anglican Eucharistic interpretation are two studies by C. W. Dugmore, *Eucharistic Doctrine in England from Hooker to Waterland* and *The Mass and the English Reformers*. The second deals with the period under study here, and is valuable in its fairness to Cranmer; but it underestimates the impact on Cranmer of the Continental Reformers and the Continental exiles in England. Gregory Dix's *The Shape of the Liturgy* also is least satisfactory in its treatment of the Continental Reformers.

[2] The Reformers had agreed on the first 14 articles, disagreeing only on the 15th. Cf. the "Relatio Rodolphi Collini de Colloquio Marburgensi," in *Zwinglii Opera*, ed. M. Schuler and J. Schulthess, IV, 173-82. Most of it is given in B. J. Kidd, *Documents illustrative of the Continental Reformation*, pp. 247-54; W. Köhler, *Das Marburger Religions-Gesprach: Versuch einer Rekonstruktion*, Schriften des Verein für Reformationsgeschichte, no. 148.

[3] Paradoxically the Oxford Movement of the nineteenth century, which owed as much to Newman's preaching as to the 90 "Tracts," tended to depreciate sermons and sermon-tasters. See W. D. White's dissertation, "John Henry Newman, Anglican Preacher," Princeton University, 1968.

of Ecclesiastical Polity written against the Puritans in 1597, there
was a continuing controversy about the Eucharist, and especially
about the mode of Christ's presence in the sacrament. This may
seem to have been a very narrow theological issue at the outset, but
as the debate developed and the issues proliferated, it became a
crucial issue of faith and life as well as liturgy; one's eucharistic
beliefs were in many of the Tudor years literally a matter of life
or death.

The importance of the matter may be seen in a listing of some
of the questions raised in the controversy. What were the chief
Scriptural sources for the institution and meaning of the Eucharist? Did not John 6 and I Corinthians 11 give radically different
interpretations? How were the Dominical words *Hoc est corpus
meum* ("This is my body") to be interpreted—literally or figuratively? Did "body" refer to Christ's historical body, His resurrected
body, or the church as His extension or "Body?" If a literal interpretation was preferred, was "Transubstantiation" or "Consubstantiation" the better interpretation of Christ's presence, following
Scripture and the Fathers? Or was it even better to affirm this as a
supreme and transcendent mystery? Where is the "body" of Christ
located: on the altar, in heaven, or in the hearts of the faithful?
If it is on the altar, is the body "in" or "under" the bread and
wine, and at what point in the Liturgy does consecration take
place? Is the Eucharist a propitiatory sacrifice for the living and
the dead? But if Christ's sacrifice was complete on the Cross, what
need was there for repetition? Or was the Eucharist a memorial
banquet? Or was it the oblation of the church with thanksgiving?
What were the eschatological dimensions of the Eucharist? Is
consecration effected by Christ the Word made flesh as the priest
uses His words of institution, as was commonly held in the West,
or was it, as in the East, effected by the agency of the re-creating
Holy Spirit? Was faith essential to the reception of the Eucharist,
or would a wicked man really be partaking of the body of Christ?
Who were the chief exponents of the various interpretations of the
modality of Christ's sacramental presence, and were their philosophical presuppositions nominalist or idealist? Was there a direct
ratio between the more conservative the doctrine of the Eucharist
and its more frequent reception? To be precise, did a transubstantiatory interpretation lead to more frequent reception of the Holy
Communion than a Zwinglian or Memorialist interpretation of the

Lord's Supper? These were some of the major theological issues raised by the Eucharistic controversy.

It should not be forgotten, however, that in the midst of the prolonged and bitter struggle the contending parties were not merely discussing doctrines. They were fighting for a way of religion as life. Complete tolerance arose only in a state of utter indifference to the issues. It was not a matter of intellectual tastes or traditions; it was one of arriving at and defending the *truth*, and it was often the disillusioned Catholic who was the most vigorous Protestant, as in the case of the former Cistercian Hooper who became a convinced supporter of Swiss theological views when in Zurich from 1547 to 1549. Unquestionably though Gardiner was Bishop of Winchester in the same Church of England in which Hooper was Bishop of Gloucester, Gardiner's meat was Hooper's poison. Bishop Gardiner was a loyal Erastian, but not a rigid and inflexible antiquarian theologian. He acknowledged that there was a time when the Catholic church had not tied the doctrine of the Real Presence to the essential explanation of Transubstantiation. Yet for him to deny the reality of the miraculous presence of Christ in the mutation of the bread and wine into the very body and blood of Christ was to deny the omnipotence of God,[4] to refuse the most straightforward explanation of Christ's words at the Last Supper, and to reject the marvellous consolation the faithful had received for more than a millennium. Hooper's explanation of the meaning of the Sacrament as a picture of the Divine benevolence, as proof of God's promises, and as a sign of Christian allegiance, must have seemed the sorry reduction of the chief sacrament to a sermon illustration. Yet that high Transubstantiatory evaluation of the Mass seemed to John Hooper the rankest idolatry and a superstition only explicable in terms of priestcraft exploiting the credulity of simple people.[5]

[4] Gardiner, in replying to Cranmer, stated that a miracle had been flattened to a mere memento in Cranmer's Eucharistic doctrine, and presented the latter's reductionistic view ironically: "So as in the ordinance of this Supper after this [Cranmer's] understanding, Christ shewed not his omnipotency, but only that he loved us, and would be remembered of us." From *Writings and Disputations of Thomas Cranmer . . . relative to the Sacrament of the Lord's Supper*, ed. J. E. Cox (Parker Society, Cambridge, 1844), p. 16. Henceforth this will be referred to as Cranmer's *Lord's Supper*.

[5] The vehement Hooper, who in Zurich had become the friend of Bullinger, Zwingli's successor, was irrational on the subject of transubstantiation, which he believed was sheer idolatry. This explains, if it does not condone, the offensiveness of his writing, as follows: "The mother of this idolatry was Rome, and the father unknown. A bastard is this transubstantiation doubtless. Lanfrancus [Archbishop of Canterbury in 1020] that enemy of truth and true religion . . . with others begat this wicked woman transubstantiation." *The Early Writings of John Hooper*, ed. Samuel Carr (Parker Society, Cambridge, 1843), pp. 117-18.

It was the same exhilarating sense of deliverance, of recognition that iconoclasm is God's demand for cleansing the temple of superstition, that was so admirably illustrated in the engraving John Day added to the heading of the chapter on the reign of Edward VI in the expanded folio edition of John Foxe's *The Acts and Monuments of Martyrs* (1583).[6] For Hooper it was the cleansing of the temple to root out the Mass books, monstrances, chalices, hosts, all the vestments, and the stone altars. For Gardiner it was to remove the only justification for having a temple; for what was a temple without an altar of sacrifice, an empty shell from which religion had vanished? So soul-deep were the differences revealed in the Eucharistic controversy.

The study of the Eucharistic controversy will, of course, center on England and the theologians of the Church of England. It will also involve several forays to the continent of Europe, to consider the Eucharistic thought of Luther,[7] Zwingli,[8] and Calvin,[9] as well as Oecolampadius.[10] It is equally important to examine the sacramental views of three men whom Archbishop Cranmer invited to England to advise him in furthering the reformation of the English church: Martin Bucer,[11] a former Dominican and influential mediating theologian who was appointed regius professor of Divinity at Cambridge in 1549; Peter Martyr Vermigli,[12] formerly abbot of the Augustinians at Spoleto and regius professor of Divinity at Oxford; and John à Lasco,[13] Polish nobleman, former Catholic bishop, the friend of Hooper, who after visiting England at Cran-

---

6 *Acts and Monuments*, p. 1294. See also the reproduction in this book.

7 See Luther, *Werke* (Weimar, 1883ff., ed. J.C.F. Knaake and others), Vols. 6 and 8; Yngve Brilioth, *Eucharistic Faith and Practice* (1956); Leonhard Fendt, *Der Lutherische Gottesdienst der 16 Jahrhundert* (Munich, 1923); J. J. Pelikan, "Luther and the Liturgy" in *More about Luther: Martin Luther Lectures*, Vol. 2 (Decorah, Iowa, 1958); Vilmos Vajta, *Luther on Worship* (Philadelphia, 1958).

8 *Huldreich Zwinglis samtliche Werke: Corpus Reformatorum*, ed. E. Egli and G. Finsler, 13 vols., IV; Brilioth, *Eucharistic Faith*; C. C. Richardson, *Zwingli and Cranmer on the Eucharist*; F. Schmidt-Clausing, *Zwingli als Liturgiker*.

9 The standard edition of Calvin's *Opera*, in 59 volumes, is Baum, Cunitz, Reuss, Lobstein, and Erichson, *Corpus Reformatorum*; see vols. 29-77. See also E. Doumergue, *Jean Calvin*, II; and Ronald S. Wallace, *Calvin's Doctrine of the Word and Sacrament*.

10 See E. Staehelin, *Das theologische Lebenswerk Johannes Oekolampads*, Quellen und Forschunger zur Reformationsgeschichte, vol. XXI; *Die evangelischen deutschen Messen bis zu Luthers Deutscher Messe*, ed. Julius Smend, pp. 213-19, for Oecolampadius' *Form und Gestalt das Herren Nachtmal*, tr. and ed. Bard Thompson, *Liturgies of the Western Church*, pp. 211-15.

11 See G. J. Van de Poll, *Martin Bucer's Liturgical Ideas*.

12 See Joseph C. McLelland, *The Visible Words of God; An Exposition of the Sacramental Theology of Peter Martyr Vermigli, A.D. 1500-1562*.

13 See K.A.R. Kruscke, *Johannes à Lasco und der Sacramentsstreit*, Studien zur Geschichte der Theologie und der Kirche, vol. VII; John à Lasco, *Works*, ed. A. Kuyper, 2 vols.

mer's request returned in 1550 to be the superintendent of the foreign Protestant congregations in England. Furthermore, it will be impossible to understand the development of Anglican Eucharistic thought unless it is contrasted with the Catholic understanding of the Mass as expounded by St. John Fisher, the martyred bishop of Rochester and friend of Sir Thomas More, and the two bishops, Stephen Gardiner and Cuthbert Tunstall, with whom Cranmer was engaged in a literary duel. As we look for the sources of Ridley's influence on Cranmer, it will be necessary to review the thought of Ratramnus of Corbie, to whom, incidentally, both Gardiner and Ridley appealed as they produced different interpretations of his famous treatise, *De Corpore et Sanguine*.[14] Nor must the sacramental writings of the earliest English Reformers be forgotten, especially those of John Frith and William Tyndale.[15]

1. Four Eucharistic Theories

Before entering the thicket of details about the Eucharistic controversy, it may prove helpful to survey the landscape, recognising that the English divines had a choice of four alternative explanations of the presence of Christ in the Eucharist. These may be labelled in the most general fashion as: Transubstantiation; the Real Presence (sometimes affirmed as Consubstantiation, but occasionally without any theory, merely affirming the mystery of Christ's corporal presence as an inexplicable mystery); Virtualism; and Memorialism.

Since the year 1215 Catholics have described their explanation of the mysterious presence of Christ in the Eucharist by the term "Transubstantiation." This scholastic explanation makes a distinction between the underlying "substance," or essence, of every material object, and its "accidents," or superficial, sensible qualities. It is asserted that at the moment of consecration the substances of the bread and wine change into the body and blood of Christ, but that the accidents remain unaltered. Since substance is imperceptible, the change is apprehended by faith, and is, in the strictest sense, as the Mass says, a *mysterium fidei*; it is effected by the omnipotence of God. Because the accidents do not change, the

---

[14] Ratramnus, *Opera*, vol. CXXI, 11-346, *Patrologia Latina*, ed. J. P. Migne. Ratramnus was opposed to the carnal view of the Eucharist held by Paschasius Radbertus.

[15] Cf. *The Whole Workes of W. Tyndall, John Frith, and Doct. Barnes, three worthy Martyrs, and principall teachers of this Churche of England* (1572); William Clebsch, *England's Earliest Protestants, 1520-1535*, pp. 88f., 109ff., 126ff., 133ff.

senses cannot detect any difference after the mutation. We may note that Cranmer had two difficulties with this theory. As a nominalist he found it impossible to think of the accidents of any material object existing apart from its substance. Thus to be told this was the real miracle denied that it was any explanation at all.[16] Furthermore, Cranmer found it difficult to accept the view that a body may exist simultaneously in many places, as is required by the view that there is the body and blood of Christ on every Catholic altar, for this was contradictory to empirical knowledge of the nature of bodies and to Scripture and the Apostles' Creed which located Christ's body in heaven.

Luther's highly original Eucharistic doctrine tries to preserve the strength of the Catholic position without its weaknesses. Often termed "Consubstantiation" because of its partial indebtedness to William of Occam's formulation, it yet moved beyond Occam's formulation of an *esse definitive* to affirm an *esse repletive*, totally rejecting an *esse circumscriptive*. Luther affirmed the identity of Christ's body in heaven with that in the sacrament, through his doctrine of the ubiquity of Christ's body which fills all things. According to Luther, at the consecration the minister did not make something present which was not there before (which was the Nominalist view). Rather, he only made manifest what in all other cases is hidden. The differences between Catholic, Lutheran, and other Protestant views are clear in the following statement of Cyril Richardson: "Where in Catholicism the body is a substance separate from other substances and made accessible in a sacrament; and while in much Protestantism the "body" is an inaccessible object, irrelevant at once to faith and to this world, so that to "eat it" means only to have faith in the Passion, in Luther it is that which underlies all reality, gives it its being, and supports it as the creation of God."[17] According to Luther, then, Christ is present in the Sacrament substantially and in His human nature. For a while Cranmer's "Commonplace Book" found this view attractive, but in time he came to deny any substantial or corporeal unity, and this doctrine of Luther's failed to meet the second of Cranmer's difficulties on the subject of ubiquitarianism. The result was that Luther's

---

16 Cranmer's *Lord's Supper*, p. 45: "And although all the accidents, both of the bread and wine remaine still, yet, say they, the same accidents be in no manner of thing, but hang alone in the air, without anything to stay them upon."

17 See Richardson's review of C. W. Dugmore's *The Mass and the English Reformers*, in *Journal of Theological Studies*, New Series, vol. XVI, pt. 2 (October 1965), 436.

explanation in the long run seemed to make the mystery even more baffling.[18]

The third account of the meaning of Holy Communion was Zwingli's, commonly known as Memorialism, a view more often caricatured than expounded. It was quite Protestant in considering the body of Christ to be located and circumscribed in heaven and in affirming that it is by His divinity and not in His humanity that Christ is present in the Lord's Supper. In no sense can Christ be substantially present in the consecrated elements. On the contrary, he can be known only by the mind or consciousness or faith of the participants in Holy Communion. The result was that there is in Zwingli no distinctive *Eucharistic* presence of Jesus Christ. Therefore, in substance the value of the Supper is that it helps to communicate faith, as does preaching. According to Cyril Richardson, "The elements" were, for Zwingli, "the reminders of a past redemption, not vehicles of a present grace."[19] There were three essential components of Zwingli's teaching: (1) the Lord's Supper is an *eucharistia*—an act of thanksgiving for the sacrificial act of Christ on the Cross here remembered; (2) it is a *synaxis*—a gathering together on the part of the church as the body of Christ to express the unity of the Christian community and its loyalty, of which the Lord's Supper is the badge; and (3) it is a prop or aid to faith.

Two lesser points in Zwingli's thinking should also be mentioned. *Hoc est corpus meum* means *hoc significat corpus meum*, on the analogy of Exodus 12:11, where the paschal lamb is eaten as a Passover, but is not itself a Passover. Second, Chapter 6 of St. John's Gospel was interpreted as the eating of Christ's flesh to be understood always as believing in Christ's Passion. The usual criticism of the Zwinglian idea is to call it, with Dom Gregory Dix, the doctrine of the "Real Absence"; that is, to argue that for Zwingli the Communion is a historical tribute and that its accent is on the past not the present. This, however, is to do less than justice to Zwingli's recognition that Christ is made contemporary by a rich doctrine of faith, not as consciousness of Christ, so much as a being grasped by God through the Holy Spirit. It is the overly subjective, even solipsistic interpretation of faith in Zwingli that

[18] It was similar considerations that caused some Anglican theologians (of whom Queen Elizabeth was possibly first) to affirm the "Real Presence" as an ultimately inexplicable mystery; the Benedictine theologians of Maria Laach Abbey have made the same point in the present century. For the latter see Horton Davies, *Worship and Theology in England*: Vol. 5: *The Ecumenical Century, 1900-1965*, Chap. 1, esp. pp. 26-31.

[19] C. C. Richardson, *Cranmer and Zwingli on the Eucharist*, p. 19.

makes his doctrine seem so thin. On the other hand, the truth in the criticism is that for Zwingli the Sacraments were not *signa exhibitiva*, signs that exhibit and communicate what is present, but *signa representativa*, signs that represent, symbolise, or commemorate what is absent.[20]

The fourth theory was what has been called, a little inadequately, "Virtualism," that is, the belief that while the bread and wine continue to exist unchanged after the Consecration, yet the faithful communicant receives together with the elements the virtue or power of the body and blood of Christ.[21] This was an attempt to preserve the positive values of Transubstantiation and of the doctrine of the "Real Presence," without metaphysical explanations that baffle the intelligence, on the one hand, and to avoid the reductionism of the Zwinglian Memorialist views on the other. It was espoused by Bucer and by Calvin, both of whom tried to produce in this way a halfway point between Luther and Zwingli; it was believed that this was in fact achieved in the Sacramental articles of the *Consensus Tigurinus* of 1549, in which Bullinger, Zwingli's successor, joined with Calvin. Calvin's *De Coena Domini*, written in 1540, was translated into English by Coverdale;[22] it showed that Calvin was much nearer to Luther's views than to Zwingli's on the significance of the Eucharist. This view was popularised in the Church of England by Hooker in his *Laws of Ecclesiastical Polity*.

Calvin defined a sacrament as "an external sign, by which the Lord seals on our consciences His promises of goodwill towards us in order to sustain the weakness of our faith, and we in our turn testify our piety towards Him, both before Himself and before angels as well as men."[23] The sacraments are seals of the word, confirmations of the Divine promises, and signs of union with the body of Christ.[24] Calvin, says Ronald S. Wallace, "cannot get away

---

20 C. H. Smyth, *Cranmer and the Reformation under Edward VI*, p. 20, draws attention to the fact that the Reformed Church in Zurich came under the guiding influence of such distinguished Hebraists as Leo Jud, Pellican, and Bibliander, who were deeply impressed by the correspondence between the Old Testament signs and the New Testament sacraments, particularly the links between circumcision and baptism, and between the Passover and the Lord's Supper. Curiously, however, in regard to the latter, they did not stress the element of sacrifice common to both the Passover and the Lord's Supper, but did emphasize unduly the commemorative aspect also common to both.

21 *The Oxford Dictionary of the Christian Church*, ed. F. L. Cross, p. 1,425.

22 Coverdale's translation of Calvin was entitled *A faythful and most godly treatise concernynye [sic] the sacrament*, which Pollard and Redgrave's *Short Title Catalogue* lists as being published probably in 1549 by Day.

23 *Institutes* IV: 14:i.   24 *Ibid.*, 17:i.

from the fact that our bodies as well as our souls are involved in this union, and that the sacrament of the Lord's Supper especially testifies that the flesh of Christ, the body in which He lived and died is also involved."[25] The point to be emphasized is that Calvin believed Christ does not merely give us the benefits of His death and resurrection, but the very body in which He suffered and rose again.[26] How does the body of Christ located at the right hand of God become our food in the Eucharist here on earth? The answer is: "Though it seems an incredible thing that the flesh of Christ, while at such a distance from us in respect of space, should be food for us, let us recall how far the secret virtue [*virtus arcana*] of the Holy Spirit surpasses all our conceptions and how foolish it is to desire to measure its immensity by our feeble capacity. Therefore what our mind does not comprehend, let faith conceive, viz. that the Spirit really unites things separated by grace."[27] Calvin rejected the Catholic and Lutheran views of the presence of Christ because they would not allow the signs to be true signs or the promises to be true promises, for, to him, Transubstantiation and Consubstantiation turn the signs of the body and blood of Christ, the bread and wine, into the body and blood of Christ. He also rejected the opposite view, that to eat is only to have faith, because this endangers the reality of the presence of Christ and thus denies the fulfillment of God's promise that the signs are not empty but show forth the promises. Calvin, without doubt, believed that it was the powerful activity of the Holy Spirit that made possible the participation of the flesh of Christ in the Supper.[28] Finally, Calvin never let it be forgotten that the mystery of our union with Christ is an eschatological reality to be completed fully in the life beyond death for our flesh as well as our spirits.[29] The defect of this theory is that it also posits a miracle to cross the abyss between two incompatible statements.

It will be apparent in our consideration of the Eucharistic controversy that a central figure such as Cranmer will, on growing dis-

[25] Ronald S. Wallace, *Calvin's Doctrine of the Word and Sacraments*, p. 151.
[26] Cf. Calvin's *Commentary* on I Corinthians 11:24, *Corpus Reformatorum* edition, vol. 39, p. 487; and Wallace, *Calvin's Doctrine*.
[27] *Institutes* IV: 17: x.
[28] François Wendel, *Calvin*, pp. 350ff., says that Calvin's two statements that seemingly cannot be reconciled, namely the presence of the flesh of Christ in the sacrament and the location of Christ's body in heaven, are in fact compatible; it is the Holy Spirit that makes possible the participation in the body of Christ without its being enclosed in the elements.
[29] See Wallace, *Calvin's Doctrine*, chap. 13, for a full explanation of Calvin's teaching, notable for both fidelity and clarity.

illusioned with the Catholic theory, appear to turn to the Lutheran view first and thereafter to the Calvinist or Zwinglian views, which of the two latter being a point of contention between interpreters. Also, it seems clear that there was a danger that a Virtualist theory would fall into a Memorialist theory, because of the greater ease of defending the latter on empirical and rational grounds, even if it seems to do less than justice to tradition and Christian experience.

It may, however, be convenient at this point in the discussion, to list the important affirmations commonly made by mediating and radical Protestant theologies. Firstly, all agreed in their opposition to the Catholic doctrine of Transubstantiation and the corporeal presence of Christ in the Eucharist as being contrary to the nature of sacraments as signs (rather than the things signified), and contrary to Scripture. Secondly, all stressed that Christ is made present by the action of the Holy Spirit. Thirdly, all asserted that the communion of the faithful is confirmed and strengthened by sharing in the Lord's Supper. Fourthly, the sacrament is in a primary sense the memorial of Christ's death and resurrection. Fifthly, there was agreement that the elements are powerful signs eliciting from the faithful the remembrance of Christ's work and consequently a confession of the forgiveness of sins through Christ's Passion. Sixthly, all affirmed that the sacrament is received by faith in the heart or conscience of the believer. Finally, some Protestants would emphasize that the presence of Christ is not limited to the sacraments, and that it is through the moving of the mind to recall the work of Christ that a sharing in Christ is effectual.[30]

2. The Catholic Case: Fisher, Tunstall, and Gardiner

Henry VIII's *Assertio Septem Sacramentorum*, written against Luther's reduction of the seven Catholic sacraments to three (Penance,[31] Baptism and the Eucharist) in *The Babylonian Captivity of the Church* (1520) won for the king the title of *Fidei Defensor*[32] (Defender of the Faith) from a grateful Pope, a title

---

30 This perceptive summary appears on p. 36 of Gordon E. Pruett, "Thomas Cranmer and the Eucharistic Controversy in the Reformation," unpub. Ph.D. thesis, which I had the privilege of directing and to which I am much indebted in this chapter.

31 Luther initially accepted Penance as a sacrament, because of his conviction that the forgiveness of sins was the first gift of the Gospel, but later gave it up because it did not conform to his threefold requirement that a Sacrament of the Gospel must be (1) clearly instituted by Christ; (2) have a divine promise attached to it; and (3) have a confirmatory sign. Only Baptism and the Eucharist met these triple requirements.

32 Leo X gave the title on October 11, 1521.

which continued incongruously to appear obscurely abbreviated as "Fid. Def." above the heads of sovereigns on English coinage within recent memory. The work is, however, more of chronological than of intrinsic importance. The Eucharistic treatise of John Fisher, friend of Erasmus, Chancellor of the University of Cambridge, Bishop of Rochester, and Catholic martyr, is of far greater importance.[33] In this work entitled, *De Veritate Corporis et Sanguinis Christi in Eucharistia*, published in 1527, Fisher defended the Catholic doctrine against the criticism of Oecolampadius.[34]

According to the most recent biographer of John Fisher, *De Veritate* is his "theological masterpiece."[35] It is in the prefaces to each of the five books that Fisher's freshness may be found. In the books themselves there is the usual catena of citations from the Church Fathers. The first preface attempts to prove the Real Presence from the unanimity of the Catholics as contrasted with the disagreements of their adversaries, as can be seen in the mutual slaughter of the Lutherans in the Peasants' Revolt and in the pride in the minds of Luther and Oecolampadius which has caused their disagreements. In the second preface Fisher demonstrates the values of the Sacrifice of the Mass from the varying names given to it, such as The Sacrament of the Lord's Body and Blood, the Mysteries, the Synaxis or Communion, the Eucharist or Thanksgiving, the Sacrifice, the Bread, the Lord's Food, the Viaticum, the Mystical Blessing, the Banquet, and the Exemplar or Antitype. These names show the wealth of the meanings of this sacrament and the folly of conceiving it as consisting of bread and wine only, as well as the sense in calling it a figure, for the bread and wine were signs of the realities, the real body and real blood of Christ. The third preface tries to confirm the faith of readers by enumerating 14 points of corroboration:

> The reality of Christ's Body in the Eucharist is proved by the most clear words of the same Christ. 2. The reality is supported by the immensity of Christ's love. 3. The same is

[33] Henry VIII had John Fisher executed on June 22, 1535, while his friend, Sir Thomas More, suffered a similar fate on July 6, 1535. Both were canonised by Pope Pius XI.

[34] Oecolampadius (1482-1531), the German Reformer, first introduced Reformed principles to Basel in 1522. At Marburg in 1529 he defended the Eucharistic doctrine of Zwingli, but in his more developed thought on the sacrament he anticipated Calvin. His sacramental theology will be considered in the thought of John Frith, his English disciple, in the next section of this chapter.

[35] Edward Surtz, S.J., *The Works and Days of John Fisher*, p. 337.

corroborated by the consent of the Fathers. 4. It is confirmed by Christ's promises. 5. It is established by many Councils. 6. It is demonstrated by innumerable miracles. 7. It is defended by revelations most worthy of credence. 8. Christ's protection against heresies has never been wanting. 9. Persons who frequently receive the Sacrament advance in every virtue. 10. Persons who have idle opinions about it meet with ill fortune. 11. Enemies of this Sacrament have no clear Scripture on their side. 12. They produce the solid testimony of no orthodox writer. 13. They can present no miracles and no revelations. 14. Dissenters from the common belief of the Church mutually cut one another's throat.[36]

The cumulative effect of these considerations is great, except, of course, that arguments 9 to 14 merely state negatively what arguments 1 and 3-7 state positively.

The fourth preface argues for a belief in the "Real Presence" on the basis of the consent of the Fathers during 15 centuries. The fifth preface shows Fisher bending all his energies to prove that the sixth chapter of St. John's Gospel refers to the Eucharist, from the unanimous testimony of the Fathers for such an interpretation, the necessity for bodily eating (*corporalis esus*) as well as for faith, and from the words of Christ Himself seen in their scriptural contexts.

Fisher, in fact, had a formidable array of weapons; the strongest was the inherent improbability of any teaching at variance with tradition, with 1,500 years to support it, being right or of the central tradition of the church having been in error for so long. He was also astute in seeing where the argument of Oecolampadius was weakest, namely in his assertion that the New Testament promises eternal life to the man "who eats my body" and to the man "who believes in Me," and that therefore there is no need for the Sacrament on the part of the man who has faith. Fisher replied that no general faith is adequate, but a special one which believes what is proposed to it and also does what is commanded. Further, not only simple faith but the sacraments are essential to the Christian life, because this is the unanimous patristic testimony, because faith dissolves unless strengthened by the sacraments, and because without receiving the sacraments, there can be no certainty about the possession of a living faith.

[36] Tr. Edward Surtz, S.J., in *ibid.*, p. 341.

The second Catholic treatise to be considered was written by Bishop Cuthbert Tunstall of Durham, a Henrician who had sympathised with Catherine of Aragon but whose position became impossible under Edward VI. Tunstall was eventually imprisoned in his house in London, where in 1551 he wrote *De Corporis et Sanguinis Domini in Eucharistia.* The manuscript was taken by his great-nephew Bernard Gilpin to Paris where it was published in 1554 by Michael Vascosanus. Gilpin's biographer recounts several incidents purporting to show that Tunstall disliked too rigid a definition of the mode of Christ's presence in the Eucharist, regarding this as a mystery that transcended human language. He reports Gilpin as saying of Tunstall: "I remember that Bishop *Tonstall* often tolde me that Pope *Innocent* the third had done very unadvisedly, in that he made the opinion of Transubstantiation an Article of faith: seeing that in former times it was free to holde or refuse that opinion. . . . Moreover the Bishop tolde me that he did not doubt but that himselfe, if he had beene in that Councell, could have prevailed with the Pope to have let that businesse alone."[37]

As one might expect, this treatise affirms the corporeal presence as the plain intention of Scripture and as the consentient tradition of the church—the two main arguments of Catholic apologists during the period.[38] The first deals with the three explanations:[39]

But how the bread, which before the consecration was common bread, by the ineffable sanctification of the Spirit passed into His Body, the most learned of the ancients regarded as something inscrutable. . . . Moreover, before Innocent III, Bishop of Rome, who presided at the Lateran Council, it seemed to those scrutinising more curiously that it could be brought about in three ways: Some thinking that together with the bread, or in the bread, the body of Christ is present, as fire in a mass of iron, which mode Luther seems to have followed: Others that the bread is reduced to nothing or corrupted: Others that the substance of bread is trans-

[37] George Carleton, *Life of Bernard Gilpin*, p. 42.
[38] C. W. Dugmore, *The Mass and the English Reformers*, p. 152. However, the treatise contains two books: the first sets forth the Scriptural evidence for the Real Presence; it attempts to answer the Protestant criticisms against the Catholic sacramental doctrine; the second consists largely of an armory of patristic proof texts to support the orthodox interpretation of the Eucharist.
[39] Here the careful translation of Fr. Edward Quinn has been used. It will be found for these central passages in his article on "Bishop Tunstall's Treatise on the Holy Eucharist," in *The Downside Review*, vol. 51, no. 148 (October 1933), 674ff.

muted into the substance of Christ's body: which mode Innocent followed, and rejected the other views at that Council, although to those investigating more curiously not fewer but rather more miracles seem to arise than in the modes rejected by him. But it seemed to those who were present with Innocent at that Council that all miracles give way before the omnipotence of God, to which nothing is impossible, because this mode seemed to them most in agreement with these words of Christ, 'This is my body, this is my blood.' For John Scotus in the fourth book of Sentences, the eleventh distinction, the third question, citing Innocent, says there were three opinions: One that the bread remains, and yet together with it is truly the body of Christ: Another that the bread does not remain, and yet it is not converted, but ceases to be, either through annihilation, or through resolution into matter, or through corruption into some other thing: A third, that bread is transubstantiated into the body and wine into the blood. But each of these opinions sought to maintain in common that one thing, that there truly is the body of Christ, because to deny that is clearly against faith.[40]

Tunstall asks whether it was right of the church to choose one of these three explanations as most in accordance with the institution of the Eucharist by Christ, and to reject the other two. His conclusion is that the church wisely decided to affirm Transubstantiation: "I think it right on a matter of this kind, because the Church is the pillar of truth, that her firm judgment should be

[40] The original reads: "Caeterum quo modo panis qui ante consecrationem erat communis, ineffabili spiritus sanctificatione transiret in corpus ejus, veterum doctissimi quique inscrutabile existimaverunt. . . . Porro ante Innocentium tertium Romanum episcopum, qui in Lateranensi concilio praesedit, tribus modis id posse fieri curiosius scrutantibus visum est: Aliis existimantibus una cum pane, vel in pane Christi corpus adesse, veluti ignem in ferri massa, quem modum Lutherus secutus videtur: Aliis panem in nihilum redigi, vel corrumpi: Aliis substantiam panis transmutari in substantiam corporis Christi: quem modum secutus Innocentius, reliquos modos in eo concilio reiecit, quanvis miracula non pauciora, imo vero plura quam in reliquis reiectis ab eo modis, oriri curiosius investigantibus videantur. Sed Dei omnipotentiae, cui nihil est impossibile, miracula cuncta cedere, his qui cum Innocentio in eo consilio interfuerunt visum esse, quod is modus maxime cum verbis hisce Christi, Hoc est corpus meum, hic est sanguis meus, congruere illis visus est. Nam Joanne Scotus libro quarto Sententiarum, distinctione undecima, quaestione tertia, recitando Innocentium, ait tres fuisse opiniones: Una, quod panis manet, & tamen cum ipso vere est corpus Christi: Alia, quod panis non manet, & tamen non convertitur, sed desinit esse, vel per annihilationem, vel per resolutionem in materiam, vel per corruptionem in aliud: Tertia, quod panis transubstantiatur in corpus, et vinum in sanguinem. Quaelibet autem istarum voluit istud commune salvare, quod ibi vere est corpus Christi, quia negare est plane contra fidem."

observed. Moreover, those who publicly contend that this mode of Transubstantiation is to be entirely rejected because the word 'Transubstantiation' is not in Scripture show themselves to be too prejudiced in their judgment. As if Christ could not effect in that way what He wills, from whose omnipotence and the operations of the Holy Spirit they seem totally to detract by their assertion. And they themselves fall into a much greater sin of audacity than that which they bring up against the Lateran Council. Since indeed if we believe them, neither Christ nor the Holy Spirit could bring it about that the bread should pass into the substance of the body of Christ."[41] The weakest point of the argument is to accuse the Protestants of lacking faith in the omnipotence of God. They did not deny that God could perform a Eucharistic miracle; they asked whether it was consonant with the revelation of God given by Christ in the wilderness when He refused to win men by miracles and when He taught that even if a man came back from the dead to warn Dives's relations, still they would not believe. Nevertheless, Tunstall by his tender devotion to the person of Christ[42] and by his long view of history, is an attractive apologist, and not least so in his lack of rigidity.

If Fisher was the most learned writer in Tudor Catholic sacramental theology, and Tunstall the most open-minded, Gardiner is the shrewdest controversialist. The chief purpose of his *Explication and Assertion of the true Catholic Faith, touching the most blessed Sacrament of the Altar with confutation of a booke written agaynst the same* (1551), as the latter part of the title obscurely indicates, was to confute Cranmer's recent treatise which, though claiming to be orthodox, was actually heterodox.

[41] The original reads: "Iustum existimo ut de eiusmodi, quia ecclesia columna est veritatis, firmum eius observeretur iudicium. Caeterum qui palam contendunt illum modum transubstantionis reiiciendum penitus esse, quod vocabulum transubstantionis in spiritus in scripturis non reperiatur, nimis praefacti iudicii sese ostendunt. Quasi vero Christus eo modo, illus quod vult efficere non posset: cuius omnipotentiae & spiritus sancti operationi, in totum detrahere sua assertione videntur. Et in multo maius audaciae crimen incidunt ipsi, quam quod Lateranensi concilio obiicunt; quando quidem si illis credimus, nec Christus, nec spiritus sanctus id efficere possit ut panis in corporis Christi substantiam transeat." The passages in notes 40 and 41 are taken from *De Veritate Corporis et Sanguinis Domini in Eucharistia*, fols. 45-47.

[42] This is finely expressed in the words in which the help of Christ is sought in writing this treatise: "Tuo igitur Christo auxilio freti obsecramus, ut luminae gratiae tuae nos illustres, et rem destinatam tuto sine formidine aggrediemur ut corporis et sanguinis tui veritatem in eucharistia esse, ex ipsis tuis verbis evincamus, atque huic veritati adversantium perversa dogmata, e fidelium circum ventorum mentibus te favente, sine quo nihil possumus, eradicare concemur. Vale." His unusual qualities are also seen in his *Certain Godly and Devout Prayers*, an example of which is cited in the next chapter, on Catholic worship.

Gardiner shrewdly points out, as I will emphasize in Chapter V on the Anglican Prayer Books, that there were five places in the 1549 Prayer Book that Cranmer took the chief responsibility for, which imply a Catholic doctrine of the Eucharist. Yet Cranmer contradicted this earlier teaching in his treatise, *A Defence of the True and Catholic Doctrine of the Sacrament of the Body and Blood of our Saviour Christ* (1550).

Gardiner begins by demonstrating how heterodox Cranmer is in denying the doctrine of the Real Presence of Christ, not only Transubstantiation, and by denying that wicked men eat and drink the body and blood of Christ. He points out that Luther, Bucer, and Melancthon accepted the doctrine of the Real Presence and berated Oecolampadius for denying it.[43] Gardiner then turns to consider the statement of Cranmer, that Christ's risen body is in heaven and cannot be on the altar. Gardiner agrees with the former part of the statement but cannot accept the consequence Cranmer draws from it. Gardiner affirms that the church acknowledges Christ's body in heaven and also in the sacrament, "not by shifting of place but by the determination of his will."[44]

Gardiner admits that there is an ambiguity in speaking of the "corporal" presence of Christ in the Sacrament. It can indeed be said that Christ is corporally present in the Sacrament, but if the term "corporally" referred to the manner of the presence, "then we should say Christes body were present after a corporall maner, which we say not, but in a spirituall maner, and therefore not locally, nor by maner of quantitie, but in such a maner as God only knoweth. . . ."[45]

The next issue Gardiner dealt with was Cranmer's contention that Christ intended the words "This is my body" to be figuratively and not literally understood. Gardiner replies that Christ made bread his body by calling it so, just as he made water wine by calling it so at Cana of Galilee. He answers the argument that when Augustine and other Fathers of the church use some passages referring to the literal presence and other passages that are clearly figurative in their reference to the presence, the latter usage is to conceal the mystery from the profane.[46] In fact, only six persons have held the figurative doctrine of the presence in the Eucharist (which Cranmer calls Catholic): Ratramnus, Beren-

---

[43] *De Veritate Corporis*, sig. 6-6v.
[44] *Ibid.*, sig. 20.
[45] *Ibid.*, sig. 38 v. See also J. A. Muller, *Stephen Gardiner and the Tudor Reaction*, pp. 210-12.
[46] *Ibid.*, sig. 41 v.-42.

garius, Wycliff, Oecolampadius, Swinglius, and Vadianus. Gardiner rather contemptuously refused to include Peter Martyr in the list since he was not learned, or Cranmer since Cranmer followed Peter Martyr.[47] The remainder of the third book of this treatise is spent hunting through the fathers in whom Cranmer found a metaphorical meaning in regard to Christ's sacramental presence, to insist that its natural meaning in each context is a literal one.

In Book Five Gardiner scores a clever debating point. He says that Cranmer has been greatly concerned with the eating and drinking of Christ's body and blood, and has been attempting to prove that the evil do not receive the body and blood of Christ. But if this is only a metaphorical eating, why waste so much useless energy on the matter?

In Book Five Gardiner defends Transubstantiation as the true doctrine of Christ's mysteries. He accuses Cranmer of error in asserting that faith is to be confirmed by reason, because it transcends reason and "natural operation," being the gift of God and the operation of the omnipotent God.[48] Nor can he accept Cranmer's argument against Transubstantiation on the grounds that there cannot be a change in the substance without a change in the accidents, for there is in all objects an inwardness. It is not, says Gardiner, by the evidence of our senses that we affirm the resurrection of the body, but only by faith.[49] Gardiner argues that as the divinity of Christ who was both God and man operates invisibly in his soul, so the invisible "soul" of the bread and wine operates through their bodily elements and thus "this Sacrament [is] the image of the principall mystery of Christes person."[50]

Gardiner observes that the Fathers used many equivalents for Transubstantiation, including "transition," "mutation," "trans-elementation," and "conversion."[51] Yet this parallel between the two natures in Christ's person and in the Sacrament cannot be pressed too far, for in the two natures of Christ there was not mutation but only assumption of the human nature into the divine nature.

Gardiner's biggest guns are trained on the six major criticisms Cranmer makes of the Catholic doctrine; it is here that his ingenuity is impressive. Cranmer argued that the Catholics, when asked what is eaten in the Sacrament, have to assert that this refers only

[47] *Ibid.*, sig. 74 r and v. [48] *Ibid.*, sig. 97 v.-98.
[49] *Ibid.*, sig. 105. [50] *Ibid.*, sig. 122.
[51] *Ibid.*, sig. 113.

to the accidents and that these have been "broken, eaten, dronken, chawed and swalowed without any substaunce at all."[52] He anticipates every logical objection that Cranmer can raise in his answer: "But where this auctor [Gardiner always refers in this indirect way to Cranmer] saith that nothyng can be answered to be brokē but the accidētes: yes verely, for in tyme of contēciō, as this is to him that would aske, What is brokē, I would in other termes answere thus, That thou seest is broken. And thē if he would aske further, What that is, I would tell him the visible matter of the Sacrament, under whiche, is present Invisibly the substance of the most precious body of Christ, if he will aske yet further, Is that bodye of Christ broken, I will say no. For I am lerned in fayth, that that glorious body nowe impassible can not be broken, or divided, and therefore it is holy in every parte of that is broken."[53] To Cranmer's second criticism, that Transubstantiation is contrary to all experience in asserting that accidents persist although the substance has changed, Gardiner replies that faith believes in the almighty power of God and that this is a mystery transcending human comprehension.[54] Cranmer's third criticism, that the Papists make substance without accidents and accidents without substance, is rather weakly rebutted by accusing Cranmer of jesting about serious matters.[55]

Cranmer's fourth empirical criticism of the metaphysics of Transubstantiation is that the Catholics assert that there is a vacuum where substances of the bread and wine were, but it is known that nature abhors a vacuum. Gardiner replies that it is quantity not substance that fills space. Here Gardiner scores a debating point, but he follows it up by admitting that ordinarily there is no quantity without substance, but that this is extraordinary because of a divine miracle.[56] Cranmer's next objection is that Catholics are not ashamed to admit that the substance is made of accidents when bread moulders or becomes worm-ridden, or wine sours. Gardiner speculates that the corruption of accidents may create new accidents, but affirms that the *prima materia* (common to both wine and vinegar) does not change.[57] Cranmer's last argument is to ridicule the Catholics for affirming that "substance is nourished without substance by accidentes onely, if it chaūce any catt, mouse, dogge, or any other thinge to eat the

52 *Ibid.*, sig. 133.
54 *Ibid.*, sig. 135 v.-136.
56 *Ibid.*, sig. 136 v.-137.

53 *Ibid.*, sig. 134 v.
55 *Ibid.*, sig. 136 r. and v.
57 *Ibid.*, sig. 138.

Sacramentall bread or drink the Sacramentall wyne."[58] Gardiner denies that vermin devouring any hosts do in fact violate Christ's precious body. He retorts in some heat that Cranmer's doctrine is an absurdity in declaring that Christ's body is actually only in heaven, that the bread only remains bread, and yet the figure should be called by the name of the reality.

It is in his criticism and attempted confutation of Cranmer's fifth book that the profoundest issue was raised: How can the all-sufficient oblation of Christ for the sins of the world made on Calvary be repeated, as Catholics claim it is, on their altars, without derogating from its sufficiency and finality? Here Gardiner replies that Scripture, while asserting that the work of Christ on the Cross was perfect, yet asserts that it is often to be remembered in the church and exhibited in such a way that it may be fed upon, this sacrifice is a propitiation for the sins of the world, as the Council of Nicea, the teachings of St. John Chrysostom, and the exposition of Peter Lombard testify.

In the course of this debate with Cranmer, Gardiner made some good points, but he fell back all too frequently for good dialectics on the mystery of the sacrament, the omnipotence of God, and the relatively unanimous consent of tradition. These were all good churchly arguments, but since the first was not denied by Protestants it might seem irrelevant to them to insist on mystery when the issue was: which is the best explanation of the mystery available? Nor did Protestants deny the omnipotence of God; they questioned whether this was His method in the Sacrament. Third, since tradition may consist of human inventions and corruptions, Protestants could not see that tradition settled anything. It is interesting, however, that so great was the fear of change in this era of intellectual and religious revolution, that even the Protestants had to appeal in their arguments to the "pure" and "primitive" church—in short, to an earlier, uncorrupted "tradition." Perhaps the best tributes to the effectiveness of Gardiner's arguments were two. The first was that Cranmer reissued his original treatise with changes to attempt to meet Gardiner's charges under the revealing title, *An Answer to a Crafty and Sophistical Cavillation devised by Stephen Gardiner.* The second was the radical reorganization of the doctrine and the words of the revised Anglican Prayer Book of 1552, in which every passage smacking of a Catholic interpretation—at the five points at which Gardiner discovered orthodox Catholicism—was wholly excised.[59]

58 *Ibid.*, sig. 138 r. and v.
59 For convenience the five places where Gardiner found Catholic doctrine in the

3. *The Teaching of the Early English Reformers*

The contrast between the sacramental theology of the Catholics and that of the early English Reformers is striking. If the Catholics believed in the corporal presence, the Protestants rejected it outright as idolatry and superstition, and tended to follow the example of Zwingli or Oecolampadius in their Memorialism. They were sacramental radicals. This section is given over to John Frith, William Tyndale, John Hooper, and Thomas Becon, all of whom endured exile and imprisonment for their beliefs, and the first three of whom were martyred for their convictions.

John Frith (1503-1533) was imprisoned for heresy in 1528, but escaped to Marburg where he assisted Tyndale in his translation of the Pentateuch and Jonah into English. Returning to England in 1532 he was arrested after a book he had prepared for a friend, and not for publication, got into the hands of Thomas More, and he was condemned to death for denying purgatory and transubstantiation.[60] *A booke made by Iohn Frith prisoner in the tower of London* (1533)[61] reads freshly; while it expounds the views of Oecolampadius,[62] it does so with dialectical brilliance and verve.

His principal contention was that the Eucharist involves "the spirituall and necessarie eating of his body & bloud, which is not received with the teth and bellye, but with the cares and faith."[63] He offered two 'proofs' that it is not an essential article of faith to believe in the presence of the natural body of Christ in the Sacrament. Many did, in fact, receive it and believe it to their damnation. By contrast, says Frith, "it is not his presence in the bread that can save me, but his presence in my hart through faith in his blood, which hath washed out my sinnes and pacified ye fathers wrath toward me."[64] His second argument was that the same faith will save his contemporaries as saved the patriarchs before Christ's incarnation, but they were not bound under pain of damnation to

---

First Prayer Book of Edward VI were: (1) in the words of administration; (2) in a post-communion rubric; (3) in the commemoration of the dead in the intercessions; (4) in the prayer of consecration and particularly in the invocation of the Holy Spirit (*epiklesis*); and (5) in the rubric requiring kneeling while saying the prayer of humble access. Each of these claims is examined in detail in Chap. V.

[60] There is a sympathetic account of Frith in Clebsch, *England's Earliest Protestants*, who claims that "Frith displayed the finest mind, the most winsome wit, and the boldest spirit among the men who wrote theology in English between 1520 and 1535" (p. 78).

[61] It was published as part of *Workes of Tyndall*.

[62] Frith borrowed from two works of Oecolampadius: *De Genuina verborum domini, Hoc est corpus meum, iuxta vetustissimos authores, expositione liber* (1525) and *Quid veteres senserint de Sacramento eucharistiae* (1530).

[63] *Ibid.*, p. 107 v.   [64] *Ibid.*, p. 108.

believe this, nor were his contemporaries.[65] The manna and water given to Moses and the Israelites are the same as the bread and wine are to those in the Christian dispensation.[66] If, however, faith is enough, then it could be objected, why institute a sacrament?

Frith maintained that there are three reasons for instituting the sacraments. The first is a sociological consideration: to be "knit in fellowship" it is essential to have visible tokens of sacraments or signs or badges of belonging. "And there is no difference betweene a signe or a badge and a Sacrament, but that the Sacrament figureth a holy thyng, and a signe or a badge doth signifie a worldly thing."[67] Second, there is a pedagogical foundation for the sacraments: "that they may be a meane to bryng us unto faythe, and to imprint it the deeper in us, for it doth customably the more move a man to believe, when he perceiveth the thyng expressed to diverse senses at once."[68] The third reason for the institution of sacraments was that "they that have received these blessed tydinges and worde of health, do love to publishe this felicitie unto other men."[69] Apart from the Gospel to which the sign testifies, the Sacrament is meaningless. For, "if a man wot not what it meaneth, and seeketh the health in the sacrament and outward signe, thē may he wel be likened unto a fond fellow, which when he is very drye, and an honest man, shew him an alepole and tell him that there is good ale inough, would go and sucke the alepole, trusting to get drinke out of it, and so to quench his thyrste."[70]

Frith continues by denying a mutation in the bread by the testimony of Scripture (especially I Corinthians 10 and 11, Acts 2, and Luke 22) by the evidence of experience since mouldering of bread proves that bread remains after the prayer of consecration, and by the witness of such doctors of the church as Gelasius, Origen, Nestorius, Bede, and Augustine.[71] His fourth argument for denying the corporal presence of Christ in the Eucharist was that Christ's body according to his manhood must be "in one place, heaven, and so not in the Sacrament."[72]

Frith then brings forward six considerations to convince his readers that the doctrine of the corporal presence in the Eucharist is unreasonable. Each sacrament is, in the first place, a sign of a holy reality, and so cannot be the holy thing itself. Therefore

[65] *Ibid.*, p. 109.
[66] *Ibid.*, p. 109 v. in the margin.
[67] *Ibid.*, p. 112.
[68] *Ibid.*, p. 112 v.
[69] *Ibid.*
[70] *Ibid.*, p. 112 v.-p. 113.
[71] *Ibid.*, pp. 117-25.
[72] *Ibid.*, p. 140.

the Sacrament is the representation of the bodily presence of Christ, not the corporal presence itself.[73] Second, it is not necessary for the flesh of Christ to be in the Sacrament if it is the remembrance, not the flesh, that profits the Christian.[74] Third, if the end is superior to the cause, and the end of the Sacrament is the remembrance of Christ's body, and if the Sacrament is his natural body, then the remembrance is better than the body itself.[75] Fourth, if the soul eats what the angels in heaven eat, and the angels' meat is "the joy and delectation they have of God and his glory," so the soul similarly eats through faith the body of Christ which is in heaven, and this is a spiritual not a corporal manducation. In the fifth place, just as the bread is Christ's body, so it is claimed, so the breaking of Christ's body would be his death; but on Maundy Thursday the breaking of the bread at the Last Supper was not the breaking of his body, but only a representation of it, and so with the Sacrament.[76] The sixth rather recondite argument from silence allows Frith to argue that since the apostles were Jews, to whom it was forbidden to eat or drink human or animal blood, any doctrine of the corporal presence of Christ in the Apostolic church would have been such an affront as to have excluded all Jews from the Christian faith.[77] His conclusion, on the nature of consecration, was: "And so I say that it is ever cōsecrated in hys hart that beleveth, though the Priest consecrate it not. . . . For except thou know what is meant therby and beleve, gevyng thankes for hys body breakyng & bloudshedyng, it can not profite thee."[78] Frith ended as he began, by insisting that it is the spirituality of the rite that matters, not the external sign, he was utterly opposed to the notion that grace is quantifiable.

The value of Frith's teaching was that it stressed the importance of the divine promise of grace and reinterpreted the meaning of grace. Like Luther, Frith understands grace to be the benevolence and mercy of God. He did not, with the scholastics, think of it as a quasi-physical essence that is infused into the soul of the recipient. This conception of grace was capable of revolutionising the role of the sacraments and the power of the church, but Frith did not draw the political and social implications of a concept which would reduce the power of the priesthood and take the sting out of

[73] *Ibid.*, p. 143. [74] *Ibid.*
[75] *Ibid.* [76] *Ibid.*, p. 143 r. and v.
[77] *Ibid.*, p. 143 v. [78] *Ibid.*, p. 153.

the interdict.[79] Frith did, however, criticise the *ex opere operato* conveyance of grace. It is infrequently realised by Protestant critics of this concept that the Catholic church had two strong pastoral reasons for affirming this concept, namely, that it stressed the objectivity of grace which was entirely appropriate if the omnipotence of God was being considered, and it reassured the faithful that unworthy priests did not obstruct the passage of grace. Frith felt that it was not the omnipotence but the generous mercy of God which was His outstanding characteristic as revealed in the Gospel, the graciousness, rather than the justice of God. He thought the church was acting as if it had a corner on the power of God, and was behaving as if it had control over grace and could almost automate it.[80]

William Tyndale's contribution to the debate was *A brief declaration of the Sacraments*.[81] He conceived of the sacraments as seals of the divine covenant made with men, so that circumcision foreshadows Baptism, and the Passover the Lord's Supper. He was chiefly aware of the pedagogical value of the sacraments, "And

[79] William Clebsch visualises dramatically (*England's Earliest Protestants*, pp. 128-29) the socio-political effects of the changed conception of grace as applied to the sacraments: "To portray sacramental grace as issuing from God's favour to the believer's faith, rather than from the empowered priesthood, to the sacramental stuff, undermined the whole religious system by which Latin Christianity had extended its influence over the common and public life of Western Europe. Without an *ex opere operato* Sacrament of Baptism the whole ideal of Christendom as a social unity, entered by being engrafted into a uniform religion, would inevitably erode. If the visible Church was not identifiable with God's elect, the weapon of excommunication would lose its power to produce conformity in religion, society, and politics. If the power to transmute bread and wine into the veritable food and drink of salvation were stripped from the Church, it must forfeit the power of the interdict and of the ban which held the temporal sword in bondage to the spiritual." The fragmentation into competing nationalisms had begun long before the Reformation, the Renaissance had a further disruptive effect, and, Canossa, at which the spiritual claims to dominate the temporal were dramatically proved, was 500 years in the past. For these reasons I believe the claims for the changed concept of grace, though important, are exaggerated.

[80] Foxe seemed to think that Frith influenced the development of Cranmer's Eucharistic thought directly, and helped to lead Cranmer to the views that were expressed in the prayer-book of 1552. This is, however, highly improbable for two reasons. The first is that in 1533 Cranmer was a believer in the corporal presence in the Eucharist. Also, it is likelier that Cranmer read Oecolampadius than Frith because of the wide reputation of the scholarly Reformer of Basel. Secondly, Cranmer could have found the emphases and interpretations he was seeking in the Eucharistic thought of Ridley and of the Ratramnus which both Frith and Ridley claimed as the partial progenitor of their thoughts. Frith refers to Bertram, i.e. Ratramnus, in *Workes of Tyndall, Frith and Barnes* (1572), p. 141 r. and v.

[81] It can be found in *Workes of Tyndall*, pp. 436-52; and in *Doctrinal Treatises and Introductions to different portions of the Holy Scriptures by William Tyndall, Martyr 1536*, ed. Henry Walter, pp. 345-89. References are to the former volume, except in note 86, where the reference is to the second volume.

hereof ye see, that our sacraments are bodies of stories only; and that there is none other virtue in them than to testify and exhibit to the senses and understanding the covenants and promises made in Christ's blood."[82] Tyndale says there are three contemporary explanations of the *Hoc est corpus meum*. The Romanists, he insists, "compel us to believe, under pain of damnation, that the bread and wine are changed into the very body and blood of Christ really: as the water at Cana Galilee was turned into very wine."[83] The Lutherans said: "We be not bound to believe that bread and wine are changed: but only that his body and blood are there presently."[84] The Reformed churches said: "We be bound by these words only to believe that Christ's body was broken, and his blood shed for the remission of our sins; and that there is no other satisfaction for sin than the death and passion of Christ."[85] Tyndale recognized that the Lord's Supper had, besides a pedagogical, a sociological value. His own view was strictly Memorialist.

How simple Tyndale's notion of the early Eucharist was may be seen in the manner in which he sets forth his detailed proposal for the restoration of the Lord's Supper. He pictured a minister gathering his flock together for a service in which the didactic interest was so dominant that it almost drove devotion away. Apart from a lengthy explanatory sermon on the doctrine of the atonement and on the meaning of the sacramental signs, there are several exhortations, and few prayers. Tyndale saw forgiveness of sins as the primary gift of the Gospel, and insisted that the congregation approach the holy table with genuine forgiveness in their hearts and a determination to follow Christ in love. The readings were to include I Corinthians 11 and John 6. The rest can be told in Tyndale's own words:

> These with such lyke preparations and exhortations had, I would every man present should professe the Articles of our fayth openly in our mother toung, and confesse his sinnes secretly unto God, praying intierly that hee would now vouchsafe to have mercy upon hym, receive his prayer, glewe hys hart unto hym by fayth and love, encrease hys fayth, geve hym grace to forgyve and to love hys neighbour as him selfe, to garnish hys lyfe with purenes and innocency, and to confirme hym in all goodnes and vertue. Then againe it behoveth the curate to warne and exhorte every man deepely

[82] *Doctrinal Treatises*, p. 358. [83] *Ibid.*, p. 366.
[84] *Ibid.*, p. 367. [85] *Ibid.*

to consider and expende with hym selfe, the signification &
substaūce of this Sacrament, so that he sit not down an hipo-
crite and a dissembler, sith God is searcher of harte and
raines, thoughtes and affectes: and see that he come not to the
holy table of the Lorde without that fayth whiche he professed
at hys Baptisme, and also that love which the Sacrament
preacheth and testifieth unto hys hart, lest hee now found
guilty of the body and bloud of the Lord . . . receive his own
damnation. And here let every man fall downe uppon hys
knees saying secretly with all devotion their *pater noster* in
English, theyr curate as example kneelyng downe before
them, which done let hym take the bread and eft the wyne
in the sight of the people hearing him with a loude voyce with
gravitie, and after a Christen religious reverence rehearsing
distinctly ye wordes of the Lordes Supper in the mother
toung. And thē distribute it to the ministers, which taking the
bread with great reverence, will devide it to the congregation
every man breakyng and reaching it forth to his next neighbour and member of the mistike body of Christ, other ministers folowyng with the cuppes powring forth & dealing them
the wyne, all together thus beyng now partakers of one bread
and one cuppe, the thyng therby signified and preached
printed fast in their hartes.[86]

Ultimately the Eucharist was for Tyndale the most vivid of all
visual aids for the mental communication of the Gospel, through
which the costliness of redemption and its chief benefit, forgiveness of sins, and the unity of love which should characterise the
Christian community were most dramatically exhibited.

John Hooper was another Zwinglian among the early English
Reformers, which is not surprising since he had spent from 1547
to 1549 in Zwingli's city of Zurich as a young man. He refused
to believe that man is saved by sacraments, for this "doth derogate
the mercy of God, as though His holy Spirit could not be carried
by faith into the sorrowful and penitent conscience except it rid
always in a chariot and external sacrament."[87] His succinct definition of the Lord's Supper was: "It is a ceremony instituted by
Christ, to confirm and manifest our society and communion in his

[86] *Workes of Tyndall*, pp. 477 v.-478, which is a separate statement of Tyndale's, not in his treatise on the Sacrament, on *The Restorying of the Lordes
Supper.*
[87] *The Early Writings of John Hooper*, ed. Samuel Carr, p. 131.

body and blood, until he come to judgment."[88] Sacraments, for Hooper, do not convey grace not otherwise obtainable; they proclaim the Gospel visually: "The sacraments be as visible words offered unto the eyes and other senses, as the sweet sound of the word to the ear, and the Holy Ghost to the heart." Each of the two sacraments "teach and confirm none other thing than that the mercy of God saveth the faithful and believers." In short, the sacraments were "seals of God's promises in Christ."[89] In another place Hooper confirmed this definition, but went on to say that the sacraments were not *signa nuda*, empty signs, which is precisely what opponents of the Zwinglian doctrine had always maintained them to be. Hooper's fullest statement was: "I believe also the holy sacraments (which are the second mark or badge of the true church) to be the signs of reconciliation and great atonement made between God and us through Jesus Christ. They are seals of the Lord's promises, and are outward and visible pledges and gages of inward faith, and are in number only twain, that is to say, baptism and the holy supper of the Lord. The two which are not void and empty signs, but full; that is to say, they are not only signs whereby something is signified, but also they are such signs as do exhibit and give the thing that they signify indeed, as by God's help we will declare hereafter."[90] However much Bishop Hooper protested that the sacraments, in his interpretation, were channels of grace, yet so strong was his Memorialism that it is difficult to accept the statement. With characteristic clarity he asserted, "I believe that the holy supper of the Lord is not a sacrifice, but only a remembrance and commemoration of this holy sacrifice of Jesus Christ. . . ."[91]

Hooper was highly critical of the idea that there could be any spiritual benefit from the corporeal presence of Christ in the Eucharist, since Christ had taught his disciples, as recorded in John 6, that "the flesh profiteth nothing," and had assured them, regarding his forthcoming return to God the Father: "It is expedient that I go away." Hooper's inevitable conclusion was: "We must therefore lift up our minds to heaven [where Christ's ascended body now is] when we feel ourselves oppressed with the burden of sin,

[88] *Ibid.*, p. 175.

[89] *Ibid.*, p. 513.

[90] *Later Writings of Bishop Hooper, together with his Letters and other Pieces*, ed. Charles Nevinson, p. 45.

[91] *Ibid.*, p. 32. Characteristically the next sentence asserts that the Mass is an invention of man, a sacrifice of Anti-Christ, and "a stinking and infected sepulchre."

and there by faith apprehend and receive the body of Christ slain and killed, and his precious blood shed, for our offences: and so by faith apply the virtue, efficacy, and strength of the merit of Christ to our souls, and by that means quit ourselves from the danger, damnation, and curse of God." He adduced a powerful citation in confirmation of this interpretation: "And St. Augustine saith, *Ut quid paras dentem et ventrem? Crede et manducasti*: 'Why preparest thou the teeth and belly? Believe and thou hast eaten.' "[92] So convinced was he of the fact that the assertion of the corporal presence of Christ in the Sacrament is idolatry, that in a sermon preached at the court of King Edward VI he urged the magistrates to require all communicants to receive standing or sitting, but not kneeling. He preferred the posture of sitting, for it expressed the fact that Christ had come, "that should quiet and put at rest both body and soul."[93] Hooper was insensitive in ridiculing the convictions of others with which he disagreed.[94] Yet he was the most vigorous exponent of Memorialism in England on both biblical and rational grounds; he was, apparently, without a trace of mysticism.

The last of the present group of early English reformers also occasionally confused sacredness with insensitivity and truth with tactlessness. The opinions of Thomas Becon are of unusual interest because Cranmer chose him to be his chaplain. Becon's *Catechism*[95] provides a full treatment of the Protestant understanding of the Lord's Supper and an extensive critique of the Catholic Mass. The dominant emphasis was on the Lord's Supper as a banquet, a spiritual food in which "Christ Jesus the Son of God witnesseth that he is the living bread, wherewith our souls are fed unto everlasting life."[96] The Lord's Supper was instituted as a remembrance of Christ's death, as a sign of the unity and concord among Christians,

[92] *Early Writings*, p. 530.

[93] *Ibid.*, p. 537. This anticipated the protest Knox would make against kneeling for the reception of holy communion before the same august auditory, and which would result in the hurried inclusion of the "black rubric" in the prayer-book of 1552.

[94] See his satirical remark that the consecration in the Mass is achieved at the final syllable *-um* in the recital of the words, *Hic est corpus meum* (*Early Writings*, p. 525). In this respect he was no better than the presumably Protestant persons who gave a dead cat a monk's tonsure and put a white disc in his mouth, resembling a host, which was later hung on the pulpit at St. Paul's Cross, or Weston, Catholic Dean of Westminster, who referred to the Protestant communion table as "an oyster-board." See Millar MacLure, *The Paul's Cross Sermons, 1534-1632*, p. 51.

[95] References are to the Parker Society edition of Becon's *Catechism*, ed. J. Ayre.

[96] *Ibid.*, p. 229.

as a means of confirming and strengthening faith, as a proclamation that Christ's body and blood are the true nourishment of the faithful soul, and to move Christians to be thankful for the death of Christ and for its benefits and fruits.[97]

Becon discovered 11 "errors" in the Roman Catholic Mass. It is a private banquet, not a common meal.[98] It is a communion in one kind, not two.[99] It changes a commemorative into a propitiatory sacrifice.[100] The sacrament which should be a meal is reserved in boxes and pixes.[101] There is idolatry in turning the sacrament into a "gazing stock."[102] The service is not in the vernacular, but in Latin.[103] The Romanists allow noncommunicants to be present as spectators.[104] Celebrations are rare, usually once a year.[105] The 9th, 10th, and 11th errors in the Mass are doctrinal in character, consisting of Transubstantiation, the corporal presence, and the assertion that the ungodly and the wicked, not only the godly and faithful, consume the body and blood of Christ in the Sacrament.[106]

Clearly the early English reformers were much more acute in their criticisms of the Catholic Mass in doctrine and practice than they were successful in elaborating a Protestant understanding of Holy Communion which was not a reductionistic Memorialism. It was to take considerable time before there would be developed a third alternative to the corporal presence affirmed by Catholics and Lutherans, on the one hand, and the failure within Zwinglian terms to define a presence of Christ in the Sacrament any different from its modality in the preaching of the word, on the other. Unquestionably this is what not only Cranmer, but Bucer and Calvin, were groping for in the 1540s and 1550s. In this search the researches of Nicholas Ridley, Bonner's successor as bishop of London, are important in their own right, as well as in their impact on the development of Archbishop Cranmer's thought.[107]

4. The Sacramental Theology of Cranmer's Advisors

It appears that it was in 1545 that Ridley read Ratramnus's *On the Body and Blood of the Lord* which held that the Sacrament

[97] *Ibid.*, pp. 231-32. [98] *Ibid.*, pp. 238-40.
[99] *Ibid.*, pp. 240-45. [100] *Ibid.*, pp. 245-51.
[101] *Ibid.*, pp. 251-53. [102] *Ibid.*, p. 253.
[103] *Ibid.*, pp. 253-55. [104] *Ibid.*, pp. 255-57.
[105] *Ibid.*, pp. 257-60. [106] *Ibid.*, pp. 260-97.
[107] Brooks, the Marian Bishop of Gloucester, said, when speaking to Ridley at Ridley's trial in Oxford in 1554: "Latimer leaneth to Cranmer, Cranmer to Ridley, and Ridley to his own wit." Cf. Ridley's, *A Brief Declaration of the Lord's Supper* (written while he was imprisoned in 1544), ed. H.G.C. Moule, p. ix.

was a figure of the body and blood rather than the veritable body and blood of Christ after the consecration. Ridley was relieved that a ninth-century monk could hold such a view without any official condemnation during his lifetime. Ratramnus's treatise was reprinted in Cologne in 1531 by Johann Präl at the instigation of Oecolampadius, and Leo Jud translated it into German. Ridley himself had referred to Ratramnus as "Bertram" in the Debate in the House of Lords in 1548, as well as in the Oxford Disputation of 1554. "This Bertram," said Ridley, "was the first that pulled me by the ear and that brought me from the common error of the Romish church and caused me to search more diligently and exactly both the Scriptures and the writings of the old ecclesiastical fathers in this matter."[108]

After Ridley has set forth the accounts of the Last Supper in the three synoptic Gospels and in the Pauline account in I Corinthians 11, he says that those who accuse him of making the Sacrament only "a bare signe or figure" which represents Christ "none otherwise than the Ivy Bush doth represent the wine in a tavern, or as a vile person gorgeously apparelled may represent a king or a prince in a play,"[109] seriously misinterpret his teaching.

The central issue in the controversy was: "As, whether there be any Transubstantiation of the bread, or no: any corporal and carnal presence of Christ's substance, or no. Whether adoration (due only unto God) is to be done unto the Sacrament, or no? and whether Christ's body be there offered in deed unto the heavenly Father by the priest or no: and whether the evil man receiveth the natural body of Christ or no." But, as Ridley was careful to show, all five questions depended on one question, "which is, What is the matter of the Sacrament: Whether is it the natural substance of bread, or the natural substance of Christ's own body?"[110] If the answer to the second part of the question is in the affirmative, then Transubstantiation must be affirmed, the Sacrament may rightly be adored, the priest does indeed offer up Christ's body to the Father, and both good and evil alike receive, with different effects, the veritable body and blood of Christ. But if it were found that the natural substance of bread is the material substance (for there cannot be two substances occupying one space) in the Sacrament, then it must follow there is no such thing as Transubstantiation and the natural substance of Christ's human nature which he took

108 *Ibid.*, p. 200.   109 *Ibid.*, p. 104.
100 *Ibid.*, p. 106.

of the Virgin Mary is in heaven where He now reigns in glory, and is not here enclosed under the form of bread.[111] Thus were the logical alternatives posed with a clarity rare in so complex and obscure a controversy.

What was the mode of Christ's presence in the Eucharist in Ridley's thought? Briefly, he denied the presence of Christ's body in the natural substance of his human and assumed nature, but granted the presence of the same by grace. He affirmed that "the substance of the natural body and blood of Christ is only remaining in heaven and so shall be" until the Day of Judgment. He also affirmed that because the human nature of Christ is joined with the divine nature in Christ, the second person of the Trinity, therefore "it hath not only life in itself, but is also able to give and doth believe on his name. . . ." This happens "By grace (I say) that is, by the gift of this life . . . and the properties of the same, meet for our pilgrimage here upon earth, the same body of Christ is here present with us."[112] He clarified the concept with the analogy of the sun, which is in heaven, yet is present on earth by "his beams, light and natural influence, where it shineth upon the earth. For God's Word and his Sacraments be, as it were, the beams of Christ, which is *Sol iustitiae*, the Sun of righteousness."[113]

Ridley went on to support his arguments with scriptural and patristic evidence, as all the other controversialists did. We need not follow him, especially in the drearier part of his pilgrimage. He did, however, make some shrewd points. One was the assertion that he could find another Transubstantiation if that is what is wanted, and he discovered it in the statement of Christ: "This cup is the New Testament in my blood." If "This cup" is a figure of speech, so is "This is my body." Another telling point was that in the Gospels of Matthew and Mark, Christ speaks of the fruit of the vine as his cup, and quite plainly so. Thus the natural substance of the wine remains still, which ought not to be the case if Transubstantiation is to be believed.[114] Transubstantiation destroys the nature of the Sacrament by denying the sign. Furthermore, it denies the truth of the human nature of Christ, making those who assert it into Monophysite heretics, a thrust he must particularly have enjoyed, being so often at the other end of that particular sword. The positive values of the Lord's Supper were three, according to the statement of Ridley in the Cambridge Debate of 1549:

111 *Ibid.*, pp. 106-108.　　112 *Ibid.*, pp. 110-11.
113 *Ibid.*, p. 111.　　114 *Ibid.*, pp. 126-27.

unity, nutrition, and conversion.[115] The first two are easily explained. The unity is the unity of the mystical body of Christ, his people, which is sealed and expressed in the Lord's Supper. The nutrition is the spiritual feeding of the soul by the spiritual body of Christ apprehended by faith. The conversion is not any substantial change in the nature of the bread and wine, but only in the use to which they are put, which is a sacred and not a common use. The conversion of use was what Cranmer and Peter Martyr termed a "sacramental mutation," a disturbingly ambiguous phrase.

This is the best-argued treatise on the Eucharist from the Protestant side in sixteenth-century England. It is notable for the logical presentation of the two major explanations and of their consequences. The writing is always pellucid, the illustrations both pertinent and illuminating. One can imagine how very attractive Cranmer found Ridley's clear explanations of complex issues and his presentation of a Eucharistic theory which attempted to be a true middle way between Transubstantiation and Memorialism. Not less impressive was Ridley's determination to put his convictions into practice however strong the conservative opposition might be. He was the first Protestant bishop in England to insist that all altars in his diocese of London be replaced by Communion tables. We can hear through the arches of the long years the clarity and firmness of his voice as he explains his reasons: "For the use of an altar is to make sacrifice upon it; the use of a table is to serve for men to eat upon. Now, when we come to the Lord's board, what do we come for? to sacrifice Christ again, and to crucify him again, or to feed upon him that was once only crucified and offered up for us? If we come to feed upon him, spiritually to eat his body, and spiritually to eat his blood (which is the true use of the Lord's Supper), then no man can deny but the form of a table is more meet for the Lord's board, than the form of an altar."[116]

Cranmer was anxious to get the best Protestant scholars in Europe to advise him, and a succession of them found haven in England. In 1547 Peter Martyr, Tremellius, and Ochino reached England. The next year John à Lasco, Martin Bucer, and Paul Fagius came. Others were Utenhove of Ghent, Dryander from

---

[115] For an account of this important debate see Jasper Ridley, *Nicholas Ridley: A Biography*, pp. 179-80.

[116] *The Works of Nicholas Ridley, D.D. Sometime Lord Bishop of London, Martyr, 1555*, ed. Henry Christmas, p. 322.

Spain, Micronius from Switzerland, Valerandus from the Netherlands, and Alexander and Veron from France. Cranmer found positions for all of them, but he made particular friends of Bucer, Peter Martyr, and John à Lasco, each of whom had decided views on the Eucharist.

Bucer[117] seems to have had the greatest influence on Cranmer, but after Bucer's death on February 28, 1551, Martyr and à Lasco came into their own. Bucer's influence was to be expected, since he had not only advised Archbishop Hermann Wied on the reformation of the archdiocese of Cologne, but, after the death of Zwingli in 1531, he was the leader of the Reformed Churches in Switzerland and southern Germany. His greatest strength was his persistently irenic concern to unite the parties within Protestantism, but which carried with it the weakness of seeming to make too many diplomatic concessions.[118] For example, he was not content to let the chasm between the Saxons and the Swiss that had opened at Marburg in 1529 remain open. In gaining from the obdurate Luther the Wittenberg Concord of 1536,[119] which accepted Consubstantiation but conceded that while the unworthy did receive the body of Christ yet the unholy did not, he found that the Swiss refused the Concord. Since Cranmer's hope was exactly the same—getting the maximum of concord among Anglicans—Bucer's stay in England was a source of great satisfaction to Cranmer. Moreover, his influence was to continue in those Cambridge theologians who were proud to be his friends and attend his immensely popular lectures, such as Parker, Sandys, and Grindal, the great statesmen-prelates of the Elizabethan church.[120]

What, then, was Bucer's Eucharistic theology? If we are to take the Strassburg Liturgy in 1539 as the standard of his practice (which, incidentally, shows an increasing conservatism as contrasted with his earlier sacramental theology in the *Grund und Ursach* of 1524[121]), then Bucer held a doctrine of the spiritual real presence of Christ in the Eucharist. Whenever Holy Communion

[117] See *Opera*, ed. R. Stupperich, *Scripta Anglicana fere omnia a C. Huberto collecta*, esp. the *Censura*; A. E. Harvey, *Martin Bucer in England*; W. Pauck, *Das Reich Gottes auf Erden. Eine Untersuchung zu Butzers De Regno Christi*; C. Hopf, *Martin Bucer and the English Reformation*; and G. J. Van de Poll, *Martin Bucer's Liturgical Ideas*.
[118] C. H. Smyth, *Cranmer and the Reformation under Edward VI*, p. 153, calls Bucer, "the Trimmer of Protestantism."
[119] *The New Schaff-Herzog Encyclopedia of Religious Knowledge*, vol. II, 323a.
[120] Smyth, *Cranmer and Reformation*, p. 173.
[121] See Bard Thompson's introduction to Bucer's Strassburg Liturgy in *Liturgies of the Western Church*, pp. 161-66.

was scheduled the minister was required to append to his sermon an exhortation which set forth in the simplest terms Bucer's doctrine, including the words: "The Lord truly offers and gives His holy and sanctifying body and blood to us in the Holy Supper, with the visible things of bread and wine, through the ministry of the church."[122] The rite has three alternative Great Prayers, replacing the Roman Canon; the emphasis in each of them was on the consecration of the persons, not the elements, that by the power of the Holy Spirit they might receive Christ, the true nourishment of the soul in the sacrament. The first prayer contained the following petition: "And grant us, O Lord and Father, that with true faith we may keep this Supper of they dear Son, our Lord Jesus, as he hath ordained it, so that we verily receive and enjoy the true communion of His body and blood, of our Saviour Himself, who is the only saving bread of heaven. In this holy sacrament, He wishes to offer and give Himself so that He may live in us, and we in Him, being members of His body and serving thee fruitfully in every way to the common edification of thy Church. . . ."[123] The second prayer included the petition ". . . grant that we may now accept with entire longing and devotion His goodness and gift, and with right faith receive and enjoy His true body and true blood, yea, Himself our Saviour, true God and man, the only true bread of heaven: That we may live no more our sinful and depraved life, but that He in us and we in Him may live His holy, blessed and eternal life, being verily the partakers of the true and eternal testament, the covenant of grace. . . ."[124] Here we can discern the distinction between the Bucerian and Zwinglian doctrines of the Lord's Supper. Bucer believed that as the body eats the bread and drinks the wine, so does the soul eat the spiritual body of Christ, the soul's true bread, and that this spiritual presence of Christ is vivifying and hallowing, to eternal life. Zwingli, on the other hand, held that bread and wine are only figures or metaphors for the body of Christ now in heaven. Sacraments for Zwingli are *signa nuda*, but for Bucer *signa exhibitiva*, conveying what they represent. Bucer's doctrine is most clearly seen in Cranmer's disputation at Oxford in 1555, when the beleaguered archbishop claimed Tertullian's statement that: *Nutritur corpus pane symbolico, anima corpore Christi*: that is, 'Our flesh is nourished by symbolical or sacramental bread, but our soul is nourished with

[122] *Ibid.*, p. 165.   [123] *Ibid.*, p. 173.
[124] *Ibid.*, pp. 176-77.

the body of Christ.' "[125] A moment later Cranmer expounded his own view: "The body is nourished both with the sacrament and with the body of Christ: with the sacrament to a temporal life; with the body of Christ to eternal life."[126] It was clearly the doctrine that underlay the Holy Communion rite in the 1549 prayer-book.[127]

Peter Martyr was the great friend of Bucer, for whom Bucer had obtained the chair of theology at Strassburg. The clearest brief statement of his belief in the spiritual presence and in spiritual manducation in the Eucharist is that which was made in a speech during the Cambridge Disputation in May 1549:

> D. Martyr. I answer: when we receive the sacrament faith-fullie, two kinds of eatings are there, and also two sorts of bread. For the receiving of the bodie of Christ, which we have by faith, is called a metaphoricall eating: even as the bodie of Christ which we receive, is a metaphoricall bread. There also have we an eating of the sacramentall signes; the which is a proper eating; and true bread was given for a signe: and so in sixt of *Iohn*, there is mention onlie of metaphoricall eating, and of metaphoricall bread: but in the supper of the Lord, wherein he communicated with his apostles, there was had a proper eating; and true bread was given for a signe; and so in the supper was given both sorts of bread, even naturall and metaphoricall: and both sorts of eating is performed; to wit, both a naturall eating in signes, and also a metaphoricall, as touching the bodie of Christ, which we receive by faith. . . . But as for the words; *Take ye, and eate ye,* I saie, that they must thus be understood: As ye receive this bread, and eat it with your bodie; so receive ye my bodie by faith, and with the mind, that ye may be strengthened thereby in stead of meate.[128]

The fullest statement of Peter Martyr's doctrine is found in Udall's translation of the detailed report of the Oxford Debate, entitled *A discourse or traictise of Petur Martyr Vermilla Florētine, the*

[125] *Writings and Disputations of Thomas Cranmer,* ed. J. E. Cox, p. 420. There is only one difference between Bucer and Cranmer. Bucer maintained that the body of Christ is spiritually present in the Eucharist and is spiritually received by the souls of worthy believers. Cranmer, however, taught that Christ's body is not present in the substance of the sacraments, but in the action of administration.
[126] *Ibid.,* pp. 420-21.
[127] This is more fully treated in Chap. v, with reference to Bucer's criticisms in the *Censura* of the 1549 prayer-book.
[128] Cited in Smyth, *Cranmer and Reformation,* p. 129.

publyque reader of divinitee in the Universitee of Oxford, wherein he openly declareth his whole and determinate iudgemente concernygne the Sacramente of the Lordes Supper in the sayde Universitee. It is mainly concerned with finding 49 arguments against Transubstantiation, and much less so in providing clear explanations of the meaning of 'spiritual eating.' It has one concluding statement on the matter: "For we have saied and doe confirme that these materiall sygnes dooe moste truely sygnyfye, represente, and exhibite unto us the bodye of Chryste, to bee eaten: howbeit it is spiritually, that is, wyth the mouth of the solle to bee eaten, and not of the bodye."[129] His most original contribution was his interpretation of Paul's doctrine of incorporation in Christ in Ephesians. This incorporation takes place in communicating in the sacrament, and by faith, but we become part, not of the incarnate body of Christ, but are conjoined with the flesh of the risen Christ.[130]

What is of chief importance, however, are those four features of sacramental doctrine common to Martyr and Bucer and which provide the lineaments of Cranmer's doctrine of the Eucharist: The first is the assertion that Christ's body and blood are in some sense in the Sacrament and not merely in a metaphorical or figurative manner. Secondly, that presence is apprehended only by those who have faith. It is the sanctifying and revivifying power of the Holy Spirit who stirs up such faith. The same Holy Spirit, in the fourth place, effects the conjunction between the believers and the body of Christ, transforming the elements from common into sacramental usage and bringing about the realities of which the bread and wine are signs. Only on the fifth point is there disagreement, concerning the consecration. Martyr joins the action of the Holy Spirit in effecting the "sacramental mutation" with the words of consecration; Bucer believes increasingly that it is people, not things that are consecrated. So (as his *Censura* on the 1549 prayerbook indicates), he wished to avoid a technical consecratory formula in the Holy Communion.

The third member of Cranmer's foreign advisers in England was John à Lasco.[131] It is just possible that his influence may have turned Bucer as well as Martyr toward a more radically Memorialist doctrine of the sacrament which found expression in the Sec-

129 Fol. cix.
130 *Ibid.*, fol. cviii v.-cix.
131 See John à Lasco, *Brevis ac dilucida de sacramentis ecclesiae Christi tractatio* (1552).

ond Prayer Book of 1552.[132] It is not necessary to rehearse the Memorialist type of sacramental doctrine, but only to recall that à Lasco reemphasizes that the Eucharist is a present reminder of a past redemption, a visual aid to the preaching of the Gospel, a badge of the loyalty of Christians gathered in community, and a confirmation and pledge of the reliability of God's promises. It is a figure or representation of the historic crucifixion, not a communication of the body of Christ. Above all, the metaphor of "eating" at the Lord's Supper is consistently interpreted as believing in the passion of Christ. His Nominalism rejected the connections the realist can see between universals in the mind and in the object considered and so makes a corporal identification with Christ impossible. On the contrary, since mind and body were a dichotomy for à Lasco, faith has to be concentrated in the mind and consciousness, however objective it may be in character. John à Lasco had nothing new to say, but what he said he said with vigor and clarity; a reductionist position is always more impressive in attack than in exposition. But by the same token, it seems to make short work of any mediating viewpoint, such as was held by Bucer and Cranmer in the 1549 prayer-book.

5. Cranmer's Eucharistic Theology

The development of Cranmer's thought as seen in the Eucharistic rites of the 1549 and the 1552 prayer-books will be considered later.[133] My immediate purpose is to illuminate its facets by two approaches: through the pen war with Gardiner and by comparing it with the teaching of Zwingli and Calvin, with whom, in its last stage, it had the closest affinities.

Cranmer will remain a mysterious figure and his teaching unclear, partly because his Erastianism, with its conviction of the divine right of kings caused him to try to meet the differing needs of two successive monarchs and their changing moods, partly because his work being incorporated in two prayer-books and a book of homilies is anonymous, and partly because he was surrounded by advisers with differing theological viewpoints. It is, however, probable that in his tenure of the Archbishopric of Canterbury he held three different views of the modality of Christ's presence in the Eucharist. The first was Transubstantiation, which

132 The probable influence of à Lasco on the 1552 prayer-book is examined in Chap. v.
133 See Chap. v.

he began to question in 1538.[134] The second was a doctrine of the Real Presence which was close to Lutheran views.[135] The third— which he came to accept under the influence of Ridley in 1546 and to which he had openly committed himself by the time the third edition of the translation he had made of the catechism of Justas Jonas appeared—was a more radical doctrine. It is not yet wholly clear *how* radical this doctrine was, whether sacramentarian like the Lollards and Zwingli, or virtualist like Calvin.[136] Cyril Richardson holds the former, and Peter Brooks and C. W. Dugmore the latter view.[137]

Cranmer wrote his own full exposition of Eucharistic doctrine in the *Defence of the True and Catholic Doctrine of the Sacrament of the Body and Blood of Our Saviour Christ* in 1550.[138] This provoked Stephen Gardiner's rejoinder, already considered, namely, *An Explication of the True Catholic Faith touching the Most Blessed Sacrament of the Altar* (1551), and a rebuttal from Cranmer—*A Crafty and Sophistical Cavillation devised by M. Stephen Gardiner . . . against the True and Godly Doctrine of the Most Holy Sacrament . . . with an Answer unto the same . . .* (1551). Cranmer's first full-length treatise on the Eucharist, in

[134] In a letter of August 15, 1538 Cranmer wrote to Cromwell about Adam Damplip of Calais, accused of denying transubstantiation while affirming the real presence, that, "herein I think he taught the truth." Cranmer, *Miscellaneous Writings and Letters*, ed. J. E. Cox, p. 375.

[135] It is difficult to evaluate Cranmer's Commonplace Book in which citations from different authors are given over a period of time. It does seem important, however, to mention two excerpts from an anti-Zwinglian tract of 1527 by Luther, entitled, *Das dies Wort Christi noch fest stehen Widder die Schwermgeister* included in the Commonplace Book and given in Peter Brooks, *Thomas Cranmer's Doctrine of the Eucharist*, pp. 30-35; and the fact that there are no citations from the sacramentarians. From this one might hazard two highly tentative conclusions. The first is, Cranmer was more anxious to refute than to espouse sacramentarianism, the second, that he had felt the sting of some of the Zwinglian criticisms of the doctrine of the Real Presence and was eager to have these doubts removed.

[136] Jasper Ridley, *Thomas Cranmer*, p. 256, inclines to the opinion that it was neither Ridley nor à Lasco who converted Cranmer to a new Eucharistic doctrine, but that "perhaps he was really converted by Henry and the times." This could mean that Henry's death in 1547 and Somerset's Protectorship over the young Edward VI encouraged Cranmer to develop or divulge an already developed Protestant direction, yet with characteristic caution. D. G. Selwyn, in "A Neglected Edition of Cranmer's Catechism," *Journal of Theological Studies*, New Series, xv (April 1964), 84, believes that Cranmer had three different doctrines of the Eucharist.

[137] Richardson's views are expounded in *Zwingli and Cranmer; Anglican Theological Review*, vol. 47, no. 3 (1965), 308-13; and *Journal of Theological Studies*, New Series, XVI, pt. 2 (October 1965), 421-37. Brooks' views appear in *Thomas Cranmer's Doctrine of the Eucharist*; and Dugmore, *Mass and English Reformers*.

[138] See *The Remains of Thomas Cranmer*, ed. H. Jenkyns, vol. 2, for the text of the *Defence of the True and Catholic Doctrine of the Body and Blood of Our Saviour Christ*. The first reference is to p. 297.

five parts and almost 200 pages long, had as its aim to show how and why the Eucharist was instituted from Scripture and the church Fathers and to confute the four major "errors" of the Catholic interpretation. The first part, sadly recalling that the sacrament which was ordained to create concord was the occasion of variance and discord, declared that the Eucharist is the spiritual food of the soul, was ordained to confirm faith, expresses the unity of Christ's mystical body, and is spiritually eaten with the heart, not the teeth. It concluded by enumerating the "four principal errors of the Papists."[139] The first was Transubstantiation; the second was the corporal presence of Christ in the Sacrament; the third was the assertion that evil men drink the very body and blood of Christ as do godly men; and the fourth was the claim that the priests "offer Christ every day for remission of sin, and distribute by their masses the merits of Christ's passion."[140] The remaining four parts of the book consist of the attempt to refute these four "errors" from Scripture and the Fathers; in it the positive teaching of Cranmer is also to be found.

Transubstantiation was "against the word of God, against nature, against reason, and against all our senses" as also "against the faith and doctrine of the old authors of Christ's Church. . . ."[141] Cranmer's nominalism[142] becomes apparent in his argument that Transubstantiation is against reason, for the Nominalists insisted that what gives being to an entity is its individual characteristics, so that if the accidents of a substance are negated then its being is also. "And most of all," wrote Cranmer, "it is against the nature of accidents to be in nothing. For the definition of accidents is to be in some substance, so that if they be, they must needs be in something. And if they be in nothing, then they be not."[143] Thus the assertion that the accidents of bread and wine remain after the consecration, but that the substance changes to the body and blood of Christ, as Transubstantiation maintains, was plainly unacceptable to any Nominalist.

Cranmer found equally unacceptable (in part three) the assertion that there is a corporal presence of Christ in the Eucharist. He referred to Christ's words in John 6: "It is the spirit that giveth

139 *Ibid.*, marginal gloss, p. 308.
140 *Ibid.*, pp. 308-12.
141 *Ibid.*, p. 320.
142 For Cranmer's philosophical presuppositions see the debate between E. G. McGee, who thinks of Cranmer as a nominalist, and W. J. Courtenay who denies it, in the *Harvard Theological Review*, vol. 57 (July and October 1964).
143 *Remains of Cranmer*, II, 318.

life, the flesh availeth nothing. The words which I spake unto you, be spirit and life." These, he says are not to be understood, "that we shall eat Christ with our teeth grossly and carnally, but that we shall spiritually and ghostly with our faith eat him, being carnally absent from us in heaven; and in such wise as Abraham and other holy fathers did eat him, many years before he was incarnate and born."[144] The Dominical words, "This is my body" are to be interpreted figuratively.[145] At this point it is worth citing an earlier explanation of what Cranmer means by spiritual feeding on Christ: "And the true eating and drinking of the said body and blood of Christ, is with a constant and lively faith to believe, that Christ gave his body and shed his blood upon the cross for us, and that he doth so join and incorporate himself to us, that he is our head, and we his members, and flesh of his flesh, and bone of his bones, having him dwelling in us, and we in him. And herein standeth the whole effect and strength of this sacrament. And this faith God worketh inwardly in our hearts by his Holy Spirit, and confirmeth the same outwardly to our ears by the hearing of his word, and to our other senses by eating and drinking of the sacramental bread and wine in his holy Supper."[146] This is the central thrust of the latest development in Cranmer's teaching. It is important to recognise that while Cranmer denied the corporal presence of Christ, since his risen body is in heaven, he affirmed the spiritual eating of the body and blood of Christ as conveying all the benefits of Christ's death and resurrection to the soul; this is the objective work of the Holy Spirit, who through the preaching of the word as well as in the sacraments creates faith and revivifies the Christian and the church.

In the fourth part of the treatise Cranmer insists that only godly persons eat the spiritual food of the sacrament, since faith must be the mouth of the soul. "For unto the faithful, Christ is at his own holy table present with his mighty Spirit and grace, and is of them more fruitfully received, than if corporally they should receive him bodily present."[147] He cites St. Augustine to prove that the unworthy do not eat the true bread of Christ.

In Part 5 Cranmer denies that the priests offer Christ anew on their altars. His first reason is that the sacrifice on the Cross was once and for all times sufficient to make satisfaction for the sins of the past and of the future. To argue that the Mass is a sacrifice

144 *Ibid.*, p. 378. 145 *Ibid.*, p. 381.
146 *Ibid.*, p. 306. 147 *Ibid.*, pp. 437-38.

propitiatory is to say "that Christ's sacrifice were not sufficient for the remission of our sins; or else that his sacrifice should hang upon the sacrifice of a priest."[148] Thus the true nature of the sacrifice in the Lord's Supper is that this is the grateful remembrance of Christ's sacrifice. It is also a sacrifice of praise, made by the people themselves. "But the humble confession of all penitent hearts, their knowledging of Christ's benefits, their thanksgiving for the same, their faith and consolation in Christ, their humble submission and obedience to God's will and commandments, is a sacrifice of laud and praise, accepted and allowed of God no less than the sacrifice of the priest."[149]

Leaving Cranmer's treatise, we may now consider his cluster of Eucharistic concepts. On the primary issue of the character of Christ's body there was no question to him but that this is an empirical object, circumscribed and located in heaven and therefore absent from this world, the sacrament, and the participants in the sacrament.

As a strictly logical consequence of the denial of any link between Christ's body in heaven and the Eucharistic elements, Christ can only be present in the Eucharist by the power of His divinity and must be absent in His flesh or humanity. Clearly Cranmer rejected any possibility of developing a doctrine through which the Body of Christ could have empirical and mystical qualities, thus conjoining a location in heaven and a pervasive, substantial presence on earth. The Occamist and Lutheran possibilities were thereby rejected.

With equal rigor Cranmer insisted that the bread and wine were and are only bread and wine. They are not in essence converted into the body and blood of Christ, as in Transubstantiation, nor do they accompany the body and blood, as in Consubstantiation. The bread and wine as self-enclosed, impermeable objects of the Nominalist tradition, they are incapable of receiving sanctification and so cannot be vehicles of grace. The elements are, of course, instruments of God's working and "promise, signify, and exhibit" the grace of the Sacrament. Yet what is accomplished, is done in us and not in the sacraments. As Richardson correctly remarks: "Nature is split from categories of consciousness and personality and cannot participate in the holy," and so, "the whole fluid, mystical, substantial way of thinking whereby the divine can impregnate the natural is overthrown in the interests of emphasis

---

[148] *Ibid.*, p. 452.   [149] *Ibid.*, p. 459.

on conscious faith and of relations between God and man in terms of personal encounter."[150]

The manducation of Christ's body must be a figure, a metaphor, for Cranmer. He freely acknowledged that the patriarchs of the Old Testament spiritually fed on Christ before He had a human body at all, and that it is possible to eat Christ's body spiritually after He has gone to heaven. Thus eating the body is possible without any reference to the Body at all. Preaching and mediation on the Passion are equally effective ways of eating Christ's body spiritually. Clearly, then, what happens in the Sacrament is altogether of the same character as what happens in preaching and meditation. The only difference is that the Sacrament provides visible signs to remind Christians of their duty. Spiritual feasting is then believing, exercising faith in the Passion of Christ.

The difficulty in explaining Cranmer with any certainty is caused by the residual ambiguities in his expressions of his Eucharistic doctrine. One mystery in Cranmer is that although he had by his denial of corporal presence reduced manducation of Christ's body to the act of believing, yet he insisted that Christ truly, substantially, and naturally dwells in the communicants and they in Christ. Thus we have the paradox that Cranmer often spoke of the double indwelling, yet at the same time insisted on the absence of Christ's body. Perhaps what we have here, as Richardson suggests, is an unresolved tension between Realist and Nominalist notions in Cranmer.[151]

For Cranmer, "incorporation" of the believers with Christ meant having faith in Christ's Passion, the presence of Christ in His divinity, and the virtue or grace that comes from Christ to us in our believing. Here Cranmer's teaching diverged considerably from Bucer's. Bucer regarded the virtue of the body as substantially identical with it. Cranmer repudiated such a view in the words: "Doth not all men know, that of everything the virtue is one and the substance another?" The unreformed Nominalist in Cranmer made him insist that the beams of the sun were not of the same substance as the sun itself. He failed to see that a true incorporation into the humanity of Christ, and not a mere metaphor, requires considerable modifications of the concept of the nature of the body of Christ. It can no longer be conceived of as merely an absent object; it must be thought of in dynamic terms, as a mystical

---

[150] *Journal of Theological Studies*, XVII, pt. 2, p. 427.
[151] *Ibid.*, p. 429.

substance which is accessible to the believer. This was the core of both Patristic and Scholastic thinking on the issue. Cranmer was inconsistent, not only in trying to unite Nominalist and Realist notions of the body of Christ, but also in allowing that through the Incarnation and Atonement there was a participation of the believer in Christ's flesh and that this was taught by the Sacrament but not effected by it.

There are strengths and weaknesses in Cranmer's Eucharistic doctrine. The strengths are chiefly the Biblical insights and recoveries: the recognition that sacraments are signs, seals of the divine promises; the importance of the interiority of faith; and the link between Christ and the application of His benefits to the believers is the power of the Holy Spirit, the giver of new life. As we have seen, Cranmer's basic weakness was his circumscribed notion of "Body" and his overcorrective stress on "spiritual eating" by faith. It was partly his unwillingness to admit, "that Christ can be present in fact by the Spirit and the sacrament, and not merely by figure, when there is no faithful reception."[152] Bromiley, while making such a criticism, was careful to point out that Cranmer was more than a Memorialist or a supporter of the view that the Sacrament is merely a symbolical reenactment of the crucifixion. The unity of Christians is with the crucified and risen Christ by the power of the Holy Spirit, but it is not a natural or organic unity. It is the Holy Spirit who creates faith, so that the presence of Christ in the signs is appropriated by faith, but never created by faith, as some critics of his subjectivity suppose it to be. It must be remembered that the term "spiritual" in Cranmer's writing and thought is not a vague noun: it has the vigor and precision of the New Testament concept of life in the Spirit as contrasted with the dying life of the world. Thus Cranmer's phrase "spiritual eating" was no equivalent for figurative eating, but for the nourishment of the soul to a life beyond life, in short, a life with new dimensions of quality as well as of extension, eternal life.

If at this stage one were asked whether Cranmer was more of a Zwinglian than a Calvinist, then he might point to the fact that the 42 Articles of the Church of England were in the process of being formulated concurrently with the preparation of the prayer-book of 1552, and that they spoke of the sacraments as "effectual signs," which convey what they signify. This would make the Eucharistic doctrine more Calvinist than Zwinglian, which, after all, is what

152 G. W. Bromiley, *Thomas Cranmer, Theologian*, p. 75.

should be expected 21 years after the death of Zwingli and two years after the *Consensus Tigurinus*.

The problem may, however, be more fully resolved after a comparison of Cranmer's teaching with that of Zwingli and Calvin. Rightly or wrongly Cranmer has been assumed to be "Zwinglian" or "sacramentarian" in the last phase of his development by such scholars as C. H. Smyth and Gregory Dix, so the assertion must be examined. Undoubtedly there is close agreement between Cranmer and Zwingli. Both denied that the Eucharist is a participation in the substance of the body of Christ.[153] Both asserted that persons and not the elements participate in the Holy.[154] Christ is present in the Eucharist by his divine, not his human, nature.[155] Christ's body is in one place, not ubiquitous, and that one place is heaven. Moreover, even if it were naturally present in the Eucharist, it would be of no avail since the flesh profits nothing, according to a Johannine logion of Jesus.[156] The distinction, if not dichotomy, between flesh and spirit, as between mind and body, is strongly stressed; this can readily be seen in the fact that Cranmer put the text, John 6:13 on the title page of the *Defence*, and "the flesh profiteth nothing" was Zwingli's favorite motto. Cranmer identified "eating the body" with believing in the Passion of Christ as effecting atonement, and so did Zwingli. Because both Cranmer and Zwingli are Nominalists, they are eager to attack Transubstantiation and to deny the abstract possibility of any participation in the concept of substance, and this causes the abyss between spirit and flesh in their sacramental thought. Quite inconsistently, however, Cranmer affirmed a substantial unity between Christ and human nature in the Incarnation, while denying it in the Eucharist.

There are, however, important differences between the thought of Cranmer and Zwingli. Both had been Catholic priests, but Cranmer had a higher evaluation of the Eucharist, since he desired a weekly Communion, for the Eucharist was the pledge of the presence of Christ. For Zwingli it was sufficient to have a quarterly Communion. Also, Cranmer interpreted the Eucharist as a *sigillum Verbi*, a seal of the Word of God and His promise to men. Zwingli

153 Cf. Cranmer, *Writings and Disputations*, pp. 96, 186; and Richardson, *Zwingli and Cranmer on the Eucharist*, pp. 20-21.

154 Cf. Cranmer, *Writings and Disputations*, p. 11 and *Miscellaneous Writings and Letters*, 413ff.; and Zwingli, in Richardson, *Zwingli and Cranmer* (1949).

155 Cf. Cranmer, *Writings and Disputations*, p. 186; and Zwingli, in Richardson, *Zwingli and Cranmer*.

156 Cranmer, *Writings and Disputations*, pp. 363, 416; and Zwingli, in Richardson, *Zwingli and Cranmer*.

held the same view, but was more impressed by the fact that the Eucharist was the public pledge of Christians, their avowal of their determination to live in the obedience of faith that should characterise disciples.[157] Moreover, Cranmer gives the elements and the act of eating greater importance than Zwingli does. The question then arises: was Cranmer willing to affirm that the "signs" in the Sacrament actually confer what they signify? In short, was Cranmer a virtualist or dynamic receptionist, with closer affinities to Calvin or Bucer than to Zwingli?

Cranmer undoubtedly would have agreed with Calvin that in the sacrament God seals his love to men through the recalling of Christ's sacrifice and that men testify their piety toward God in their participation in the sacrament. He would also share Calvin's view that Transubstantiation and Consubstantiation are to be rejected because they turn the signs into the realities and therefore cease to be true sacraments. Furthermore, he would see that the faith by which the sacrament is appropriated is objective as well as subjective and that the Holy Spirit of God is the link between Christ and His people.

There are passages of Cranmer where he sounds very like Calvin: ". . . although Christ in his human nature, substantially, really, corporally, naturally, and sensibly be present with his Father in heaven, yet sacramentally and spiritually he is here present."[158] Yet if we analyse that very sentence, the second part is Calvinist, but the first part is not. Calvin believed that the flesh of Christ,[159] as well as His divine power, and our bodies and spirits are mysteriously united in the Eucharist. Cranmer did not hold the latter view. More characteristic of Cranmer is the following citation: "I say (according to God's word and the doctrine of the old writers) that Christ is present in the sacrament, as they teach also, that he is present in his word, when he worketh mightily in the hearts of the hearers."[160] As Gordon Pruett has suggested, in the light of this citation, we are justified only in speaking of "Cranmer's doctrine of the presence in the Christian life." This, of course, includes Christ's presence in the Lord's Supper. This diffused rather than focussed or concentrated presence of Christ in the Christian life was, for Cranmer, a consequence of the theologian's belief that

---

157 Cranmer, *Writings and Disputations*, p. 398; and Zwingli, in Richardson, *Zwingli and Cranmer*, p. 21.
158 *Writings and Disputations*, p. 47.
159 Calvin, *Institutes* 4:17:32; see also note 25 above.
160 *Writings and Disputations*, p. 11; and Bromiley, *Cranmer*, p. 72.

Christ is influential through his divine, and not his human, nature. Yet when one is ready to conclude that for the latest expression of Cranmer's thought all thought of the sense of Christ being present in his humanity, flesh, or *body*, is ruled out, one recalls these words: "And therefore in the book of the holy communion we do not pray that the creatures of bread and wine may be the body and blood of Christ; but that they may be unto us the body and blood of Christ; that is to say, that we may so eat them, and drink them, that we may be partakers of his body crucified, and of his blood shed for our redemption."[161] This is much more than Zwingli would have said. We know that Cranmer frequently and quite explicitly rejected the notion that the sacraments are bare signs. Yet it is less than Calvin had affirmed. On the other hand, the acceptance of the "black rubric" in the Prayer Book of 1552 is more consonant with a Zwinglian than a Calvinist interpretation.

Disappointing as this conclusion may be, after following the winding trail of Cranmer's thought, it is appropriate to recall that Cranmer did more than write a treatise on the Eucharist. He left the Church of England with a copy of the Great Bible in every parish and copies of the Book of Common Prayer in almost every pew. F. E. Hutchinson's tribute to Cranmer is just: "*lex orandi*, the regulation of public worship together with a wide circulation of the Bible, would do more to affect the character of the religion of English people for years to come than definitions of doctrine reflecting the controversies of the time. Here Cranmer was at his best and rendered his greatest service to the Church of England."[162]

6. *Eucharistic Doctrine after Cranmer*

Cranmer's influence, both in its affirmations and negations, lived on. In the 42 Articles of 1553, the negation was foremost: "a faithful man ought not to believe . . . the real and bodily presence (as they term it) of Christ's flesh and blood." By 1563 the 28th of the 39 Articles declared more positively: "The body of Christ is given, taken, and eaten in the Supper only after an heavenly and spiritual manner. And the mean whereby the body of Christ is received and eaten in the Supper is faith."

Jewel followed Cranmer in his assertions that there are three distinctions between Anglican and Catholic doctrine: "first, that we put a difference between the sign and the thing itself that is

161 *Writings and Disputations*, p. 271.
162 *Cranmer and the English Reformation*, p. 104.

signified. Secondly, that we seek Christ above in heaven, and imagine not him to be bodily on earth. Thirdly, that the body of Christ is to be eaten with faith only, and none other wise."[163] Cranmer would have been delighted with Jewel's explanation that the sacrament was eaten "with the mouth of faith" and equally with his denial of any corporal presence of Christ because Christ's body was in heaven: "the sacrament-bread is bread, it is not the body of Christ: the body of Christ is flesh, it is no bread. The bread is beneath: the body is above. The bread is on the table: the body is in heaven. The bread is in the mouth: the body in the heart. The bread feedeth the outward man: the body feedeth the inward man. . . . Such a difference is there between the bread, which is a sacrament of the body, and the body of Christ itself."[164] With Cyprian, Jewel affirmed that the sacramental food is *cibus mentis, non ventris*.[165] The same distinction between the sacrament and the thing signified by it was made succinctly two decades later by Edmund Grindal: "Christ did eat the sacrament with the apostles: ergo, the sacrament is not Christ."[166]

Edwin Sandys, who became Archbishop of York in 1576, wrote of a spiritual eating, as contrasted with the physical eating, which Transubstantiation presupposes: "In this sacrament there are two things, a visible sign and an invisible grace: there is a visible sacramental sign of bread and wine, and there is the thing and matter signified, namely the body and blood of Christ. . . . Thy teeth shall not do him violence, neither thy stomach contain his glorious body. Thy faith must reach up into heaven. By faith he is seen, by faith he is touched, by faith he is digested. Spiritually by faith we feed upon Christ, when we steadfastly believe that his body was broken and his blood shed for us upon the cross; by which sacrifice, offered once for all, as sufficient for all, our sins were freely remitted, blotted out, and washed away. This is our heavenly bread, our spiritual food. This doth strengthen our souls and cheer our hearts."[167]

The fullest and freshest treatment of the Eucharist in the Eliza-

---

[163] Jewel was answering the criticism of the Catholic writer, Harding, that what is given in the Eucharist must exist in the Eucharist to be given, and cannot, therefore, reside exclusively in heaven. See Jewel, *Works*, I, 449.

[164] Jewel, *Works*, II, 1,121; also see I, 9, 12.

[165] *Ibid.*, II, 1,121.

[166] *The Remains of Edmund Grindal, D.D.*, ed. William Nicholson, p. 43, cited in C. W. Dugmore, *Eucharistic Theology in England from Hooker to Waterland*, pp. 9-10.

[167] Sandys, *Sermons*, ed. J. Ayre, 2 vols., I, 88-89.

bethan age was that of Richard Hooker, though he too moved in the tracks laid down by Thomas Cranmer. Although he acknowledged that "the fruit of the Eucharist is the participation of the body and blood of Christ," he rejected Transubstantiation and Consubstantiation with the statement: "there is no sentence of Holy Scripture which saith that we cannot by this sacrament be made partakers of his body and blood except they be first contained in the sacrament, or the sacrament converted into them."[168] Hooker was plainly tired of unprofitable and "fierce contentions" on the explanations of the Eucharist, and insisted that they were irrelevant from a practical standpoint, "because our participation of Christ in this sacrament dependeth on the co-operation of his omnipotent power which maketh it his body and blood to us, whether with change or without alteration of the element. . . ."[169] Hooker expounded his own theory of instrumentality as if interpreted by Christ, thus: *"this hallowed food, through concurrence of divine power, is in verity and truth, unto faithful receivers, instrumentally a cause of that mystical participation, whereby as I make myself wholly theirs, so I give them in hand an actual possession of all such saving grace as my sacrificed body can yield, and as their souls do presently need, this is to them and in them my body."*[170] The real presence of Christ is to be sought, according to Hooker, not in things but in persons, not in consecrated elements but in consecrated persons receiving grace through faith.[171]

I may appropriately end this chapter, which has asked so many questions and dealt with so many differences where there may not always have appeared to be distinctions, with the positive appraisal of Hooker, ecstatically speaking of the communicant's experience of the Eucharist: "They are things wonderful which he feeleth, great which he seeth and unheard of which he uttereth, whose soul is possessed of this Paschal Lamb and made joyful in the strength of this new wine, this bread hath in it more than the substance which our eyes behold, this cup hallowed with solemn benediction availeth to the endless life and welfare both of soul and body, in that it serveth as well for a medicine to heal our infirmities and purge our sins as for a sacrifice of thanksgiving; with touching it sanctifieth, it enlighteneth with belief, it truly conformeth us unto the image of Jesus Christ; what these elements are in them-

[168] *Ecclesiastical Polity*, ed. Keble, v, lxvii, 6, p. 353.
[169] *Ibid.*   [170] *Ibid.*, v, lxvii, 12, p. 359.
[171] *Ibid.*, v, lxviii, 1, p. 363.

selves it skilleth not, it is enough that to me which take them they are the body and blood of Christ, his promise in witness hereof sufficeth, his word he knoweth which way to accomplish; why should any cogitation possess the mind of a faithful communicant but this, O my God thou art true, O my Soul thou art happy!"[172]

[172] Three criticisms may be offered of Hooker's Eucharistic thought: piety is no substitute for theological clarity; he had an inadequate doctrine of sacrifice; and Hooker, like Cranmer, did not draw the Eucharistic implications of his admirable doctrine of the Incarnation. It has been stated best by L. S. Thornton in *Richard Hooker, a Study of his Theology*, pp. 87-88: "The principle of the Incarnation is that things Divine and heavenly are actually embodied in things earthly and visible in such sense that they are no longer two entities but one; that one entity however has two aspects and belongs to two orders, the heavenly and the earthly. This is the truth both as regard to the Church and the Eucharist which Hooker missed because he had another presupposition in his mind."

PART TWO:

THE LITURGICAL ALTERNATIVES

CHAPTER IV

CATHOLIC WORSHIP

To RECOUNT the history of sixteenth-century Catholic worship in England is to walk along a heavily shadowed path on which the gleams of light are few, and those are of a wintry and ominous glint. Henry VIII might keep up the outward panoply of worship, as he insisted on the overdone royal pageantry of the Field of the Cloth of Gold. It became an increasingly hollow ceremonialism, however, for the dissolution of the monasteries was bringing to an end the great traditions of spirituality fostered by the monks, friars, and nuns, and the all too common "bare, ruined choirs" of the religious houses were mute testimony to a new secularity.

A new spring (though short-lived, as it turned out) seemed possible to faithful Catholics when Reginald Pole, Cardinal and kinsman of Henry VIII, was appointed one of the three presidents of the Council of Trent in 1545, and four years later was almost elected Pope. Another sign of the same hope was the important role in the Council played by Thomas Goldwell, bishop of St. Asaph, a Theatine who had experienced in this revolutionary group the disciplined fervour of the counterreformation, and who was the protagonist of a renovated liturgy and disciplined spirituality. The hope was, in fact, achieved with the accession of Mary to the throne of England and the submission of England to the spiritual authority of the Papacy in December 1554. But Mary and Pole, then the Archbishop of Canterbury, died on the same day four years later to the great relief of Protestants and most of England. The last desperate rays of the sun setting on a dominant English Catholicism were seen in the brilliant and doomed gallantry of the Jesuit priests, like Campion, who invaded England in 1580 and after, and by their agonies and dying testimonies encouraged vacillating Recusants (those who refused to worship according to the Book of Common Prayer) to stand firm in the faith. The story falls into three major parts: pre-Elizabethan Catholic worship; the Tridentine revaluation of worship and the part played in it by Englishmen; and the largely secret and wholly valiant worship of the Elizabethan Recusants.

127

1. *On the Eve of the Reformation*

Despite the Lollard insistence on simplicity in religion and worship, and the Erasmian criticisms of the superstition of the people and the avarice of the clergy, yet the Latin Mass, according to the Salisbury Use, especially when celebrated on a great festival in a medieval cathedral or by the entire community of a prestigious abbey, was an impressive act of corporate adoration of the living God. It is almost impossible to describe what the loss of such a liturgy must have meant to a nation that had accepted for a thousand years its reinforcement of faith and spirituality. It is easier to appreciate the visible aesthetic than the invisible spiritual loss. The austere sublimity of the Gregorian chanting, the sunlight fractured through the rose window or dyed by the lights of the vast West window, the images, all testified to a faith that made wayfaring men the companions of the saints, and provided a fit context of *son et lumière* for the miracle by which the God of the Incarnation made a new epiphany in the Mass, to provide iron rations for the warrior-pilgrims of the church militant. How splendidly did the church through the sacraments, despite the worldliness of its prelates and the somnolence of its priests, surround the main stages of the life of the plain man and woman with glimpses of transcendental glory that transfigured their dull diurnal routine! How unimaginably bleak and bare and colour-blind must the simple and didactic services of incipient Protestantism have seemed to the same Darby and Joan, when the iconoclastic excitements of sermons, which appealed to brain but not to heart or imagination, had worn off! Sacramentals abolished; sacraments reduced from seven to two and these explained so as to lose their sense of the numinous. Images in glass and stone smashed and replaced by the ubiquitous royal coat of arms. Ceremonies such as lustrations, elevations, prostrations, and the like, as well as exorcisms and dirges and tolling of bells, were prohibited as superstitious. Vestments were radically reduced in numbers and ornateness. The magic and mystery, as well as the majesty of worship had been dissipated. What was left must have seemed vacuous and uninspired, mere penny-plain, whitewashed religion.

It is only too easy, in the tradition of Pugin and Ruskin, to re-create a sentimentalised, rose-coloured Middle Ages in which donors are always godly and artists are inspired allegorists.[1] G. G.

---

[1] One is reminded of Dean Inge's remark about Ruskin, that he knew about everything in the medieval cathedral except the altar.

Coulton has rightly questioned whether many of the donors were not trying to bribe their way into heaven, and making a good end to what had been a wicked life, and whether the designers were not repeating superstitious legends in place of the authentic narratives of the Gospel.[2] A. G. Dickens asks similar questions: "Where lay communities genuinely supplied the means, did they not pay more regard to pretty fictions about the saints than to the searching demands of Pauline theology? When actually Christocentric, did not popular art concentrate unduly upon such favourite themes as the Passion and the Nativity, to the neglect of a fuller and more balanced presentation of Christ's life and teaching?"[3] When allowance is made for mixed motives, one can only regret the greedy iconoclasm which led to the loss of so many medieval art treasures through the melting down of reliquaries, images, and metal-encrusted Gospel covers. The lost of monastic libraries was also a cultural and spiritual deprivation. Some gains to education resulted from the spoliation of the monasteries, through the foundation or reendowment of grammar schools in the cathedral cities and the establishment of Christ Church in Oxford and Trinity in Cambridge, which dwarfed the older foundations, and the creation of new chairs in both universities.[4] The vast number of empty ecclesiastical buildings in London testified to the disappearance of a way of life; the new architectural landscape was secular. There were no new foundations to train men in the culture and discipline of the spiritual life,[5] nor any hospices for the poor. In this spiritual drought the twin streams of spirituality and hospitality had almost dried up.

While there must have been immorality in the monasteries, it was to Cromwell's advantage to exaggerate this to provide an excuse for pillaging them. If God made the country and man made the town, as Cowper believed, and a succession of prophets like Amos anticipated him in this judgment, then it is in the country that we would expect the familiar piety of Catholicism to linger longest, undisturbed by the new learning and newer ambition of

2 *Art and the Reformation*, chaps. 19, 20.

3 *The English Reformation*, p. 10.

4 *Ibid.*, p. 150. At Cambridge in 1540 and Oxford in 1546, Henry VIII established regius professorship in divinity, medicine, law, Hebrew, and Greek.

5 One must not forget Archbishop David Mathew's warning that "the state of the monasteries during their last years was less encouraging than is suggested by the Gasquettinted sunset. In country parishes however there were in many cases a deep constant attachment between the parishioners and their priests. . . . The Christian ages and the sacramental life were behind them constantly supporting." *Catholicism in England*, p. 2.

the cities and towns. Countrymen awaited with faith the regularity of God's blessing on the seasons; for them religion was as inevitable as the march of the seasons. It is not surprising that it was the areas farthest from London that proved truest to Catholicism, both in the Pilgrimage of Grace of 1536 which protested against the dissolution of the monasteries and the Western Uprising of 1549 which protested against the abolition of the Latin Mass with the demand that all the people use the First English Prayer Book of 1549. Church and state were interlocked; there was no reason to think that the anti-Lutheran Henry VIII would bring the great house of Catholicism tumbling down merely to secure his throne and line by the hope of an heir from another wife and to fill his depleted coffers with the proceeds from the deprived monasteries. Wittenberg and its rebellious priests seemed as remote as Doomsday.

Yet in 1534 England passed the Act of Supremacy which cut England off from the Pope and Western Catholic Christendom, and which declared: "Be it enacted by the authority of the present Parliament that the King our sovereign Lord, his heirs and successors, Kings of this realm, shall be taken, accepted, and reputed the only Supreme Head in earth of the Church of England, called *Ecclesia Anglicana*."[6] The following summer there were the martyrdoms of John Fisher, Bishop of Rochester, and of Thomas More.

The proof of the discontent among the common people was their dissatisfaction with the new English prayer book. The people's deep sense of loss in being prevented from using the Catholic ritual and ceremonial was evident in the Western Uprising. The uprising began when the parishioners of Sampford Courtenay in Devon succeeded in persuading their priest to say the Mass publicly and in Latin in open defiance of the law, to the great approbation of the common people in the neighbourhood. There was a suspicion of the rich who stood to benefit most from Protestantism. An old lady, forbidden to say her beads in public in the village of St. Mary's Clyst, informed the other villagers: "Ye needs must leave beads now; no more holy bread for ye, nor holy water. It is all gone from us or to go, or the gentlemen will burn your houses over your heads."[7] Whether or not the discontent was fomented by dis-

6 Henry Gee and W. J. Hardy, *Documents Illustrative of English Church History*, pp. 243-44.
7 Mathew, *Catholicism in England*, p. 13.

possessed priests, as seems likely, the nature of the protest could plainly be seen in the articles put forward by the Devon and Cornwall men, all wanting to turn the clock back. The conservatism of these countrymen was untouched by the influence of the counterreformation, for their third request was: "We will have the Mass in Latin as was before and celebrated by the priest without any man or woman communicating with him." They also desired to adore the Reserved Sacrament; wanted the Communion, as far as partaking of the consecrated wafer was concerned, only once a year, at Easter; desired to have Baptism administered on weekdays as well as Sundays; and obviously missed the Catholic ceremonies, as the 7th article indicates: "we will have holy bread and holy water made every Sunday, palms and ashes at the times appointed and accustomed, images to be set up again in every church, and all other ancient ceremonies used by our mother the Church." They found the new liturgy impersonal and deficient in the sense of the Communion of the Saints, hence they begged: "we will have every preacher in his sermon and every priest at his Mass pray specially by name for the souls in purgatory as our forefathers did." Perhaps the most naive of all the articles was the 10th, which offered a lefthanded compliment to Protestants as "men of the Book." It read: "we will have the Bible and all other books of Scripture in English to be called in again; for we be informed that otherwise the clergy shall not of long time confound the heretics." The entire document was the touching manifestation of a spiritual vacuum in the English countryside.[8]

2. *Piety in Mary's Reign*

The tragedy of Queen Mary's reign is that she is "Bloody Mary" to all Protestants because her burning of heretics in Smithfield scorched itself into the English memory through the work of the great martyrologist, Foxe. To Catholics she was "Good Queen Marye," as in the long and undistinguished, but devout, daily prayer she composed and was retained in the Bedingfield Papers in the "Sydenham Prayer Book." This prolix prayer breathes the spirit of an utter dedication to God, narrow and proper rather than deep, legalistic rather than lyrical; it evinces a total commitment to the ways of the Catholic church.[9]

[8] *Troubles connected with the Prayer Book of 1549*, ed. Nicholas Pocock, New Series, xxxvii, pp. 148f.
[9] Catholic Record Society Publication No. 7, entitled, *Miscellanea, VI, Bedingfield Papers*, etc., pp. 23-27.

What a deep and genuine Catholic piety was like in Queen Mary's days may be ascertained from the account of the life of Magdalen Dacre, who was a maid of honour at Mary's court and later married the Viscount Montague. Her father was a man of wit and daring. When Henry VIII told Lord Dacre that Parliament had given consent for the king to become head of the English church, and sought his opinion, Magdalen's father replied, "Hereafter, then, when your Majesty offendeth, you may absolve yourself."[10] It is related in her biography that the court of England was then "a school of virtue, a nursery of purity, a mansion of piety," and that "the Queen herself did shine as the moon in all kinds of virtue, whose praises all histories do record."[11] Magdalen Dacre used to spend several hours in devout meditation before the crucifix in the chapel either before others had risen from their beds or long after they had gone to bed, so sincere was her devotion to God. Even in the reign of Protestant Elizabeth, Magdalen had Mass said twice a week for the repose of her husband's soul, and she herself said the Office of the Dead for him. She maintained three priests in her household and often had as many as 80 persons dependent on her for sustenance.[12] She visited the poor in their cottages and sent them medicines, food, wood, or money.[13] The strict regimen of her spiritual life in the time of the Renaissance deserves to be recorded, for it was unaccompanied by any Pharisaic superiority or censoriousness.

> For the most part she did every day hear three masses, and more would willingly have heard if she might, and such was her affection to this divine sacrifice that, when upon any occasion in the winter it was said before day, she in that cold and unseasonable time could not contain herself in her bed, but rising endured not to be absent from that heavenly sacrifice.

> In her private devotions she did every day say three offices, that is of the Blessed Virgin, of the Holy Ghost, and of the Holy Cross, whereto she added at least three rosaries, the Jesus Psalter, the fifteen prayers of St. Brigit, which because they began with O are commonly called her fifteen Oes, and the common litanies, and, finally, sometimes the Office of the

[10] A. C. Southern, ed., *An Elizabethan Recusant House, comprising the Life of the Lady Magdalen, Viscountess Montague*, p. 7.
[11] *Ibid.*, p. 12.   [12] *Ibid.*, p. 39.
[13] *Ibid.*, p. 40.

Dead. Which prayers, when in her infirmity she could not say herself, she procured to be said by others, distributing to every one a part.[14]

If such is the private image of Marian piety, what was the public view of the restoration of the Catholic faith and worship in Queen Mary's reign? The first Parliament in the new reign abolished the second Edwardian Prayer Book of 1552, reestablished the Mass, the Catholic ordination service, and the remainder of the Catholic sacramental ritual. Laws abolishing certain holy days and days of fasting and a law for putting away images and certain books were repealed, but only after eight days of stormy debate and by a vote of 270 votes to 80 on the third reading in the House of Commons.[15] The episcopate was purged of married and heretical bishops, and the old Henricians and the new appointees proceeded to dismiss about one-eighth of the clergy who had married and would not give up their wives.[16] The most difficult task was to persuade Parliament to acknowledge the spiritual supremacy of the Apostolic See and the abolition of the spiritual jurisdiction of the sovereign over the English Church. This delicate matter was accomplished early in 1554 in the long Second Act of Repeal, which begins by acknowledging that England had "swerved from the obedience of the Apostolic See, and declined from the unity of Christ's church," and which restored the ecclesiastical status quo of 1529. Insofar as Parliament could legislate the return of England to the Catholic faith, it had been accomplished, but a series of factors made a complete return impossible.

The new nobility and gentry were dependent on the Henrician dissolution of the monasteries, and would not think of giving up their lands to the church, nor, indeed, was the queen prepared to endanger her throne by asking for their return. Thus it was a church shorn of most of its temporalities that returned to power. Furthermore, the effect of a monarch as head of the Church of England had been to diminish the power of the archbishops and bishops, not to mention the clergy, almost all of whom had shown themselves to be compliant to the monarch's demands and in no

[14] *Ibid.*, p. 47.
[15] Philip Hughes, *Rome and the Counter-Reformation in England*, p. 67. For the Act of Repeal see Gee and Hardy, *Documents of English Church History*, pp. 377-80.
[16] W. H. Frere, *The Marian Reaction*, chap. 3, where it is claimed that 1,057 deprivations of the clergy for marriage out of approximately 8,000 incumbents can be traced in the documents.

sense the successors of Thomas à Becket. The queen seemed to be more Spanish than English in her style of life, and her consort was, of course, an unpopular Spaniard. Her chief spiritual advisor, Cardinal Pole, the Archbishop of Canterbury, was a man of great spirituality and scholarship, but of little diplomatic ability. It was excessively myopic of Mary to revive the hated statute *De Heretico Comburendo* in 1554, for the English were a tolerant people and would not forgive her the deaths of so many valiant men of another faith in so short a reign.[17] Whatever the reasons, Mary had time to make enemies, and she lacked the popularity and the length of reign to make friends on the grand scale. It would have been wiser if she had vowed, like her sister and successor Elizabeth, "not to make windows in men's minds."

The new liturgical requirements were made clear in the instructions Mary issued to the archbishops and bishops on March 4, 1554.[18] The traditional procession before the High Mass on Sundays and festival days was to be restored, and the old Litany of the Saints in Latin was to displace the new Litany of the Book of Common Prayer in English. The days of fasting and the holy days abolished in the reign of Edward VI were reinstituted, together with the ancient ceremonies associated with Ash Wednesday, Palm Sunday, and the last three days of Holy Week. The Catholic rite alone was to be used for Baptism and Confirmation.

The rest of the intentions for Marian liturgical practice can be derived from a consideration of the remarkably careful articles and injunctions prepared for Bishop Bonner's prolonged visitation of the diocese of London in the autumn of 1554.[19] His aim was to restore in all their plenitude the old religious customs as well as the full Catholic liturgy. Detailed directions were given for the replacement of service books, vestments, and church furniture presumably destroyed by the iconoclasts of the prior reign. The previously unemployed craftsmen must have been working overtime in

[17] The situation was well characterised by the Venetian ambassador, *Venetian Calendar*, VI, 1,074-75: "With the exception of a few very pious Catholics, none of whom, however, are under thirty-five years of age, all the rest make this show of recantation, yet do not effectually resume the Catholic faith, and on the first opportunity would be more than ever ready and determined to the unrestrained life previously led by them. . . . They discharge their duty as subject to their prince by living as he lives, believing what he believes, and in short doing whatever he commands, making use of it for external show to avoid incurring his displeasure rather than from any internal zeal; for they would do the like by the Mahometan or Jewish creed."
[18] Printed in Gee and Hardy, *Documents of English Church History*, pp. 380-83.
[19] See W. H. Frere and W. M. Kennedy, *Visitation Articles and Injunctions of the Period of the Reformation*, Alcuin Club Collections xv, vol. 2, pp. 331-72.

the first years of Mary's reign as they provided stone altars for wooden Communion tables, carved the calvaries for rood-lofts and statues, and made pyxes. A Catholic handbook of doctrine and piety for the least instructed of his clergy was prepared by Bonner under the title, *A profitable and necessary doctrine with certain homilies adjoined thereto.* In the simplest terms it explained the Creed, Sacraments, Commandments, *Our Father, Hail Mary*, seven capital sins, and Beatitudes.

There is one interesting sidelight on both the infrequency with which the chief Sacrament of the Eucharist was celebrated, and on the conservative way in which it was administered. It shows that the new insights of the Council of Trent had not yet reached the Bishop of London, who, during the sequestration of Cranmer, was virtually the Archbishop of Canterbury. He did not think it proper to exhort the priests to invite the people to communicate at Mass more frequently. Indeed, he implied that this is the less necessary precisely because bread is blessed at the Mass and given to the people as a sacramental. It was ironical that he mentioned that the bread was a reminder of unity, and "that all Christian people be one mystical body of Christ, like as the bread is made of many grains, and yet but one loaf," when it is the sacrament of incorporation which best testifies to that unity. Almost incidentally he suggested that "the said holy bread is to put us also in remembrance of the housel, and the receiving of the most Blessed Body and Blood of our Saviour Jesus Christ, which the people in the beginning of Christ's church did oftener receive than they do use now in these days to do."[20]

How conservative and insulated England had become through being cut off from the revivifying counterreformation piety of the Council of Trent can be seen by comparing Bonner's attitude to the Sacrament, with the statement of the Fathers at Trent, made only three years earlier: "This holy synod, with fatherly concern, warns, exhorts, begs, and beseeches by the loving heart of our God, that all and singular, who count themselves of the Christian name, come together in harmony now at any rate, in this sign of unity, this bond of love, this symbol of agreement, and that, mindful of the great majesty and understanding love of Our Lord Jesus Christ who gave His beloved life as the price of our salvation, and His flesh for our eating, all Christians will believe and reverence

---

20 Philip Hughes, *The Reformation in England*, III, 75, comments on Bonner's merely historic interest in days when Holy Communion was received frequently.

these sacred mysteries of His body and blood with a faith so constant and so firm, with such devoutness of spirit, with such dutiful worship, that they may be able frequently to receive this supersubstantial bread, and that it may truly be to them the life of their soul, and an everlasting health of mind, made strong by whose strength they may be qualified to arrive, from the journeying of this woeful pilgrimage, at our fatherland which is heaven."[21]

Pole's reforming concern is seen in the plans for a restored Catholicism in England. It is visible in the decisions of the great national council he summoned and over which, as Papal legate, he presided during the winter of 1555-1556. Every anniversary of St. Andrew's Day 1554, the memory of the day of England's reconciliation with Rome, was to be gratefully recalled at a special celebration with a solemn procession and a sermon explaining the reason for the festival. Pole's hand can be seen in the decree against absentee bishops and in another, insisting that it was the personal responsibility of the bishop to preach and catechise. The spirit of the counterreformation and of Pole is also seen in the bishops' requirement to erect diocesan seminaries to train candidates for the priesthood and to provide (until a preaching priesthood was trained) bands of itinerant preachers to tour the dioceses. In four years the death of the queen and the cardinal ruined the entire plan. Pole's wider influence was felt through the work of the Council of Trent.

3. Englishmen and the Counterreformation

The two most eminent Englishmen at the Council of Trent were the Cardinal President, Reginald Pole, and the Theatine, Thomas Goldwell, who was bishop of St. Asaph; Goldwell was very interested in the reform of the liturgy.

Pole was Cardinal Deacon with the title of Santa Maria in Cosmedin. His mother was niece to Edward IV, reigning at the time of her birth, first cousin to his successor, Edward V, niece to Richard III, and first cousin to Elizabeth, the consort of Henry VII, as well as dear friend of Catherine of Aragon. He was schooled by the Carthusian monks at Sheen, and went on to Magdalen College, Oxford, for six years, which was still in the fervour of its founding by William Waynefleet. In 1519 Pole went to the University of Padua where he became adept in Renaissance learning. Even

---

[21] October 11, 1551; Session XIII; Decretum, c. viii, De usu admirabilis huius sacramenti.

Bembo praised him in 1522 as "perhaps the most virtuous, erudite
. . . grave young man in Italy at the present time."[22] Many years
as a diplomat for the Pope, combined with an unassailable purity
of life and modesty of manner and an eagerness for the reform of
the church, excellently equipped him to be one of the three presidents of the Council of Trent when it resumed in 1545. When the
canon of Sacred Scripture was under preliminary consideration, as
a proponent of reform, it was Pole who argued that the Council
should examine the claims of each separate book for inclusion in
the Canon, for the Lutherans had argued that the Catholics were
afraid to look too closely into their origins. When the discussion
moved to the adoption of an official version of the Bible, Pole made
the interesting suggestion that the Hebrew and Greek texts be
added to the Vulgate lest the Catholic church should seem to be
limited to the Latin culture.[23] If a man is known by his friends,
then we may judge that Pole's great friend Cardinal Contarini
spoke for him when attempting to construct a theory of justification that would provide a way of moving closer to Lutheran views.[24]
Pole and Contarini were later accused of heresy by the inquisitorial
Pope, Paul IV, a charge that shadowed Pole's last years. The last
irony was that he, who had such good reason to fear the machina-
tions of the politicians because of the way his ancestors and near
relations had been slaughtered because of their closeness to an
insecure throne and who had therefore been inclined to put his
complete trust in God, should find God's man, the Pope, treach-
erous, too.

Goldwell, the other Englishman at Trent, was the last survivor
of the ancient English hierarchy, who lived to be 84. This bishop
of St. Asaph in remote Wales went abroad during the first months
of Henry VIII's anti-Catholic measures and met Pole, with whom
he was thereafter to be closely associated, in Padua in 1532. From
1538 to 1547 Goldwell was in charge of the hospice for English
pilgrims in Rome. In 1547 he joined the new reformist order of
the Theatines, one of the greatest of the new counterreformation
orders (like the Jesuits), founded in 1524. Aiming to combat
the worldliness and general laxity of the clergy, the Theatines
demanded an oath of poverty to apply communally as well as personally. Its members went to work among the poor in the great

---

[22] Cited Hughes, *Reformation in England*, p. 32.
[23] *Ibid.*, pp. 41-42.
[24] See the article on Pole by Herbert Thurston, S.J., in *Catholic Encyclopedia*,
v, 335-37.

cities and were dependent on the providence of God as they went about preaching, catechising, and administering the sacraments. Their founders were Count Gaetano da Tiene (St. Cajetan) and Gian Pietro Carafa, bishop of Chieti and the future Paul IV. They revolutionised the image of the clergy and the church in the areas where they worked, but their numbers were select and small. Their own lives were austere, but their churches were splendid tributes to the Divine Majesty. Goldwell joined them when he was about 47, a remarkable tribute to his willpower and dedication.[25]

From time to time Goldwell would leave his pastoral duties for other tasks, as when he was given leave in 1549-1550 to attend Cardinal Pole at the conclave to elect a successor to Paul III. When he had the great privilege of conducting the Vespers of the Council of Trent on June 25, 1561, the vigil of St. Vigilius,[26] the protector of the Council, his great friend Pole had been dead for three years. Since the Theatines were greatly concerned with the improvement of the Breviary and the Missal, having been encouraged by Pope Clement VII in 1529 to devote their energies to the task, and as Paul IV was also a Theatine, Goldwell was greatly concerned with liturgical renewal. Paul IV granted permission to the Theatines to adopt certain minor alterations in worship, but he did not believe that these matters had been sufficiently discussed by the representatives of the entire Catholic church to make them universally binding on the faithful. When the Council of Trent met anew under Pope Pius IV, the matter was referred by His Holiness to the fathers of the Council. At their request he sent them the annotations made by Paul IV and preserved in the Theatine archives. The Council entrusted the work of revision to a Commission, of which Goldwell, being a Theatine, was appointed a member. The Council of Trent did not have time to complete its liturgical work. The Pontiff then appointed a special Congregation at Rome, of which Goldwell was a member, to continue the work of revision;[27] the work was concluded on July 9, 1568 with the provision for a uniform Roman Missal and Breviary which became the undeviating standard of Catholic worship throughout the world until the second Vatican Council initiated the far-reaching changes of the present day.

Goldwell received many honours, including the presidency of

---

[25] T. E. Bridgett and T. F. Knox, *The True Story of the Catholic Hierarchy deposed by Queen Elizabeth with fuller memoirs of its last two survivors*, pp. 214-15.

[26] *Ibid.*, p. 226.   [27] *Ibid.*, pp. 238-39.

the General Chapter of the Order of Theatines in 1566, 1567, and 1572. St. Pius V appointed him to the office of Vicar (or representative of the Cardinal Archpriest) in the St. John Lateran Church in Rome; and in 1574 Cardinal Savelli made him his suffragan, or vicegerent. Among Goldwell's duties was the responsibility for administering the sacrament of Holy Orders to the many persons from many lands who desired to be ordained priests in Rome. The *Pontifical* Goldwell used has several corrections and annotations in his own handwriting. After his death the Congregation of Rites made considerable use of it in the correction of the old Roman *Pontifical*, and many emendations were adopted to conform with Goldwell's wishes. This was his final liturgical legacy to the Catholic church, in whose renewal he had shared.[28] The Catholics of England could take great pride in the contributions of Pole and Goldwell to the renewed spirituality and worship of the international Roman Catholic church.

4. The Case for Catholic Liturgical Reform

Having observed, however briefly, the two major English representatives among the fathers of the Council of Trent, it is important to consider the overall liturgical work of the Council, as it was to determine the pattern of English Catholic worship for four centuries to come.[29]

It is impossible to understand the reasons for the Tridentine reforms of the liturgy apart from some minimal knowledge of the morphology of the medieval Mass. It is now customary for liturgiologists to describe the medieval Mass as the "allegorical" or "dramatic" mass, by which is intended the view that the holy banquet had been turned into a mysterious spectacle, the chief moment of which was the elevation of the Host by the celebrating priest. As Latin became increasingly unfamiliar to the humble believers who assisted at the Mass, they came to believe that the

[28] *Ibid.*, pp. 247-48.
[29] The major volumes consulted for this and the following section in this chapter are: *Concilium Tridentinum. Diariorum, Actorum, Epistolarum, Tractatuum,* henceforth *Conc. Trid. Nova collectio edidit Societas Goerresiana;* Hermanus A. P. Schmidt, *Introductio in Liturgiam Occidentalem;* Joseph A. Jungmann, *Missarum Sollemnia; eine genetische Erklärung der römischen Messe;* and the English edition, *The Mass of the Roman Rite: Its Origins and Development (Missarum Sollemnia),* tr. Francis A. Brunner, 2 vols. I have also had the benefit of the advice of a learned monk and friend, Father Reinhold Theisen, O.S.B., the author of *Mass Liturgy and the Council of Trent,* which has proved invaluable. An authoritative article is H. Jedin's "Das Konzil von Trient und die Reform des Romischen Messbuches," *Liturgisches Leben,* VI, 30-66.

Mass could be interpreted as a complete representation of the Passion of Christ, so that the arrival of the priest at the altar symbolized the capture of Christ, the *Confiteor* the standing before Annas and Caiaphas, etc., or, that the whole life of Christ was figuratively represented in the mass and hence was to be understood as divided into the 40 works of Christ's life or the 33 years of his life.[30] Like human nature, the liturgy could not leave well enough alone, so imagination went to work during the long spells of unintelligibility to create elaborate and unhistorical fancies.[31] Popular devotion was no longer concentrated on incorporation into the body of Christ or on fellowship with Christ and His church. As the immensely popular legend of the Middle Ages, the Holy Grail, makes clear, the people's desire was to glimpse the celestial mystery of the miracle. To see the sacred Host became the principal, if not the exclusive, concern of worshippers in the High Middle Ages. In the cities the people ran from one church to another in the hope of witnessing the miracle often. In England it was not enough for the priest to elevate the Host, for the people cried, "Higher, Sir John, higher!"[32] They insisted on glass chalices[33] in order to have ocular demonstration of the miracle effected by consecration. They believed that a glimpse of the Host would spare them from blindness, or poverty, or death on that day on which they had seen it. In all these ways, the total action of the Eucharist was sacrificed to a single moment of epiphany; the sense of the community created by the Holy Communion was sacrificed to an individual egotism. The result was that all sense of the Mass as a *sacrifice* for the living and the dead, as a *communion* with God and His people, or as a *banquet* was lost. Clearly some reinterpretation was necessary, not to mention changes in the mode of celebration.

Another valid reason for reconsideration was the trenchant criticism of the Mass made by the Protestant Reformers. These were contained in Luther's *Von der Winkelmesse und Pfaffenweihe* and

[30] Schmidt, *Introductio*, p. 367.
[31] *Ibid.* "Fine Medii Aevi allegoria Missae irretitur suis propriis imaginationibus et pervenit usque ad insolitissimas inventiones, ne dicamus ad stultitias."
[32] *Ibid.*, p. 269. "In Anglia eveniebat, quod sacerdoti qui S. Hostiam non satis alte elevebat, acclamabatur: 'Altius, Domine Joannes, altius.' " See also E. Dumoutet, *Le désir de voir l'Hostie*, for the growing importance of the "elevation" for the people. So great had the mania for miracles become, and so deep was the conviction of the "instant" benefits available from the Mass, that there was an enormous growth of "altarists" whose only duty it was to celebrate the Mass and say the Office. In 1521, for example, Strassburg had 120 Mass foundations, and in England Henry VIII suppressed 2,374 chantries just before his death. Jungmann, *Missarum Sollemnia*, I, 130.
[33] Schmidt, *Introductio*, p. 269.

Calvin's *Petit Traicté de la Saincte Cène*. The Reformers had three main objections to the private Mass.[34] They objected to its sacrificial character. Luther, for example, argued that the Mass was not a *sacrificium*, but a *beneficium*—not a sacrifice renewed by men but the thankful recognition by men of a gift of God. They particularly criticised the failure to administer Communion to the laity, when the priest receives alone, as a departure from the example and institution of Christ and a denial of the fellowship of Christ of which this sacrament is the renewal of incorporation. They believed that the Mass is essentially a Communion service and that where there is no reception by the faithful there is no Communion. They insisted that the custom of the private Mass lacked apostolic authority and was introduced by Gregory the Great. The Council eventually limited the conditions on which the private Mass was permitted.

In other matters there was considerable agreement between Protestants and Catholics, such as on the need for a simplification of ceremonies and for greater intelligibility in worship with vernacular preaching (the Protestants would also have said a vernacular rite was necessary), and a necessity for greater lay participation in worship. These criticisms, whether accepted or rejected, required the attention of the fathers at Trent.

A third reason for liturgical revision at the Council of Trent was the prevalence of abuses in the celebration of the Mass. An objective survey of such abuses in contemporary practice was ordered on 20 July, 1562.[35] The seven members appointed prepared their work, but because the fathers were growing increasingly weary of the Council it was only summarized instead of being presented in full. Several canons were proposed for the correction of the abuses,[36] and, of course, there was a definitive decree "On what must be observed and avoided in the celebration of the Mass." The decree exhorted the bishops to eliminate avarice, irreverence, and superstition as they might find them in the celebration of Mass

34 See Theisen, *Mass Liturgy*, p. 76. The most recent treatment of Calvin's sacramental teaching is Kilian McDonnell, O.S.B., *John Calvin, the Church, and the Eucharist*, which is invaluable in showing that Calvin's teaching was elaborated in contradistinction to that of the Roman Catholic Church at every point. Calvin declared in the *Petit Traicté de la Saincte Cène* (*Opera Selecta*, I, 519f.): "wherever there is no breaking of bread for the communion of the faithful, it is only a false and deceitful imitation of the Lord's Supper."

35 *Conc. Trid.*, VIII, 721.

36 Schmidt, *Introductio*, p. 370. Schmidt prints a list of the abuses (pp. 371-77), of the corrective canons (pp. 378-80), and the decretum (pp. 380-81).

in their own dioceses. In this compendious fashion the commission summarised the three major abuses they found.

Avarice is present whenever the Mass is sold for a price or celebrated only for the sake of profit. Bishop Perez de Ayala, for example, believed that the offering of Masses for many intentions gained more money for the celebrants but rendered Masses less efficacious in the popular mind. On another occasion de Ayala insisted, "the sacrifice of the Mass is offered for sins and for the appeasement of God's anger; not for human affairs such as a successful business deal, fertility, good sailing or the recovery of stolen goods."[37]

Irreverence was common, and could be attributed to celebrations by ignorant and unlettered priests. The Council fathers observed that sincerity was of great importance, and in an earlier draft had criticised those who pronounced the sacred words like actors on a stage; it was recommended that celebrants speak gravely and clearly.[38] They also condemned excessive slowness and celerity in speech. Others behaved more like amateur actors, gesticulating inappropriately, as by moving their heads in the shape of a cross, bowing so low in their veneration of the sacred species that their hair becomes enveloped in it, or exaggeratedly striking their breasts at the *Agnus Dei* or at the *Lord, I am not Worthy*.[39]

The third abuse was superstition. This was particularly attendant on the celebration of series of Masses which led to foolish expectations of material benefits. Erasmus had criticised it in his *Liber de sarcienda ecclesiae*[40] and Luther in his *De captivitate babylonica*.[41] They particularly objected to the idea that the material benefits desired, such as good hunting or good business, could be guaranteed by a multiplicity of Masses. Another superstition took the form of celebrating with a predetermined number of candles, such as 12 for the apostles or 7 for the Mass of the Holy Spirit, because they were believed to render the votive Masses more efficacious. The Council criticised the abuses, along with the "dry Mass"—that is, a rite in which the consecration and communion were missing, but in which the blessed Sacrament reserved in a ciborium is exhibited to the people for adoration.[42] Although the Commission proposed the abolition of the dry Mass, 12 fathers

[37] Theisen, "Reform of the Mass Liturgy and the Council of Trent," in *Worship*, vol. 40 (1966), no. 9, pp. 565-83; this particular reference is to p. 569.
[38] *Conc. Trid.*, VIII, 919, 10-13.
[39] *Ibid.*, pp. 14-27.
[40] Theisen, "Reform of Mass Liturgy," p. 572.
[41] Luther, *Werke*, vol. 6, p. 519.
[42] Theisen, "Reform of Mass Liturgy," pp. 573f.

spoke against it as a useful means of stimulating the devotion of priests, so the matter was shelved.

A final reason for the reconsideration of worship at the Council of Trent was the need for greater uniformity of celebrations of worship and for less corrupt texts. Nicholas of Cusa, bishop of Brixen, provided a practical example of reform by requiring in 1453 and 1455 that all the Mass books in his diocese should be assembled at certain centres and corrected from one perfect example.[43] The truth was, the Mass books had in many ways become a "tangled jungle."[44] This could easily happen, as Theisen explains, because in the Middle Ages all types of votive Masses, sequences, prefaces, and Mass series came to be included in the "Missals," as new formularies and texts were composed to suit the needs of the people. It was easy for this to take place; the Missals were copied by hand, with no supervision by ecclesiastical authority.[45] It was, in fact, Thomas Campeggio, bishop of Feltria, who was the first of the Tridentine fathers to recommend the provision of an authorised uniform Missal. In 1546 he urged that a purged Missal be prepared that would be identical for all churches.[46] If there was chaos in the late medieval period, there was more in the Reformation because many priests had taken the law into their own hands and abbreviated or added as they saw fit. In Austria some priests omitted even the Canon of the Mass.[47]

On the matter of purged texts, one of the most far-reaching proposals came from the progressive leader of the Augustinians, Jerome Seripando, who suggested that only the words of holy Scripture should be used in the Missal and Breviary.[48] He was apparently anxious to eliminate the apocryphal and poorly devised sequences and prefaces, and to remove the legendary accounts of some of the saints. The commission on worship was also concerned that the Council should consider some unauthorized accretions to the Canon of the Mass including such expressions as *Hostia immaculata, calix salutaris* at the Offertory and the prayers added at the commingling.[49] There was the further complication, that different dioceses had their own patron saints and appropriate

43 Jungmann, *Missarum Sollemnia*, I, 131-32.
44 *Ibid.*, p. 133.
45 Theisen, "Reform of Mass Liturgy," p. 577. See also Jungmann, *Missarum Sollemnia*, I, 133.
46 *Conc. Trid.*, V, 25, 8-10.
47 Jungmann, *Missarum Sollemnia*, I, 134.
48 *Conc. Trid.*, V, 26, 42-43.
49 Jungmann, *Missarum Sollemnia*, I, 134.

Masses. In short, there was a desperate need for consistency in the celebration of the Mass, the lack of which was bewildering to the priests and the people.

These four factors—the Gothic transmutation of the Mass and its continuing influence on the sixteenth century; the Protestant criticisms; the necessity to avoid the abuses of avarice, irreverence, and superstition in the celebration of the Mass; and the demand for the unification of rites and texts—cumulatively accounted for the attention paid by the Tridentine fathers to the reform of the liturgy.

5. The Character and Consequences of Trent's Liturgical Reforms

If the Gothic Mass can be characterised as "dramatic" or "allegorical," the Mass that was finally made uniform by the fathers at Trent may be termed the "rubrical" Mass.[50] The term rightly suggests the end of individualism, eccentricity, and arbitrary variety, and the arrival of the codification of the liturgy. Such a description of the liturgy codified in 1570 indicates that the primary concern of the fathers at the Council of Trent was protective rather than creative.

Certainly in regard to the Canon of the Mass, the Mass ceremonies, and the acceptance of the private Mass (within certain carefully stated conditions), the work of the Council was defensive. Theisen has described its work as "defensive, on the one hand, purifying and immobilizing on the other."[51] It was in the most literal sense distinguished by an attitude of anti-Protestantism or counterreformation. Individual reformists such as Seripando were too far ahead of their times, in suggesting that a Biblical norm would purify the worship of the church and remove the reasons for the defection to Protestantism, and it was not likely that an Augustinian, even the general of the Order, could convince conservatives who would not forget that Luther himself had been an Augustinian monk.

The Council was aware of the defects in the celebration of the liturgy. Jedin has referred to the report of the commission of seven bishops appointed in the later sessions to provide a survey of current practise as "the most comprehensive accumulation of reform ideas."[52] The fathers were even made aware of some defects

---

[50] Schmidt, Introductio, p. 382: "Nunc rubricistae, h.e.iuris liturgici periti, primas partes habent. Nunc semper quaestio est de codificatione liturgicae."
[51] Theisen, Mass Liturgy, p. 111.
[52] Jedin, "Das Konzil," vi, 34-35.

in the Canon of the Mass during this time. There was, for example, the omission of the words *quod pro vobis tradetur* (I Corinthians 11:24) and *quod pro vobis datur* (Luke 22:19) in the consecration of the bread. There was also the insertion of the words *mysterium fidei* in the words of consecration over the chalice. There were severe grammatical difficulties in the prayer *Communicantes*. Similarly the recitation of the Canon and of the consecratory formulas aloud, instead of secretly, were demanded by many.[53] The Council also considered the many ceremonies that might be thought of as cluttering the essential simplicity of the rite, as its members had criticised, and condemned such abuses as avarice, superstition, and irreverence in celebrating the Mass.

Why, then, did the Council not incorporate these changes by revising the ceremonies and the Canon? Why was a great opportunity missed? First, the Council saw, as its chief task, the defense and preservation of the liturgy as the supreme channel of the sanctification of the church, a church severely challenged by the Protestants who had particularly directed their criticisms at the Canon.[54] To have changed the liturgy would have amounted to admitting serious error on the part of the church.[55] The Protestants were cast in the role of innovators, the Catholics as conservers. Since many of the significant reforms had been first suggested by the Protestants, it seemed like a confirmation of heresy to agree to them.

A second and important reason for conservatism was that historical studies of the liturgy were only at their very beginnings at this time. It was, says Theisen, the common opinion of the time that St. Peter had instituted the way of celebrating Mass and that St. James had set this down in writing.[56] The intervening centuries have added greatly to the store of early liturgical texts, which were

53 Theisen, *Mass Liturgy*, pp. 111-12.
54 Luther was vitriolic in speaking of the Canon in the preface to his *Formula Missae et Communionis pro Ecclesia Wittembergensis*: ". . . the Canon, that mangled and abominable thing gathered from much filth and scum. Then the Mass began to be a sacrifice; the Offertories and paid-for prayers were added; then Sequences and Proses were inserted in the *Sanctus* and the *Gloria in excelsis*. Then the Mass began to be a priestly monopoly, exhausting the wealth of the whole world, deluging the whole earth like a vast desert with rich, lazy, powerful celibates. Then came Masses for the dead, for travellers, for riches, and who can name the titles alone for which the Mass was made a sacrifice?" (Bard Thompson, *Liturgies of the Western Church*, p. 108.) Luther's anger caused him to mix his metaphors badly!
55 Bishop Frederick Nausea of Vienna (as Theisen points out in *Mass Liturgy*, p. 112) admitted as much in a speech before the Council in January 1552. See *Conc. Trid.*, VII, 48ff.
56 *Mass Liturgy*, p. 39.

unavailable in the sixteenth century, and historical method has determined the histories of various families of liturgies in East and West which were not even divined by Protestant or Catholic scholars in the controversial period under study here. Whatever the reasons, however, the Tridentine attitude to liturgy was essentially protective and defensive.

As for the liturgical achievements of the Council, although there was a strong demand for it in the Council, the Council only provided the impetus for an unified and purified Missal. But many proposals and ideas of reform suggested by the Council fathers were acknowledged by the post-Tridentine Commission that prepared the Missal of Pope Pius V, published in 1570. Unquestionably the Council provided the momentum that resulted in the production of major authorized liturgical texts in the years immediately following the conclusion of the Council of Trent in 1563: the *Breviarium Romanum* (1568) and the *Missale Romanum* (1570). In 1596 the *Pontificale Romanum* came out, the *Ceremoniale Episcoporum* in 1600, and in 1614 the *Rituale Romanum*. Second only to the publication of the first two of these volumes was the establishment of a new Congregation of Rites in 1588, to which disputed questions could be referred, and whose usual reply, it has been said was: Observe the rubrics![57]

The Council itself did not undertake the task of reforming the Missal. By a decree of the 25th session, the task of reforming Breviary and Missal was left to the Pope. In 1564 Pius IV created a commission for this purpose which was enlarged by his successor, Pius V. There are no official reports of the Commission available, but its major achievement was the *Missale Romanum ex decreto ss. Concilii Tridentini restitutum, Pii V. Pont. Max. iussu editum*, which was made binding on almost the entire Catholic church by a bull of July 14, 1570. The ideal of the Commission was to return "to the liturgy of the city of Rome, and indeed, the liturgy of that city as it was in former times."[58] The bull indicates that the scholars of the Commission had consulted the oldest and most uncorrupt texts available and that they believed they had brought the Missal to the pristine norm of the rite of the holy Fathers.[59] They had cleared away several accretions, but had not produced a rite of any patristic norm—assuming there was such

---

[57] Jungmann, *Missarum Sollemnia*, I, 140.
[58] *Ibid.*, p. 136.
[59] The exact wording of the bull is "ad pristinam sanctorum Patrum normam ac ritum."

a single rite. Nor had they produced an early Roman rite, for that was not marked by ceremonial elaboration or diffuseness of ritual, since "the genius of the Roman rite is marked by simplicity, practicality, a great sobriety and self-control, gravity and dignity."[60]

What were the characteristics of the new Roman Missal? The most important fact was that a uniform Mass rite was now imposed on the whole church. There were Council fathers who had wanted each bishop to have the liberty to allow variations, but they were greatly outnumbered by those who felt that the primary need was for order in the contemporary chaos.[61] There is plenty of evidence of pruning. The later Middle Ages had greatly cluttered the church year with saints' feasts. These were so reduced in the new Missal that there were only about 150 days, excluding octaves, free of feasts. Only those feasts were retained which were kept in Rome itself up to the end of the eleventh century.[62] The innumerable later feasts introduced chiefly under Franciscan influence from the thirteenth century on were nearly all eliminated; and there were few saints of countries other than Italy still commemorated in the Missal. On the other hand, memorial days for the Greek Fathers were introduced into the Missal, indicating the newer appreciation of the church fathers. This was later to be expressed in Bernini's *Cathedra Petri* in St. Peter's, which depicts Saints Athanasius and Chrysostom, together with two Latin bishops, as supporters of the Petrine throne and of the implied infallibility of the teaching office of the Pope, under the guidance of the Holy Spirit, symbolised by the dove in effulgent rays.

Pruning was also evident in the ruthless excision of sequences which had flourished in other Mass books. The sequences were a luxuriant modern growth that had never taken hold in Italy. The sentimental Marian insertions in the *Gloria in excelsis*, then widely accepted, were deleted.

A decided innovation lay in the fact that the same commission had prepared both the Breviary and the Missal. There was now agreement between the two basic books of the Catholic church's worship as to the calendar and the choice of collects and Gospels for each Mass.

[60] See Edmund Bishop's superb essay on the early Roman Mass, "The Genius of the Roman Rite," in *Liturgica Historica, Papers on the Liturgy and Religious Life of the Western Church*, p. 12.

[61] Schmidt, *Introductio*, I, 380, states that the Spanish and Portuguese bishops sponsored a uniform rite, while the French bishops argued for diocesan variety.

[62] Jungmann, *Missarum Sollemnia*, I, 136.

Furthermore, the rubrics were unified and codified and printed in the preface of the new Missal. Nearly all of them were lifted bodily from the *Ordo Missae* of John Burchard of Strassburg, a papal master of ceremonies; the book had appeared in 1502 and had circulated widely since that time. The codification of rubrics, coupled with the work of the Congregation of Rites, greatly reduced the opportunities for irreverence or idiosyncrasy.

Happily, the Council and the church, despite some conservative criticism, did not curb the rich polyphonic music that originated during this period. Palestrina was working closely with the influential Theatines, and the compositions of the Fleming, Jacques de Kerle, which were sung at the last sessions of the Council of Trent, seem to have convinced the wavering that the new music could be used to glorify God.[63] The road to the future was left open for great masterpieces of music in this mode.

The chief advantage of the new Missal, then, was its almost universal prescription, which thus enabled it to be increasingly the organ of the unity of the church's teaching on faith, morals, and devotion. Deviations by addition or omission were henceforth severely dealt with; the only rites allowed to vary from the new Roman Missal were those of churches which could prove a tradition of 200 years for their own use. The dioceses that took advantage of the permission included Milan, Toledo, Trier, Cologne, Liège, Braga, and Lyons.[64]

Obviously it was a great advantage to the church to have greater uniformity in the celebration of the Roman Rite, for the new rite would reflect the order, clarity, and stability of the Catholic church, menaced alike by the passionate subjectivity of Gothic and Protestant minds. Exaggerations and arbitrarinesses, squalor and irreverence, were now excluded. Concern for a better form for the liturgy, more elegantly expressed than hitherto and controlled by the Congregation of Rites, was an admirable ideal. There was, however, one grave defect in the entire conception of fixation: it excluded the development which is natural to the life of the Catholic church. If, as Jungmann suggests, the advantage was that "all arbitrary meandering to one side or another was cut off, all

[63] Schmidt, *Introductio*, p. 382: ". . . quia in ipso Concilio patres pluries musicam polyphonicam audierant, ad conclusionem pervenerant, artem polyphonicam esse aedificantem, devotam ac convenientem, ergo non improbandam, dummodo omne lascivum ac impurum excluderetur."
[64] Jungmann, *Missarum Sollemnia*, I, 138.

floods prevented, and a safe, regular and useful flow assured," yet the price to be paid was "that the beautiful river valley now lay barren and the forces of further evolution were often channelled into the narrow bed of a very inadequate devotional life, instead of gathering strength for new forms of liturgical expression."[65]

The Baroque culture might give the Roman Rite glorious polyphonic music, and the Mass could be celebrated in churches the architecture of which tried to bring heaven down to earth with *trompe l'oeil* painted domes with phalanxes of angels providing a new Jacob's ladder of worship, and there could be censings to delight the nose and candles to please the eye, and sermons to fascinate the ear—but all this was merely external. The radical failure, the legacy of which has lasted to our own day, was the inability to give the people a sense of participating in worship. The repudiation of the vernacular translations of the Roman Rite and the insistence on a *Roman* rite without diocesan variations, and the legalism that ensued in this "rubrical" Mass, created an abyss between priests and people. The result was that while the aesthetically splended Mass went on, the people were occupying themselves with devotional exercises that were not based on the liturgical texts, for these were not translated for the common people; in a very short time the devotion of the masses was not based on the Mass at all, but became extraliturgical. It was an ironical consequence of a reform of the Liturgy designed to avoid deviations, that it should have encouraged unliturgical devotions in reaction to the externality of a fixed and inalterable liturgy, artificially made interesting as a theatrical spectacle by Baroque adventitious aids. No amount of pomp and circumstance could disguise the seriousness of the alienation of the people from participation in the Messianic banquet of their Lord.[66] Happily the Second Vatican Council has opened the dam, and the waters of liturgical freedom now flow again.

Having looked at the international movement in Catholic worship (and the contribution of two Englishmen to it before and during Queen Mary's reign), we are now ready to survey the worship of English Catholics during the long reign of Queen Elizabeth, from 1558 to 1603.

[65] *Ibid.*, pp. 140-41.
[66] Father Louis Bouyer, a French oratorian and former Protestant minister, presents a devastating critique of Baroque liturgy in his *Life and Liturgy*.

6. *Elizabethan Recusant Worship*

The worship of the Catholics in Elizabethan England reflected the propitiousness or hostility of the government toward them. From 1559 to 1573 they did not find it unduly difficult to exercise their faith or to worship. The situation deteriorated in 1570 when the bull, *Regnans in Excelsis*, which excommunicated Queen Elizabeth, forced her Catholic subjects to choose between their religion and their country. From 1574 to 1583 there was a serious attempt to revive English Catholic life with the training of Jesuits for the English Mission in Douai, which had been opened as a seminary for English priests in 1568. After 1584 there was an acceptance of the decline of Catholicism into a cultured and quiescent minority, with the occasional raising of false hopes by political plots, from the Gunpowder Plot to the final disillusionment brought on by the failure of the Pretender. This, the third phase, was to be a long one in Catholic life, lasting from 1584 until 1829, when the Catholic Emancipation Act gave Catholics the same political and religious rights as all other denominations of Christians in England.[67]

Some indication of the disadvantages Catholics were under in the second half of the sixteenth century may be gained from the consideration of the penal ecclesiastical legislation passed by the Parliaments of Elizabeth. The Act of Uniformity of 1559 made illegal in the churches of England the use of any service book other than the Book of Common Prayer; it required every person to attend his parish church on Sundays and holy days, "upon pain of punishment by the censures of the Church, and also upon pain that every person so offending shall forfeit for every such offence twelve pence, to be levied . . . to the use of the poor in the same parish. . . ."[68] In 1585, however, it was found necessary to enact legislation against Jesuits and Seminarists. The Act[69] declared that any Jesuit or Catholic seminarist who came to England was to be considered guilty of high treason, and anyone helping such persons shall be "adjudged a felon, without benefit of clergy, and suffer death, lose, and forfeit, as in the case of one attainted of felony."[70] The same act made it a crime to send any child overseas for edu-

---

[67] This division of the reign of Elizabeth into three periods corresponding to the varying character of Catholic activities and attitudes is suggested by W. R. Trimble, *The Catholic Laity in Elizabethan England.*
[68] Gee and Hardy, *Documents of Church History*, p. 403.
[69] 27 Elizabeth, Cap. 2. *Ibid.*, pp. 485-92.
[70] *Ibid.*, p. 487.

cation, without permission, and £100 had to be paid in each case.[71] Even to withhold information about a Jesuit in the realm was punishable by a fine of 200 marks.[72] Finally—and this is an unintended testimony to the astonishing endurance of Catholics in the old faith—the Act Against Recusants was passed in 1593. Its provisions included the demand that every person over 16 and a Roman Catholic ("popish recusant" is the term in the Act) convicted of not attending a place of common prayer to hear divine service there, should stay in his home and not move outside a radius of five miles from it, and "shall lose and forfeit all his and their goods and chattels, and shall also forfeit to the queen's majesty all the lands, tenements, and hereditaments, and all the rents and annuities of every such person so doing or offending, during the life of the same offender."[73] The laws became increasingly severe against the Catholics until they were financially crippling for the nobility and gentry and mortally dangerous for the daring Jesuits and secular priests who ministered to them. Evelyn Waugh's novel *Edmund Campion* gives a poignant account of the furtive priests living in the shadows continual fear of the queen's pursuivants during the latter part of the Elizabethan period.

It is perhaps easier to criticise than to understand the motives of those Catholics who in the early years of Elizabeth's reign apparently attended Anglican worship without affronting their consciences. J. H. Pollen may well be right in suggesting that many of them were simply "waiting for another change of the royal whim, of which they had experienced so many. Three creed-compelling sovereigns had died in 11 years, and the reigning queen was far from enjoying strong popular support. Or again, she might marry a Catholic, for there was as yet no Protestant prince who was her peer."[74] On the other hand, Elizabeth's was a forced Catholicism in the reign of Mary, her sister; she could not have forgotten how her people hated the Catholic *autos-da-fé* at which the Protestants were burned in Smithfield, nor was she unaware of the fact that to the Pope her father's marriage to Anne Boleyn, her mother, was illicit, and therefore she was illegitimate. Quite apart from the fines which must have made the Laodicean and lukewarm Catholics frequent Anglican sanctuaries, the more committed Catholics could have argued truly enough that attendance at morning and evening prayer according to the Book of Common Prayer was

71 *Ibid.*, p. 488.   72 *Ibid.*, p. 491.   73 *Ibid.*, p. 501.
74 Pollen, *The English Catholics in the Reign of Queen Elizabeth*, p. 94.

not attending Mass (for they recognised a doctrinal world of difference between the Roman Mass and the English Communion service), although the prayers were for the most part a translation from the Breviary and the Missal.

Whatever the reasons or excuses, Catholics seem to have attended Anglican services at least minimally. What is of greater interest, however, is the information that certain Catholic priests found it possible to square their consciences to permit them to celebrate Anglican Communions and Roman Masses in succession. This may be called the period of convenient Catholic compliance, and those who took advantage of these arrangements might be termed Anglican papists! Nicholas Sanders in *The Rise and Growth of the Anglican Schism*, first published in 1585, gives the facts: "At the same time they had Mass secretly in their own houses by those very priests who in church publicly celebrated the spurious liturgy, and sometimes by others who had not defiled themselves with heresy; and very often in those disastrous times they were on one and the same day partakers . . . of the Blessed Eucharist and the Calvinistic supper. . . . Yea, what is still more marvellous and more sad, sometimes the Priest saying Mass at home for the sake of those Catholics whom he knew to be desirous of them, carried about his Hosts consecrated according to the rite of the church, with which he communicated them at the very time in which he was giving to other Catholics more careless about the faith, the bread prepared for them, according to the heretical rite."[75] Allen, a future cardinal, also attests to the truth of this practise, of which he was a witness and which he condemned, since it involved Catholics becoming "partakers often on the same day (oh! horrible impiety) of the chalice of the Lord and the chalice of devils."[76]

In London in the earlier years of Elizabeth's reign many Catholics attended Mass at the Spanish embassy chapel. In December 1567 some Englishmen who had heard Mass at the chapel were questioned by the government and a few were imprisoned. As a result of severe remonstrations the ambassador, Don Guzman de Silva, agreed to exclude Englishmen from his chapel in the future.[77] Two years earlier the ambassador had stated that the authorities

[75] The quotation is from the 1877 edition, with introduction and notes by D. Lewis, pp. 266-67.
[76] *Douay Diaries*, p. xxii, cited p. xxx of T. Law's introduction to Laurence Vaux, *A Catechisme or Christian Doctrine*.
[77] *Calendar, State Papers, Spanish*, I, 686.

were so confused by the chaos of the differing liturgical practises in the Anglican churches (mirrored by the Vestiarian Controversy) that Catholics could celebrate their own form of worship with less danger; consequently "Mass is much celebrated in secret, and many people confess and communicate most devoutly [even in London], which is quite common in other parts of the country."[78]

It appears that during the first decade of Elizabeth's reign the ecclesiastical penal laws were not strictly applied. The laxity would account for the varying degrees of nonconformity. The result was that "some failed to come to church; some, it would seem, worshipped according to the prescribed rites, but absented themselves from the sermon; others did not receive Communion."[79]

The slackness in prosecution of the Catholics, especially where they were numerous and of good social position, must also account for the persistence of Catholic religious customs in an officially Protestant England, which were so frequently reported to the bishops and as frequently warned against in their visitation articles and injunctions.

[78] *Ibid.*, I, 418.

[79] Trimble, *Catholic Laity*, p. 14. There was even a fourth alternative—to say Catholic prayers during the Anglican service. "There be many in the Diocese of Chichester, which bring to the church with them the popish Latin primers and use to pray upon them all the time when the Lessons be a reading and in the time of the Litany. . . ." *Disorders in the Diocese of Chichester Contrary to the Queen's Majesty's Injunctions*, P.R.O., S.P., 12, vol. 60, no. 71, cited in Hughes, *Reformation in England*, III, 248.

A fifth ruse was employed by lords of the manor—to allow their church to go to ruin; thus there would be no church from the worship of which they could dissent and be delated as Recusants. Augustus Jessopp, *One Generation of a Norfolk House*, p. 205, describes the method: "It was easy to reduce the fabric to a ruinous condition in any out-of-the-way village where the lord of the manor was all but supreme, where he was resident and the parson was not; accordingly a systematic destruction of the churches in Norfolk commenced and went on to an extent that may well amaze us; foremost among these was the church at Bowthorpe. It was convenient to have a clergyman of the new school coming and using the new Prayer-Book, and reporting absences at the bishop's visitation, therefore Mr. Yaxley, the lord of the manor, 'converted [the church] to a barne, and the steeple to a dove house,' and Mr. Waldegrave could no more be returned as 'not keeping his church.' It could hardly be expected, however, that the family would live like heathens, and it was in houses of this kind that the Missionaries [Jesuit priests] found an eager welcome."

Finally, a common trick was to refuse attendance at Anglican communions by pretending to be out of charity with one's neighbors or with the parson. Christopher Boreman of South Newington declared in 1584, "that he is not in love and charity with his neighbours"; Philip Wakeridge of Newton Valence in Hampshire, who absented himself from communion for a year, declared in court that the incumbent "Mr. Stanlie . . . is his adversary." Some were successful in their recusancy, notably John Downes of Babingley, of whom it was reported in 1597: "He is a notorious recusant and obstinately refuseth to be partaker with the Church of England. He hath not repayred to church this xx years." A Tindal Hart, *The Man in the Pew, 1558-1660*, pp. 181-82.

All of these Catholic customs were forbidden in Elizabeth's Injunctions of 1559 in which the clergy were instructed as to the subjects of their sermons, including what to approve and what to condemn. The third injunction forbade such "works devised by man's fantasies," as "wandering to pilgrimages, setting up of candles [presumably before images or relics], praying upon beads or such like superstition."[80] The 23rd injunction required the removal and destruction of "all shrines, coverings of shrines, all tables, candlesticks, trindals, and rolls of wax, pictures, paintings, and all other monuments of feigned miracles, pilgrimages, idolatry, and superstition, so that there remain no memory of the same in walls, glass windows, or elsewhere within their churches and houses."[81] Against the background of the determination of the queen and her bishops to extirpate Catholic customs, their survival was remarkable.

Bells were still rung on All Saints' day and on All Souls' day in honour of the saints and to remind parishioners of their duty to pray for their holy dead. Bell-ringing on these two days was prohibited in Norfolk in 1569, "as a superstitious ceremony used to the maintenance of popery, or praying for the dead."[82] The sacring bell continued to be sounded in the Midland counties and in the north.[83] The bringing of candles to church on Candlemass day so that they might be blessed and carried in procession was a custom maintained in the Catholic north well into Elizabeth's reign[84] As late as 1584 parishioners of Churchill in Oxfordshire were presented for keeping copes and relics. The indictment was: "there was abowt a vi or vii yeares agoe in the custodie of one Wm Kerrie three copes or vestments, one of velvit and ii of silke and two crosses, of wch they tooke one of the crosses and put it to the belfounders in Oxford to make there sauncebell withall, wch said reliques have remaynid in the custodie of this respondent ever synce undefaced untill they by chaunce came to light of late."[85]

A persistent use of the Rosary must have continued among the country people in remote places for many years, both because England's devotion to the Virgin Mary was renowned in Catholic days when every large church had a chapel in her honor and the

80 Gee and Hardy, *Documents of Church History*, pp. 419-20.
81 *Ibid.*, p. 428.
82 W.P.M. Kennedy, *Parish Life under Queen Elizabeth*, p. 117.
83 *State Papers, Domestic*, vol. 36, 41 (21).
84 Kennedy, *Parish Life*, p. 118.
85 *The Archdeacon's Court, 1584*, ed. E. R. Brinkworth, 2 vols., II, 200, 201, 205, cited in A. Tindal Hart, *The Country Clergy in Elizabethan and Stuart Times*, p. 36.

Rosary was one of the most common forms of devotion. Also, as late as 1571 all the clergy and churchwardens of the northern counties are requested to see that none of their parishioners used or had beads. At the same time the people were forbidden "superstitiously to make upon themselves the *sign of the cross* when they first enter into any church to pray."[86] The tenaciously held religious customs of many generations are not easily uprooted.

Yorkshire had many faithful Catholics as well as "loyal Lancashire." The Visitation returns of 1567 indicted clergy for many forbidden practises such as making wafers for Holy Communion with the impression of the crucifixion on them, retaining holy water stoops, or keeping images and tabernacles. The Rector of Roos, Nicholas Coke, even managed to preserve his rood-loft intact with all its images of the crucified Christ, the weeping Virgin Mother, and St. John.[87] Prohibited as these devout customs were, the faith had been too dearly purchased by the lives of Catholic saints and martyrs from Saints Thomas More and John Fisher up to the sacrifices of Edmund Campion and Robert Southwell, for it to be easily extinguished by government decree. But as government pressure increased and its agents grew more persistent, Catholics had to celebrate more and more secret Masses.

7. Clandestine Masses

From 1570 on, Recusants, if discovered at a Catholic Mass, were heavily fined; their priests were imprisoned and often hung, drawn, and quartered at Tyburn. The wonder is not that many conformed to Anglicanism, but that so many priests and laity remained committed and courageous Catholics.

The stiffening of Catholic resistance was undoubtedly aided by the example of the Elizabethan martyrs. There were 189 priests and layfolk, men and women, put to death for their faith between January 4, 1570 and February 17, 1603. Of the 189, 126 were priests, 111 of them secular clergy, one Benedictine, one Dominican, two Franciscans, and 11 Jesuits.[88] Their spirit was eloquently expressed in the challenge of the Jesuit Campion to the Privy Council: "Many innocent hands are lifted up to heaven for you daily by those English students, whose posterity shall never die, which beyond seas, gathering virtue and sufficient knowledge for

---

[86] Kennedy, *Parish Life*, p. 118.
[87] J. S. Purvis, ed., *Tudor Parish Documents of the Diocese of York*, p. 31; see also pp. 152, 162 for further evidence of the continuation of Catholic customs.
[88] Philip Hughes, *Rome and the Counter-Reformation in England*, p. 240.

the purpose, are determined never to give you over, but either to win you heaven, or to die upon your pikes."[89] Campion reminded his judges in his statement before receiving the sentence of death that he and martyrs like him were no treasonable foreigners, but inheritors of a proud English traditon: "In condemning us, you condemn all your own ancestors—all the ancient priests, bishops and kings—all that was once the glory of England, the island of saints, and the most devoted child of the see of Peter. . . . To be condemned with these lights—not of England only, but of the world—by their degenerate ancestors, is both gladness and glory to us."[90]

The Catholic apologists found it necessary, especially after 1570, to try to persuade the government that Catholics were recusants or emigrants into the Low Countries for spiritual and not political reasons. At the same time, they hoped that such writings would reinforce wavering Catholics in their determination not to bow the knee in the House of Rimmon in England. William Allen's reasons for recusancy or emigration in 1581 are as follows:

> The universal lacke then of the soveraine Sacraments cath-olikely, ministred, without which the soule of man dieth, as body doth without corporal foode: this cōstrainte to the con-trarie services, whereby men perish everlastingly: this intol-erable othe repugnāt to God, the Church, her Ma.ties hon-our, and al mens cōsciences: and the daily dangers, dis-graces, vexations, feares, imprisonments, empoverishments, despites, which they must suffer: and the railings and blas-phemies against Gods Sacraments, Saincts, Ministers, and al holies, which they are forced to heare in our Countrie: are the only causes, most deere Sirs, or (if we may be so bold and if our Lord permitte this declaration to come to her M.ties reading) most gratious Soveraine, why so many of us are departed out of our natural Countrie, and so absent our selves so long from that place where we had our being, birth, and bringing up through God, and which we desire to serve with al the offices of our life and death: onely craving corre-spondence of the same, as true and natural children of their parents.[91]

[89] Philip Caraman, *The Other Face, Catholic Life under Elizabeth*, p. 117.
[90] *Ibid.*, p. 227.
[91] *An Apologie and true declaration of the institution and endevours of the two English Colleges* (1581), sig. B3v.-5r.

In 1601 the Jesuit leader Robert Persons wrote *A brief Discourse containing certain reasons why Catholikes refuse to go to Church*. The main reasons were: the pollution of heresy; the scandal or stumbling-block this was to true believers; the fact that going or not going to church "is made a signe now in *England* distinctive between religion and religion, that is, betwixte a Catholike and a Schismatike";[92] and because it encouraged schism. Because such risks were taken to celebrate and assist at the clandestine Masses, they were marked by great fervor. Father Robert Persons reported, "No one is to be found in these parts to complain that services last too long. Nay, if at any time Mass fails, to last nearly a whole hour, this is not much to the taste of many of them. If six, eight, or even more services are held on the same day and in the same place, which happens not infrequently when priests are holding meetings among themselves, the same congregation will be present at all of them."[93] The letters of Father Persons provide the fullest account of how the Douai-trained priests were welcomed by Catholics, the secret places in which they celebrated worship, the ruses used to put government agents off the trail, the thoroughness and relentlessness of the government's search for the recusants' gatherings for worship, and the effects the martyrdoms had on the constancy of Catholics' commitment. His letters, in short, give the most complete, vivid, and accurate account of what the religious life of the Recusants, a *gens lucifuga*,[94] as Newman was to call them, was like in those days. For this reason, I will make full use of them.

In 1580 Persons and other members of the English Mission passed through most of the counties of England, preaching and administering the sacraments, "in almost every gentleman and nobleman's house that we passed by, whether he himself was a Catholic or no, if he had any Catholics in the house." They would enter as acquaintances of relations of some person in the house or as friends of some gentleman accompanying them. After the usual greetings they were shown to their quarters which were always in a retired part of the house. There they "put on priestly dress, conferred secretly and generally late at night with the Cath-

92 *Ibid.*, sig. C8v. See the similar views expressed in Thomas Stapleton's *Apologia pro rege Catholico Philippo II* (1592).
93 Letter of August 1581, in *Letters and Memorials of Father Persons, S.J.*, no. 39, vol. I, ed. L. Hicks, p. 46.
94 This crepuscular term was used by Newman in his sermon, "The Second Spring," in *Sermons preached on various occasions*.

THE LITURGICAL ALTERNATIVES

olics of the household and those who would come from outside, and heard their Confessions. Next morning Mass was said, followed by a Sermon or Exhortation."[95]

The fear in which the Recusants and their fugitive priests lived is conveyed in a letter describing the ruses that were resorted to for the avoidance of the pursuivants: "It is the custom of the Catholics themselves [Persons wrote] to take to the woods and thickets, to ditches and holes even, for concealment, when their houses are broken into at night. Sometimes when we are sitting at table quite cheerfully, conversing familiarly about matters of faith and piety . . . if it happens that someone rings at the front door a little insistently, so that he can be put down as an official, immediately, like deer that have heard the voice of hunters and prick their ears and become alert, all stand to attention, stop eating, and commend themselves to God in the briefest of prayers. . . . It can truly be said of them that they carry their lives always in their hands."[96] Father Robert Southwell gave an equally vivid description of the searches of the pursuivants for their human prey: "Their serches are very many and severe. Their serve God, as on Sondaies, holy daies, Easter, Christmas, Whitsontide [Pentecost] and such very great feastes. They come ether in the night or early in the morning, or much about dinner time; and ever seeke their opportunities when Catholikes are or would be best occupied or are likely to be worst provided, or looke for nothing. Their manner of searching is to come with a troupe of men to the house as though they come to fight a field. They beset the house on every side, then they rush in and ransacke every corner—even women's beds and bosomes—with such insolent behaviour that their villanies in this kind are half a martyrdome."[97]

---

[95] *Letters and Memorials*, p. xxiii.
[96] *Ibid.*, p. 86, letter of August 1581.
[97] *The Letters and Despatches of Richard Verstegan*, ed. A. G. Petti, no. 52 (1959), 7. The letter is probably the work of Southwell, the Jesuit priest and poet, author of the nativity poem, "The Burning Babe." It was written in December 1591. Southwell lived hourly with the thought of death, and his poem, "I dye alive," ends with the quatrain:

> Not where I breath, but where I love, I live;
> Not where I love, but where I am, I die;
> The life I wish, must future glory give,
> The death I feele in present daungers lye.

(Christobel M. Hood, *The Book of Robert Southwell*, p. 116.) For an account of pursuivants hammering at the door at 5 a.m. on a day in October 1591 while Father Southwell was celebrating Mass and Friars Garnet, Gerard, and Oldcorne

To these general accounts may be contrasted a description of the "Little Rome" which Lady Magdalen, Viscountess Montague, maintained in Elizabeth's reign at her country house. Her biographer states: "She built a chapel in her house (which in such a persecution was to be admired) and there placed a very fair altar of stone, whereto she made an ascent with steps and enclosed it with rails, and, to have everything comfortable, she built a choir for singers and set up a pulpit for the priests, which perhaps is not seen in all England besides. Here almost every week was a sermon made, and on solemn feasts the sacrifice of the Mass was celebrated with singing and musical instruments, and sometimes also with deacon and subdeacon. And such was the concourse and resort of Catholics, that sometimes there were 120 together, and 60 communicants at a time had the benefit of the Blessed Sacrament. And such was the number of Catholics resident in her house and the multitude and note of such as repaired thither, that even the heretics, to the eternal glory of the name of the Lady Magdalen, gave it the title 'Little Rome.' "[98]

To have been apprehended as a priest, even in the early days of the reign before the screw of torture was fully turned, must have been an ignominious and horrifying experience. Stow reports such an incident on September 8, 1562, when a priest was seized while saying Mass at Lady Cary's house in Fetter Lane, London. His captors were the Bishop of Ely's men, by whom the "priest was violently taken and led, as ten times worse than a traitor, through Holborn, Newgate market and Cheapside to the Counter at the stocks called the Poultry, with all his ornaments on him as he was ravished from Mass, with his Mass-book and his porttoys [Breviary] borne before him, and the chalice with the pax[99] and all other things, as much as might make rude people to wonder upon him. And the number of people was exceeding great that followed him, mocking, deriding, cursing, and wishing evil to him, as some to have set him on the pillory, some to have him hanged, some hanged and quartered, some to have him burnt, some to have

---

were in the same house, see *John Gerard, the Autobiography of an Elizabethan*, pp. 41-42.

The contribution of Nicholas Owen, a renowned Jesuit, who is said to have helped Gerard escape, should not be forgotten. He designed famous "priests holes," or hiding places, for recusant priests at Hindlip Hall. *D.N.B.*, XLII, 433.

[98] *An Elizabethan Recusant House, comprising the Life of the Lady Magdalen, Viscountess Montague (1538-1608)*, ed. A. C. Southern, p. 43.

[99] A pax was a tablet with a representation of the crucifixion, which was kissed by the officiating priest and the people at Mass.

him torn in pieces and all his favourers, with as much violence as the devil could write. . . ."[100]

In the midst of Persons' anguish in having to report to the Superior of the English College in Rome the death of other members of the Jesuit Order, he rejoiced that the martyrdoms were strengthening the Catholic faith. "Finally, it cannot be told," he writes, "and far less believed, unless we saw it with our own eyes, how much good their death has brought about. All with one voice, our enemies as well as ourselves, declare that if their lives had been prolonged to their hundredth year they could not have benefited their cause as much as has their short life[101] but glorious death."[102] The result was that Masses were being celebrated in London with greater frequency than for many years and were better attended and listened to than ever. So loyal were the Catholic recusants that when they were chased out of one house, they celebrated Mass in another; "when they are dragged to the prisons, they find a way to perform the holy sacrifice there also."[103] The supreme consolation for those being put to death for their faith— despite public abuse, private torture, and horrible consummation of their witness on earth—was the assurance of wearing the crown of martyrdom which the risen Christ himself would place upon their victorious brows. This is the only explanation of the following record of the reaction of Catholics who watched the martyrdom of Thomas Maxfield: "Whilst all these things were in acting [Maxfield was being hung. drawn and quartered before their eyes] wee Catholikes, who had before implored for him the divine assistance, did now beseech him, as being arived [sic] at the haven of glorie, to render by his prayers the eternal Maiestie propicius."[104] It was a practical response, which in its sheer matter-of-factness witnessed profoundly to the Catholic belief in the Communion of Saints. Martyrdom is the ultimate and climactic gift of adoration and worship of the living God. It was given unstintingly by the Catholics of Elizabethan England.

8. The Faithful Remnant Prepared for Endurance

Tempting as it is to conclude this chapter at the high point of martyrdom, the sad truth is that those responsible for the spiritual

100 Stow Memoranda, p. 121, cited in Caraman, The Other Face, p. 42.

101 Hughes, p. 242, points out that 85 of the 126 priest martyrs had not been ordained more than five years before their martyrdom and that 28 had been ordained less than two years.

102 Letters and Memorials, p. 133; letter of March 1, 1582.

103 Ibid.

104 Miscellanea, III, no. 3 (1906), 46.

welfare of English Catholics knew that martyrdom was impossible for all except the choicest few. Preparation had to be made for the long-term future, which did not realistically include the dream of England as a nation restored to obedience to Rome. Rather it envisioned a faithful remnant who had to be prepared for endurance in a climate at worst of hostility, at best of indifference.

Harding, one of the ablest English Catholic apologists, saw that the situation had gone beyond the hope of convincing Englishmen through polemical writing. He believed the time had come to try to make converts through devotional books. Richard Hopkins, in his *Of Prayer and Meditation*, claimed that it was Harding who "perswaded me earnestlie to translate some of those Spanishe bookes into our Englishe tounge, affirming that more spirituall profite wolde undoubtedlie ensewe thereby to the gayning of Christian sowles [since they were much more successful in disposing] the common peoples myndes to the feare, love, and service of almightie God."[105] Hopkins took Harding's advice, to the great advantage of English spirituality. *Of Prayer and Meditation* is a translation of Luis de Granada's *Libro de la Oracion y Meditacion* (1554); Hopkins also translated de Granada's *Memorial de la Vida Christiana* in 1586. Both were highly successful and may well have inspired the religious poetry of Crashaw and Vaughan in the next century.[106] The fiery, intense mysticism of de Granada, now available for the first time in English, seemed most suitable for English Catholics suffering from persecution.

If Harding was disillusioned by crabbed controversialism, so was Persons by the linking of recusancy with political plotting. As he was about to leave Rheims with Edmund Campion for England, Allen, the leader of the English Mission in Rome, informed them of the papal expedition to Ireland of 1580, which was ultimately directed against England. Persons said: "We were heartily sorry. . . . As we could not remedy the matter, and as our consciences were clear, we resolved, through evil report or good report, to go on with the purely spiritual action we had in hand; and if God destined any of us to suffer under a wrong title [that is, as political traitors] it was only what He had done and would be no loss to us, but rather gain in God's eyes who knew the truth, and for whose sake alone we had undertaken this enterprise."[107] Since both

---

[105] *Of Prayer, and Meditation* (1582), sig. A6v, cited in A. C. Southern, *Elizabethan Recusant Prose, 1559-1582*, p. 181.
[106] An opinion of E. Allison Peers, *Spanish Mysticism*, p. 20.
[107] Richard Simpson, *Life of Edmund Campion*, p. 146.

controversial and political approaches were dubious, it was better to concentrate on spiritual means of witnessing.

Hotheads would continue to believe in political action even after the crushing of the Spanish Armada had doomed any hope of Catholic Spain's conversion of England by the sword. Wiser heads, however, realised that Catholics were too few to have prevented the disestablishment of Catholicism at the beginning of the reign, and powerless in the cities, where middle-class mercantile interests were Protestant, as were the majority of the nobles and the higher gentry. In addition, the most influential force for conformity was the almost sacrosanct respect of all classes for the crown, so that the upper classes "by training, and intellectual formation and self-interest obeyed the royal will, and the lower classes by long conditioning followed the leadership of those above them.[108]

In reading Laurence Vaux's *A Catechisme or Christian Doctrine* (and *The Use and Meaning of the Holy Ceremonies of Gods Church* that appears with it in all English editions) one has the sense that this exiled English priest was providing his minute account of Catholic ritual and ceremonial because it was in danger of being forgotten, and it might keep alive hope in the minds of the recusants. It went through eight editions from 1567 to 1605, and was the manual of the English exiles in Louvain. It is probably because a full ritual and ceremonial of Catholic worship in England had virtually disappeared in Elizabeth's reign that we have this marvellously detailed description of worship as the recusants would have celebrated divine worship had they been allowed to do so. Baptism's ceremonies are detailed. They include exorcism, the signation of the Cross, the placing of salt in the child's mouth, the putting of spittle in the child's ears and nose, the anointing the child's breast and back with oil, the pouring of holy chrism on the head, and the use of the candle. The meaning of each is given in Vaux. The explanation for the anointing with oil is: "The childe is anoynted upon the breast with holy Oyle, to signifie: that the holy Ghoste should alwaies dwell in that harte and breast by faith and charitie," and, "The child is anoynted upon the backe with holy Oyle, to signifie the yoke of our Lord, which is sweete and light."[109] Holy Orders and Extreme Unction are similarly described in great detail.

[108] Trimble, *Catholic Laity*, p. 266.
[109] *A Catechisme or Christian Doctrine*, ed. Thomas G. Law, n.s., vol. IV (reprint of the 1583 edition), pp. 55-56.

The fullest and longest description is of the celebration of the Mass, or Sacrament of the Altar. There is a rather fanciful allegorical interpretation of the celebrant's vestments, which is of interest. The amice signifies the cloth on Christ's face when he was buffeted by the Jews, the alb the white cloak which Christ wore when Herod returned him to Pilate, the girdle represents the scourge, the "Favell" on the left arm the cord with which Christ was bound, and the stole stands for the other ropes. The upper vestment represents the derisive purple garment put upon Christ as a king.[110] The various actions of the priest are heavily allegorised as representations of various incidents in the narrative of the passion and death of Christ. The preparation of the elements is symbolically interpreted, too: "Bread & wine are then brought to the Priest at the Altar, to the ende he may do with thē as Christ in his last supper did, when he was now going to his deth. The Chalice betokeneth the grave: the white corporace betokenethe the white sheete, wherein Ioseph did fold Christes body, whē it was layed into the grave: & the paten representeth the stone wherewith the grave was covered."[111] After explanations of the priest's secret prayer and the salutations at the beginning of the great prayer of intercession, Vaux reaches the act of consecration, described and explained thus: "[he] cometh at the laste to take Christes person upon him, saying in his name and power over the breade. This is my body and over the wine. This is my bloude &c. By whiche wordes no faithfull man doubteth, but that Christes body and bloud are made really present under the forme of bread and wine, In token of which beleefe the priest lyfteth up the holy Sacrament, to put us in remembrance, how Christ was exalted upon the Crosse for us, and the people adore with godly honor the selfe body and bloud, which dyed, and was shed for us. And the in wordes also the Priest beseecheth, the said body and bloud of Christe being most acceptable to God, in his owne nature, to be accepted also of God in respect of the Church, which being yet sinfull, adventureth to handell and to offer suche preciouse giftes. And anone the faithfull soules are cōmended also unto God, to the end no members of the Churche may be omitted of the Churche in the cōmon sacrifice which toucheth the whole body of the Churche."[112]

The Mass concludes with the Lord's Prayer (curiously this usage is allegorised as a prayer of seven petitions calling to mind the seven "words" of Christ on the Cross), the kiss of peace, and if

110 *Ibid.*, pp. 89-90.   111 *Ibid.*, p. 91.   112 *Ibid.*, p. 92.

the priest communicates alone, "if none other be prepared there unto (as Christ upon the Crosse ended his owne Sacrifice alone) or if other be ready, they receave also with the priest even as Christ at his Supper gave his Sacrament to others also."[113] The persistence of the allegorical interpretation of the Mass (the pedigree is by Durandus out of Dionysius) is striking at this late date. While there is some recovery of the Tridentine insistence on the centrality of the sacrifice offered by Christ, together with the whole church, (the head with the members), there is more of the Gothic allegorical than of the Tridentine rubrical Mass in Vaux's interpretation.

What is unmistakable is that the Mass was the chief means of grace for recusant Priest and people alike. It enabled them to enter deeply into the sufferings of Christ and so bear their own burdens with less complaint; and it offered them the medicine of immortality. For John Rastell, writing in 1564, there was a deep spiritual consolation in the sacrament, which was not to be interpreted in a grossly carnal manner, for Christ is now at the right hand of God the Father. He advised his readers not to let their thoughts and desires rest in Christ's flesh only, but "goe hyer by your faith, and cōsider that blessed sowle of his, so chast, patient, wise, charitable, bright, glorious, and yet hyer and hyer, in to the heavens, and above all heavens, beholding and wondering, how the maker of them all, whom thowsand thowsandes, and ten hundred thousand thousands do wayte upon: ys present here for us, to be receaved of us, and to incorporate us in to hym selve."[114]

113 *Ibid.*, p. 93.
114 *A Confutation of a Sermon* (1564), sig. LI, cited in Southern, *Elizabethan Recusant Prose*, p. 87.

CHAPTER V

ANGLICAN WORSHIP:
THE PRAYER-BOOKS

As CATHOLIC worship is mediated by the Missal and the
Breviary, so is Anglican worship maintained and renewed by
the Book of Common Prayer. The Roman rite, at least until
the Second Vatican Council, was an international liturgy magis-
terially celebrated in the Latin tongue; the Anglican rite, which
has great affinities with the Roman rite, has always been celebrated in the English vernacular, and with the spread of English
colonists in the Western world it has become an international liturgy. On all hands it is acknowledged as the only surviving
sixteenth-century liturgy in continuous use, and as the bond of
union of the Anglican communion. According to Gregory Dix, the
various Anglican rites in their diverse forms probably serve some
20 million people,[1] and may well have influenced some of the 150
million of the different Protestant churches of the world in their
ordering of worship.[2] The Book of Common Prayer is the book
of both priest and people, uniquely so, and has continued to be an
impressive handbook of devotion and discipleship for four centuries. Within its wide and charitable bounds High Churchman,
Broad Churchman, and Low Churchman or Evangelical have
found ample freedom and adequate spiritual nourishment. It is a
book which has raided the devotions of Judaism as well as Christianity, and is a treasury that borrows from Eastern Orthodoxy as
well as from Western Catholicism (Roman and Gallican), and
from German and Swiss, as well as indigenously English, Protestantism. Its dignity, sobriety, compactness, and practicality make
it a peculiarly English vehicle of devotion. To understand something of its history is an introduction to the spirit of the Church of
England. If Protestantism is best understood in its Confessions
such as the Heidelberg (or Puritanism in the Shorter Westminster
Confession) and in the vigorous application of the Word of God
to the contemporary situation in the pulpit, Anglicanism is only
imperfectly understood in the Books of Homilies or in the Thirty-

---

[1] *The Shape of the Liturgy*, p. 613.
[2] For Anglican liturgical influence on Protestant forms of worship see W. D.
Maxwell, *The Book of Common Prayer and the Worship of Non-Anglican
Churches*.

Nine Articles, but supremely in the Book of Common Prayer and in the celebration of Holy Communion. It is for this reason that the chapters on Anglican worship concentrate on the successive revisions of the Book of Common Prayer in 1549, 1552, and 1559 and on the immense importance of Holy Communion and the prolonged controversies as to the mode of Christ's presence in that central Sacrament, while not ignoring the pulpit.

1. *The Preparation for Liturgical Reform*

The great achievement of Anglicanism is, of course, the First Prayer Book of King Edward VI of 1549. This was no sudden achievement of Archbishop Cranmer and his advisers, but the slow growth of various earlier experiments in England and overseas, and notably issuing in the production of an English *Litany* and of an English *Communion Order* before the publication of the climactic *Book of Common Prayer*.

Thomas Cranmer, whose name will always be associated with the first English Prayer Books, achieved fame as the supporter of Henry VIII's claim to the right to divorce his Spanish queen in 1529, and was employed by the monarch as his ambassador in Germany in 1532, where Cranmer first made contact with the Protestant vernacular rites of Lutheranism in Nuremberg and other centres.[3] The next year he was made Archbishop of Canterbury, but, although Germany had apparently convinced him of the importance of vernacular services, he had to wait 12 years before he could put these convictions into practice, with the publication of the Litany in English.

With so arbitrary a master as Henry VIII, Cranmer had to walk warily during this reign and to conceal his embryonic Protestantism. Some slight indications of his future reformist attitude can be gleaned from certain changes made in the church services during Henry's reign. There was a significant reduction between 1536 and 1541 in the number of holy days to be observed. A "Sarum" Breviary was issued which omitted the title of the Pope and was to that minimal extent "reformed." In the Convocation on February 21, 1543 Cranmer announced there would be a sweeping reform of all the service books; on the same day there was intro-

---

[3] Cranmer married Margaret, the niece of the German Reformer, Osiander. He was largely responsible for drafting the Brandenburg-Nuremberg Church Order of 1533, the prototype of the famous Church Order of Cologne, which influenced the First Book of Common Prayer of 1549 and *The Order of the Communion* of the previous year.

duced an English reading from the Bible at both matins and vespers on every Sunday and holy day in every church. All this was but the tip of the iceberg. Below an even more fundamental revolution was being evolved in the scholarly mind of the archbishop, in his plan to reduce the Breviary to two hours only.

Cranmer's first revolutionary achievement in the reign of Henry VIII was the issue of an English Litany on June 18, 1544 under the unwieldy title, *An Exhortacion unto prayer, thoughte mete by the Kinges maiestie, and his clergy, to be read to the people in euery church afore processyons. Also a Letanie with suffrages to be said or song in the tyme of the said processyons.*[4] In 1544 Henry VIII was at war with Scotland and France. On June 11, 1544 he wrote to Cranmer declaring his determination to have "general processions in all cities, towns, churches, and parishes of this realm, said and sung with such reverence and devotion as appertaineth, forasmuch as heretofore the people, partly for lack of good instruction and calling, partly for that they understood no part of such prayers and suffrages as were used to be sung and said, have used to come very slackly to the procession . . . we have set forth certain godly prayers or suffrages in our native English tongue."[5]

It appears that Cranmer took a patriotic opportunity, as the king conceived it, for some moderate measure of Reformation. Its importance was that it became soon afterwards the sole authorized form of service for processions; it is the oldest single part of the Book of Common Prayer.[6] It was derived from many sources, including the Sarum Litanies, Luther's Litany in its Roman form, the York Litany, possibly the Litany of Brixen, and certainly the Liturgy of Constantinople, from which is derived St. Chrysostom's prayer.[7] It could hardly be more critical of the Papacy than it is, with the scurrilously honest petition for delivery "from the tyranny of the bishop of Rome, and all his detestable enormities."[8] Its dominating characteristic is a concentration on the mediatorial power of the Holy Trinity and a diminished emphasis on the mediation of the

[4] It is conveniently reprinted as an appendix (pp. 564ff.) to *Private Prayers put forth by authority during the reign of Queen Elizabeth*, ed. W. K. Clay; or in F. E. Brightman, *The English Rite*, 2 vols., I, 174-90.

[5] Cranmer, *Miscellaneous Writings and Letters*, ed. J. E. Cox, p. 494.

[6] See Brightman, *English Rite*, I, lviii-lxviii.

[7] The "Prayer of Chrysostom" is an English version of the Latin translation of the Orthodox Liturgy, namely, the *D. Liturgia S. Ioannis Chrysostomi* (Venice, 1528). See Dowden, *The Workmanship of the Prayer Book*, pp. 227ff.

[8] *Private Prayers set forth by authority during the reign of Queen Elizabeth*, p. 572.

saints, though the Virgin Mary is mentioned by name and there is a general reference to saints, angels, and patriarchs. It was probably as close an approach to Protestantism as Cranmer dared make at the time.[9]

The accession of the boy King Edward VI on January 28, 1547 gave Cranmer the opportunity for religious reformation. Somerset, the Lord Protector of the realm, was a convinced Protestant. The Protector and the archbishop issued a series of *Royal Injunctions* to be administered in a general visitation of all the dioceses by 30 visitors. The *Injunctions* made a series of minor but important changes in the church service. Injunction 22 required that in addition to the existing English lesson at matins and evensong, the Epistle and Gospel at parish high Mass were also to be read in the vernacular. The 24th Injunction insisted that the Litany was in the future to be sung kneeling and not in procession so that the veneration of shrines on the way might be prohibited. Provision was also made for a *Book of Homilies* to be distributed by the visitors to each church, which were to be read each Sunday. The *Homilies*, according to T. M. Parker, "expounded a moderate version of justification by faith only and tacitly ignored such subjects as the Eucharist and other Sacraments, while attacking in strong terms popular Catholic customs regarded by the Reformers as superstitious."[10] On the 27th of January 1548 the Council ordered the abolition of the proper ceremonies for Candlemas, Ash Wednesday, and Palm Sunday. Those special to Good Friday were soon abolished, as well as the use of holy water and holy bread. While this was happening it seems that there was official encouragement of the wholesale smashing of images that took place. All foreshadowed the future simplification of the English liturgy and its total rendering into English.

Furthermore, there were several experimental services in the vernacular during the first two years of Edward VI's reign.[11] The first experiment took place on East Monday, April 11, 1547, when Compline in English was sung in the Royal Chapel.[12] The

---

[9] Brightman, *English Rite*, I, lxvii, observes, rightly that "the Litany is enough to prove he [Cranmer] had an extraordinary power of absorbing and improving other people's work."

[10] *The English Reformation to 1558*, p. 125.

[11] Fully described from the relevant manuscript evidence in the British Museum and the Bodleian Library in "Edwardine Vernacular Services before the First Prayer Book," in Walter Howard Frere, *A Collection of Papers on Liturgical and Historical Subjects*, ed. J. H. Arnold and E.G.P. Wyatt, Alcuin Club Collections, no. 35 (1940), pp. 5-21.

[12] F. A. Gasquet and E. Bishop, *Edward VI and the Book of Common Prayer*, p. 58.

service already existed in several versions, as in Bishop Hilsey's *Manual*, Henry VIII's Primer, and in the older primers. The significance in 1547 was its use in the Royal Chapel, which in Edward VI's reign was something of a national liturgical laboratory. A second important experiment took place on November 4, 1547, at the opening Mass of Parliament and Convocation, when the *Gloria in excelsis*, *Credo*, and *Agnus Dei* were sung in English. Six months later, according to Wriothesley, "Poule's quire with divers other parishes in London sung all the service in English, both Mattens, Masse & Evensonge"; on May 12, 1548, on the anniversary of the death of Henry VII, which was kept at Westminster Abbey, the Mass was "song all in English with the consecration of the Sacrament also spoken in English."[13] On September 4, 1548 Somerset sent a letter to the authorities of the University of Cambridge, ordering them "in their colleges, chapels, or other churches to use one uniform order, rite and ceremonies in the Mass, Matins, and Evensong, and all divine service in the same to be said or sung such as is presently used in the King's Majesty's Chapel" pending further changes.[14] On September 9 Robert Ferrar was consecrated Bishop of St. David's; at the Eucharist both the administration and the consecration of the Sacrament were in English. Nor should it be forgotten that the *Litany* had been available since 1544 in English; its use was required by the *Royal Injunctions*. In these ways the English vernacular was making liturgical inroads. The final barriers seemed to be yielding when even the Eucharist was celebrated in English, though admittedly the consecration of a bishop was a special and far from regular occasion. As Cranmer's *Litany* had been the first step in the vernacular walk, so *The Order of the Communion*[15] was the penultimate step in the same path.

On December 17, 1547 an act was passed in Parliament, "Against such as unreverently speak against the Sacrament of the Altar, and of the receiving thereof under two kinds." As a directory for the priest *The Order of the Communion* was issued on March 8, 1547-48. It comprised the English devotions, provided solely for communicants, which were to be interpolated in the middle of the Latin Mass, which now had the lessons in English. The time was almost ripe for the provision of the Canon of the Mass in

[13] Charles Wriothesley, *A Chronicle of England*, ed. W. D. Hamilton, 2 vols., new series, xi, pts. i and ii, 1875-77), ii, 2.
[14] Gasquet and Bishop, *Edward VI*, p. 147.
[15] Printed as a facsimile of the British Museum copy, C.25, f.15, in H. A. Wilson, *The Order of Communion, 1548*, Henry Bradshaw Society, vol. 34, 1908.

English, and last of all for the formula of Consecration in English. These were the last bastions of the Latin rite, the retention of which seemed synonymous with the maintenance of mystery and the unity of all Catholic worshippers. Thus the English *Order of the Communion* provided the communicant with the intelligibility necessary to full sharing in the rite, while reserving to the celebrant the privilege of conducting the worship in the venerable ecclesiastical language. The authorship of the *Order*, as of the First Book of Common Prayer which followed a year later, remains obscure. The letter of the Council addressed to the bishops hardly lifts the veil in affirming that the book was compiled by "sundry of his majesty's most grave and well learned prelates and other learned men in the Scripture."[16] Like the *Litany*, the *Order* shares the distinction of being subsequently incorporated into the Book of Common Prayer. This was Cranmer's most ambitious liturgical project hitherto.

The *Order* was the form of administration to be used in the Mass, immediately following the celebrant's Communion, and was inserted in the traditional Latin Mass. It included an exhortation to the communicants, with a warning and an invitation, and followed with confession, absolution, and the "comfortable words," together with a prayer before Communion, the words of administration, and a Blessing.

Its major sources are the Holy Scriptures, the Greek rite, the medieval commonplaces, the traditional Roman rite, and the *Pia Deliberatio*, or Church Order, of Cologne, prepared for Hermann von Wied, Prince-Archbishop of Cologne, by Martin Bucer; it in turn, as I have indicated, was based on the work of Osiander in the Brandenburg-Nuremberg Church Order. The *Pia Deliberatio*, a distinctively Lutheran Church Order, influenced the *Order of the Communion* in providing suggestions for the exhortation, the confession (in part), the "comfortable words," and the clauses added to the end of the formula of absolution and to the form of administration.[17] In view of the importance of the words of administration it is interesting to learn that they were adapted from the Sarum form of Communion for the Sick, which is: *Corpus domini nostri iesu christi custodiat corpus tuum et animam tuam in vitam aeternam.* This form is partly reduplicated and partly redistributed

16 Edward Cardwell, *Documentary Annals of the Reformed Church of England*, 2 vols., I, 61.
17 Brightman, *English Rite*, I, lxxiii.

so that "body" may answer to "corpus" and "soul" to "sanguis," and is expanded with "which was geven for thee" and "which was shed for thee." The relevant rubrics and words of administration read:

> And when he doth deliuer the Sacrament of the body of Christe, he shall say to euery one, these wordes followyng.
> The bodye of oure Lorde Jesus Christ, which was geuen for the, preserue thy body unto euerlastyng life.
> And the priest deliuering the Sacrament of the bloud, and geuing euery one to drinke once and nomore, shal saye.
> The blud of oure Lorde Jesus Christ, which was shed for the, preserue thy soule unto euerlastyng life.[18]

These words are typical in their conflation of Scripture, the Sarum Use of the Mass, and the German Lutheran church orders of Brandenburg-Nuremberg of 1533 and Cologne of 1543.[19]

The Sarum Use and the orders were a series of open experiments in liturgy from the *Litany* of 1544 to the *Order of the Communion* of 1548. In his study and in his mind, however, Archbishop Cranmer was conducting other liturgical experiments.[20] The experiments were entirely concerned with the simplification of the Daily Office, and almost certainly never went beyond Cranmer's study or their ideas beyond the minds of his confidants and fellow-liturgists in Henrician and Edwardine days. Frere believes the experiments "belong to the earlier stages of development, the first probably being anterior to the accession of Edward VI, and the second not long subsequent to it."[21] Brightman, on the other hand, believes that the first literary experiment belongs to the last years of Henry VIII, and the second one probably from 1547.[22] The manuscript itself introduces a complication, since it places the earlier draft

---

18 Wilson, *Order of Communion*, sig. C.j. verso.
19 The general theological character of the *Order* is described by W. Jardine Grisbrooke as follows: "It contains little positively offensive to the upholders of the old faith, but is clearly a product of the new. . . ." *Studia Liturgica*, Vol. I, No. 3, p. 150.
20 See Cranmer's *Liturgical Projects*, ed. J. Wickham Legg, Henry Bradshaw Society, vol. 50, 1915. These volumes were brought to public notice by Gasquet and Bishop after they had evaluated MS. Reg. 7 B iv. in the British Museum in their book, *Edward VI and the Book of Common Prayer*. The manuscript is discussed in Appendix i, its chief contents described and summarised in Appendices ii-iv, and the character of the experiments considered in Chaps. 2 and 4. Legg, *Edward VI*, p. ix, assigns the second draft to about 1543-47.
21 *Papers*, p. 5.
22 *English Rite*, I, lxxviii. Brightman appears to follow Legg's dating.

second. More important, however, than the dates are the contents of the two experiments.

The first includes all the canonical hours from matins to compline, derives its material almost wholly from the Sarum Breviary, and yet follows very closely the structure of the second recension of Cardinal Francisco Quiñones' revised *Breviarium Romanum* which was first published by Paul III in 1535, the second recension following a year later.[23] It reduced the readings from the lives of the saints to a bare minimum, so that the entire Psalter might be read in a week and nearly all the Bible in a year. Furthermore, the different Hours were made almost equal in length. In addition, the Breviary was made more compact by eliminating the antiphons, versicles, and responses. This immensely popular work of reconstruction went through a hundred editions before it was proscribed by Pope Paul IV in 1558.

The second literary experiment is even more interesting, because here the Hours were reduced to two, matins and evensong. The reduction was due in part, according to the explanation, because the existing arrangement required far to much unnecessary repetition, and partly because the ancient distribution of the Hours of the day had become obsolete; in practice the services were accumulated or concentrated at two hours in the day. The schema is introduced by a preface which is reproduced in the main from the first recension of Quiñones. This particular experiment foreshadowed Cranmer's work in the First and Second Prayer Books, where Sunday worship consisted of matins and evensong (which the minister was also to say daily in his parish church) with the Order for Holy Communion.

Both Offices begin with the Lord's Prayer in English, followed by *Domine labia* at matins and *Deus in adiutorium* at evensong, with the ensuing *Gloria Patri*, etc. Then follows the Hymn and three Psalms, then the Lord's Prayer in English. At matins there were three lessons (at evensong, two), followed by *Te Deum* at matins and *Magnificat* at evensong. A fourth lesson was to be read at matins on Sundays, great festivals and saints' days, and lessons were always to be read in English and from the pulpit, that they might be audible and intelligible. After *Benedictus* at matins and *Magnificat* at evensong there followed *Dominus vobiscum*, the

[23] Quiñones' breviary of 1535 was reprinted by J. Wickham Legg, and the second recension of 1536, also edited by Legg, appeared in two volumes, as Henry Bradshaw Society publications in 1908 and 1912.

collect of the day, and *Benedicamus Domino*, to which the response was *Laudemus et superexaltemus nomen eius in saecula. Amen.*

In both public experiments in worship, from the *Litany* to the *Order of the Communion*, as in the special services in the Royal Chapel and in St. Paul's Cathedral in London and at Westminster Abbey, as well as in the literary experiments in his study, Cranmer had been preparing the Church of England to take the momentous step of relinquishing the Roman Liturgy, which had sustained the devotion of English Christians for almost 10 centuries. It had been the exclusive medium and channel of the nation's worship from the last decade of the sixth century—when Augustine of Canterbury (as he was to be known) and his monks landed on the island of Thanet and thence came to the shores of Kent, where their singing of the Gregorian chant charmed the ancient English and made them anxious to glorify God in the Latin Liturgy—down to the sixteenth century. No wonder that the step to give it up was taken warily, that its abandonment was the cause of riots, that the interruption of their liturgical habits[24] made the English, who were renowned for their devoutness in the fifteenth century, antipathetic to a coerced national form of worship, which excluded the most conscientious Christians, whether Roman Catholic recusants or Puritan pietists.

2. *The First English Prayer Book (1549)*

It is curious that the First Prayer Book, which along with the King James translation of the Bible is one of the great religious landmarks of the English spirit, has resisted the most assiduous efforts of scholars to penetrate the obscurity of its joint authorship. It is well known that its chief author was Cranmer, impelled thereto as much by duty as studious inclination. But it is not known who his fellow compilers were, though there are some likely candidates. It is known that in September 1548 a group of bishops and divines were assembled in Chertsey and (probably during the king's stay on September 22-23) at Windsor, in order to settle

---

24 Dix, *Shape of Liturgy*, pp. 686-87, writes: "With an inexcusable suddenness, between a Saturday night and a Monday morning at Pentecost 1549, the English liturgical tradition of nearly a thousand years was altogether overturned. Churchgoing never really recovered from that shock. Measures of compulsion kept the churches reasonably full in the reign of Edward VI and the earlier half of Elizabeth's. But voluntary, and above all weekday, churchgoing—on the popularity of which most fifteenth century travellers had remarked—virtually disappeared." See also Ridley, *Works*, p. 60, for a reluctant confirmation that Catholic was much higher than Protestant church attendance.

outstanding liturgical questions and provide "a uniform order of prayer."[25] It is also known that five bishops and four divines took part in the consecration of Ferrar to the see of St. David's on September 9, 1548. It is a natural supposition that those who took part in the consecration were those who were conveniently near because they were the same group who were concerned with the reform of the Liturgy. If this supposition is correct, their names are: Archbishop Cranmer, Bishops Ridley of Rochester, Holbeach of Lincoln, Thirlby of Westminster, and Goodrich of Ely, together with Drs. May, Dean of St. Paul's, Haynes, Dean of Exeter, Robertson, afterwards Dean of Durham, and Redman, Master of Trinity College, Cambridge. An additional piece of evidence is that at the service the consecration and administration of Holy Communion was performed in English.[26] The first Edwardine Act of Uniformity, passed January 21, 1549, states that the king, on the advice of the Lord Protector and his council, "has appointed the Archbishop of Canterbury, and certain of the most learned and discreet bishops, and other learned men of the realm" on the basis of Scripture and the usages of the primitive church to "draw and make one convenient and meet order, rite, and fashion of common and open prayer and administration."[27]

What were the purposes of the new Book of Common Prayer as defined in the first Act of Uniformity of Edward VI and in the preface of the Prayer Book itself?[28] The Act suggests that a primary consideration was the general confusion in the kingdom over the use of different forms of Roman rite, namely, "the Use of Sarum, of York, of Bangor, and of Lincoln" and of more recent varieties of "forms and fashions."[29] From the Council's standpoint it is most desirable, the Act goes on to maintain, that "a uniform quiet and godly order should be had."[30] The preface to the First Prayer Book also holds, though only as the fourth reason, that

[25] *Grey Friars of London Chronicle*, ed. J. G. Nichols, Camden Society, vol. liii, p. 56.

[26] Proctor and Frere, *A New History of the Book of Common Prayer*, pp. 44-45. See also British Museum *Catalogue of an Exhibition commemorating the four hundredth anniversary of the introduction of the Book of Common Prayer*, pp. 19f.

[27] Gee and Hardy, *Documents illustrative of English Church History*, p. 359.

[28] Further insight into the intentions of Cranmer and the other compilers of the First English Prayer Book of 1549 is provided by the proclamation prefixed to the *Order of the Communion* of the previous year. It expressed a determination, "to trauell for the reformation & setting furthe of suche godly orders as maye bee moste to godes glory, the edifying of our subiectes, and for the advauncemente of true religion," so that, "a uniform quiet and godly order should be had. . . ."

[29] Gee and Hardy, *Documents*, p. 358.

[30] *Ibid.*, p. 359.

unity will come from a uniformly prescribed liturgy: "And where heretofore, there hath been great diuersitie in saying and synging in churches within this realme: some folowyng Salsbury use, some Herford use, some the use of Bangor, some of Yorke, and some of Lincolne: Now from hēcefurthe, all the whole realme shall haue but one use."[31]

Unquestionably the notion that the "Uses" of the medieval cathedrals had created the confusion is greatly exaggerated, though the innovators were the likelier culprits. In fact, the medieval "Uses" were not alternatives to the Roman rite, but minor alterations and adaptations of the same rite in accordance with the usages of the great cathedral churches, such as York, Hereford and Salisbury. The employment of the "Uses" was not necessarily rigid; there was some variety in ceremonial details, and some difference with reference to lessons or Psalms, even sometimes with regard to collects and antiphons. "English worship in the pre-Reformation period nevertheless exhibited a general unity of pattern so that the habitual attendant at the services of Salisbury cathedral was entirely at home in the services of York minster."[32] The great "inconveniences" claimed to result from a diversity of medieval "Uses" had in fact been seriously mitigated by a previous decision at the Convocation of Canterbury in 1542 to prohibit the use of any but the Salisbury Breviary in the Southern Province.

The first reason given in the preface to the First Prayer Book— and it is a substantial one—for the issuance of the Book of Common Prayer was that a Scriptural and primitive form of worship should be available in the vernacular English. The Latin service was not understood by the people, and thus contradicted St. Paul's demand that worship should be with the spirit and with the understanding. As a result, "the seruice in this Churche of England (these many yeares) hath been read in Latin to the people, whiche they understoode not; so that they haue heard with theyr eares onely; and their hartes, spirite, and minde, haue not been edified thereby."[33] The sincerity of this motive is hardly subject to question when considering the slow and steady experiments that we have seen Cranmer engaging in with this paramount aim in mind.

31 All references to the 1549 prayer-book will be to the Page Collection copy in the Henry E. Huntington Library, which was printed by Edward Whitechurche in May. This reference is to Aiiv.

32 *The book of common prayer of the Churche of Englande: its making and revisions M.D.xlix—M.D.clxi. set forth in eighty illustrations, with Introduction and Notes*, ed. E. C. Ratcliff, p. 9.

33 Aii.

Cranmer and his advisors were equally concerned with providing a rite that was Biblical in inspiration and content, uncluttered by saints' lives and legends, and in which the arrangement of the lections was such that it enabled the books of the Bible to be read consecutively. In this revision they claimed they were returning to patristic use: "But these many yeares passed this Godly and decent ordre of the auncient fathers, hath bee so altered, broken, and neglected, by planting in uncertein stories, Legēdes, Respondes, Verses, vaine repeticions, Commemoracions, and Synodalles, that commonly when any boke of the Bible was begon: before three or foure Chapiters were read out, all the rest were unread."[34] In consequence, now "is ordeyned nothyng to be read, but the very pure worde of God, the holy scriptures, or that which is euidently grounded upon the same,[35] and that in suche a language and ordre, as is moste easy and plain for the understandyng, bothe of the readers and hearers."[36]

A third consideration was that of convenience. It is almost impossible to imagine what a revolution it was to provide a single service book that was adequate for all regular occasions and many special occasions. Previously the parish had had to provide a considerable number of service books for the use of the priest. First, he needed the *Ordinal* or *Pica* or *Pie*, from which he would discover what service he should recite in the Canonical Office on any given day. The preface to the first Book of Common Prayer refers to "the nōbre and hardnes of the rules called the pie, and the manifolde chaunginges of the seruice, was the cause, it to turne the boke onely, was so hard and intricate a matter, that many times, there was more busines to fynd out what should be read, then to read it when it was founde out."[37] In addition to the *Ordinal* the priest needed a *Missal* or Mass-book. A *Processional* was also required, this being a collection of psalms, anthems, litanies, and prayers which were appointed to be sung or said in processions before the celebration of the principal Mass of the day on Sundays and holy days. For the occasional rites such as Baptism, Matrimony, Churchings of Women, Visitation of the Sick, Burials, and Blessings over food, the priest needed a *Manual*. For his own Canonical Office, he required a *Breviary*. Churches with choirs of singers needed at least one *Graduale* (in English "Grayle") containing the

34 *Ibid.*
35 Presumably this means exhortations based on doctrine derived from the Bible.
36 Aii v.   37 *Ibid.*

musical portions of the Mass and an *Antiphonary* for the musical parts of the Breviary. Furthermore, a bishop needed a *Pontifical* for ordaining priests and deacons, confirming children, and consecrating a church, although this was his personal book. Thus the parishioners were responsible for providing a minimum of six and a maximum of eight or nine service books for their priest. The inconvenience of using so many books must have been irritating in the extreme and disruptive of dignity and reverence in worship.

There was another great convenience that the single vernacular, Biblically-based, uniform Book of Common Prayer brought. It is that for the first time in the history of a nation a vernacular prayer book was made available with traditional and contemporary resources which was both the priest's and the people's common book. In the four centuries in which it has been used, it has served as a resource for family and private prayers as well as for public and corporate worship, and has stimulated a splendid and singular lay devotion of which some outstanding examples are the Ferrars of Little Gidding in the seventeenth century and Susannah Wesley in the eighteenth. Another of its by-products may well be the lack of any serious anti-clericalism in English history, because the married English clergyman in his rectory or vicarage was closely in touch with the life of his parishioners; he could never be thought of as celebrating a mysterious and magical rite in a foreign tongue which gave him social distance and distrust; he was freed from the temptation of any sacerdotal airs and graces. Debatable as the latter is, however, the main point is that the Book of *Common* Prayer was a form of devotion which bound clergy and congregation together because priest and people alike shared it and had their copies of it, because the responses were a kind of "conversation" between them in worship, and because both parties shared a common appreciation for a rite that spread a Biblical faith that attempted to be intelligible in its expression, was rationally organised in its structure, and one that linked the Jewish and Christian centuries in felicitous phraseology and musical cadence.

In sum, then, we can see in the first English prayer-book three distinguishing characteristics that made it unique. It was a vernacular rite aimed at edification and presupposing good order and intelligibility. It was a single book to be used by priest and people alike. It was a uniform rite aimed at the religious unification of the English people. Furthermore, it was imposed by the political power of England, with inhibitory penalties. The Act of Uniformity required the use of *The Booke of the Common Prayer and Administracion of the Sacramentes and Other Rites and Ceremonies of the Churche, after the Use of the Churche of England* on and after Whitsunday, June 9, 1549.

The 1549 prayer-book is divided into 14 sections. First comes the Preface with the reasons for introducing the Prayer Book in English. There follows a Calendar for Psalms and Lessons. Then comes the Order for matins and evensong throughout the year. The next and most substantial section of the book (amounting to two-thirds of it) contains the Introits, Collects, Epistles, and Gospels to be used at the celebration of Holy Communion throughout the year, with proper Psalms and lessons for various feasts and days. The fifth section provides the Order for "The Supper of the Lorde and holy Communion, commonly called the Masse." The following sections contain the services of Baptism (public and private), Confirmation (including a catechism for children), Matrimony, for the Visitation of the Sick and Communion of the Sick. The 10th section provides a burial service and the 11th a form for the Purification of Women after childbirth. The 12th part contains Scripture and prayers suitable for Ash Wednesday. The 13th and 14th sections are practically extended rubrics, the former dealing with Ceremonies omitted and retained, and the final section providing notes "for the more plain explicaciō, and decent ministracion of thinges cōteined in this boke." The First Prayer Book provided a comprehensive, condensed, pruned and purified as well as primitive, liturgy for England.

3. A Predominantly Catholic or Protestant Rite?

Opinions differ as to how conservative or radical a revision of Catholic worship the 1549 prayer-book was, as can be seen from the carefully expressed opinions of three Anglicans. Bishop Gibson maintained that "it was an honest attempt to get rid of mediaeval corruptions and to go back to what was primitive and Catholic."[38] To Dom Gregory Dix, Cranmer's aim, more ambiguously and subtly in 1549 and more openly in 1552, was to replace the Catholic understanding of the sacrifice of the Mass with a commemoration of Calvary not distinguishable from Zwinglian "Memorialism," and therefore a radical Protestantism.[39] E. C. Ratcliff asserts that

[38] *The First and Second Prayer Books of King Edward VI*, ed. E.C.S. Gibson, Introduction, p. x.
[39] Dix, *Shape of the Liturgy*, pp. 656f., 670f.

"Cranmer in 1546 at Ridley's persuasion abandoned the Catholic belief that by consecration the eucharistic bread and wine become the Body and Blood of Christ."[40] What is clear in this crepuscular situation is that the government wanted to give the impression that this was a simplified and purified version of the Roman Rite in English dress. So much at least can be gathered from the reply given to the Western rebels who had called the new Prayer Book Service a "Christmas game" and wanted their old Mass back. They were, in effect, told that they were receiving the Mass. The government's words were: "It seemeth to you a new service, and indeed it is none other but the old: the selfsame words in English which were in Latin, saving a few things taken out."[41]

Some clue and answer to the question of the theological conservatism or radicalism of the 1549 prayer-book may be provided by a consideration of the sources Cranmer and his advisers drew on. The single greatest source for the new rite was the Holy Scriptures; it is interesting that the Psalms, all Lessons, including the Gospels (with a single exception),[42] and the Epistles are taken from the "Great Bible" which was issued in 1539 and revised in 1540 and is a composite Tyndale-Coverdale version. That, of itself, would indicate a strong Protestant bias in the rite, as also would the preference for the vernacular. On the other hand, there are strongly traditional Catholic sources that went into its making. The Latin rite, according to the Use of Sarum,[43] had a considerable influence also. It provided the structure of the different offices and (with the exception of the Lessons) suggested the various passages of Scripture to be used in the services. It also supplied most of the rest of the content of the services, with the important exceptions of the didactic and hortatory materials.

If we look at the latter element, which is strong, then the scales are tipped on the Protestant side, for this type of contemporary material comes almost exclusively from the Lutheran *Kirchenordnungen*, though it is difficult to pinpoint an exact source in many cases.[44] The Church Orders showing the clearest influence on the

40 *First and Second Prayer Books*, p. 10.
41 *The Acts and Monuments of John Foxe*, ed. George Townsend, 8 vols., v, p. 734.
42 The Gospel at baptism.
43 See *The Use of Sarum*, ed. W. H. Frere, 2 vols. and *The Sarum Missal*, ed. J. Wickham Legg.
44 Brightman, *English Rite*, I, lxxx, warns that the contributions of the Lutheran Church Orders are considerable but not strictly measurable, "since similarity between the books apart from actual quotation does not of necessity imply—and

prayer-book are those of Electoral Brandenburg of 1540 (largely the work of Osiander) and Cologne of 1543 (largely the work of Bucer). The ideas of the didactic and hortatory passages in the prayer-book were borrowed from these two Orders, and occasionally the very words either in the Latin or, in the case of the latter, in the English translations that appeared conveniently in 1547 and 1548. There was also an influence in practice from the same Lutheran sources, particularly when this involved a modification of the traditional customs in worship. It can be seen in providing a direction for worshippers to offer money at the celebration of Mass or Holy Communion, in the prayers to be offered at the holy table in a dry Mass (*missa sicca*),[45] in the way in which declarations are made in marriages, in the use of Mark 10:13ff. as a Baptismal lection, and in the recognition of private houses, in some conditions, as a suitable locale for Baptisms. The same influence is evident in the adoption of new customs, such as the separating of communicants from the general congregation at Mass, the use of the litany throughout the year and not only at special seasons like Lent and Rogationtide, and the communicating of the sick directly from the altar.[46] Here again the preponderance of Protestant and therefore of radical influence seems to be definitely decided.

Nonetheless, other and more traditional sources for the first prayer-book can be found. The influence of Eastern Orthodoxy can be found in two spots: in the inclusion of an invocation of the Holy Spirit, or an *epiklesis*, in the prayer preceding the recital of the Institution narrative, and in the Prayer of St. Chrysostom. There is also a probable influence from the Mozarabic rite. Traditional influence can also be seen in certain Continental Catholic revisionary documents. The first and second recensions of the Breviary of Cardinal Quiñones left their mark on Cranmer's simplification and purification of the Offices of matins and evensong and on the Lectionary and Calendar as a whole. The *Encheiridion* and the *Antididagma*, both coming from the Catholic party in the reforming diocese of Cologne, also had an impact.[47]

where the similarity is one of omission there is no means of knowing—that the one has borrowed from the other."

[45] See J. Dowden, *Further Studies in the Prayer Book*, pp. 186ff.

[46] These are detailed in Brightman, *English Rite*, I, lxxx-lxxxi. The claims of German Lutheran influence on the Book of Common Prayer, although exaggerated, are set forth in H. E. Jacobs, *The Lutheran Movement in England*, Chaps. 17-23.

[47] These influences are shown in detail in Brightman, *English Rite*, II, 690, 692, 694, 734, 778.

Finally, it should be remembered that the first Book of Common Prayer incorporated the *Litany* of 1544 and *The Order of the Communion* of 1548. The former was derived from Catholic and Lutheran sources, as well as from Eastern Orthodoxy; the latter derived from the same sources, although its dependence was greatest on the *Simplex ac Pia Deliberatio*—the Cologne Church Order.

My conclusion, in reference to the sources used by Cranmer and his advisors, is deliberately inconclusive. That is, I still see a considerable and wary conservatism, bent on preserving all that could be preserved from tradition that was consonant with Biblical directives and the customs of the primitive church. At the same time I observe an increasing trend toward the vernacular and an approximation to Lutheran *Kirchenordnungen*.[48] This doubtful conclusion leads to another inquiry, and that is, was the language of the First English Prayer Book *deliberately* and consciously ambiguous in the hope of placating Catholics, while pleasing more radical Protestants?

Are we to place any credence in the charge that the traditional language of the 1549 prayer-book was a deliberate equivocation, as is suggested by E. C. Ratcliff,[49] Jardine Grisbrook,[50] and T. M. Parker?[51] Was the traditional language employed for the purpose of creating a comprehensive rite wide enough to include both Catholics and Protestants? Or was it a camouflage for a liturgical revolution, that without it might have led to widespread civil war? Intriguing as these suggestions are, it is improbable that such a wary scholar as Cranmer would plot a theological or liturgical revolution, so different is this from all the previous evidence we have for his methods. Furthermore, it is extremely unlikely that a radical Communion doctrine would be introduced into a uniform liturgy intended to produce—as the Act of Uniformity insists—

---

[48] Jardine Grisbrook, *Studia Liturgica*, affirms that many Anglicans tended to maintain the Catholic character of the first prayer-book only by comparison with the more radical nature of the second, but that they forgot (a) what was omitted from the Catholic rite in the 1549 formulary and (b) were ignorant of what was common to the 1549 rite and the Continental Lutheran Church Orders. Gasquet and Bishop (*Edward VI*, p. 184), make a similar observation.

[49] Ratcliff, *Book Common Prayer*, p. 15, refers to "the large equivocal element in the Book of 1549."

[50] Grisbrook, *Studia Liturgica*, p. 158, suggests that Cranmer disguised a novel Eucharistic doctrine in traditional language to avoid the probability of widespread civil war.

[51] Parker, *English Reformation*, p. 130, insists that the conservatism of the Eucharistic doctrine is more apparent than real, arguing that "this is an ingenious essay in ambiguity," since, while the prayers do not deny Catholic doctrine, yet a careful use of the words "would enable a Protestant to use the service with a good conscience."

"so godly order and quiet in this realm."[52] The suggestion seems to me to be too sinister to comport what is known of Cranmer's character.

On what, then is the charge of deliberate equivocation based? Ratcliff rightly points out that contemporaries interpreted the doctrinal trend of the 1549 book differently. To the radical Bishop Hooper of Gloucester and Worcester it was so redolent of Sarum as to be "very defective . . . and in some respects indeed manifestly impious."[53] On the other hand, the immovably Catholic Princess Mary would not admit the rite into her chapel so convinced was she of its departure from the Catholic norm. It is significant, however, that the Henrician Catholic, Stephen Gardiner, found it to be so traditional in its phraseology that he believed it was compatible with a Catholic interpretation. In fact, Gardiner maintained that there were five places in the Communion Order where he detected Catholic Eucharistic doctrine: (1) the words of administration and the use of the terms 'body' and 'blood'; (2) the rubric which requires the bread to be broken into at least two pieces, and comments, "menne muste not thynke lesse to be receyued in parte then in the whole, but in eache of them the whole body of our sauiour Jesu Christ";[54] (3) the *epiklesis* implies a mutation of the elements; (4) the Prayer for the Church[55] and especially the commendation of the dead presupposes a "Mass propitiatory"; and (5) the central petition of the Prayer of Humble Access, which begs: "Graunt us therefore (gracious lorde) so to eate the fleshe of thy dere sonne Jesus Christ, and to drynke his bloud in these holy Misteries, that wee maye continually dwell in hym and he in us, that our synful bodyes may bee made cleane by his body, and our soules washed through his most precious bloud."[56]

Gardiner, a former Master of Trinity College, Cambridge and a devotee of the New Learning, had for a time believed it was not necessary to tie the doctrine of the corporal Presence of Christ in the Sacrament to the Aristotelian metaphysics with which it was defended and explained in the theory of Transubstantiation. It

[52] Gee and Hardy, *Documents*, p. 360.
[53] *Original Letters relative to the English Reformation*, ed. Hastings Robinson, I, 79.
[54] For his life see J. A. Muller, *Stephen Gardiner and the Tudor Reaction*. Gardiner's Eucharistic views are contained in the *Explication and Assertion of the True and Catholic Faith of the Blessed Sacrament of the Altar*, which, together with Cranmer's response to it, *Answer unto a Crafty and Sophisticall Cavillation devised by Stephen Gardiner*, are to be found in Cranmer, *Works*, I, 10-365.
[55] The Book of Common Prayer (1549), fol. cxxi.
[56] *Ibid.*, fol. cxviii.

is not generally known that Cranmer for a while accepted a similar belief or that he constantly repudiated the charge that he taught a nude Memorialism (such as Zwingli is charged with upholding),[57] and that he was firmly convinced of the real and spiritual presence of Christ in the Eucharist. Cranmer denied the *inclusio localis* because it was contrary to the evidence of the senses, destroyed the nature of the sign in the sacrament by the mutation of bread, and contradicted the assertion of the Scriptures and the Apostles' Creed that the body of Christ was in heaven. At the same time he believed Christ was spiritually present, objectively so, because it was the power of the Holy Spirit that conveyed the benefits of His Cross and Resurrection to the believer, who received Christ in the heart by faith.

It has been a common mistake to assume that there was no fourth alternative open to Cranmer besides the Catholic, Lutheran, and Zwinglian. There was, in fact, a fourth available possibility in Virtualism, the Eucharistic doctrine according to which, while the bread and wine remain unchanged after the consecration, the faithful communicants receive with the elements the *virtue* or power of the Body and Blood of Christ. This was the view of the Eucharist affirmed by Martin Bucer, Henry Bullinger, Peter Martyr, and John Calvin. It has been argued at length by C. W. Dugmore in *The Mass and the English Reformers* and more recently by Peter Brooks in *Thomas Cranmer's Doctrine of the Eucharist*, that Cranmer's was a high Calvinist doctrine.

Cranmer shared with Calvin several negative and positive points of Eucharistic doctrine. Both denied that the Eucharist is in the medieval sense a propitiatory sacrifice. Both denied Transubstantiation for the same reasons that it is destructive of the "sign" in the Sacrament and contrary to empirical experience. Both affirmed that the body of Christ is located in heaven and only in heaven. Both had what Brooks calls a "sursum corda" approach; they asserted that the hearts of the believers have to be lifted to heaven there to feed on Christ spiritually. Such participation was declared possible through the power of the Holy Spirit. Both affirmed fur-

---

[57] The late Dom Gregory Dix, that learned and witty liturgiologist, has widely spread the notion that Cranmer's was a Eucharistic doctrine of the "Real Absence," and reiterates the view in his *Shape of the Liturgy* (pp. 656, 659) that Cranmer taught a Zwinglian doctrine. He is, however, unjust to Zwingli in accusing him of teaching subjectivity. As Cyril Richardson rightly insists, for Zwingli "the objectivity is not understood in substantial categories but in mental and personal ones." *Journal of Theological Studies*, New Series, vol. 16, pt. 2 (October 1965), 436.

ther that the "signs" in the Lord's Supper are neither empty nor nude, but deliver what God has promised to attach to them. Calvin and Cranmer denied that there is any feeding of the ungodly in the Eucharist (*manducatio impiorum*) since the Sacrament works from faith to faith. Calvin would also have agreed with Cranmer's claim that as the bread and wine feed the body of the believer, so does the spiritual presence of Christ feed the soul of the believer.

In these ways Cranmer's doctrine seemed closer to Calvin's than Zwingli's. My own judgment is that Dugmore and Brooks have made a convincing case, particularly in the light of three considerations: (1) If Cranmer was a Zwinglian, why didn't he more patently and unambiguously use Zwinglian language instead of the conservative language he does? I reject the thesis that accuses him of using ambiguous language for political purposes. (2) In 1548 and 1552 the Eucharistic debate and issues had long since passed the positions affirmed by Zwingli, who had died in 1531; it seems therefore anachronistic to call Cranmer's views Zwinglian 17 or 20 years after Zwingli's death. This is particularly so when Bullinger, Zwingli's successor in Zurich, had helped make the Eucharistic doctrine of Zurich agree more closely with that of Calvin in the *Consensus Tigurinus* of 1549. (3) Calvin provides in his doctrine of the Holy Spirit an admirable account of the dynamics of the Eucharist in a way that preserves its objectivity.

Yet as Cyril Richardson has shown,[58] there were still two obstacles in the path of easy acceptance of Cranmer's doctrine as unequivocally Calvinist. The first was that Calvin uses substantialist language in speaking of manducation frequently and apparently deliberately, whereas there is only one clear case of such usage in Cranmer. Calvin intended us to understand that there is in the Sacrament a true participation in Christ's flesh or human nature. This Cranmer denies; he affirmed that the participation is through the divine nature and power, mediated through the Holy Spirit. Cranmer, therefore, stands with Zwingli in drawing a strict line between spirit and substance. The second difficulty in seeing Cranmer as a consistent Calvinist in Eucharistic doctrine derived from his agreement to accept the "black rubric" in the Second Prayer Book, which utterly excluded "anye reall and essencial

---

[58] Cyril Richardson's powerful statement of the view that Cranmer's Eucharistic doctrine is Zwinglian was first stated in *Zwingli and Cranmer on the Eucharist*; it has recently been elaborated over against the views of Dugmore in *The Journal of Theological Studies*, New Series, vol. 16, pt. 2 (October 1965), 421-37; and against the views of Brooks in *The Anglican Theological Review*, vol. 47, no. 3 (1965), 308-13.

presence there beeyng of Christ's naturall fleshe and bloude." This would appear to be conclusive proof of his Zwinglianism but for two other considerations. Since this was apparently foisted on Cranmer at the last minute when the Second Prayer Book was already coming off the printing presses and at the royal request, after the beseeching of King Edward by his chaplain, John Knox, it can hardly be regarded as the free expression of Cranmer's Eucharistic teaching. Nor, unless Knox deviated in his Eucharistic beliefs from Calvin (an unlikely supposition of one who was so close a disciple), can we regard this doctrine as un-Calvinist.

Furthermore, however close to Calvin or Zwingli Cranmer's Eucharistic beliefs were, it must be noted that Cranmer and Zwingli differed in their evaluation of the importance of the Eucharist. Cranmer, like Calvin, desired a weekly Eucharist, whereas Zwingli settled for a quarterly Eucharist. On balance, then, I think Cranmer moved from a Catholic through a Lutheran to a Calvinist or Virtualist doctrine of the Eucharist, and that the final stage was accompanied by the strong influence on him of Nicholas Ridley, relying on the Nominalism he found in Radbertus. Cranmer, it must be insisted, affirmed that by the power of the Holy Spirit, the true consecratory agent in the sacrament, Christ with all the benefits of his passion and resurrection was spiritually present at the Lord's Table, and that this was known in the hearts of believers by the interior testimony of faith. Faith did not create the presence —that would be blasphemy. Rather it confirmed the presence through the power of the Holy Spirit. Cranmer would undoubtedly have agreed with the statement made by his mentor, Ridley, in the Cambridge Debate of 1549. There Ridley stated that the three practical benefits of the Eucharist were unity, nutrition, and conversion.

If this view of Cranmer's Eucharistic teaching is accepted, then, there is no need to describe Cranmer as a crypto-Zwinglian hiding behind a quasi-Catholic mask of language, for his was a mediating, not an iconoclastic, position. Possibly Cranmer was inconsistent, even confused. It may be that he was groping toward a clearer expression of his sacramental theology. He may even have been adumbrating a mediating position which tried to conserve the Catholic value of the Eucharist, while abandoning its anachronistic metaphysical underpinnings. But he was assuredly not an *éminence grise*, nor was he an archiepiscopal Rasputin at Edward's court, however intriguing such a hypothesis may be.

There is, however, one further argument to be considered, which seems to shadow Cranmer's integrity by those who use it. It is the claim of Ratcliff[59] and Dix[60] that the rite of 1549 was merely a trial balloon, an experiment that could be revised at the first opportunity. It was conceived as a kind of liturgical litmus paper to test exactly how much Protestant acidity Englishmen could stomach, with the intention of giving them a larger dose next time. This view is bolstered by the evidence in a letter written by Martin Bucer and Paul Fagius on April 26, 1549 to the ministers of Strassburg, reporting on a visit they had paid the previous day to the Archbishop of Canterbury. Referring to the proposed prayer-book, presumably then in the process of being printed since it was to be introduced at Pentecost that year, they wrote: "We hear that some concesions have been made both to a respect for antiquity, and to the infirmity of the present age; such, for instance, as the vestments commonly used in the sacrament of the eucharist, and the use of candles: so also in regard to the commemoration of the dead, and the use of chrism; for we know not to what extent or in what sort it prevails. They affirm that there is no superstition in these things, and that they are only to be retained for a time, lest the people, not having yet learned Christ, should be deterred by too extensive innovations from embracing his religion, and that rather they may be won over."[61]

As a comment on this excerpt, it can be said that Cranmer's welcome to the strangers of distinction may have led them to believe that he was more firmly of their opinion than was in fact the case. But, at all events, it is notable that the reasons given for the conservative nature of the revision, were two and not merely one. There was the typically English desire to keep the tradition of the past, as far as possible; and there was the sensible idea that reform should be undertaken slowly, so that, using the scriptural analogy, babes in the new faith should be given milk before they are ready for the strong meat of the Gospel. Moreover, Bucer and Fagius called for no changes in the actions, words, or theology of the first Prayer Book, but only in the ceremonial with which it was celebrated, which they thought might be changed. Furthermore, they explicitly stated that the leaders of the English church "affirm that there is

[59] *Ibid.*, p. 15.
[60] *Ibid.*, p. 658. Dix, however, relies on F. M. Powicke, *The Reformation in England*, pp. 89f.
[61] *Original Letters relative to the English Reformation*, ed. Hastings Robinson, 2 vols., II, 535-36.

no superstition in these things." My conclusion is that there is no indication here of a proposal for a revised Prayer Book to follow the Book of 1549, only for a possible future modification of ceremonial. I find no evidence of any equivocation by Cranmer.

In short, I attribute the apparent confusion in the rite which has caused Catholics and Protestants to interpret it diversely, as due to several possible causes in combination, none of them amounting to the sinister purpose of saying one thing and meaning another. First, it should be remembered that, however important Cranmer was, he was working with a committee, and further that there was a considerable disagreement between the similar Cranmerian and Ridleian views and those of highly placed Bishops Bonner, Gardiner, Tunstall, Day, Thirlby, and others.[62] Next, Cranmer had not fully clarified his own views, which appeared to have moved from a Lutheran to a Calvinist direction and which were intermediate between Transubstantiation and Memorialism. Third, since the aim of producing this uniform rite was to end the innovations of the ultra-Protestants and to prevent the recrudescence of the more crassly magical popular Catholic view of the Mass, its intention clearly was mediatorial and not iconoclastic. Finally, *festina lente* was precisely the imperative of gradualism that Cranmer had followed throughout his career, of which the First Prayer Book was the continuance of a policy that had produced the ever-increasing Protestantism of the *Litany* followed by *The Order of the Communion*, and the further stage of which would be the prayer-book of 1552. It was not a radical or Zwinglian type of Protestantism, but a more conservative or mediating Protestant doctrine of the Eucharist that was expressed clearly or darkly in the Communion order of the first Prayer Book.

After these preliminary discussions we are in a position now to estimate how Protestant the 1549 prayer-book was by comparing the Communion rite in it with the Roman Catholic Mass in the

---

62 The lack of agreement of the bishops in their interpretation is carefully analysed in Gasquet and Bishop, *Edward VI*, pp. 129-39. This is also the theme of the letter written by Peter Martyr to Bucer, dated from Oxford, December 26, 1548: "The other matter which distresses me not a little is this, that there is so much contention among our people about the Eucharist that every corner is full of it. And even in the supreme council of the state in which matters relating to religion are daily brought forward, there is so much disrupting of the bishops among themselves and with others, as I think was never heard before." There is also a later, interesting reference to the new respect that Cranmer had won as a theologian, who had hitherto been regarded as an administrator, but had now "shewn himself so mighty a theologian against them [the 'popish party'] as they would rather not have proof of, and they are compelled . . . to acknowledge his learning and power and dexterity in debate." *Ibid.*, pp. 469-70.

Sarum Use. In this comparison I will be concerned not only with words but also significant actions, whether included or omitted.

The first half of the Communion service of 1549 closely followed the medieval rite; Introit, *Kyrie*, *Gloria*, collects, Epistle, Gospel, and Creed were retained in the same order as in the Catholic Mass. The preparatory prayers were fewer, however, and an Erastian flavor appears in the following of the collect for the day with the collect for the king. The Gradual, which in the Sarum rite was accompanied by the priest's ceremonial preparation of the chalice and paten, was entirely omitted. Notice that the term "Mass" is the third alternative permissible title after "Lord's Supper" and "Holy Communion" and that it will disappear altogether in 1552, also that the fourth rubric in the introduction to the service allows the ceremonial cope to be substituted for the sacrificial chasuble, which Grisbrook described as "a discreet but powerful blow at the old sacrificial doctrine at the very outset."[63]

The most notable deviations from the Sarum Rite come after the Creed and sermon, or homily, for the character of the immediately following Offertory was greatly altered. In the ancient and medieval liturgy the bread and wine were offered to God as gifts of the people in readiness for the sacrifice. But in the 1549 rite there was only an offering of alms at this point, and they were not placed on the altar but in the "poor mennes box." The offertory sentences had no Eucharistic reference; the preparation of the elements takes place without any ceremony. One can only conclude that Cranmer, like Luther, wished to rid his rite of any suggestion of oblation or sacrifice.[64]

If the character of the Offertory was changed, the Canon of the Mass was so altered as to be wholly unrecognisable, after the *Sursum corda*, the Preface, and the *Sanctus*. It consisted of eight parts.[65] The first was a long intercession for the church, king, ministers of state, clergy, those in adversity, and, after a short thanksgiving, for the example of the saints and for the dead. If the commemoration of the dead and mention of the saints is a Catholic item, it is important to notice the omission of any pleading of the sacrifice or of the resurrection of Christ in connection with the intercession. The second section includes a commemoration of the

63 Grisbrooke, *Studia Liturgica*, p. 154.
64 Gasquet and Bishop do not overstate the case (*Edward VI*, p. 196): "It will therefore appear that the ancient ritual oblation with the whole of which the idea of sacrifice was so intimately associated, was swept away."
65 The entire Canon can be found in fols. cxv-cxvii.

death of Christ and of the institution of the Last Supper, but it excludes any pleading of the sacrifice of Christ in the Eucharist. The third section is a petition for the consecration of the elements: "with thy holy spirite and worde, vouchsafe to blesse and sanctifie these thy gyftes, and creatures of bread and wyne, that they maie be unto us the bodye and bloude of thy moste derely beloued sonne Jesus Christe." Here we have an interesting combination of Eastern and Western consecratory forms, since the Eastern *epiklesis* regards the Holy Spirit as the agent of consecration, while the Western formula regards Christ the word of God as consecrating through the repetition of his words in the Institution narrative. Even if there is a strong stress on the subjective, "that they may be unto us,"[66] there is a strong stress on the objective power of God the Holy Spirit and God the Son, which seems to be wholly orthodox, though perhaps unduly anxious in providing a double insurance in a hybrid Eastern-Western formula of consecration.

The fourth section of the Canon is the Narrative of Institution from I Corinthians 11, which is common to Catholic and Protestant Communions, but with the significant rubric: *These wordes before rehersed are to be saied, turning still to the Altar, without any eleuacion, or shewing the Sacrament to the people.*[67] What Cranmer gave with one hand—the affirmation of the corporal presence in the consecratory formula—he seemed immediately to take away with the other by the abolition of the ceremony of elevation. The fifth section is the *anamnesis*, "the memoryall whyche thy sonne hath wylled us to make, hauyng in remembraunce his blessed passion, mightie resurreccyon, and gloryous ascencion," which seems to be so worded as to avoid any reference to any offering or sacrifice of the body and blood of Christ or of the bread and wine as people's oblations. The sixth part of the Canon is a petition for the acceptance of "this our Sacrifice of praise and thankes geuing." The seventh part is an offering of "our selfe [*sic*], oure soules, and bodies, to be a reasonable, holy, and liuely sacrifice unto thee," that the communicants may be incorporated into Christ. The eighth and final section of the Canon is the petition that the "bounden duetie and seruice" of those present may be accepted and that their prayers may be brought up to the Divine Majesty.

[66] Day, bishop of Chichester, refused his assent to the book because, among other reasons, the words immediately following the *epiklesis*, "that they maie be unto us," were not changed to read "that they maie be *made* unto us the bodye and bloude of . . . Jesus Christe." See Gasquet and Bishop, *Edward VI*, pp. 131-32.

[67] *Ibid.*, fol. cxvi.

The Canon, or Prayer of Consecration, concluded, there follow the Lord's Prayer and the *Pax Domini*. From there the deviation from the Sarum Rite is almost total. Although the *Agnus Dei* is retained it is removed until the people's Communion. After a reminder that Christ the Paschal Lamb is offered up, there is an exhortation to repentence and charity, and an invitation to partake of the Sacrament after making a general confession. The absolution follows, then the "Comfortable Words" of Christ and the "Prayer of Humble Access." The Communion is delivered in the following words: "The Body of oure Lorde Jesus Christe which was geuen for thee, preserue thy bodye and soule unto euerlasting lyfe," and "The bloud of our Lorde Jesus Christ which was shed for thee preserue thy bodye and soule unto euerlastyng lyfe."[68] As was mentioned, the *Agnus Dei* in English is sung while the people communicate. The Communion Order approaches the end with a section known as the "Post-Communion," which includes sentences of Scripture stressing the ethical duties of Christians, since sanctification is the proper expression of gratitude for redemption, a prayer of thanksgiving for "the spirituall foode of the most precious body and bloud of thy sonne, our saviour Jesus Christ" and the concluding Blessing.

What conclusion are we, at last, to draw from this Communion Rite which is an amalgam of ancient Catholic and modern Lutheran forms, with an Eastern *epiklesis* added for good measure? It is clear that in the absence of the elements of oblation and of sacrifice and with the obvious intention to do away with any occasion of adoration of the elements (explicitly denied in the rubric forbidding elevation) that the corporal presence is denied. Even the force of the *epiklesis* tending in the direction of a corporal presence is weakened if not nullified by two considerations. The first is the strongly subjective modification introduced by the phrase "be unto us," which was seen by the conservative bishop Thirlby of Westminster and which he proposed to rectify by changing the phrase to read "be made unto us," which was hotly rejected by Cranmer and his associates. The second consideration is that the emphasis on a "spiritual presence," dear to Cranmer's heart since 1546, would be strengthened by an *epiklesis* invoking the power of the Holy Spirit, and the final prayer of thanksgiving reinforces this view with its gratitude for "*spirituall* foode." If it can be established as at least probable that no corporal presence is intended, then this

68 *Ibid.*, fol. cxviii v.

cannot be considered a Catholic rite, however much its structure, order, and terminology recall a partially Catholic model. Nor, since the German Lutherans of the north in this period believed in the corporal presence, though not in Transubstantiation (a belief sometimes unhappily termed "Consubstantiation"), can this Order be considered a Lutheran Rite, despite its indebtedness in so many details to various German *kirchenordnungen*, and despite the fact that Gasquet and Bishop, who have made the most thoroughgoing comparative analysis of the rite, considered it to be Lutheran.[69] The only alternative left is to call it a Reformed rite, even if its phraseology and contents show an immediately greater indebtedness to Roman and Lutheran than to Reformed sources. But to which section of the Reformed church does its theology of the Eucharist belong? Is it Zwinglian and Memorialist, as Gregory Dix believed?[70] It is impossible to accept this view, because Zwingli had been dead 18 years and his successor Bullinger had helped frame a more mediating doctrine of the Eucharist. It was this view that was held, with only minor variations, by Bucer, Peter Martyr, Pullain, à Lasco (all foreigners befriended by Cranmer and living in England), and the greatest Reformed theologian of them all—Calvin—whose sacramental views had been fully formulated as early as 1540, four years after the publication of the first edition of his *Institutio Christianae Religionis*. It was a Eucharistic doctrine that affirmed that *in the sacramental action* Christ was present to grant forgiveness and eternal life through His humanity and that, although the body of Christ was in heaven, by the power of the Holy Spirit He was present through transformation, and in the sacrament united His elect to one another.[71]

[69] *Ibid.*, p. 195: "And even if it were not an ascertained fact that, during the years when it was in preparation, Cranmer was under the influence of his Lutheran friends, the testimony of the book itself would be sufficient to prove beyond a doubt that it was conceived and drawn up after the Lutheran pattern."
[70] *Op. cit.*, pp. 656, 659.
[71] The best recent treatments of Calvin's sacramental teaching are Ronald S. Wallace, *Calvin's Doctrine of the Word and Sacraments*; and Kilian McDonnell, *Calvin, the Church, and the Eucharist*. Calvin's *Petit Traicté de la cène*, first appearing in 1540, was translated into English by Miles Coverdale and published probably in 1549 (according to the *Short Title Catalogue*, item 4412.1) as *A Faythful and moste Godly treatyse concernynge the most Sacred Sacrament of the blessed body and bloude of our sauior Christ, compiled by John Caluine*; it was reprinted in *Writings and Translations of Myles Coverdale, Bishop of Exeter*, ed. G. Pearson, pp. 434-66. Calvin's doctrine in its distinctive emphasis may be gathered from two brief citations from this treatise. First he writes: "The bread is not unworthily called the body; forasmuch as it doth not only represent it unto us, but also bring unto us the same thing . . ." (p. 440). Second, he approves Zwingli and Oecolampadius for having criticised a gross carnal presence of Christ

Much has been said about the indebtedness of Cranmer and his associates to many and varied sources, but what of the values of this rite? Was it no improvement over the ancient Roman rite? One improvement it clearly had, apart from the other advantages of a vernacular liturgy and of a book which is equally for the use of priest and people: the Roman Canon, unlike the liturgies of the East, were insufficient in intercessory prayers. This was remedied in 1549 by the comprehensive intercessions of the Prayer, "for the whole estate of Christ's Church," which follows the singing of the *Sanctus*. Furthermore, while early liturgies were called "Eucharists" in token of the large element of thanksgiving in them, this was not prominent in the Latin Canon. The defect was rectified in the 1549 book, both in the Prayer of Consecration and in the noble post-Communion Prayer of Thanksgiving. Furthermore, the emphasis is strong on the need for the offering of "ourselfe, our soules, and bodies" in union with Christ's supreme offering now being commemorated. Perhaps one can even mark as the distinctive characteristic of the Anglican rite the sobering demand that devotion should lead to the duty to serve the community. It seems as if Anglicanism was from the first afraid of that romanticism which could so easily drown ethics in aesthetics; so at the conclusion of the service of Holy Communion there is the disenchanting series of Scriptural reminders that the service of God must issue in the service of men.

Moving from content to style, it is widely recognised that Cranmer was not only a highly skilful editor who could meld the most heterogeneous sources into a polished literary product, but that he

---

in the sacrament, but condemns them for having "omitted to declare what presence of Christ in the supper we ought to believe, and what communion of his body and blood is there received: insomuch that Luther supposed them willing to leave nought else but the bare signs, void of the spiritual substance" (pp. 463-64). Calvin's point is that to the seals of God's promises, which the sacramental signs are, the verity is also added. Also see Wallace, *Calvin's Doctrine*, pp. 151-53; and Calvin's *Institutes* IV:17:iii-x.

If the question is asked, how did Cranmer have access to Calvin's thought? it can be answered that John Knox was a minister at Berwick-on-Tweed during Edward VI's reign, that by 1549 the great *Institutio* was 13 years old, and that Coverdale had already published an English translation of Calvin's *Short Treatise on the Sacrament*, together with a Calvinist Church Order for Denmark and parts of Germany, and, most important, à Lasco and Pullain were using Calvinist orders of worship for the churches of the strangers of which they were superintendent ministers in England. In fact, there is clear verbal dependence on Pullain's Order of worship in the third collect in the Communion Order of 1549 which is to be used after the Offertory when there is no Communion to follow; it is possible that there are verbal echoes of the three prayers of thanksgiving after Communion in the Calvinist Danish order, already referred to, in the post-Communion Prayer of Thanksgiving, since all four are concerned with spiritual eating and drinking.

could write with an ear sensitive to the cadences of the English language. Sometimes he even improved on the lapidary elegance of the Roman collects he was translating. For example, the second collect (for Peace) which is used at matins, renders the Sarum original which begins "Deus auctor pacis et amator, quem nosse vivere, cui servire, regnare est . . ." as "O God, which art the author of peace and louer of concorde, in knowledge of whom standeth oure eternall life, whose seruice is perfect fredome. . . ." Another example of a happy translation from an equally felicitous Latin original can be found in the first collect after the offertory, to be used when Communion was not celebrated. The Latin phrase *inter omnes vias et vitae huius varietatis* is rendered as "emonge all the chaunges and chaunces of this mortall life."[72] These are freer but far more felicitous translations than more exact ones would have been. The fact is that Cranmer was much more than a facile manipulator of scissors and paste; he was, in several instances, a creator.[73] His were the then new collects for Eastertide and those for several of the evangelists, as well as the admirable prayer of thanksgiving after Communion, which is an excellent summary of the varied meanings of Holy Communion, as thanksgiving, mystery, grace, incorporation with Christ, fellowship in the Church, or anticipation of the Kingdom of God. The prayer is, in addition, most happy in making the transition from the mystery of the sanctuary to the morality of "good works" in the workaday world. Another masterly prayer of Cranmer's was the famous "Prayer of Humble Access" which combines reverence and tenderness in the approach to God, with sense and without sentimentality.[74]

Unquestionably the literary quality of the First Book of Common Prayer is high. But while the language of the time was vigorous and rich, there was also a tendency toward prolixity and vaguely worded sentences, along with a Latinity that could easily become cumbersome if not firmly controlled. As it was, Cranmer may be said to have created a superb vehicle of liturgical prose, which

[72] Book of Common Prayer, fol. iiii and cxx.

[73] Another example of Cranmer's improving on a collect in the original Latin is that for the Fourth Sunday after Easter. But in justice, it must be admitted that Cranmer on the rare occasion could perpetrate an unhappy construction, as in the confusion of the following petition in the second collect at evensong (the offending words are italicised): "Geue unto thy servauntes that peace which the world cannot geue; *that both* our hartes may be sette to obey thy commaundementes, *and also that by thee we* being defended from the feare of our enemies, may passe our time in rest and quietnesse."

[74] See the comments of Massey H. Shepherd, Jr. on these two major prayers in *The Oxford American Prayer Book Commentary*, pp. 82-84.

met simultaneously, as Stella Brook has defined them, "the workaday but important requirements of ease of articulation, and the need to create aural effects of sonority and dignity and rhythmic balance."[75] In addition, the language gave shape to profound thought and aspiration.

In concluding my consideration of the contents and style of the 1549 prayer-book, I may have left the impression that in structure, diction, and sources used the volume was fairly conservative; that view may be reinforced as we consider its successor, the more radical prayer-book of 1552. In fact, however, as the revision of other services in the 1549 book show, it was a volume more distinguished by innovation than conservatism. In the two daily offices of matins and evensong all invocations of the saints and all allusions to their merits and intercession were excised; also, the Calendar removed the names of all but a few "scriptural" saints. In the office of Baptism the impressive secondary ceremonies such as the use of salt, spittle, oil and candle are done away with, with the exception of the exorcism before and the signation of the cross and the anointing with chrism after Baptism. The new rite of Confirmation removed the anointing with chrism which all previous rites in Christendom had considered essential. Further, there are whole areas of life untouched by the new book, such as the lack of any formulas for the blessing of sanctuaries or of their contents. Nor was there any form for the commendation of the dying to God, which must have been a grievous loss to an elderly person who might have received great comfort from the rich medieval rite. After all, the central rite of Mass or Holy Communion had been changed from the offering of a sacrifice for the living and the dead to a commemoration of the sacrifice of Christ offered on the cross for the world and a renewal of spiritual fellowship with Him and with living Christians. Change, to mid-sixteenth-century worshippers, was more obvious than continuity.[76] That was clearly shown by the popular reaction to the introduction of the Book of Common Prayer of 1549.

4. The Reception of the 1549 Prayer-book

The publication of the First English Book of Common Prayer generated three responses, varying from acceptance through tem-

---

[75] The Language of the Book of Common Prayer, p. 122.

[76] On this issue I find the summary judgment of F. E. Brightman (English Rite, I, lxxxii), that "Rite and Ceremony are simplified," inadequate. W. Jardine

porary acquiescence to dissatisfaction. Since the book was introduced by an Act of Uniformity which carried heavy fines for noncompliance, it had to be accepted willy-nilly by most of the country. In some cases it may even have been warmly accepted. This is the impression given by Dryander in a letter of June 5, 1549 to Bullinger in Strassburg: "A book has now been published a month or two back, which the English churches received with the greatest satisfaction." Whatever reservations Dryander thought Bullinger may have had about some ceremonies retained in the book, he was anxious that Bullinger realize what an advance it represented, so he added: "Meanwhile this reformation must not be counted lightly of; in this kingdom especially, where there existed heretofore in the public formularies of doctrine true popery without the name."[77]

There is more evidence that the Continental divines who had recently been invited to England by Cranmer and certain Englishmen in the Reformed cities of Strassburg or Zurich were acquiescent, and not enthusiastic about the Prayer Book of 1549. Their attitude was that this was only the important first instalment of a reformation of worship that must proceed further. For them the 1549 book was important only as an interim rite.[78] John Butler, writing on February 16, 1550 to Thomas Blaurer, obviously considered the prayer-book a halfway house: "The affairs of religion are now, through the mercy of God, in a more favourable position, considering the state of infancy and rudeness of our nation. Baptism, for instance, and the Lord's Supper, are celebrated with sufficient propriety, only that some blemishes in respect to certain ceremonies, such for instance as the splendour of the vestments, have not yet been done away with."[79] Richard Hilles, writing to Bullinger on June 4, 1549, gave the impression that Cranmer is more anxious to please the German than the Swiss divines, and that Bucer, who had recently come to Cambridge, might keep him conservative. "Thus," he wrote, "our bishops and governors seem, for the present at least, to be acting rightly; while, for the preservation of the public peace, they afford no cause of offence to the

---

Grisbrook's statement, pp. 152-53, is more convincing when he affirms that "the majority of the ceremonies with which the Englishman approached his Maker were not simplified at all—they were abolished."

[77] *Original Letters*, ed. Hastings Robinson, 2 vols., I, 350-51.

[78] This is clear in a joint letter from Martin Bucer and Paul Fagius to the Strassburg ministers, April 26, 1549. *Original Letters*, II, 534-37.

[79] *Ibid.*, p. 635.

Lutherans, pay attention to your very learned German divines, submit their judgment to them, and also retain some popish ceremonies."[80] Thus the opinion of Protestant scholars in England seemed to be acceptance for lack of anything better.

But it was Bucer, whom Hilles had thought to be in the conservative school, who had written the fullest and most careful evaluation of the 1549 book in the *Censura Martini Buceri super libro sacrorum, seu ordinationis Ecclesiae atque ministerii ecclesiasticii in regno Angliae.*[81] His most serious criticism was a dislike for the very concept of the consecration of things, including the water in Baptism and the bread and wine in the Communion service. He believed the signs of the sacraments were potent only while in sacramental use.[82] He would abolish in the Communion Order the invocation of the Holy Spirit, the manual acts accompanying the recital of the Institution narrative, and the sign of the cross in Baptism. He would also abolish the exorcism and chrism in Baptism and the oil for the sick. Bucer was highly critical of mass vestments and of ceremonial signs without Biblical warrant. He argued the need of many more homilies to be written for the instruction of clergy and people, and rejected several traditional customs, among them the use of the choir for divine service as perpetuating an unfortunate distinction between ministers and laity and because it was difficult to hear the minister leading worship from the choir. He disliked the use of wafer bread instead of ordinary bread for the Eucharist, condemned the placing of the sacrament in the mouths rather than in the hands of communicants, and would abolish as superstitious all prayers for the dead.[83] Hooper was the most critical of the Protestants: "I am so much offended with that book, and that not without abundant reason, that if it be not corrected, I neither can nor will communicate with the church in the administration of the supper."[84]

A serious Catholic criticism was that of the deposed Bishop of

---

[80] *Ibid.*, I, 266.

[81] This comprises pp. 456ff. of *Scripta Anglicana*, published in Basel in 1577. For the liturgical work of Bucer see Jan van de Poll, *Martin Bucer's Liturgical Ideas*.

[82] "Nonnulli eam sibi fingunt superstitionem, ut existiment nephas esse, si quid ex pane et vino communicationis ea peracta supersit, pati id in usum venire vulgarem; quasi pani huic et vino insit per se aliquid nominis aut sancti etiam extra communicationis usum." *Ibid.*, p. 464.

[83] These were, of course, only the major negative criticisms in a book of 28 chapters, which was laudatory of many of Cranmer's innovations. The main points are summarised in Proctor and Frere, *New History*, pp. 73ff.

[84] Letter of March 27, 1550, *Original Letters*, I, 79.

Winchester, Stephen Gardiner. Gardiner accused Cranmer of believing a radical Eucharistic doctrine which Cranmer had concealed in the 1549 Prayer Book. Gardiner, tongue in cheek, took great delight in interpreting the doctrine in the most Catholic way possible.[85] Indeed, if one wished to find the most compelling reason for revising the 1549 book soon after it appeared, it would be to silence Gardiner's criticism of inconsistency on the part of Cranmer. The criticism appeared in *The Explication of the True and Catholic Faith of the Blessed Sacrament of the Altar*, summarised earlier in this chapter. The contention was that in five separate places in the Communion Order a Catholic interpretation could be put on the words and actions. Gardiner strongly believed that the Dominical words, "This is my Body" should be accepted by the faithful as literal truth, and if Christ be the eternal Son of God, the miracle is not only possible it is probable. If the corporal Presence transcends man's wit, then reason must bow to the superiority of faith which transcends it. "I know by faith Christ to be present but the particularity how he is present, more than I am assured he is truly present, and therefore in substance present, I cannot tell. . . ."[86] Holding such views, it was inevitable that Gardiner would either retain the 1549 rite in order to prevent matters becoming worse or hope for a return to the Sarum Use, possibly in a vernacular rite.

The most vehement critics of the 1549 book were the Cornish and Devonian rebels who revolted in the Western Uprising of 1549, and who were finally put down by soldiers. They had been clearly instructed by priests, as was evident in their demands, but the statement of the demands was a significant pointer to what they felt was lacking in the first prayer book. There were 16 demands. Four were crucial—the 4th, 7th, the 9th (which marked their acute sense of loss), and the 8th, which expressed an acute dislike for the new Prayer Book. They missed the adoration of the reserved Sacrament: "(4) Item, we will have the Sacrament hang over the high altar and there to be worshipped as it was wont to be, and they which will not consent we will have them die like heretics against the holy Catholic faith." They missed the colorful ceremonies and images of the familiar medieval rite: "(7) Item,

---

[85] As an exercise in Catholic ingenuity it foreshadowed Newman's interpretation of the Thirty-Nine Articles in Tract 90.

[86] Muller, p. 210. Gardiner held that the 1549 book "is well termed not distant from the Catholic faith." *Writings and Disputations of Thomas Cranmer relative to the Lord's Supper*, p. 92.

we will have holy bread and holy water made every Sunday, palms and ashes at the times appointed and accustomed, images to be set up again in every church and all other ancient, old ceremonies used heretofore by our mother the holy Church." They particularly missed a sense of unity with their recent ancestors, for the repose of whose souls Masses and prayers might be said: "(9) Item, we will have every preacher in his sermon and every priest at his Mass pray specially by name for the souls in purgatory as our forefathers did." All of this was a protest against the barrenness and impersonality of the new Anglican rite, and the sheer didacticism which reached the top of the mind, but did not penetrate through the senses, as ancient rites and profound symbolism had. The rebels felt the new prayer book was only a shadow, or a mimicry, of the ancient Mass, and they said so: "(8) Item, we will not receive the new service because it is like a Christmas game, but we will have our old service of matins, mass, evensong, and procession [the Litany] in Latin, not in English, as it was before. And so we, Cornishmen (whereof certen of us understand no English) utterly refuse this new English."[87] Elsewhere the dissatisfaction among adherents of the older faith was deep, though less articulate.

Some indications of the difficulty the government had in forcing the people to accept the new prayer-book can be seen in the extraordinary measures that were taken to make the prayer-books acceptable, for example, to the universities. Early in May 1549 Ridley was an official visitor to the University of Cambridge, where he led an inquisition into the religious life and services in the colleges. On Sunday, May 26, he ordered six altars to be removed from the chapel,[88] and presided over a public disputation in which two propositions were to be affirmed. The first was that Transubstantiation could not be confirmed by the Scriptures or the writings of the first 10 centuries; the second was that the only oblation in the Lord's Supper was the giving of thanks and commemoration of Christ's death.[89] The purge was effective only for a short time, since during Pentecost in 1550 Bucer wrote to Calvin: "by far the greater number of the Cambridge fellows are either the most bitter papists or profligate epicureans"; many parochial clergy recited the service so "that the people have no more understanding of the mystery of Christ than if the Latin instead of the vulgar tongue

[87] *Troubles connected with the Prayer Book of 1549*, ed. N. Pocock, New Series, 37, pp. 148f.
[88] Charles Henry Cooper, *Annals of Cambridge*, 3 vols., II, 28.
[89] Proctor and Frere, *New History*, pp. 212-13.

were still in use."[90] Nor was the success of the visitor to Oxford, Holbeach, bishop of Lincoln, any greater. Bucer was distressed by the account he read of the Acts of the Disputation at Oxford, which he had received from Peter Martyr. He felt that most people reading the report, "will be entirely of the opinion that you assert that Christ is altogether absent from the Supper and that the only presence is that of his power and spirit."[91] Oxford seemed to be slipping back into the old faith more rapidly than Cambridge. John Stumphius, a disciple of Bullinger, informed him by letter on February 28, 1550 that, "those cruel beasts the Romanists, with which Oxford abounds, are now beginning to triumph over the downfall of our duke [Somerset], the overthrow of our gospel at its last gasp, and the restoration of their darling the mass, as though they had already obtained a complete victory."[92] Stumphius informed Bullinger on November 12, 1550 that, "the Oxford men, who have been hitherto accustomed to do so, are still pertinaciously sticking in the mud of popery."[93] Even the cruelest measures had failed to achieve loyalty to the prayer-book in Oxford. John ab Ulmis reported from Oxford on August 7, 1549 that there had been a rebellion in Oxfordshire; "the Oxfordshire papists are at last reduced to order, many of them having been apprehended, and some gibbeted, and their heads fastened to the walls."[94]

The most dramatic exhibition of the division of conviction that the introduction of the new prayer-book had created was in St. Paul's cathedral in London, where Bishop Edmund Bonner, who disliked the book, postponed celebrating worship according to its requirements in ritual and ceremonial as long as possible. His Dean, William May, however, introduced the new book before he was legally required to do so,[95] and rejoiced in the iconoclasm it encouraged. On March 17, 1549, the second Sunday in Lent, following a sermon by Coverdale, the Dean "commanded the Sacrament at the high altar to be pulled down."[96] Although all chantry

---

[90] *Original Letters*, II, 546-47.
[91] *Scripta Anglicana*, p. 549. So alarmed was Bucer that he urged Peter Martyr, if his conscience permitted, to secure a chance to alter the report to affirm the reality of Christ's presence in the sacrament.
[92] *Original Letters*, II, 464.
[93] *Ibid.*, II, 467-68.
[94] *Ibid.*, 391. These riots, as well as being occasioned in part by religious reaction, were partly caused by dissatisfaction with the enclosures of formerly monastic lands.
[95] Wriothesley, *Chronicle of England*, II, 9, states that "Paul's choir and divers parishes in London began the use after the new books in the beginning of Lent."
[96] *Grey Friars of London Chronicle*, p. 58.

priests had been dismissed, Mass continued to be said in St. Paul's in private chapels; on June 24 the Council, considering it unsuitable that the cathedrals of St. Paul's and St. Peter's, Westminster, should continue to disregard the Act of Uniformity, wrote letters to both Bonner and Thirlby commanding them to discontinue the deviations from the new book.[97] Meanwhile Cranmer entered the fray, and on Sunday, July 21, 1549, set an example of the new simplicity; he "did the office himself in a cope and no vestment, nor mitre, nor cross, but a cross staff was borne afore him, with two priests of Paul's for deacon and subdeacon with albs and tunicles, the dean of Paul's following him in his surplice."[98] Wriothesley records that the archbishop, wearing a silk cap instead of a mitre, "gave the communion himself unto eight persons of the said church."[99] Again, on Saturday, August 10, 1549, Cranmer went to St. Paul's and preached on the Western Uprising, accusing Popish priests of fomenting the revolt. On the same day Bonner was summoned before the Council; injunctions were delivered which accused him of celebrating Mass in the old days in person, but now of seldom or never celebrating Communion according to the new order.[100] The Council compelled Bonner to celebrate the Communion in his cathedral in mid-August, where he "did the office at Paul's both at the procession and the communion, discreetly and sadly."[101] He was also required to preach to a vast auditory at Paul's Cross, where he maintained with all his might the corporeal presence in the Lord's Supper."[102] A doomed man, he made a final public protest on his last free Sunday, October 15, when the preacher at St. Paul's vehemently criticized the doctrines of Transubstantiation and corporal presence, by leaving the church before the sermon was over to show his abhorrence of such heretical doctrines. He was committed to the Marshalsea prison on September 20 and deprived of his bishopric on October 1, 1549.

Considerable opposition to the Prayer Book must have been expected from the beginning. How otherwise could it be undergirded by an Act of Uniformity *insisting* that the new rite be used, with penalties for a failure to use it? Indeed, the prescience of the Council is seen in their prohibition in the Act of "any interludes, plays, songs, rhymes or by other open words in derogation, deprav-

[97] Strype, *Ecclesiastical Memorials*, 3 vols., II, 210-11.
[98] *Grey Friars of London Chronicle*, p. 60.
[99] *Ibid.*, II, 16.   [100] Foxe, *op. cit.*, V, 762.
[101] *Grey Friars of London Chronicle*, p. 62.
[102] Foxe, *Acts and Monuments*, V, 750.

ing or displaying of the same book; or of anything contained therein."[103] It is unfortunate that their collective wisdom did not enable them to see that a compromise book that was neither truly Catholic nor consistently Protestant, instead of pleasing both religious persuasions, would alienate both. The general unpopularity of the prayer-book ensured that its life would be a brief one—only three years.[104]

5. The Second Prayer Book of Edward VI

There were three reasons for the Second Prayer Book of 1552. First, Cranmer had been accused of duplicity (or inconsistency) by Gardiner for holding a more radical doctrine of the Eucharist in his *Defence of the True and Catholick Doctrine of the Sacrament* (1550) than he had dared to express in the First Prayer Book. Cranmer's honour was impugned, and the implication of Gardiner's accusation was obvious: if rite and theology were indeed at odds, let Cranmer create a new rite which would be a clearer mirror of his Eucharistic theology than the vague rite of 1549 was, which, in its obscurity, could still be interpreted in a Catholic way. It is significant that when the 1549 Book came to be revised, at every point where Gardiner had detected residual Catholicism the words were expunged; even the Canon itself was so rearranged as to exclude the remotest possibility of its being interpreted as a propitiatory sacrifice for the living and the dead.

Second, there were several who were urging Cranmer to proceed to produce a genuinely reformed formulary of worship. Peter Martyr's brief, and Martin Bucer's exhaustive, criticisms—as well as the examples of the liturgies devised on the model of Geneva by the superintendents of the "churches of the strangers," John à Lasco and Valérand Pullain, not to mention the urgent appeals of Nicholas Ridley—were pushing Cranmer in the direction of the Swiss.

Third, the more precipitate English members of the Reforming

---

103 Gee and Hardy, *Documents*, p. 362.

104 Yet its usefulness was much greater than might be supposed. It lived on partly in the revised rite of 1552 and in its reappropriation in 1559. It was regarded as a Eucharistic norm by English and Scottish High Churchmen. The Scots produced the mislabeled "Laudian Liturgy" of 1637 (cf. Gordon Donaldson, *The Making of the Scottish Prayer Book of 1637*), and were influenced by the 1549 prayer-book, as was the English revision of 1661 and as were the Non-Jurors from 1688. The Scottish Episcopal Church bishops consecrated Samuel Seabury the first bishop of the Protestant Episcopal Church of the United States on the condition that the American church would use the Scottish Communion Order, which owed more to the Prayer Book of 1549 than to its successor of 1552.

party, notably Ridley who moved from Rochester to London in April 1550, and Hooper who was consecrated bishop of Gloucester on March 8, 1551 after scrupling the vestments, were wholly dissatisfied with the rubrics of the 1549 book. Ridley, first on his own authority and later with the backing of the Council, had demanded that all altars be removed in the churches of his diocese, insisting that altars were appropriate only for the renewal of a sacrifice, whereas holy tables befitted the celebration of the reformed rite of the Lord's Supper. Hooper was also eager to sweep away the "altars of Baal," along with all vestments and ceremonies associated with the old sacrificing priests. There were also great practical difficulties involved in these ceremonial changes, for which the old rubrics were unable to provide any guidance. The people were used to kneeling before the altar at Communion, but what was the appropriate posture in front of a holy table? The priest had previously been told to stand before the middle of the altar fixed at the east end of the choir, but where should he stand at a movable table placed either at the entrance to the chancel or even in the nave of the church?[105] To compound the rubrical problems, it was well known that conservative priests had continued to offer what was in effect an English Mass by retaining the old ceremonial, including many signs and gestures not comprehensively excluded by the rubrics of 1549.[106] Cumulatively these considerations built up a head of reforming pressure which was finally released in 1552. Also, the political fall of Somerset and his replacement by the radical Protestant Northumberland, a man of violence, facilitated the changes.

Now the only difficulty remaining was the government's need to explain its about-face and how a good rite it had authorised was being abolished after only three years' use. In fact, the First Prayer Book was eulogised in the opening paragraph of the second Act of Uniformity, but it was added that the sensuality and ignorance of God on the part of so many who refused to attend parochial worship made a second reformation necessary, a revision of the First Prayer Book to make it, "more earnest and fit to stir Christian people to the true honouring of Almighty God."[107] To this end the

---

[105] Proctor and Frere, *New History*, pp. 69-70.

[106] *Ibid.*, p. 64. Bucer early in 1551 wrote: "Some turn the prescribed form of service into a mere papistical abuse." Cited Gasquet and Bishop, *Edward VI*, p. 269.

[107] Gee and Hardy, *Documents*, pp. 370-71. These reasons do not square with the eulogy of the First Prayer Book in the opening paragraph of the Act, which

First Prayer Book was perused, explained, and perfected in the Second Prayer Book, and an Ordinal was added.

Differences between the two books will be examined first in the Occasional Offices, then in the two Daily Offices, and, finally, in the Communion Order. Almost all that survived from the ancient baptismal office in the 1549 order was abandoned; the exorcism, the white vesture put on the child as a token of innocency after baptism, the anointing, and the triple renunciation and profession were reduced to a single renunciation and profession, while the sign of the cross was made on the child's forehead but no longer also on its breast. The service of Confirmation in 1552 eliminated the bishop's signing each child on the forehead with the cross (which was appropriately foreshadowed by the signation of the cross at baptism, retained in the 1552 order of Baptism and inconsistently excluded from the Order of Confirmation), but keeps, as it must, his laying his hand upon the head of each child. However, an excellent, brief new prayer was introduced: "Defende, O lord, this child with thy heauenly grace, that he may continue thine for euer, and dayly encrease in thy holy spirite more and more, until he come unto thy euerlastyng kyngdom."[108] In the burial service of 1549 allowance was made for "the celebration of Holy Communion when there is a burial of the dead." This was eliminated from the Burial Order of the 1552 book. References to an intermediate state are removed and a lengthy service is reduced without any theological loss. One prayer, thanking God for the deliverance of the deceased person from this world, ends much more felicitously in the revised version: "beseching thee, that it maye please thee of thy gracious goodnesse, shortely to accomplyssh the noumbre of thyne electe, and to hasten thy kingdome, that we with this our brother, and al other departed in the true faith of thy holy name, maye haue our perfect consummacion and blisse, both in body and soule, in thy eternal and euerlastyng glory." The main differences are the elimination of Catholic vestiges in doctrine or practice, the reduction of ceremonies, the abbreviation of unduly long forms, and the improvement of language for the sake of clarity. The additions are generally of a didactic or hortatory character.

---

speaks of it as "a very godly order" in the vernacular, "agreeable to the word of God and the primitive Church, very comfortable to all good people desiring to live in Christian conversation . . ." (*ibid.*, p. 369). If it was so admirable what was the need of revision?

[108] All references to the 1552 prayer-book are to the Grafton impression of August 1552 (S.T.C. 16283) in the Huntington Library. This reference is to fol. 112.

Matins and evensong were retitled "morning prayer" and "evening prayer." Each of the Daily Offices had a lengthy introduction of a pentitential character prefixed to it, that consisted of a selection of 12 scriptural sentences, a short exhortation to the confession of sins, which gave an admirable summary of the reasons for worship,[109] after which followed the general confession and absolution. While it might be argued that the introduction to the Daily Offices was excessively introspective, especially when some of the exaggerated affirmations of the General Confession are remembered, and too pentitential for all times and seasons. Nonetheless, it served as a useful preparation for worship and as a transition from the street to the sanctuary.

The most radical revision and reconstruction was undertaken on the Order for the "Lordes Supper or Holye Communion." It is noted that there was a change in the title and that "commonly called the Masse" was eliminated. The term "offertory" was removed, and the terms "table" and "Lord's table" were substituted for "altar." The rubrics were radically changed. The curious and complicated rubric referring to the provision of bread and wine by families, together with a communicant from each family in turn, was eliminated. No direction was given as to the time when the bread and wine should be placed on the table. Ordinary bread was to be used instead of the unleavened bread. The rubric directing the minister to take sufficient bread and wine for the number of communicants disappeared. Bucer had called attention to the way many priests inclined their heads over the bread and wine during the prayer of consecration as if they longed for a mutation of them, and particularly asked "that the little black crosses and the rubric about taking the bread and wine into the hands should be removed from the book as well as the prayer for the blessing and sanctifying the bread and wine."[110] All these requests were fulfilled. The previous provision that a priest in a cope should on Wednesdays and Fridays say the first part of the Communion office was abolished. It was now directed that the table for the Communion stand in the body of the church and that the minister place himself at the north end of the table. Finally, a long rubric on kneeling, thereafter commonly called the "black rubric," was issued as a royal proclamation through the insistence of John Knox after some copies of the

<hr/>

[109] These were: "humbly to knowledge our synnes before God," "to rendre thanckes . . . ," "to sette furth hys moste worthie praise, to heare his mose holy worde, and to aske those thynges whiche be requisite and necessarie, as wel for the bodye as the soule." 1552 Prayer Book, fol. 1 verso.
[110] *Censura*, p. 472.

revised book had already been printed.[111] The rubric denied that kneeling at the Sacrament was not to be interpreted as if "any adoracion is doone, or oughte to bee doone, eyther unto the Sacramentall bread or wyne there bodily receyued, or unto anye reall and essencial presence there beeyng of Christ's naturall fleshe and bloude."[112] All vestments were abolished except for the episcopal rochet and surplice.

It should also be noted that "low Mass" became the norm in the revision, as singing was discouraged. The Introit, the response to the Announcement of the Gospel, the *Osanna* and *Benedictus*, the "Peace of the Lord" and "Christ our Paschal Lamb," the *Agnus Dei*, and the "Postcommunion" were omitted. All that was left that was required to be sung was the Epistle, Gospel, and *Gloria in excelsis*. Proof that "low Mass" was the new model can be seen in the fact that the Epistle and Gospel were assigned to the priest alone.

The most significant changes in the Communion rite, however, were changes in the order of the items, as may be seen in the diagram on Communion order, which compares the orders of 1549 and 1552. Items whose order was changed are italicised.

Communion Order

1549	1552
Sermon	Sermon
	Offertory
	Intercession for living
Exhortation	Exhortation
	Penitential preparation
Offertory	
Preface and Sanctus	Preface and Sanctus
	Prayer of Humble Access
Intercession for living	
Consecration	Consecration
	Communion
Oblation	
Lord's Prayer	Lord's Prayer
Penitential preparation	
Prayer of Humble Access	
Communion	
Thanksgiving	*Oblation*, or Thanksgiving

[111] For its history see R. W. Dixon, *History of the Church of England*, 6 vols., III, 475ff.

[112] 1552 Prayer Book, fol. 102.

In comparing the first with the second Communion office it is quite obvious that the order of 1549 fairly closely resembled the shape and order of the Sarum Use of the medieval Mass, while the 1552 order was so altered as to bear no shadow of a semblance with the order of the Mass. It seems as if Cranmer was determined to eliminate all those parts of the 1549 rite that Gardiner had claimed were patently orthodox Catholic. The intercession was removed from any connection with the Consecration, and the Prayer for the Dead was omitted; no one could argue this was intended as a propitiatory sacrifice for the living and the dead.

The prayer for the sanctification of the elements, which Gardiner had approved as orthodox, was replaced by a form designed, as Bucer wished, to avoid the supposedly superstitious notion of consecration. But in the attempt to avoid Romanism it had unwittingly become more Romanist than ever, for it was perpetuating the medieval scholastic belief that consecration came through the repetition of the Dominical words in the Institution narrative! The Prayer of Humble Access ("We do not presume to come to this thy table"), which previously came immediately before the Consecration (though the term was not used in 1549 or 1552), where it could not be referred to the Eucharistic elements and from the phrase "in these holy misteries," was expunged. The *Benedictus* was deleted from the *Sanctus* presumably because it might imply the most realistic corporal descent into the elements. The very brief *anamnesis* of 1549 was entirely abolished, and sadly abbreviated remnants at the end of the 1549 Canon were reworded and made an alternative to the prayer of thanksgiving which was also revised slightly. Even the Lord's Prayer was placed after the Communion, possibly lest no one should misinterpret the petition for daily bread as having any Eucharistic reference. The last criticism of Gardiner was met with the formulas of administration, which were highly subjective and strongly Memorialist. The words at the delivery of the wine, for example, were: "Drynke this in remembraunce that Christes bloude was shedde for the, and be thankful."[113] There was no direction for the preparation of the elements, presumably because this might lead to the reintroduction of the old conception of offertory which Cranmer was eager to eliminate from the rite. Even more disturbing was the absence of directions for fraction and libation. Indeed, the fear of superstition had itself become a superstition with Cranmer, so that almost every

[113] *Ibid.*, fol. 100.

element of the numinous was excluded from the rite. Typical, for instance, was the insistence that ordinary bread rather than wafers was to be used for the Communion, and that the minister might take home whatever was left over.

It is easier to criticise this rite than to defend it. It was anathema to High Churchmen, while it was perhaps excessively eulogised by Low Churchmen and ultra-Protestant evangelicals in the Anglican fold. The rite tried to model itself more on the New Testament account of the Last Supper than on ancient traditions. Also, what is often called an iconoclastic oddity, its transference of the *Gloria in excelsis* as a climactic post-Communion thanksgiving may have been a stroke of genius.[114] Since in a true Eucharist in which the people of God really participate in the eschatological banquet, the act of communion is the climax; all that usually follows afterwards is necessarily an anti-climax. Cranmer, however, prevented such a lowering of the spirits by this lyrical act of superabundant praise. Finally, if Cranmer's theological outlook, with its increasing Protestantism, is accepted, the Prayer Book of 1552 may indeed be—in words of Gregory Dix—"not a disordered attempt at a catholic rite, but the only effective attempt ever made to give liturgical expression to the doctrine of 'justification by faith alone.' "[115] Cranmer had in fact restructured the Offertory to conform with his view that, "the humble confession of all penitent hearts, their knowledging of Christ's benefits, their thanksgiving for the same, their faith and consolation in Christ, their humble submission . . . to God's will and commandments, is a sacrifice of laud and praise."[116] This is, however, far from being a typical assessment of the 1552 rite. More common is the view of Proctor and Frere, that it was an "ill-starred book" and that with it "English religion reached its low water mark."[117]

Some light is thrown on the character of the 1552 Book of Common Prayer by a consideration of the sources Cranmer and his unknown collaborators used. Since the book was a revision, naturally enough the First Prayer Book was the primary source. Bucer's careful and detailed *Censura* was another important source; while his conservative theological ideas proved unacceptable, about two-

---

[114] See *Walter Howard Frere: A Collection of his papers on Liturgical and Historical Subjects*, ed. J. H. Arnold and E.G.P. Wyatt, Alcuin Club Collections, no. 35, p. 193.

[115] *Shape of Liturgy*, p. 672.

[116] *Defence of the True and Catholick Doctrine of the Sacrament*, in *The Remains of Thomas Cranmer*, ed. Henry Jenkyns, 4 vols., ii, 459.

[117] *New History*, p. 85.

thirds of his objections were satisfied.[118] His friend, Peter Martyr, a former Augustinian canon and Regius Professor of Divinity at Oxford, made an independent critique of the 1549 book on the basis of a partial translation of it into Latin by Cheke; but when he saw the fuller work of Bucer he concurred with all of Bucer's criticism. He did, however, object to the rubric permitting the reservation of the Sacrament for communicating the sick.[119] All Gardiner's commendations of places in the rite where a Catholic interpretation of the 1549 rite was possible were taken as criticisms, and, as we have seen, were entirely altered by rephrasing or repositioning.

Another possible source of influence must be considered, where influence is hard to detect. This influence, if proven, demonstrates that Calvinist rather than Lutheran rites were looked to for models of the 1552 book. The first was the *Liturgia Sacra seu Ritus Ministerii in ecclesia peregrinorum profugorum propter Euangelium Christi Argentinae. Adiecta est ad finem breuis Apologia pro hac Liturgia per Valerandum Pollanum Flandrum*, published in London on February 23, 1551. This is the liturgy that was used by the French and Flemish congregation which had migrated from Strassburg and had been given the use of the buildings of Glastonbury Abbey by the Protector, Somerset. The interest of the liturgy is that it had connections with both Bucer and Calvin. Pullain had, in fact, succeeded Calvin as the minister (with one intervening link) of the *ecclesiola gallicana*, and Calvin had translated and paraphrased for the French reformed congregation in Strassburg the service which Bucer had developed for the German Reformed congregation in Strassburg.[120] Considering that the Archbishop of Canterbury had to oversee the "churches of the strangers" in the realm and that he was deeply involved in liturgical revision, it would be surprising if Cranmer had not scrutinised the liturgy more out of delight than from a sense of duty.

Proctor and Frere conclude that the similarity between the 1552 Prayer Book and the *Liturgia Sacra* (and also à Lasco's *Forma ac Ratio tota ecclesiastici Ministerii in peregrinorum, potissimum uero Germanorum Ecclesia*, a rite prepared for the Emden refugees worshipping in Austin Friars in London, and which originated

118 Brightman, *English Rite*, book I, p. cxlv.
119 *Ibid.*, p. cxliv.
120 For the history of this rite see A. Erichson, *Die Calvinische und die Alt-strassburgische Gottesdienstordnung*; and E. Doumergue, *Jean Calvin*, 7 vols., II, 494f.

with Farel, the first Genevan reformer and colleague of Calvin)[121] is collateral, not lineal.[122] But this assessment is difficult to maintain if one examines the Latin of Pullain and the English of Cranmer side by side in additions that were made for the first time in the rite of 1552.

The real novelty in the Daily Offices was the penitential introduction that was added. It is precisely in the general confession and in the following absolution that the interdependence of ideas and words can be shown. Pullain's *tuas leges sanctissimas assidue transgredimur* was exactly paralleled by "we haue offended against thy holy lawes"; Pullain's *agnoscimus . . . peccatores esse nos miseros* becomes "us miserable offendours." Furthermore, it may well be that the inclusion of the recitation by the minister of the Decalogue in the Communion Office, to which the people made a response, originated in Pullain's *Liturgia Sacra*, or in another Swiss Church Order of Calvinist provenance through the example of Bishop Hooper of Gloucester and Worcester. What *is* clear is that the final petition of the people, which includes the *Kyrie* and reads, "Lorde, haue mercye upō us, & write al these thy lawes in our hartes, we beseche the," could easily have come from Pullain's *dignare cordibus nostris eam ito tuo spirito inscribere.*[123] It is possible that the words in the absolution in the Daily Offices, referring to God—"whiche desireth not the deathe of a synner, but rather that he maye tourne from his wickednesse, and liue"—find their origin in à Lasco's *Forma ac Ratio*, where there appears *neque amplius velis mortem peccatoris sed potius convertatur et vivat.*[124] *The Pious Consultation* of Archbishop Hermann von Wied of Cologne also continued to influence the revision.

If the major influences on the 1549 rite were Roman and Lutheran, the chief impact on the 1552 rites was Reformed. This is not a revolutionary conclusion, because the iconoclastic nature of the revision is evident. What is not commonly understood, however, is that the Eucharistic theology that dominates the revised rite may well be not low Zwingli but high Calvin.[125] If the Eastern

121 The Farel directory is reprinted in J. G. Baum, *Première Liturgie des églises reformées de France de l'an 1533.*

122 Proctor and Frere, *New History*, p. 90.

123 *Ibid.*, pp. 86-88, where the relevant passages from the rite of Pullain are cited.

124 *Ibid.*, pp. 89-90, where the relevant material from the à Lasco rite can be found.

125 Among Anglicans C. W. Dugmore, *The Mass and the English Reformers*, and Peter Brooks, *Thomas Cranmer's Doctrine of the Eucharist*, are the only historians known to me who stress that Cranmer's doctrine of the Eucharist was a

invocation of the Holy Spirit was formally excluded—because Gardiner had argued that this clearly implied a mutation of the elements—it was clearly understood that the Holy Spirit made the spiritual presence of Christ in the sacrament possible, with all His benefits. It was equally understood that the same Holy Spirit created and maintained the gift of faith and united in the *vinculum caritatis* the Head and the members of the church. With such an august belief, the ceremonial could afford, like the rite itself, to be simple. It is only simple for those who have interpreted it through a reductionistic Memorialism, in which Christ is remembered as a martyr, not as the living and transforming Lord. We do well to remember that Cranmer died for an affirmation, not a negation. Although his rite of 1552 was only to endure for a year before the Marian Reaction swept it away, it was revived in 1559 by Queen Elizabeth. It, or its predecessor, has remained the sum and substance of Anglican worship ever since.

6. *The Elizabethan Prayer Book of 1559*

The third and least creative of the Tudor Books of Common Prayer was the Elizabethan Prayer Book of 1559, yet it was the one that lasted the longest. In fact, it was only modified in the most minimal way until the revision at the Restoration in 1661. It is in the Elizabethan setting that we have the best opportunity to see its long-term qualities and its inherent weaknesses.

On the accession of Elizabeth November 17, 1558, the people were readier to accept Protestantism than they had been in the days of Henry VIII or Edward VI. It was expected that the prayer-book would be restored and the Sarum Use under Mary set aside. On December 27 the queen issued a proclamation prohibiting any change in the existing order, "until consultation may be had by Parliament." As an interim arrangement an exception was made for reading the Gospel and the Epistle of the day in English, and for the recitation of the English Litany (the only vernacular form of worship permitted under Queen Mary) as said in the Queen's Chapel. The clause for deliverance, "from the tyranny of the bishop of Rome and his detestable enormities," was omitted from the printed edition of the Litany issued January 1, 1559. According

---

high Reformed doctrine, rather than a low Reformed one. These views, however, have been challenged by T. M. Parker in the *Journal of Theological Studies*, vol. 12, pt. 1 (1961) and C. C. Richardson in the same journal, vol. 16, pt. 2 (1965).

to E. C. Ratcliff,[126] there is some reason to think that two proposals were made for appropriate rites, the first for the restoration of the rite of 1549, and the second for the one similar to that used by the congregation of English emigrants in Frankfort, where a famous liturgical war[127] had broken out among the Marian exiles between the Coxians, who supported the 1552 prayer-book, and the Knoxians, who wanted a more Genevan form.

The prayer-book eventually established by the Elizabethan Act of Uniformity of April 1559 was the Second Prayer Book of 1552, but with three specific important changes. An additional table of lessons for Sundays, festivals, and holy days was provided. In the Eucharist an attempt at comprehension and inclusion of the differing viewpoints was made, which might be confusing theologically, but which was admirable in its sense of charity. The objective and subjective emphases of the Communion were combined by amalgamating the forms of administration of 1549 and 1552 so that at the delivery of the consecrated bread the words are: "The body of our Lord Jesus Christ which was geuen for thee, preserue thy body and soule into euerlasting life: and take and eate this, in remembraunce that Christ died for thee, and feede on hym in thy hearte by fayth, wyth thankes geuynge." The Litany's form was not that of 1552, but was the form used in the Royal Chapel, in which the anti-Papal clause was removed. The infamous black rubric was excised, and, although this had not been hinted at in the Act of Uniformity when the 1559 book appeared, it contained a new rubric ordering the use of the vesture worn in "the seconde yeare of the reyne of King Edward the VI." This was intended to cover the use of the Mass vestments. In practice, however, the usual vesture was the surplice, over which was worn a cope in the Queen's Chapel and in some cathedral and collegiate churches, in accordance with the rubric of 1549.

Several unauthorised changes in worship were made during Elizabeth's reign.[128] There was no uniform text of the Book of

---

[126] *The booke of common prayer of the Churche of England: its making and versions*, p. 19.

[127] *A Brief Discourse of the Troubles at Frankfort, 1554-1558*, ed. E. Arber.

[128] There was an inconsistency in requirements. The queen's *Injunctions* demanded the use of wafer bread at the Communion, "of the same fineness and fashion round, though somewhat bigger in compass and thickness, as the usual bread and wafer, heretofore named singing cakes which served for the use of the private Mass." (Frere and Kennedy, *Visitation Articles*, III, 28.) On the other hand, the prayer-book directed the use of "bread such as is usual to be eaten." *Liturgies and Occasional Forms of Prayer set forth in the Reign of Queen Elizabeth*, ed. W. K. Clay, p. 198.

Common Prayer issued. The texts printed by Jugge and Cawood, on the one hand, and Grafton, on the other, were not themselves uniform, nor were they consistent with each other. One major unauthorised change was made in the collect for St. Mark's Day (involving an inversion of the order). The new ritual "settlement" combined the Act of Uniformity, the new prayer-book, and the Royal *Injunctions* of 1559 (which repeated 26 of the 38 Injunctions of 1547, together with 29 new ones). But even these regulations were insufficient to bring order to the general ceremonial chaos, so that Archbishop Parker's *Advertisements* were issued for his own province without the formal royal consent.[129] The queen, only 18 months after coming to the throne, ordered Parker and his advisors to prepare a new table of lessons and Kalendar. In 1563 the Convocation of Canterbury sanctioned the production of a further Book of Homilies which included 20 sermons, to which another was added in 1571 by Convocation, entitled "Against Disobedience and wilful Rebellion." The latter was occasioned by the uprising in the north in November-December 1569. Haphazard as the arrangement may seem, Elizabeth, in fact, jealously preserved her prerogatives as "Supreme Governor" of the Church of England and used her bishops not only as licensers of all books printed but also to keep a tight rein on their clergy. The clergy were expected to be the maintainers of loyalty to the throne, the makers of morale, and the upholders of "degree" and stability in society as much as judges were. Although there was a shortage of preachers for the first 20 years of her reign, those who preached were carefully licensed, and muzzled if they dared to speak out of turn.[130] The

129 The precise authority for the Ornaments Rubric and its relation to the "Advertisements" of 1566 is still a controversial question. The rubric is the ruling placed in the 1559 prayer-book at the beginning of the Order for Morning and Evening Prayer, which requires that the Ornaments of the Church and Ministers shall be those in use "by the authority of the Parliament in the second year of the reign of King Edward VI." The Act of Uniformity of 1559, however, to which the prayer-book was attached, added the qualification that such use should continue "until other order shall be therein taken by the authority of the Queen's Majesty with the advice of her Commissioners appointed and authorized under the great seal of England for causes ecclesiastical, or of the Metropolitan of this realm." The only explicit statement about the ornaments is in the Prayer Book of 1549, but it was produced in the third (not the second) year of Edward VI. The Judicial Committee of the Privy Council have on two occasions ruled that Parker's "Advertisements" were the "other order" foreshadowed in the Act. This ruling has been widely disputed because the "Advertisements" had not statutory authority and because they could not be regarded as overriding the later reenactment of the rubric by Parliament in 1604 and 1662. See J. T. Micklethwaite, *The Ornaments of the Rubric.*
130 W.P.M. Kennedy, in *Elizabethan Episcopal Administration*, 3 vols., I, cvi, defines preaching as "a ministerial activity which was the most uniformly controlled of all, as it was made the active handmaiden of Elizabethan state-craft."

Liturgy was for Queen Elizabeth the way to maintain religious uniformity in her realm, a political mechanism for the manufacturing of loyalty. But she was pitiless toward those who marched to the beat of a different drummer, or those whose consciences would not allow them to worship according to the Book of Common Prayer, whether Catholic or Puritan.[131]

When the novelty of a vernacular rite had worn off, and the reading of printed homilies rather than the preaching of unpredictable sermons was the order on Sunday (as it was for the first 20 years of Elizabeth's reign), and now that the Anglican services were set off by so little ceremony and symbolism, church attendance must have been excessively dull, especially with the lack of other interesting activities on Sundays. It should be remembered that a large number of people were there by compulsion, not by choice, that the services were long and routine, and that most churches were crowded. In such circumstances it is not surprising that many should find the worship boring, or that misbehaviour and shouting, not to mention scuffling, were frequent interruptions of the service. Men talked, laughed, cleared their throats, slept and snored, and refused to stand up for the Creed or the Gospel, to turn to the East or to bow at the name of Jesus. The accusation against William Hills of the parish of Holton St. Mary in Suffolk, that he "used in tyme of devine service open and lowde speeches to the disturbance of the minister"[132] seems characteristic of the time.

Females were equally adept at interrupting worship, whether it was Mary Knights of Bythburgh in Suffolk, who "do bring maistifs to church into the stoole with her wherebie the parishioners cannot haue their seats," or Jane Buckenham, who entered church one Sunday and abused the parson, "calling him blacke sutty mowthed knave, to the greate disgrace of his callinge."[133] There is one recorded case in which two women brawled in church over their right to sit in a certain seat. The Southwell Act Books recorded for February 16, 1587 that a woman of Nottinghamshire, Joan Halome, had alleged, "that the . . . said Luce Wentworth did give the occasion of making the disturbance in the churches, for that she would not kepe her place, where she was first sett; but came ouer her backe and marred her apparel in the stall, where she was

131 E. C. Ratcliff, Book of Common Prayer, p. 19: "The Elizabethan Prayer Book was equally obnoxious to Papists and to Puritans."
132 A. Tindal Hart, The Man in the Pew, 1558-1660, p. 130.
133 Ibid.

sett. Wherappon the saide Johanne Halome did pricke the said Lucie with a pynne."[134]

We can only suppose that it was sheer monotony that drove some of the less patient members of the captive congregations to high jinks, such as those described in the visitation articles of the Elizabethan bishops. W.P.M. Kennedy does not exaggerate when he says, from a thorough examination of episcopal injunctions and visitation articles of the Elizabethan period, that "there was a disposition . . . to treat the Sunday services as fit subject for merriment, and to turn the parish church into either a parochial club, or a controversial meeting."[135] In the parishes of the huge diocese of Coventry and Lichfield the churchwardens, along with the clergy, were instructed to choose four to eight well-built men in each parish who would take an oath to maintain order during the services. These "orderlies" paced the aisles during services with a monitory white wand in their hands. If the misbehaving were impenitent or refractory, the "two honestest" of the orderlies frogmarched them to the chancel door and made them stand facing the people for 15 minutes.[136]

It doesn't seem to fit the romantic picture of "Good Queen Bess" and her golden days, but the village churches, as is apparent from the numerous visitation returns from the countryside, were in a ruinous condition. As a consequence of destruction and neglect, the churches had "damp green walls, rotting earth floors, and gaping windows."[137] The country people, always conservative, were tired of the disruptive religious changes; their spirits had been shaken by the iconoclasm of the returning Genevan exiles, the lack of energy or money on the part of clergy and churchwardens, and the sheer greed or indifference of the lay impropriators and farmers of the rectory. One did not expect color or pageantry in Elizabethan Anglicanism, for, as William Harrison wrote, "all images, shrines, tabernacles, roodlofts, and monuments of idolatrie are removed, taken down and defaced; onelie the stories in glasse windowes excepted, which for want of sufficient store of new stuffe . . . are not altogither abolished in most places at once, but by little and little suffered to decaie, that white glasse may be provided and set

---

[134] *Ibid.*, p. 131.
[135] *English Life under Queen Elizabeth*, p. 128.
[136] *Calendar of State Papers, Domestic*, XXXVI, 41 (13).
[137] A. Tindal Hart, *The Country Clergy in Elizabethan and Stuart Times, 1558-1660*, p. 40.

up in their roomes."[138] It soon reached the point where one expected a bare simplicity but was grateful to find that the church walls were newly whitewashed and that the earth floor had recently been covered with straw or rushes to make kneeling easier.

An interesting description of English religion in general and worship in particular is given in Harrison's *Description of England*, written from the standpoint of a loyal Anglican with some Puritan, but no Separatist, tendencies. Acknowledging that there was a substantial lack of pastors, so that every parish could not have one, there was also, he affirmed, a dearth of preaching, so that besides the four sermons per annum there were homilies read in church, "by the curate of meane vnderstanding."[139] His account of Sunday worship (daily offices and Communion) is worth citing in its entirety:

> And after a certeine number of psalmes read, which are limited according to the dates of the month, for morning and euening praier, we haue two lessons, wherof the first is taken out of the old testament, the second out of the new; and of these latter, that in the morning is out of the gospels, the other in the after-noone, out of some of the epistles. After morning praier also we haue the letanie and suffrages, an inuocation in mine opinion not deuised without the great assistance of the spirit of God, although manie curious minsicke persons vtterlie condemne it as superstitious, and sauoring of coniuration and sorcerie.
>
> This being doone, we proceed vnto the communion, if anie communicants be to receiue the eucharist; if not we read the decalog, epistle and gospell, with the Nicene creed (of some in derision called the drie communion), and then proceed vnto an homilie or sermon, which hath a psalme before and after it, and finallie vnto the baptisme of such infants as on euerie sabaoth daye (if occasion so require) are brought vnto the churches; and thus is the forenoone bestowed. In the afternoone likewise we meet againe, and after the psalmes and lessons ended, we haue commonlie a sermon, or at the leastwise our youth catechised by the space of an houre. And thus do we spend the sabaoth daie in good and godlie exercises,

---

[138] *Harrison's Description of England in Shakespere's Youth*, ed. F. J. Furnivall, 2 vols., I, 32.
[139] *Ibid.*, p. 28.

all doone in the vulgar tong, that each one present may heare and vnderstand the same. . . .[140]

This picture of Elizabethan worship in 1577 can be filled out from the information in the rubrics of the 1559 prayer-book and the norm that is suggested by consideration of the deviations from it recorded in episcopal visitations and injunctions. Harrison only hinted at two of the defects in the worship of his time.

The clergy were under a daily obligation to say the services of morning and evening prayer, "in a loud voice" and "in the accustomed place of the church, chapel, or chauncel, except it be otherwise determined by the ordinary of the place." When the services were public the clergyman was to ring a bell to give the well-disposed good warning. The Litany was to be said on Sundays, Wednesdays, and Fridays. Those who intended to communicate at the Lord's Supper or Holy Communion were requested to give their names to the curate, whose duty it was to warn notoriously wicked persons and forbid them to approach the Lord's table until they had openly declared their repentance and amendment, and to reject those unreconciled in a quarrel until penitent and prepared to return to amity.

The Lord's table, covered with a white linen cloth at times of Communion, was "to stand in the body of the church, or in the chancel where morning and evening prayer be appointed to be said." The "priest standing on the north side" was to begin the rite. After the offertory it was the duty of the churchwardens to gather "the devotion of the people and put the same into the poor men's box." On the offering days—Christmas, Easter, St. John the Baptist, and St. Michael—every man and woman was ordered to pay the accustomed dues. After the Nicene Creed the priest announced the holy days and fasting days of the ensuing week. On holy days, when there was no Communion, the "ante-communion" service, with the reading of a homily, was prescribed. No celebration of Holy Communion was to take place "except there be a good number to communicate with the priest according to his discretion." His discretion was restricted by the rubric "and if there be not above twenty persons in the parish of discretion to receive the Communion, yet there shall be no Communion except four, or three at the least communicate with the priest." In cathedrals or collegiate churches weekly Communion was the minimum laid down for priests and

140 *Ibid.*, pp. 29-30.

216

deacons. The bread had to be "such as is usual to be eaten at table with other meats, but the best and purest white bread that conveniently may be gotten." The bread and wine were to be paid for by the parish through the churchwardens. As Communion attendance became larger the old chalices (in which it was usual for only the priest to communicate) were ordered to be melted down to produce Communion cups with wider and deeper bowls, looking like inverted bells.[141] Communicants knelt to receive at the holy table or altar, and every parishioner was ordered to receive Communion at least three times a year, of which Easter was one.

Baptism was to be administered on Sundays and holy days, immediately following the last lesson at morning or evening prayer. Provision was to be made for godfathers and godmothers. The child was to be dipped in the font (or, if it were weak, water was to be poured on it[142]) and the forehead was to be signed with the sign of the cross. In special cases private Baptism could be arranged.

Confirmation according to the approved rite was to be administered to those who could repeat in English the Creed, the Lord's Prayer, and the Ten Commandments, and who knew the short Catechism which was printed with the Order for Confirmation in the Elizabethan Prayer Book.

Banns were to be published on three consecutive Sundays or holy days; provision was made for Communion at a marriage. An order was provided for the visitation of the sick, with an opportunity for private Confession and Absolution. Further, if opportunity and the condition of the sick person allowed, and there were sufficient friends and neighbours to join in, a private celebration of Holy Communion was possible. A burial service (in which, it may be noted, it became increasingly common to include a sermon,[143] and not only for distinguished deceased persons), a form for the Churching of Women, and a Commination against Sinners (to be used at different times of the year) concluded the

---

[141] The contrast of the old and the new is perfectly exemplified in the marginal illustration to *A Booke of Christian Prayers*, printed by John Day in 1578 (see illustrations, Pl. 11). Note at the bottom of the picture the deep, bell-shaped Protestant communion cup and the "idolatrous" Catholic chalice depicted in the lefthand margin, along with rosary beads, candlestick, monstrance, episcopal rochet and crucifix.

[142] Another illustration reproduced from *A Booke of Christian Prayers* depicts an Elizabethan baptismal ceremony. The child, wrapped in swaddling bands, is presumably weak, since the clergyman appears to have poured water over it, rather than dipped it.

[143] A. F. Herr, *The Elizabethan Sermon, A Survey and Bibliography*, p. 46.

different public services of the Church of England in Elizabethan days.[144]

Harrison's description and its amplification fail, however, to do more than mention two problems of Elizabethan worship. The first was insufficient amount of preaching. Many country parishes must have gone month after month hearing only a homily; four sermons a year, often the maximum, was a spare diet of the word of God. As Elizabeth's reign went into its third decade the number of preachers increased considerably, until by the end of the reign there were many good preachers and two outstanding ones—the brilliant Lancelot Andrewes and the popular Henry Smith. The increase in the number of sermons was undoubtedly stimulated by Puritan criticism of nonpreaching parsons as no more than "dumbe dogges," the admirable examples Puritan ministers provided of faithful expository preaching, and the excellent Puritan training schools for preaching ("prophesyings"[145]).

If there ever was an ideal of a weekly Communion, it soon lapsed. The second weakness of Anglican worship during this period was infrequent celebration of Communion. The most frequent directive from the bishops is that there must be "sufficient number of celebrations for the parishioners to receive three times in the year at the least—Easter being one."[146] In larger city or county town parishes the ideal seems to have been that of a monthly communion.[147] But there was a great abyss between the ideal and the real.

We have seen that there was a Puritan supplementation of Anglican worship by further religious exercises such as "prophesyings." Similarly there was an undoubted Catholic supplementation of worship, either by priests with strong inclinations toward older ways or by pious laity who had not ceased to believe in the religious practices of their youth. The clergy must frequently have aided their parishioners in maintaining the old devotional practises.

---

[144] For much of the material of the summary of worship according to the Book of Common Prayer in the previous five paragraphs, I am indebted to Kennedy, *Elizabethan Episcopal Administration*, I, xxxviii-xl.

[145] "Prophesyings" were meetings of ministers for the study, discussion, and exposition of scripture. The meetings served as a means of educating the clergy homiletically and for disseminating Puritan views. They originated in the Swiss and Rhenish cities (where the Marian Exiles learned their value). (See Leonard J. Trinterud, "The Origins of Puritanism," in *Church History*, XX [1951], no. 1, p. 46. They were held in England as early as 1564 and became very popular in the next decade.

[146] See Frere and Kennedy, *Visitation Articles and Injunctions*, III, 275, 307, 337.

[147] *Ibid.*, III, 167.

Some priests paid the holy table the same respect as they had the older altar, elevated the Host at the sacrament, and rang the passing bell more than was permitted. They would also be the priests who affected not to notice parishioners praying in the Lady Chapel, and who readily offered absolution to people whose consciences were not stilled by the all-too-convenient General Confession they were taught to repeat in the Book of Common Prayer.

The episcopal articles and presentment lists give us a glimpse of countrymen who hoarded the old vessels, images, and vestments in their cottages when the parish authorities wished to dispose of them, because they evoked precious memories and might even be needed again.[148] There was a deep, innate conservatism that attached people to religious objects and holy associations which could not be erased by official Protestant decree. P. M. Dawley has described these survivals: "People stole into the churches at night to pray, occasionally burning a candle stub on the feasts of Our Lady and the saints; they paused before the ruined churchyard crosses to utter the familiar intercessions. During the services they fingered their beads and could not keep their hands from the sign of the cross or penitent 'knockings' upon the breast. Through many a darkened village on the eve of All Souls' the bells of the parish tolled the forbidden remembrances of the departed, and by the time the churchwardens arrived at the church they found either the belfry ropes stilled, or a group gathered there too formidable to restrain."[149]

7. An Evaluation of Anglican Worship

There is much to criticise and admire in the Book of Common Prayer, though many of the criticisms are directed less to the book itself than the circumstances of its introduction.

The three Tudor Books of Common Prayer were a coerced formulary of worship intended for "soul control"—that is, to force the parson and people in a direction predetermined by their sovereign and Council. The blasphemous fact is (and it held wherever the doctrine of *cuius regio, ejus religio* was accepted) that Almighty God was to be honoured by a form of worship reinforced by the strongest temporal penalties. Every minister declining to use the

---

[148] One Christopher Smyth of Pateley Bridge in Yorkshire was ordered to make a declaration penance for selling a New Testament in the Tyndale translation for 10 shillings, to be paid "whan masse shalbe said within this realme." *Tudor Parish Documents of the Diocese of York*, ed. J. S. Purvis, p. 150.

[149] *John Whitgift and the English Reformation*, p. 119.

Prayer Book or using other forms of worship was subject to an ascending series of punishments, ending, for the third offense, in deprivation and life imprisonment. Any lay person depraving the book or obstructing the use of it was subject to heavy penalties. Absence from church on Sundays or holy days was punishable by a fine of 12 pence for each offense, the sum to be levied by the churchwardens for the use of the poor of the parish. The later Elizabethan Acts against Recusants and Puritans contained even stiffer penalties, including imprisonment and in extreme cases, death. Among those who used the Book of Common Prayer we are then to assume in every last parish a number of liturgical prisoners to whom the Prayer Book was anathema, either because its doctrine and ceremonies were insufficiently Catholic, or, in the case of the Puritans, for whom it was still too Catholic in retaining certain ceremonies and vestments. The Catholics believed that a memorial to Christ had replaced a sacrifice of His Body for the living and the dead, hence that the Anglican Eucharist was but the shadow of a shade. The Puritans deplored the retention in the prayer-book of such Catholic ceremonies as the signing of the cross in Baptism, the bowing at the name of Jesus, the turning to the East for the Creed, the kneeling for Communion, and the use of the ring in marriage. They also found the vestments as well as the retention of the episcopal form of government, supported by ecclesiastical courts, too great a deviation from primitive Christianity as they understood it. Moreover, Puritans and Separatists abhorred the very notion of a set form of liturgy as a "quenching of the Spirit," a denial of the Spirit-induced spontaneity of the New Covenant. Catholics and Puritans sensed in Anglican worship a lack of intensity and deep conviction. Neither group could think of bishops as *pastores pastorum*, but only as ecclesiastical superintendents of police, as the *Marprelate Tracts* so clearly show.[150] Compliant worshippers used the Book of Common Prayer; the conscientious worshippers of Roman or Puritan allegiance preferred their own more committed communities and ways of worship, but still had to attend the parish churches of the Church of England.

The English people greatly admired the mettle of Elizabeth, but found it difficult to determine her religious convictions as, at the beginning of her reign, she swayed between Catholicism and Protestantism. Jewel complained of the royal prevarication to Peter

[150] See William Pierce, *An historical introduction to the Marprelate Tracts; a chapter in the evolution of religious and civil liberty in England.*

Martyr in Zurich in a letter of April 2, 1559: "If the Queen would but banish it [the Mass] from her private chapel, the whole thing might be easily got rid of. Of such importance among us are the examples of princes."[151] Sampson, another former exile in Marian days, also wrote to Peter Martyr in 1560 of his disappointment that three recently consecrated bishops were to celebrate the Lord's Supper without a preparatory sermon, and "before the image of a crucifix, or at least not far from it, with candles, and habited in the golden vestments of the papacy. . . ."[152] Although she was "Supreme Governor" of the Church of England, Elizabeth left her Archbishop of Canterbury to undertake unpopular tasks such as enforcing Parker to issue the *Advertisements* to unify the use of vestments and ceremonial in the current rubrical anti-Nomianism. Yet in her high-handed way she had no qualms in sequestering Archbishop Grindal when he had the courage to refuse to suppress the "prophesyings" in 1577 at her behest. Since religion was essentially a matter of political expedience with the queen, it was no wonder that the bishops found it difficult to enforce her Prayer Book, especially as men and women of conscience so often turned out to be Recusants or Puritans.

In the 1559 prayer-book there was what political realists would have called "compromise" and what theologians, Catholic or Reformed, would have considered only confused thinking. The most notable liturgical example was the juxtaposition of a Catholic sentence (implying Christ's corporal presence in the Eucharist) with another (implying a doctrine of Memorialism) in the words of Administration, thus combining the 1549 and the 1552 formulas. Perhaps Jewel, the first apologist for the Church of England, was not far from the truth in seeing the "Elizabethan Settlement" as less a golden mean than a leaden mediocrity. After all, could Catholic church order be readily joined to Calvinist Articles of Faith, and a liturgy combining Catholic, Reformed, and Lutheran elements? The English Reformation, more the work of sovereigns and statesmen than of heroic reformers like Luther, Zwingli, and Calvin, has been a profound enigma to Catholics and Protestants alike: the Elizabethan Prayer Book, too, has a sphinx-like character for those who like theology and liturgy clear and consistent.

As I concluded earlier, worship according to the Book of Com-

---

[151] *Zurich Letters*, ed. Hastings Robinson, pp. 28-29.
[152] *Zurich Letters, 1558-1579*, ed. Hastings Robinson, pp. 63-64.

mon Prayer must often have seemed exceedingly boring. The disappearance of so much ceremonial splendor was an affront to the eyes and a deadening of the pictorial imagination. Strict instructions were given that "all shrines, covering of shrines, all tables, candlesticks, trindals, rolls of wax, pictures, paintings, and all other monuments of feigned miracles, pilgrimages, idolatry, and superstition" were to be completely destroyed, "so that there remain no memory of the same in walls, glasses, windows, or elsewhere, or repairing both the walls and the glasses."[153] Thus the medieval *biblia pauperum* were rudely snatched away from the illiterate, and even the literate are often grateful for "illustrations" of doctrine or duty. With these representations of divine persons or events also went many of the symbolical ceremonies which had made the ancient rites of baptism, ordination, and burial so memorable. The substitutes for these acts of aesthetic iconoclasm were the royal arms painted on boards, in which an improbably quiescent greyhound faces an incredibly tame lion as supporters, the monitory Decalogue painted on two wooden tablets, and the carved and painted tombs which the more successful Elizabethans built for the ancestor worship of the future, and which ironically were the liveliest artistic objects in the church. Nor were the popular metrical psalms which were introduced in the hobbling metre of Sternhold the musical equivalent of the Gregorian chanting, so austere in its solemnity and so apt to induce reverence. The irreverence at so many Elizabethan services is one indication of their dullness, a dullness compounded by aesthetic colour-blindness and cacophony.

A more serious and allied criticism of worship according to the prayer-book is that it was impersonal, formal, and artificial. The sense of the church as a great family divided by the narrow stream of death disappeared between the 1549 and the 1552 Prayer Books. While in the 1549 book there was a prayer for the whole state of Christ's church, in the 1552 book the prayer was restricted to the whole state of Christ's church militant here on earth; thus the sense of the communion of saints was weakened and hardly kept alive by the greatly reduced festivals of the "scriptural" saints, who, after all, had lived long ago. Nor did Anglicanism compensate for this loss, as Puritanism did, by stressing the intimate communion of the covenanted members of Christ's church, for the national church was a "mixed multitude," not a community of

[153] Kennedy, *Elizabethan Episcopal Administration*, I, 44.

"visible saints" to use a favourite, though later, Puritan term.[154] The doctrine of the church as consociation of the elect gave support and warmth to Puritan gatherings for worship.

The Puritans were conscious of an excess of decorum, dignity, and formality—not to mention legalism—in the Anglican services. They sensed a chilling social distance which, they were sure, conflicted with Christian cordiality and community. That is why the author of the *Second Admonition to Parliament* attacks the impersonality and artificiality of the Elizabethan Book of Common Prayer in 1572: "The Book is such a piece of work as it is strange we will use it, besides I cannot account it praying, as they use it commonly, but only reading or saying of prayers, even as a child that learned to read, if his lesson be a prayer, even so it is commonly a saying and reading prayers, and not praying. . . . For though they have many guises, now to kneel and now to stand, these be matters of course, and not any prick of conscience, or piercing of the heart. . . . One he kneeleth on his knees, and this way he looketh and that way he looketh, another he kneeleth himself fast asleep, another he kneeleth with such devotion, that is so far in talk, that he forgetteth to rise till his knees ache, or his talk endeth, or the service is done! And why is all this? But that there is no such praying as should touch the heart."[155]

Finally, on the negative side, there was a spirit of dignity and decorum, a sense of good order and good taste, but there was neither the lyricism that is appropriate to souls committed to the high ventures of faith in the power of the resurrection (such as inspired most of the prayers of the Eastern Orthodox rites), nor was there that massive historic sense of the world being an amphitheatre in which the saints watching in the stands encourage the earthly runner, though weary, to run the straight race (which the sanctoral cycle of the Roman rite expressed). The spirit of Anglican liturgy was rather one of reverent humility and submission than of triumphant rejoicing. It was too often introspective and penitential; its dignity was in danger of being a Laodicean substitute for enthusiastic commitment. By the same token it was a suitable form of worship for sober, pragmatic, prudent, and rational members of the social establishment, which was presumably why it has been described historically—but there are too many excep-

---

[154] J.H.A. New, *Anglican and Puritan, the Basis of their Opposition, 1558-1640*, pp. 40-41.
[155] *Puritan Manifestoes*, ed. W. H. Frere and C. E. Douglas, p. 115.

tions for the *mot* to be *juste*—as the formulary of the Tory party on its knees.

This has been a formidable series of negative criticisms against the Book of Common Prayer. It has been criticised as the instrument of religious control as put forward by the authority of a queen who equivocated in matters of faith, as a document marred by theological confusion, as the medium of a repetitive and dull form of worship, as a formulary of devotion likely to lead to a sense of impersonality in worship, and, finally, as a book spiritually tepid and admirably suited to the "top people." Yet this cannot be the last word on a historic, influential, and well-loved book of corporate devotions, for if that is all there is to be said about it, it would never have survived its detractors. It alone has survived in continuous use amid the plethora of evanescent new liturgies of the sixteenth century; those who use it today are found in many countries and they worship through it in many languages. What, then, are the qualities of the Book of Common Prayer that have given it enduring value?

First, it is salutary to recall that what can be conceived politically as a "great compromise" may be conceived liturgically as the "great comprehension." P. M. Dawley has written of the combination of the "Catholic" (1549) and "Protestant" (1552) words of the administration in the Elizabethan rite of Holy Communion, that "no single action was more important in ensuring a wide acceptance of the religious settlement."[156] In this ecumenical age, we are able to realise what an immense and honourable task was attempted by Cranmer and his associates in trying to span the Catholic and Protestant chasm with the bridge of liturgy (as also in church order by preserving the historic forms of the ministry). For our present consideration it does not matter that Catholics and Separatists excluded themselves from the Anglican Settlement, but only that the experiment of inclusiveness was made and that the Prayer Book of 1559 was its liturgical laboratory. So, in fact, although it was the national formulary of English worship, it preserved an international heritage. The Book of Common Prayer enabled Englishmen to worship God with the voices of the past and the present conjoined. As Jews had worshipped God in the Temple and Synagogue in the Psalter, as citizens of the Roman empire had worshipped Him in lapidary collects and in the Order of the Mass, as members of the Eastern Orthodox Church did in the

[156] *John Whitgift and the English Reformation*, p. 58.

prayer of St. John Chrysostom or as in the climactic moment of "invocation" in the prayer of Consecration in the 1549 rite, as the German Lutherans did in the penitential introductions to the Daily Offices, as Calvinists did in the metrical psalmody and in parts of the prayer of confession, so did English worshippers make these experiences their own through the Book of Common Prayer. The comprehensive and compassionate spirit of Cranmer—a reserved, reverent, scholarly, and tentative spirit—breathes through its entirety, and made it possible through his researches and superb editing and writing. Many countries, many centuries, march through the Book of Common Prayer to the present.

Second, the Book of Common Prayer has become the bond of unity of the Anglican communion through its many provinces throughout the world. It has managed to satisfy the spiritual needs of high churchman, broad churchman, and low churchman. They appear to have found ample sustenance and variety in this treasury of devotions and guide to spirituality. (The tragedy is that an act of Parliament is needed in England to add to it the splendid prayers that have been composed since the last legally authorised revision of over 300 years ago).

Third, the Book of Common Prayer is unique in being the first Book of *Common* Prayer, in that it has always been used both by priest and people, whereas previously priests had a different set of prayer books and the people had none, unless they were Latin scholars. Such a book has kept minister and people close in English life. It has encouraged the growth of a lay spirituality of depth. In yet another sense, the prayers of the Prayer Book are *common*— that is, they have a richly responsive construction and character, whether in the Litany, or in the versicles and responses of the Suffrages, or in the general confessions, and elsewhere. This means that there has always been in Anglican worship a continuing dialogue between priest and people, an ongoing liturgical "conversation." This can be contrasted with the tyranny of the ministerial voice in Reformed worship, for the minister's is often the only voice heard in prayer, and the congregation is mute, and often says the permitted "Amen" with little conviction. As Hooker was to point out to the Puritans, these "arrow prayers" are likelier to provoke the people to devotion than the long pastoral prayers of the Reformed Churches which are likelier to turn the church militant into the church somnolent. Not only praises, but prayers are shared in the Anglican (as in the Roman and Orthodox) rites.

The Book of Common Prayer gave a sense of the priority of prayers and sacraments over preaching, sometimes to the detriment of preaching, though there was a great post-Elizabethan Anglican tradition of preaching. But prayer was given the primacy of consideration in the Book of Common Prayer, in accordance with its title, and this is its great glory that through this model of devout spirituality it has, under God, helped to train prophets like F. D. Maurice and William Temple and saints like George Herbert and Nicholas Ferrar, William Law, and John Wesley, as well as John Henry Newman. It has also raised many men and women from the mire to the mountains of aspiration.

CHAPTER VI

ANGLICAN PREACHING

THE LITURGY has nearly always taken precedence over the sermon in the esteem of the Church of England, yet the new church slowly established an important preaching tradition which reached its zenith in the golden sermons of Andrewes and Donne early in the next century. The assigning of an important role to preaching was greatly accelerated by the Christian Humanists of Catholicism and by the Protestant Reformers on the Continent and in England.

The friends of the new learning were also the friends of preaching. John Fisher, the Catholic bishop of Rochester and vice-chancellor (from 1501) and chancellor of the University of Cambridge (from 1504), was himself a great oratorical preacher who was selected to preach the funeral sermons of Henry VII and Lady Margaret, the great benefactress of the university. Lady Margaret, probably on the advice of her confessor, John Fisher, established a preachership at Cambridge in 1504. The preacher, a resident Cambridge fellow without cure of souls, was to preach once every two years in each of 12 parishes in the dioceses of London, Ely, and Lincoln. Fisher also advised Erasmus to write his treatise on preaching, *Ecclesiastes, sive concionator Evangelicus*, and recommended that the Lady Margaret Readers should give attention to preaching. John Colet, the friend of Erasmus and More and a distinguished Greek scholar, lectured at Oxford University on the meaning of the Pauline Epistles with great critical insight and the plea for a return to the discipline of the early Church. In 1505 he became dean of St. Paul's cathedral in London, where he preached expository sermons at every festival. The same innovation was introduced by Ralph Collingwood, dean of Lichfield, who preached a weekly sermon in his cathedral.[1] Richard Fox, the bishop of Winchester, founded Corpus Christi College, Oxford in 1516 and a Readership in Greek to promote the study of the Scriptures in the original languages. In line with the same emphasis and concern was the establishment of Regius Professorships of Greek and

---

[1] F. E. Hutchinson, "The English Pulpit from Fisher to Donne," in A. W. Ward and A. R. Waller, eds., *The Cambridge History of English Literature*, vol. 4, p. 224. For Colet see E. H. Harbison, *The Christian Scholar in the Age of the Reformation*, pp. 56-67.

Hebrew at Cambridge in 1540 and at Oxford in 1546. The Reformation increased the momentum started by the Renaissance.

To the Reformation is owed the great drive toward set preaching instead of largely informal instruction. The Reformation's impact took three forms. First, in the first generation of Protestantism no man could be born a Protestant—he was *made* one. He had to be convinced of the truth of the Protestant position, which was based largely on a knowledge of the Scriptures, which was expounded in sermons, rather than on entrenched traditional Catholicism. Protestants were convinced that every man must make up his mind about his religion for himself under the eyes of his Maker. This conviction urged them to preach the word of God and instruct men in the truth. Finally, the substitution of English for Latin in the worship of the church was itself a stimulus to preaching.[2]

Anglican preaching in Tudor England moves from Latimer to Hooker through two quite different periods, the Edwardian and the Elizabethan. In the reign of Edward VI, which marked a radical Protestant change from the conservative, non-Papal Catholicism of Henry's final years, there were few preachers who appeared in the public forum; but all who did were strong personalities. They zestfully proclaimed the dominance of Scripture over tradition, and were prophetic Reformers, applying the Biblical criterion to the overthrowing of false doctrine and social injustice. Such were Latimer, England's most popular sixteenth-century preacher and a friend of the poor, and the inflexible, Savonarola-like John Hooper, imprisoned for his Vestiarian views, then freed and, like Latimer, a Protestant martyr in Mary's reign. With them should be numbered the significant, but less dominant, Lever and Bradford whose preaching was delivered with an urgent simplicity appropriate to the early days of open Protestantism in English history. Lever's and Bradford's preaching was, inevitably, iconoclastic, and therefore produced much excitement especially in court circles. Latimer and Hooper preached memorable series of sermons. Their second-generation Elizabethan successors seem, perhaps unfairly, smaller and more inhibited figures in "tuned" pulpits.

1. *Homilies*

Apart from the individual sermons of Latimer and Hooper, a new simple style of thematic preaching was being forged in Henry's

[2] Arthur Pollard, *English Sermons*, p. 8.

reign, and became public in Edward's reign in the form of model sermons, or official "homilies." The plan to issue prescribed homilies for the use of illiterate or discontented clergy was first agreed to at a convocation early in 1542, and was duly prepared within the next 12 months. The collection of 12 homilies was not, however, issued until early in the reign of Edward VI. They appeared July 31, 1547, with the authority of the council. A second book, containing another 21 homilies, was issued in 1571 under Queen Elizabeth, although it was probably completed by 1563.[3]

Bishop Ridley asserts that the *Homilies* of 1547 were so designed that some were "in commendation of the principal virtues which are commended in Scripture," and "other against the most pernicious and capital vices that useth (alas!) to reign in this realm of England."[4] In its approach the plan seemed, therefore, medieval rather than Protestant. Its Henrician context is suggested by the preponderance of ethical over theological themes.

It is generally agreed that of the 12 homilies in the first book, Cranmer wrote 5: A fruitful Exhortation to the Reading of Holy Scripture; Of the Salvation of all Mankind; Of the true and lively Faith; Of Good Works; and, probably, An Exhortation against the Fear of Death. Bishop Bonner is the reputed author of the homily, Of Christian Love and Charity, while Archdeacon John Harpsfield is believed to have written the homily, Of the Misery of all Mankind. With less confidence the homilies, Against Whoredom and Adultery, and Against Swearing and Perjury, have been attributed to Cranmer's chaplain, Thomas Becon.[5]

In these homilies there is evidence of nostalgia for the simple Christian way of life depicted in the New Testament, and of a protest against the legalities and complexities of a tradition-bound church, which A. G. Dickens has characterised as the motive power of English Protestantism.[6] It can be found in the words of Cranmer,

[3] The two books of homilies were reproduced as *Certain Sermons appointed by the Queen's Majesty to be declared and read by all Parsons, Vicars, and Curates, every Sunday and Holiday in their Churches, and by Her Grace's Advice perused and overseen for the better understanding of the Simple People.* The second edition of 1574 is the basis of G. E. Corrie's edition of the homilies, entitled, *Certain Sermons*, which appeared in Cambridge in 1850, and to which all references hereafter will be made. The 21 Elizabethan homilies were largely the work of Jewel, although Grindal wrote number 5, and probably Parker, number 17.

[4] *Works*, ed. H. Christmas, p. 400.

[5] See J. T. Tomlinson, *The Prayer Book, Articles, and Homilies*, pp. 232-35. The three other homilies were: Of the Declining from God; An Exhortation to Obedience; and Against Strife and Contention.

[6] *The English Reformation*, p. 138. "In the Bible, in the notion of a return to the original spirit of Christianity, in the rebirth of a fragment of the ancient

where the nostalgia becomes also an impatient iconoclasm; Cranmer was trying to tear down superstition and replace it with a supposedly more authentic, because more Biblical and simpler, faith:

> What man, having any judgment or learning, joined with a true zeal unto God, doth not see and lament to have entered into Christ's religion such false doctrine, superstition, idolatry, hypocrisy, and other enormities and abuses; so as by little and little, through the sour leaven thereof, the sweet bread of God's holy word hath been much hindered and laid apart? Never had the Jews, in their most blindness, so many pilgrimages unto images, nor used so much kneeling, kissing, and censing of them, as hath been used in our time. . . . Which sects and religions had so many hypocritical and feigned works in their state of religion, as they arrogantly named it, that their lamps, as they said, ran always over: able to satisfy not only for their own sins, but also for all other their benefactors, brothers and sisters of religion, as most ungodly and craftily they had persuaded the multitude of ignorant people: keeping in divers places, as it were marts or markets of merits; being full of their holy relics, images, shrines, and works of overflowing abundance ready to be sold. And all things which they had were called holy, — holy cowls, holy girdles, holy pardons, beads, holy shoes, holy rules, and all full of holiness.[7]

Even the willingness to accept the Bible as the criterion for criticism of the corrupt accretions of tradition could not be assumed. So the first homily, A fruitful exhortation to the Reading of Holy Scripture, had to refute the common notion that only the learned

---

world so infinitely more precious to Christians than the glories of Greece and the grandeurs of Rome, here lay the true strength of the Reformation. One who has never felt this nostalgia, this desire to sweep away the accretions, to cross the centuries to the homeland, can understand little of the compulsive attraction of the New Testament, even less of the limited but real successes of Protestantism, successes which cannot, any more than those of the Counter Reformation be explained away by reference to ambitious kings and greedy nobles."

[7] *Certain Sermons*, pp. 55-56. Later in the same Homily of Good Works annexed unto Faith, Cranmer denounced the superstition of the life of the "religious" with their three-fold vows, and many "papistical" superstitions and abuses, from beads, rosaries, stations, and jubilees, to fastings, indulgences, and masses satisfactory. The gravamen of his criticism is that they "were so esteemed and abused to the great prejudice of God's glory and commandments, that they were made most high and most holy things, whereby to attain to the everlasting life, or remission of sins" (p. 58).

were trained to expound its teachings: "Unto a Christian man, there can be nothing either more necessary or profitable than the knowledge of holy Scripture; forasmuch as in it is contained God's true word, setting forth his glory, and also man's duty. And there is no truth nor doctrine, necessary for our justification and everlasting salvation, but that is, or may be drawn out of that fountain or well of truth. . . . These books, therefore, ought to be much in our hands, in our eyes, in our ears, in our mouths, but most of all in our hearts. For the Scripture of God is the heavenly meat for our souls, the hearing and keeping of it maketh us blessed, sanctifieth us, and maketh us holy; it turneth our souls; it is a lantern to our feet; it is a sure, steadfast, and everlasting instrument of salvation; it giveth wisdom to the humble and holy hearts, it comforteth, maketh glad, cheereth, and cherisheth our conscience; it is a more excellent jewel, or treasure than any gold or precious stone; it is more sweet than honey or honeycomb; it is called *the best part* which Mary did choose, for it hath in it everlasting comfort."[8]

2. *Edwardian and Elizabethan Sermons Compared*

With the conviction and recommendation indicated in the foregoing, it would be astonishing if the leading Edwardian preachers were not primarily expositors of the Scripture. It is important that the first book of homilies was itself so full of the content of the Scriptures, and that it used the topical approach with simplicity. The excitement was in the novelty of scriptural teaching in the vernacular, its controversy with the old faith, and the stripping away of the accumulation of superstitious and idolatrous beliefs and practises.

The first book of homilies was issued as a standard of Biblical doctrine and preaching for the nation. It fulfilled a need, since there were few Reformed ministers with the requisite background, training, and conviction. Their numbers did not grow appreciably in Mary's reign, during which the more convinced members of the Reformed faith went to the stake or into exile. The same situation at the beginning of Elizabeth's reign made it necessary to continue to use the first book of homilies and, later, to produce a second. This time, however, as the universities turned out more and more clergymen for the (Reformed) Church of England, the books of homilies became less an incentive for preachers than a theological

8 *Ibid.*, pp. 1-3.

straitjacket, which the Puritans, along with enlightened prelates, such as Grindal, and other former Marian exiles, felt inhibited the encouragement of imaginative application of Biblical exposition.

In due course there appeared on the Elizabethan scene, not the towering prophetic figures of Edward VI's days but a large number of competent and devoted preachers, some of whom, like Henry Smith,[9] Richard Hooker, and the early Lancelot Andrewes,[10] were admirable orators of the English pulpit. The opportunities for preaching had increased considerably since Edward.

St. Paul's Cathedral provided lectures every day, and sermons on Sunday, from its large staff comprising a dean, chancellor, treasurer, five archdeacons and 30 Prebendaries. London boasted 123 parish churches.[11] Several clergymen drew large and growing congregations with their preaching ability. Stephen Egerton, of St. Anne's, Blackfriars, had many female admirers in his congregation. Richard Bancroft, later Archbishop, preached to large congregations at St. Andrew's, Holborn, where he was succeeded by the popular Dr. John King. The brilliant preacher at St. Giles, Cripplegate was Lancelot Andrewes. At Christ Church, Newgate, the Puritan minister Richard Greenham gathered crowds; Henry Smith, another Puritan, was more popular than any preceding preacher as he preached in St. Clement Dane's.[12] Such enthusiastic listeners did not always mean that the people craved the bread of life (though they often did); it meant they also delighted in the rich fruitcake which some Christian cooks provided. The unquestioned fact is that Elizabethans not only listened to sermons but increasingly read them for their pleasure and edification.[13]

The Elizabethans, like St. Paul's Athenians, were eager to hear "some new thing." It was not always soundness of doctrine or sincerity of manner that drew them, but humour, wit, even sensationalism. Controversy was sure to attract them. Indeed, the more thoughtful prelates were disturbed about the criteria by which the populace judged sermons. Archbishop Sandys referred to the popular view in one of his sermons: "The preacher is gladly heard of

[9] Smith is considered in Chap. VIII.
[10] Andrewes' important contributions to worship, spirituality, and preaching will be considered in the forthcoming vol. II of this series.
[11] Stow, *Survey of London*, ed. C. L. Kingsford, vol. 2, p. 143.
[12] See Alan F. Herr, *The Elizabethan Sermon, A Survey and Bibliography*, p. 27; and John Bruce, ed., *The Diary of John Manningham*, p. 75.
[13] Herr, *Elizabethan Sermon*, p. 27, affirms that the number of sermons published in each decade increased from nine in 1560-1570 to 69 in 1570-1580, to 113 in 1580-1590, to 140 in 1590-1600.

the people that can carp the magistrates, cut up the ministers, cry out against all order, and set all at liberty. But if he shall reprove their insolency, pride, and vanity, their monstrous apparel, their excessive feasting, their greedy covetousness, their biting usury, their halting hearts, their muttering minds, their friendly words and malicious deeds, they will fall from him then. He is a railer, he doteth, he wanteth discretion."[14] Jewel was of much the same opinion, but he said the generally poor (in his opinion) response to preaching was due to a contempt for the role of the preacher.[15] At the turn of the century both Hooker and Andrewes complained that their congregations were full of persons with itching ears. Hooker remarked that in the day when religion flourished, most men, "in the practice of their religion wearied chiefly their knees and hands, we especially our ears and tongues." Lancelot Andrewes, preaching on the text, "Be ye doers of the words and not hearers only,"[16] claimed that the teaching of St. James was particularly urgent in an age "when hearing of the Word is growen into such request, that it hath got the start of all the rest of the parts of Gods service," so that "Sermon-hearing is the *Consummatum est* of all Christianitie."[17]

The competitive pressure to interest the congregations and even more the readers of sermons can be seen in the attempt to produce intriguing sermon titles. The genesis of one such title was provided by the author in the dedication of his volume of sermons. John Stockwood of Tunbridge was ready to publish a treatise on August 20, 1589, when he recalled that August 24 would be St. Bartholomew's Day and that the famous London Fair would closely follow. Hence he wrote: "the time falling out so fitlie with the finishing of this worke, and the publishing of the same, I have geven unto it the name of a *Bartholmew Fayring*, the rather by the noveltie of the title to drawe on the multitude of people that nowe out of all places of our county repaire unto to the citie, to the better beholding and consideration of the matter cōtained in the treatise."[18]

To assume that Elizabethan sermons consisted only of Christian doctrine, Christian standards of behaviour (though this was an elastic category), and the requirements of Christian piety would be inaccurate. There was a great deal of what would today be

---

14 *Works* (chiefly sermons with miscellaneous pieces), ed. J. Ayre, Sermon 14.
15 *Works*, 4 vols., II, 1,014.   16 James 1:22.
17 Both citations are given by F. E. Hutchinson in *Cambridge History of English Literature*, vol. 4, chap. 12.
18 *Bartholmew Fayring*, sig. A3 r.

regarded as secular matters in the scope of the Elizabethan sermon, which helped to account for the growing interest of congregations in the final decades of the century. Local and international politics, the question of the royal succession, the social and economic problems of the poor, the ravages of plagues or catastrophic fires, the inconsistency of the weather, and facts every pregnant woman should know[19] were all included in sermons of the period. Such scope not only allowed for a great variety of treatment and originality in sermons, but also gave them a direct relevance to the issues of the day.

Perhaps the greatest distinction between Edwardian and Elizabethan sermons was mood and manner. The typical, early vernacular sermons had been those of the harangue, in which Becon and Latimer specialized. During Elizabeth's reign the controversial sermon had moved "from the hustings to the lecture room."[20] A new generation had arisen since the advent of the Reformation, which was used to theological controversy and interested in its technicalities. As a result, the sermons of Jewel and Hooker appealed to the intelligence in addition to the emotions of their hearers. Appeals to antiquity, supported by citations from the Fathers and Doctors of the church, take the place of appeals to prejudice.

3. Biblical Interpretation[21]

In the sixteenth century medieval exegesis had appealed to the literal, tropological, allegorical, and anagogical senses of Scripture, an approach which led to a great deal of ingenuity and hunting for hidden meanings, and which made possible the explanation of what on a consistently literal basis looked like contradictions in Scripture. Its defect was both the subjectivity it encouraged and the departure from the historical sense which stimulated the acceptance of fables in lieu of facts.

Inevitably Protestantism preferred a literal or historical to an allegorical interpretation of Scripture, except when the intention was clearly metaphorical, as when Jesus declared that he was the true vine or the door. While Protestants were as anxious as Catholics to preserve the primacy of faith, though interpreting the term differently, they also paid a long-needed tribute to reason as they

[19] See Christopher Hooke, *The Child-birth or Woman's Lecture* (1590), referred to in Herr, *Elizabethan Sermon*, p. 29.
[20] Hutchinson, "English Pulpit."
[21] Further consideration is given to this theme in chap. VIII.

expounded the historical contexts in which the servants of God spoke or acted in their own times and for the benefit of the future. J. W. Blench notes that the early English Reformers discontinued the allegorising of the figurative expressions of the poetic and prophetic books of the Old Testament favoured by the exponents of the old learning, that instead they were fond of the historical books of the Old Testament. Blench also observes that the same preachers were anxious to provide the original meaning of Biblical expressions, according to the original languages in which the Bible was written.[22] In this the English reformers were aided by the exemplars of the new learning such as Colet, Erasmus, and Stafford.[23] Latimer, for example, protested against Cardinal Pole's exegesis as being too farfetched; he argued that no preeminence for Peter is intended in Luke 5:1-7, in which Christ selected Peter's boat rather than another's as the pulpit from which He addressed the people, and from which He commanded the nets to be lowered with the resulting miraculous catch of fish. Latimer refuted the argument from common sense experience, asserting that a wherryman at Westminster bridge could tell the cardinal the natural meaning of the passage. It was obvious that Christ "knoweth that one man is able to shove the boat, but one man was not able to cast out the nets; and therefore he said in the plural number, *Laxate retia*, 'Loose your nets'; and he said in the singular number to Peter, 'Launch out the boat.' Why? Because he was able to do it. But he spake the other in the plural number, because he was not able to convey the boat, and cast out the nets too; one man could not do it. This would the wherryman say, and with better reason, than to make such a mystery of it, as no man can spy but they."[24]

The English reformers, however critical they were of Catholic allegoric glosses did not themselves always abide by the literal sense without making any accommodations to their own times or to their audiences. Hooper's series of sermons on the Book of Jonah

22 *Preaching in England in the late Fifteenth and Sixteenth Centuries: A Study of English Sermons, 1450—c. 1600*, pp. 39-40. This erudite volume is indispensable.
23 Thomas Becon said it was a common saying that "When Master Stafford read, and Master Latimer preached, then was Cambridge blessed." He added that it might be doubted whether Stafford owed more to Paul, "or that Paul, which before had so many years been foiled with the foolish fantasies and elvish expositions of certain doting doctors . . . was rather bound unto him, seeing that by his industry, labour, pain, and diligence, he seemeth of a dead man to make him alive again. . . ." (*The Jewel of Joy*, in Thomas Becon, *The Catechism*, ed. J. Ayre, p. 426.) Stafford was expert in Hebrew, Greek and Latin; he brought not only the Pauline Epistles but also the Gospels to life.
24 *Sermons*, ed. G. E. Corrie, pp. 205-206.

are a case in point. For example, it is very doubtful if a strictly historical interpretation would draw a trite moral lesson from the sleep of Jonah, "that when we think ourselves most at rest, then we be most in danger."[25] Further, Jonah emerges from the whale and the waters strikingly like a newly baptized Lutheran theologian. Latimer, too, was hardly consistent in the same way as Hooper. Quite unhistorically Latimer asserted that those labourers who criticised the Lord of the vineyard because labourers hired late in the day received the same pay as themselves, despite their longer hours, are "merit-mongers, which esteem their own work so much, that they think heaven scant sufficient to recompense their good deeds; namely for putting themselves to pain with saying of our lady's psalter, and gadding on pilgrimage and such-like trifles."[26] There was considerable flexibility in showing the relevance of the Biblical teaching to the contemporary situation of the hearers. In its way, this meant making a wax nose of Scripture, as the Catholics so often accused the Protestants of doing, if not in the interpretation, then in the application.

Anglican preachers in Elizabeth's reign, such as the Calvinist William Whitaker, the Puritans Deringe and Perkins, and Thomas Drant and John Chardon, were extremely critical of the Catholic fourfold sense of Scripture. Yet they did employ typology to some extent, particularly in their use of Old Testament figures. For example, John Foxe affirmed that Samson, in defeating the Philistines, was Christ victorious over the powers of darkness.[27] Bishop John Jewel of Salisbury used the traditional types and anti-types in seeing the water that Moses struck from the rock as a figure of the blood of Christ, the miraculous manna as a foreshadowing of the body of Christ, and the brazen serpent of Numbers 21:9 as the type, or foreshadowing of the crucifixion of Christ. In most cases, however, they are referred to as such in Scripture.[28]

But other Elizabethan preachers, particularly the witty and florid Thomas Playfere or the erudite and subtle Lancelot Andrewes, refused to be restricted by the literal sense. Playfere delighted in allegorising poetical expressions from the Psalms or the Song of Songs. A simile from the latter—"Thy two breasts are like two

[25] Hooper, *Early Writings*, ed. S. Carr, p. 454.

[26] *Remains*, ed. G. E. Corrie, p. 220. This instance of inconsistency is taken from Blench, *Preaching in England*, p. 47.

[27] *A Sermon of Christ crucified preached at Paules Crosse the Friday before Easter, commonly called Goodfryday* (1570), fol. 41 recto. See also Blench, *Preaching in England*, p. 61.

[28] See *Works*, ed. J. Ayre, 4 vols., vol. 2, 968-70.

young roes that are twins, which feed among the lilies" (4:5) —
he interpreted as the two Testaments that feed the church with the
pure milk of God's word.[29] Andrewes, following the example of
such a distinguished Latin expositor as St. Jerome, used the allegorical method to extract "spiritual" meanings from refractory
texts with great ingenuity. On the whole, however, most Elizabethan preachers insisted on the primacy of the historical meaning
of Scripture in its context.

4. Sermon Topics

The single most decisive change in interest, as one moves from
Catholic to early Protestant sermons, was from the other-worldly
to the this-worldly, though the distinction was not absolute, since
Catholic sermons were also concerned with current duty and Protestant sermons with future reward. Nonetheless, the change of
emphasis is striking.

Some important Catholic themes were either ignored or were
treated infrequently or incidentally, such as the shadowy transience
and mutability of this world contrasted with the substantial and
enduring joys of eternity or the morbid meditation on skulls, skele-
tons, and advancing decrepitude. It was a serious loss that the
loving contemplation of the Passion of Christ also virtually
disappeared.[30]

Edwardian preachers warmed to the immediate and urgent task
of Reformation in church and state, according to the divine plan
revealed in the Scriptures. They demanded that the old strongholds of superstitious doctrines and idolatrous practices be torn
down; there is the intoxication of iconoclasm in their sermons. The
really courageous critics of Catholicism in Henry's reign, like "little" Bilney who so impressed Latimer and Barnes, paid the supreme
penalty. Latimer himself waited until the time of Edward VI before
attacking Catholic doctrine directly; prior to this, he delivered
attacks on the abuses of neglecting the poor in favour of giving
opulent vestments to a church, or decorating images when the true
image of God was to be seen in the face of a poor man. Even in
1537 Latimer in his forthright Convocation sermon harshly criticised contemporary teachings on purgatory, with their assertion

29 *The whole Sermons of That Eloquent Divine of Famous Memory: Thomas
Playfere, Doctor in Divinitie* (1623), p. 230. See Blench, *Preaching in England*,
pp. 64-65.
30 In Chap. XII I suggest that this is because Protestants are less interested in
the *person* than in the *work* of Christ, and concentrate more on the Atonement
than on the Incarnation.

that imprisoned souls are utterly dependent on the assistance of the living. He insisted that such teaching was purely speculative, that it denied the mercy of the living God, and that it was founded on the greed of the "purgatory pick-purse."[31]

In Edward's reign Latimer and Hooper assaulted Catholic doctrines with vigour. Both struck out at the doctrine of the sacrifice of the Mass, which Latimer accused of evacuating the fulness and finality of Christ's offering on the cross, and which Hooper echoed by declaring that the historic crucifixion is the only propitiatory sacrifice necessary.[32] Their style was often mocking and railing, which must have been highly offensive to good Catholics, although Catholics, in turn, were equally insensitive to tender Protestant consciences. Hooper ridiculed the legalism of a formula of consecration in the Mass (as contrasted with the sovereign action of the Holy Spirit): "After this their wicked and idolatrical doctrine, this syllable (*um*) in this oration, *Hoc est corpus meum*, to say, 'This is my body,' hath all the strength and virtue to change and deify the bread! But I pray you, what syllable is it that changeth and deifieth the wine?"[33] While intellectually superior, it was not morally superior to the Protestants who referred to the Mass as a "Jacke of the boxe"[34] or who hung on Paul's Cross a dead cat with shaven crown and a white disc in its mouth.[35] If such insensitivity ruled, the Marian dean of Westminster, Hugh Weston, could hardly be blamed for describing the Protestant Communion tables set up by Bishop Ridley of London as oyster boards!

It was also interesting that Bishops Latimer and Hooper were linked in the public mind as defenders of and liberal donors to the poor. Both were critical of the corruption of judges and magistrates who took bribes. What differentiated their sermons from those of their Catholic predecessors was that their charges were uncomfortably precise and particular.[36] Latimer spoke like a sixteenth-

---

31 *Sermons*, p. 50. For other abuses Latimer criticised see *ibid.*, pp. 38-55.
32 See Latimer's *Sermons*, pp. 72-73; and Hooper's *Early Writings*, p. 500.
33 Hooper, *Early Writings*, p. 523.
34 *Chronicle of the Grey Friars of London*, ed. J. G. Nichols, p. 55.
35 Millar Maclure, *The Paul's Cross Sermons, 1534-1642*, p. 51.
36 See Latimer, *Sermons*, pp. 128, 190; and Hooper, *Early Writings*, pp. 482-83. Latimer, in his second sermon before Edward VI, appealed to the young king to hear himself the suits of the poor: "I must desire my Lord Protector's grace to hear me in this matter, that your Grace would hear poor men's suits yourself. Put them to none other to hear, let them not be delayed. The saying is now, that money is heard everywhere; if he be rich, he shall soon have an end of this matter. Others are fain to go home with weeping tears, for any help they can obtain at any judge's hand. Hear men's suits yourself, I require you in God's behalf, and put it not to the hearing of these velvet coats, these upskips" (p. 127).

century Amos at court in the name of a mightier sovereign—God. The calling of a prophet was a dangerous one, especially in Henry VIII's time. In fact, Latimer reported that he was once accused of preaching seditious doctrine in the king's presence, a charge Henry asked him to reply to. After facing his accuser Latimer ended his words to the king by saying: "But if your Grace allow me for a preacher, I would desire your grace to give me leave to discharge my conscience: give me leave to frame my doctrine according to mine audience: I had been a very dolt to have preached so at the borders of your realm, as I preach before your Grace."[37] The answer was a tribute to Latimer's honesty and courage, but its shrewd prudence contained a whiff of the danger he was in. Latimer told Henry's successor: "The poorest ploughman is in Christ equal with the greatest prince there is. Let them therefore, have sufficient to maintain them, and to find their necessaries."[38]

In Elizabeth's reign it was the Puritan Deringe alone who in 1569 dared to criticise the queen to her face. Deringe was prohibited for a time from preaching because he had told Elizabeth to take away her authority from the bishops because they would not ordain preaching ministers and tolerated pluralities and non-residence.[39] Other preachers before less august persons were, however, ready to continue the diatribe against greedy, lazy, illiterate, and incompetent parsons. This kind of jeremiad runs through the Elizabethan age, whether the critic in the pulpit was a prelate like Jewel[40] or a minister such as Pagit[41] or Bush.[42] Some plain speaking took place during the Vestiarian and Disciplinarian controversies. The sermons of Sampson and Humphrey at Paul's Cross[43] and Fulke at Great St. Mary's, Cambridge,[44] against the vestments, and the sermon of Robert Beaumont[45] for compliance with the ecclesiastical law, show that few holds were barred in the homiletical warfare. The "disciplinarians" who supported Cartwright were equally forthright in their demands and accusations. William

---

[37] *Sermons*, p. 135.
[38] *Ibid.*, p. 249; this citation was part of a plea to stop enclosures of common land.
[39] See J. W. Blench, *Preaching in England*, pp. 300-301. The offending sermon and its consequences are described in the article on Blench in the *Dictionary of National Biography*, xiv, 393.
[40] *Works*, vol. ii, 999, 1,011.
[41] Eusebius Pagit, *A Godly Sermon preached at Detford* (1586), sig. A vi v.
[42] Bush, *A Sermon preached at St. Paules Crosse on Trinity Sunday 1571*, sig. F ii r.
[43] Maclure, *Paul's Cross Sermons*, p. 205.
[44] H. B. Porter, *Reformation and Reaction in Tudor Cambridge*, p. 121.
[45] *Ibid.*

Charke, in demanding a Presbyterian parity of ministerial status from the pulpit of Great St. Mary's, Cambridge, asserted that archbishops and bishops were introduced into the church by the devil.[46] Less than a year later, Richard Crick argued for an equality in the ministry from St. Paul's Cross,[47] where Thomas Cooper,[48] bishop of Lincoln and later a butt of "Marprelate," had preached a sermon in June 1572 in reply to the *Admonition to Parliament*. While the timid or the office-seekers tuned their pulpits to the prevailing doctrine, there was some independence of judgment and freedom in its expression in the Elizabethan pulpit. That is, indeed, why Archbishop Whitgift discouraged frequent preaching: "But, if any hath said that some of those which use to preach often, by their loose, negligent, verbal and unlearned sermons, have brought the word of God into contempt, or that four godly, learned, pithy, diligent, and discreet preachers might do more good in London than forty contentious, unlearned, verbal and rash preachers, they have truly said, and their saying might well be justified."[49]

In addition to airing internal controversies from the pulpits, there were anti-Catholic polemics which continued into Elizabeth's reign. Happily the spirit in which it was conducted was generally more scholarly and less vindictive than it was under Edward. The new tone was set by Jewel's famous Challenge Sermon which he preached at Paul's Cross on November 26, 1559 and March 31, 1560, as well as at court. Jewel provided a historical and rational basis for his attack on the Catholic Mass—for the use of Latin in the Mass, its giving communion in one kind, the teaching on sacrifice, the encouragement of adoration of the sacrament, and private celebrations. After 1570, when Pope Gregory XIII issued the bull *Regnans in Excelsis* absolving Catholics from loyalty to Queen Elizabeth, and with the surreptitious introduction of seminarists into England, Protestant-Catholic relations worsened. The situation was reflected in the mocking manner of Tyrer's and Bridges' sermons which condemned Roman Catholicism as a superstitious religion of externals, for its postponement of salvation through the teaching of purgatory, and for an evasive scholastic subtlety.[50]

The Elizabethan preachers did, however, find new subjects for

---

[46] *Ibid.*, p. 141.
[47] Maclure, *Paul's Cross Sermons*, pp. 208-209.
[48] *Ibid.*, p. 308.
[49] *Defence of the Answer to the Admonition*, tractate XI, sec. 3.
[50] Ralph Tyrer's criticisms are contained in *Five godlie sermons . . .* (1602), pp. 302-304, while Bridges' are in *A Sermon preached at Paules Cross* (1571), p. 125.

their critical powers. There was a complaint that the indifference of the people had led to decay of the furnishings in the churches. In his Homily for Repairing and keeping clean of Churches, Bishop Jewel exhorted many congregations: ". . . forasmuch as your churches are scoured and swept from the sinful and superstitious filthiness, wherewith they were defiled and disfigured, do ye your parts, good people, to keep your churches comely and clean; suffer them not to be defiled with rain and weather, with dung of doves and owls, stares and choughs, and other filthiness, as it is foul and lamentable to behold in many places of this country."[51] Another theme for denunciation was the new paganism emerging through classical learning, which, it was claimed, was replacing Christianity. This occupied the attention of Bishops Curteys and Cooper. Curteys charged: "Wee buylde Castles and toures in the ayre to get us a name. So many heads, so many wittes, so many common wealths. Plato his Idaea [sic], Aristotle's felicitie, and Pythagoras numbers, trouble most men's brayns."[52] Bishop Cooper inveighed against the new "Heathenish Gentilitie, which raigneth in the hartes of godlesse persons, Atheistes, and Epicures. . . ."[53] Closely allied to the criticism—and a particular aversion of the Puritans[54] —was the disapprobation of stage plays, popular romances, and translations of novellas about love.

On the positive side, Elizabethan Anglicans found much to praise. The "Gloriana" of the poets found clerical panegyrists on "Queen's day" each November 17th, and thanked God that "the Supreme Governor of this realm . . . as well in all spiritual things and causes as temporal"[55] had maintained the peace and unity of church and state, preserving them from the perils of religious wars abroad and civil war at home.

The Church of England itself was praised in sermons of Richard Bancroft, future Archbishop of Canterbury, Richard Hooker, who wrote the finest apologia of the *via media* between Papal and Puritanical extremes that had appeared. Bancroft saw the Anglican church as the middle ground between authoritarianism and libertin-

51 *Certain Sermons appointed by the Queen's Majesty*, ed. Corrie, p. 278.
52 *A sermon preached before the Queens Maiestie at Grenewiche*, sig. Civ verso, cited in Blench, *Preaching in England*, p. 305.
53 *Certain Sermons wherein is contained the Defense of the Gospell now preached, against such Cavils and false accusations, as are objected against the Doctrine it selfe and the Preachers . . . by the friendes . . . of the Church of Rome* (1580), p. 189, cited in Blench, *Preaching in England*, p. 305.
54 These are considered in Chap. VIII.
55 An important phrase in the oath of Supremacy: Elizabeth wisely avoided calling herself "Supreme Head of the Church," as her brother and father had done.

ism in the Christian religion,[56] which was a church that combined scriptural authority with decent custom. Hooker desired to move from theory to facts, to ask whether the Church of England's ministry of the word and sacraments had not nourished its children in spirituality and the love of God, in a moving appeal: "I appeal to the conscience of every soul, that hath been truly converted by us, whether his heart were never raised up to God by our preaching; whether the words of our exhortation never wrung any tear of a penitent heart from his eyes; whether his soul never reaped any joy, any comfort, any consolation in Christ Jesus, by our sacraments, and prayers, and psalms, and thanksgiving; whether he were never bettered, but always worsed by us."[57] Hooker's strength was not only in his recognition of the rule of law and the force of custom in human societies, but in the judicious reasonableness and quiet dignity of his own spirit. In an age of contention his fairness in presenting the positions of his opponents was refreshing.

Perhaps the most important contribution Anglican preachers made during the century was the "new wholesomeness of tone" and "peculiarly fragrant spirituality" in the preaching of Hooker and Andrewes.[58] For ultimately it was recognised that the Christian gospel was not a matter of theological contention but of the application of the transforming love of God to the souls that were battered by sin, beaten by suffering, and terrified by the approach of death. This spirituality of forgiveness, the courage of faith, and the hope of the life everlasting was central to the preaching of Hooker and Andrewes, and, one might add, to the gospel of Puritanism, where the disputatiousness was more evident.

5. Sermon Structure

In the Middle Ages there were two forms of sermon construction, "ancient" and "modern." The ancient, copied from the homilies of the Church Fathers, either expounded and applied a passage of Scripture according to the order of the text or a thematic treatment of any subject according to Scripture and sense. The modern construction was more elaborate; it usually consisted of the Theme (a Scriptural text), the Exordium (which was a subordinate element

---

[56] A Sermon preached at Paules Crosse . . . Anno 1588, pp. 89-90.
[57] Works, vol. III, 679-80.
[58] Blench, Preaching in England, p. 319. The Puritan divines were greatly concerned with manners and ethics, which did not greatly concern Hooker and Andrewes. But they are equally interested in providing Christian palliatives for bruised consciences. See Chap. VIII.

of the theme), the Bidding Prayer, the Introduction of the Theme, the Division (with or without subdivisions of the topic), and finally, the Discussion. Dean Colet, Richard Taverner's *Postils* and Longland's sermons on the Penitential Psalms all favoured the ancient form. It was, however, chiefly on important public occasions such as the funeral of a prince, a confutation of heresy (e.g. Fisher's Sermon against Luther in 1521[59]), or in addresses to the clergy that the modern form was used by the early sixteenth-century Catholics.

By contrast the early Reformers discarded the modern form entirely in favour of various simpler ones. These might be topical, as the official *Book of Homilies* was, or they might be a simple exposition and application of the text, as in the sermons of John Bradford. Subtlety and erudition were avoided in the interests of making a simple and direct exposition of Christian truth. Hooper's sermons on the Book of Jonah are particularly interesting because in their use of lessons ("Doctrines") and moral applications ("Uses") from each verse, they anticipated the structure of the favourite Puritan preachers' sermons in Elizabeth's time.

Elizabethan sermons used a great variety of constructions. Perhaps most used a simplified version of the modern style, which almost reverted to the classical form. The classical was recommended by Reuchlin's *Liber Congestorum de arte praedicendi* (1503) and Erasmus's *Ecclesiastes* (1535); the form came to Elizabethan preachers by way of the treatise of Hyperius of Marburg, which was translated into English by John Ludham in 1577 as *The Practice of Preaching*. According to the treatise the essential parts of a sermon were: (a) reading of the sacred Scripture, (b) invocation, (c) exordium, (d) proposition, or division, (e) confirmation, (f) confutation, and (g) conclusion.[60]

The ancient sermon form is found in Becon's *Postils* (which are like Taverner's in expounding and applying the meaning of a summary of the Gospels and Epistles of the church's lectionary). Abbott's and King's sermons on Jonah, Deringe's on Hebrews, and Rainolds's on Obadiah and Haggai were constructed on this model.

The new Reformed method, of which Bishop Hooper's Jonah sermons were the first examples in England, was particularly

---

[59] See the analysis of this sermon's structure in Edward Surtz, *The Works and Days of John Fisher (1469-1535)*, pp. 302-306.
[60] See Blench, *Preaching in England*, p. 102.

approved by the Puritans and was "canonised" by the recommendation of William Perkins in his *Art of Prophecying*.[61]

6. *Sermon Styles*

Generally there was a great contrast between the styles of Edwardian and Elizabethan preachers. Like the Gospel, it was commending, Edwardian preaching was plain and direct. Particularly in Latimer, and a few others, it was also a racily colloquial style. The first *Book of Homilies* shares the simplicity and directness, but only occasionally the colloquialism.

Style among the Elizabethan preachers varied more widely. There was a great difference in pulpit style between the Ciceronian eloquence of Richard Hooker and the studied simplicity and concentrated directness of Walter Travers, a Puritan divine, both of whom preached at the Temple. Thomas Fuller says, "Here the pulpit spake pure Canterbury in the morning, and Geneva in the afternoon."[62] Travers expressed his views about the purposes of preaching and the imperative necessity for plainness of style forthrightly: ". . . the holie scriptures are to be expounded simplie and sincerely, and uttered with reverence. For some to shew themselves to the people to be lerned, stuffe ther sermons with divers sentences out of Philosophers, Poets, Orators, and Scholemen, and of the auncient fathers, Augustine, Hierome and others, and thos often times rehersed in greek or latin; by which pieces, sometime illfavoredly patched togither, they seeke and hunt for commendation, and to be esteemed lerned of the people, which also some doe that are unlerned."[63]

Travers was a learned man who hid his learning in his sermons. He was a senior fellow of Trinity College, Cambridge, a friend of Calvin's successor, Beza, and in 1595 he became provost of Trinity College, Dublin. William Holbrooke would have agreed with him when Travers observed: "The pulpit is not a place for a man to shewe his wit and reading in, but for plainnesse and evidence of the spirit."[64]

---

[61] The influence of Perkins is considered in Chap. VIII.

[62] *The Worthies of England*, ed. John Freeman, p. 133. Walter Travers, Cecil's candidate for the Mastership of the Temple, had been passed over in favour of Hooker. Their dignified dispute about the nature of a reformed national church was the crucible out of which the fifth book of Hooker's *Ecclesiastical Polity* came.

[63] *A full and plaine declaration of Ecclesiastical Discipline*, p. 104.

[64] Maclure, *Paul's Cross Sermons*, p. 147. Most preachers at Paul's Cross, Maclure notes, condemned the witty and extemporary extremes of preaching, preferring the plain, though carefully prepared, style (p. 146).

While English literature was delighting in euphuism, the sheer love of language for its own sake, and the brilliant joining of spiritual and sensual experience in the metaphysical love poems of John Donne, it would have been surprising if there were no preachers who revelled in wordplay and wit. Such preachers were Thomas Playfere, Lady Margaret Professor of Divinity at Cambridge, who used subtle reasoning and minute analysis and quoted from the Church Fathers and from classical poets; and John Carpenter, rector of Northleigh in Devon, who was fond of puns and heaping up similes. Two examples must suffice for a style which was in the reign of James I to achieve the acme of popularity, and which was peculiarly suitable for dazzling courtiers, as used by Lancelot Andrewes and John Donne. Typical was Playfere's: "Nay our very reason is treason, and our best affection is no better than an infection."[65] This is very like the wit in the early Shakespearian comedies, as in *Love's Labour's Lost* or in some of the passionate exaggerations in the hero's amorous protestations in *Romeo and Juliet*. Carpenter's examples boggled the imagination by leaving a confused, ornate impression, as when he said: "The polluted vessel is with water washed, the raw flesh is seasoned with salt, the dropping vine must bee pruned with a knife: the weake stomackes must have bitter wormewood; the slow Asse a whip, the heavie Oxe the goade, the idle scholler correction, and the fleshly Christian sower affliction."[66]

One could say of the varieties of Elizabethan preaching what Dryden said about Chaucer's *Canterbury Tales*—"Here is God's plenty." A contemporary preacher at Paul's Cross afforded the following description of their differences: "Some would have long texts: some short textes. Some would have Doctours, Fathers, and Councels: some call yt mans doctrine. Some would have it ordered by Logicke: some terme that mans wisdome. Some would have it polished by Rhetoricke: some call yt persuasibleness of wordes. And agayne in Rhetoricke some would have it holy eloquence, liable to the Ebrue & Greeke phrase: Some would have it proper and fittyng to the English capacitie. Some love study and learnyng in Sermons: Some allow only a sudaine motion of the spirite. Some

---

[65] *The Meane in Mourning* (originally preached in 1595), printed 1623, p. 55, cited in A. F. Herr, *The Elizabethan Sermon*, p. 102.

[66] *Remember Lots Wife* (1588), sig. C8 r., cited in Herr, *Elizabethan Sermon*, p. 102. Carpenter was also addicted to puns, as when he said, "From Lots wife is learned the lot of the wicked in this world, with a terror unto all backsliders and wicked Apostates" (sig. E1 v.).

would have all said by heart: some would have oft recourse made to the booke. Some love gestures: some no gestures. Some love long Sermons: some short Sermons. Some are coy, and can broke no Sermons at all."[67]

7. Representative Preachers: Latimer and Hooker

After a quick survey of the sermons, rather than the preachers, of the Edwardian and Elizabethan periods, a corrective can be provided by two portraits of preachers contrasting in style and approach yet complementary, namely those of Latimer and Hooker.

Latimer was the originator of popular preaching. There would be no one like him until George Whitefield or John Wesley in the eighteenth century and Charles Haddon Spurgeon in the nineteenth. He exemplified Phillips Brooks' definition of preaching, "truth through personality." Not a great scholar, in his earlier years he was an opponent of the new learning, yet he was marvellously compounded of courage and compassion. He not only produced a series of sermons on the Card; he was himself a card and character.[68] He had a great gift for vivid narration and used an incisive vocabulary, replete with colloquialisms and proverbs.

Latimer had a whimsical and kindly sense of humour, which was occasionally ironical and mordant. Himself a preaching prelate (he was consecrated Bishop of Worcester in 1535), he denounced unpreaching prelates with an arsenical application of alliteration: "But now for the fault of unpreaching prelates, methink I could guess what might be said for excusing of them. They are so troubled with lordly living, they be so placed in palaces, couched in courts, ruffling in their tents, dancing in their dominions, burdened with ambassages, pampering of their paunches, like a monk that maketh his jubilee; munching in their mangers, and moiling in their gay manors and mansions, and so troubled with loitering in their

[67] John Dyos, *A Sermon preached at Paules Crosse* (1579), sig. F3 r., cited in Maclure, *Paul's Cross Sermons*, p. 146.

[68] He introduced his listeners to a game much like whist, known as "triumph": "I will, as I said, declare unto you Christ's rule, but that shall be in Christ's cards. And whereas you are wont to celebrate Christmas in playing at cards, I intend, by God's grace, to deal unto you Christ's cards, wherein you shall perceive Christ's rule. The game we shall play at shall be called the triumph, which if it be played at, he that dealeth shall win; the players likewise shall win; and the standers and lookers upon shall do the same; insomuch that there is no man that is willing to play at this triumph with these cards, but they shall all be winners, and no losers." The ingenuity of the man, his ability to catch common interest, and the looseness and naturalness of the style, with its digressions and repetitions and grammatical imprecision, are all to be found in this excerpt from the *Sermons*, ed. Corrie, p. 8.

lordships, that they cannot attend it."[69] But irony more often yielded to anecdote, as when in his third sermon before Edward VI he told the following story: "A good fellow on a time bade another of his friends to a breakfast and said, 'If you will come, you shall be welcome; but I tell you aforehand, you shall have but slender fare: one dish and that is all.' 'What is that?' said he. 'A pudding and nothing else.' 'Marry,' said he, 'you can draw me round about the town with a pudding. These bribing magistrates and judges follow gifts faster than the fellow would follow a pudding."[70]

Latimer's humour expressed his empathy with the common people when he said: "I had rather ye should come of a naughty mind to hear the word of God for novelty, or for curiosity to hear some pastime, than to be away. I had rather ye should come as the tale is told by the gentlewoman of London: one of her neighbours met her in the street, and said, 'Mistress, whither go ye?' 'Marry,' said she, 'I am going to St. Thomas of Acres to the sermon; I could not sleep all this last night, and I am going now thither; I never failed of a good nap there.' " It was a story Latimer could tell without fear of it rebounding on him, for no one surely ever fell asleep at his lively sermons, which had the character of conversations with the congregation, often divagatory, but always lively.

For his iconoclastic utterances Latimer had developed a vocabulary as inventive in abuse as Skelton's, with some unforgettable minting of words: "flattering clawbacks," "merit-mongers," "pot-gospellers," "mingle-mangle" (compromise), "these bladder-puffed up," "wily men," "flibbergibs," "upskips," "ye brain-sick fools, ye hoddy-pecks, ye doddy-pols, ye huddes." Latimer represented the Pharisees as saying to Christ: "Master, we know that thou art Tom Truth."[71] He succeeded in holding the easily straying attention of courtiers and the interest of a royal boy of 11 years in "a nipping sermon, a rough sermon and a sharp, biting sermon."[72]

Latimer comes more alive in his sermons than any preacher in Tudor England, not only because of his impressive individuality and deep humanity, but also because of the many autobiographical references in his sermons. Some of his finest passages of illustration are drawn from his memories of early life as a yeoman's son and country boy,[73] how he was taught to shoot the bow,[74] or his con-

[69] *Sermons*, p. 67.   [70] *Ibid.*, p. 140.
[71] *Ibid.*, p. 201.
[72] Most of Latimer's phrases in this paragraph were collected by Hutchinson, *Cranmer and English Reformation*.
[73] *Sermons*, pp. 24, 101.   [74] *Ibid.*, pp. 197-98.

THE LITURGICAL ALTERNATIVES

version through Bilney, with whom he would visit prisoners in the castle.[75] He is never better able to use his countryman's knowledge than in the famous sermon, "Of the Plough."[76] He was equally moving when he spoke of the perils of a court preacher[77] or of the confusion and fear in Bilney's mind as he bore his faggot to the stake (incidentally an anticipation of Latimer's own martyrdom in 1555).[78]

For all its strength, it was not flawless preaching. Latimer was a people's preacher, not a preacher's preacher; he was certainly not a scholar's preacher, for he was critical of theological subtleties.[79] His sermons lacked the imaginative power of Fisher, and there was no poetry in them. They lacked clear structure; Latimer wandered all over the place as the fancy or the interested (or bored) look of the congregation dictated.[80]

Latimer's sermons and life show that he had two overriding concerns: to proclaim the word of God without fear or favour and to be the champion of the poor and the oppressed. As early as 1522 Cambridge University recognized his zeal in reforming ecclesiastical and social abuses by licensing him as one of only 12 preachers commissioned to preach anywhere in England. Even after he became Bishop of Worcester he continued to denounce evils in church and state. Latimer's Protestantism brought him first in Henry's time to resign his see of Worcester because his views conflicted with his sovereign's, and to the Tower in 1546; he was liberated after the accession of Edward VI and reached his zenith of popularity as court preacher. Mary's reign brought him to the Bocardo prison, with Ridley as his companion and his friend Cranmer as the silent and distant observer. Latimer's preaching was too courageous, too direct, and too compassionate ever to be mistaken for demagoguery. It was, whether in denunciation, retelling a Biblical narrative, or in exposition, despite all its delightful divagations, prophetic and popular preaching at its best. Latimer was even capable of writing a good Communion meditation, perhaps the last task one would have laid on his charitable shoulders. Here,

[75] *Ibid.*, pp. 334-45.   [76] *Ibid.*, pp. 59-78.
[77] *Ibid.*, pp. 134-35.   [78] *Ibid.*, p. 222.
[79] He dismissed "school-doctors and such fooleries" after his meeting with Bilney, and characterised the spinners of theological subtleties with the words, "as for curiouse braynes nothinge can content them."
[80] Latimer does not worry about being discursive ("I will tell you a pretty story of a friar to refresh you withal") or to make a good point even if it is out of order ("peradventure it myght come here after in better place, but yet I wyll take it, whiles it commeth to my mind.")

too, his understanding of the Gospel's relevance to the hardships of life was as practiced as ever, as when he was expounding the Lord's Supper as a Great Banquet where Christ is *epulum et hospes*: "But now ye know that where there be great dishes and delicate fare, there be commonly prepared certain sauces, which shall give men a lust to their meats: as mustard, vinegar, and such like sauces. So this feast, this costly dish, hath its sauces: but what be they? Marry, the cross, affliction, tribulation, persecution, and all manner of miseries: for, like as sauces make lusty the stomach to receive meat, so affliction stirreth up in us a desire to Christ. For when we be in quietness, we are not hungry, we care not for Christ: but when we be in tribulation, and cast in prison, then we have a desire to him; then we learn to call upon him. . . ."[81] Latimer was in the unfamiliar role of priest; he was better as a prophet.

Hooker, though lacking the immediate empathy, the capacity to project his personality, the humour, and the prophetic zeal of Latimer, was able to plumb a spirituality of greater profundity. Latimer certainly believed in the doctrine of the cross; his life was its scarlet seal. Yet he lacked the theological subtlety, the command of style, and the imagination to express it as Hooker or Andrewes could. Latimer was incapable of the concentration of thought and expression of the doctrine of the cross, in order to be able to say, as Hooker did: "Affliction is both a medicine if we sin, and a preservative that we do not."[82] Hooker shone best as a priest, as the apologist of the church and the representative of the people before God.

Hooker's sermons read better today than Latimer's, though they were preached in a very indifferent manner. Fuller said of his delivery: "He may be said to have made good music with his fiddle and stick alone, without any rosin, having neither pronunciation nor gesture to grace his matter."[83] Izaak Walton confirmed the impression: ". . . his sermons were neither long nor earnest, but uttered with a grave zeal, and an humble voice; his eyes always fixt on one place to prevent his imagination from wandering, insomuch that he seemed to study as he spake. . . ."[84] They read well because they are meaty and the skilful use of rhetoric was largely con-

[81] *Sermons*, pp. 463-64.
[82] *Works*, III, 636. See also the fourth paragraph of the sermon, "A Remedy against Sorrow and Fear," for a fuller statement of the grounds for patience in troubles.
[83] Fuller, *Worthies of England*, p. 264.
[84] Walton's *Life of Hooker*, in *Works*, ed. Keble, I, 79.

cealed. Walton rightly emphasized the reasonableness of his approach—he did "rather convince and persuade, than frighten men into piety"—and the modesty which sought for neither amusement nor self-glorification in sermons. But it is difficult to believe that his "unlearned hearers" were edified by sermons demanding such concentration. Certainly his sermons that have survived seem rather to have been prepared for the lawyers of the Temple than for the parishioners of Drayton Beauchamp, Boscombe, or Bishopsbourne churches, of which Hooker was successively rector. The clarity of his definitions, the building up of his "cases," the confirmation of his doctrines with Biblical and Patristic examples, and the rationality of his approach, confirm the impression.

It is only a *felix error* (Walton's phrase) that Hooker's sermons survived at all. It was the opposition of "Geneva" Travers (afternoon Reader at the Temple Church) that made Hooker write out for private circulation some of the sermons to which Travers had made objections. Samuel Taylor Coleridge wrote of the sermon, "The Certainty and Perpetuity of Faith in the Elect": "I can remember no other discourse that sinks into and draws up comfort from the depths of our being, below our own distinct, with the clearness and godly loving kindness of this truly evangelical and God-to-be-thanked-for-sermon."[85] Canon F. E. Hutchinson was of the opinion that Hooker's sermons on the certainty of faith, justification, and the nature of pride had "more permanent value than any sermons of the reign."[86] The historian of preaching, J. W. Blench, sees Hooker's great mission as that of "bringing peace to the bruised conscience; instead of excoriating the wounds of the soul, he brings precious unguents to soothe and heal them."[87] This was high praise, especially when it is recalled that only a fraction of Hooker's sermons have survived.

Hooker's themes were central and practical. He was the great preacher of faith—faith as the power to trust in God's promises because of the memory of His past mercies;[88] the invincibility of the faith of the elect;[89] the primacy of justification by faith, with works as consequences, not preconditions;[90] contagious faith as it

[85] See S. T. Coleridge, *Notes on English Divines*, in *Works*, IV, ed. H. N. Coleridge, *ad. loc.*

[86] "The English Pulpit from Fisher to Donne," *The Cambridge History of English Literature*, IV, chap. 12, p. 236.

[87] *Preaching in England in the late fifteenth and sixteenth centuries*, p. 319.

[88] See "The Certainty and Perpetuity of Faith in the Elect," esp. in sec. 7.

[89] *Ibid.*, sec. 6.

[90] See "A Discourse of Justification, Works, and how the Foundation of Faith is overthrown."

works through love;[91] faith's necessary edification, especially against mockers;[92] and faith strengthened by the Lord's Supper.[93] How practical and judicious he could be is evident in his consideration of whether God's elect can ever find their faith utterly to fail. "The question," he said, "is of moment; the repose and tranquillity of infinite souls doth depend on it."[94] Equally practical was his advice to the despondent in the same sermon: "of us, who is here which cannot very soberly advise his brother? Sir, you must learn to strengthen your Faith by that experience which heretofore you have had of God's great goodness to you, *Per ea quae agnoscas praestita, discas sperare promissa*, By those things which you have known performed, learn to hope for those things which are promised."[95]

Faith, though central, was not the only theme of Hooker's sermons. He preached on "The Nature of Pride" (a sermon title), and had an impressive funeral sermon called "A Remedy for Sorrow and Fear." He was always impressed by the generosity of grace, and ended a sermon on Matthew 7:7, 8 ("Ask and it shall be given you; seek and ye shall find; knock, and it shall be opened unto you") by asking his hearers to act as if they believed "it is the glory to God to give": "Let there be on our part, be no stop, and the bounty of God we know is such, that he granteth over and above our desires. Saul sought an ass, and found a Kingdom. Solomon named wisdom, and God gave Solomon wealth also, by way of surpassing. 'Thou has prevented thy servant with blessings,' saith the Prophet David. 'He asked life, and thou gavest him long life, even for ever and ever.' God a giver; 'He giveth liberally, and upraideth none in any wise': and therefore he better knoweth than we the best times, and the best means, and the best things, wherein the good of our souls consisteth."

[91] See the second Sermon on Jude, para. 32 and 33: "If there be any feeling of Christ, any drop of heavenly dew, or any spark of God's good Spirit within you, stir it up, be careful to stir it up, be careful to build and edify, first yourselves, and then your flocks, in this most holy Faith. 33. I say, first, yourselves; for he, which will set the hearts of other men with the love of Christ, must himself burn with love."

[92] See the first sermon on Jude.

[93] See the second sermon on Jude, para. 11: ". . . is not all other wine like the water of Marah, being compared to the cup which we bless? . . . Doth not he which drinketh behold plainly in this cup, that his soul is bathed in the blood of the Lamb? O beloved in our Lord and Saviour Jesus Christ, if ye will taste how sweet the Lord is, if ye will receive the King of Glory, build yourselfes." It should be noted that Hooker, advocate of reason as he is, shows his affectionate pastoral concern here and particularly in paragraph 12 of this sermon.

[94] "The Certainty and Perpetuity of Faith in the Elect," para. 6.

[95] *Ibid.*, para. 7.

Among the qualities of Hooker's sermons were the capacity to develop a central and practical theme with many Biblical illustrations and an occasional patristic citation,[96] the appeal to reason and experience,[97] the use of careful definitions,[98] his charitable tolerance,[99] and his quiet and uncontentious spirit. Allied to these was his mastery of the art of persuasion and his use of various literary devices to attain his ends.[100]

Hooker was a master of the balanced sentence. The following is an example of his architectonics, and typical in its robust common sense: "The light would never be so acceptable, were it not for that usual intercourse of darkness. Too much honey doth turn to gall; and too much joy, even spiritual, would make us wantons. Happier a great deal is that man's case, whose soul by inward desolation is humbled, than he whose heart is through abundance of spiritual delight lifted up and exalted beyond measure. Better is it sometimes to go down into the pit with him, who, beholding darkness, and bewailing the loss of inward joy and consolation, crieth from the bottom of the lowest hell, 'My God, my God, why hast thou forsaken me?' than continually to walk arm in arm with Angels, to sit as it were in Abraham's bosom, and to have no thought, no cogitation, but 'I thank my God it is not with me as it is with other men.' No, God will have them that shall walk in light to feel now and then what it is to sit in the shadow of death. A grieved spirit is therefore no argument of a faithless mind."[101] How majestically the magisterial conclusion is arrived at, yet how inevitable!

Hooker also knew how to begin and end sermons. Perhaps the best examples of each can be drawn from the sermon, "The Nature of Pride." It begins: "The nature of man, being much more delighted to be led than drawn, doth many times stubbornly resist authority, when to persuasion it easily yieldeth." It concludes by

---

96 "The Certainty and Perpetuity of Faith in the Elect" is full of such instances, including Abraham, Job, Habakkuk, Peter, John, Paul, Nathanael; and there is a reference to Sallust.

97 The first sermon on Jude, para. 15.

98 "A Remedy against Sorrow and Fear," para. 6, has this definition of fear: ". . . fear is nothing else but a perturbation of the mind, through an opinion of some imminent evil, threatening the destruction or great annoyance of our nature, which to shun it doth contract and deject itself." See also his statement of the cause of superstition in "A Discourse of Justification," sec. 23.

99 "A Discourse of Justification," sec. 27 and 39, where he argued that the Church of Rome, even in its insistence on works, did not deny the foundation of Christianity.

100 See Fritz Pützer, *Prediger des englischen Barok*, pp. 36f.

101 "The Certainty and Perpetuity of Faith in the Elect," para. 6.

agreeing, with St. Augustine, that God helps the conceited soul by withdrawing his grace and so establishes it later the more surely on Him: "Ask the very soul of Peter," demands Hooker, "and it shall undoubtedly make you itself this answer: My eager protestations, made in the glory of my ghostly strength, I am ashamed of; but those crystal tears, wherewith my sin and weakness was bewailed, have procured my endless joy; my strength hath been my ruin, and my fall my stay." These concluding paradoxes, which anticipate both Donne and Andrewes, are also reminiscent of El Greco's *St. Peter's Repentance*,[102] with its great luminous eyes, washed with "crystal tears."

Hooker, too, could create memorable similes and metaphors to fasten his lessons on the memory of his congregations. "As a loose tooth is a grief to him that eateth, so doth a wavering and unstable word in speech, that tendeth to instruction, offend."[103] This is in itself the perfect lapidary comment on Eliphaz's query: "Shall a wise man speak words of the wind?"[104] Equally apt was his vivid description of the bewilderment fear throws Christians into: "But because we are in danger, like chased birds, like doves, that seek and cannot see the resting holes that are right before them: therefore our Saviour giveth his Disciples these encouragements beforehand, that fear might never so amaze them. . . ."[105]

Not all excellences are to be found even in Hooker. The vigour of Latimer and his attack are replaced by a peace that sometimes is indistinguishable from a euphoric blandness. The brilliant surprises sprung by Donne and Andrewes are missing in him. There is probably far too much solid meat in every one of his sermons to be easily digestible by his listeners. His delivery might do for the law court, but it was extremely dull for the pulpit. His strengths were the deep respect he had for the church, the adoration and reverence for Christ and His ordinances, and the dignity he kept even in controversy, and the compassion for humanity, which, though less obtrusive, might be as real as Latimer's.

The first generation of the English reformers needed a prophet such as Latimer, with his thundering demand for obedience to the Gospel, his brusque way with opposition, his denunciation of social injustice, his racy colloquialisms, and his courage and compassion.

---

102 This painting is to be found in the Putnam Gallery of the Art Museum of San Diego.
103 "A Discourse of Justification," sec. 39.
104 Job 15:2.
105 "A Remedy against Sorrow and Fair," last para.

The second generation of Anglicans needed the balance, the attachment to the Church of England, and the sweet reasonableness and deep devotion of Hooker. The first generation needed prophetic courage to begin the task of reformation, the second, saintly sagacity for the task of consolidation. The contribution of Latimer and Hooker were therefore complementary.

CHAPTER VII

PURITAN WORSHIP[1]

THE SCOPE of Anglican worship was firmly defined by the official prayer-books of the Church of England. Puritan worship is a much more amorphous term, partly because of the various groups included within Puritanism, partly because of the change in the interests of Puritans from one decade to another.

Puritanism included loyal Anglicans using the prayer-book, yet hoping for an improved formulary of worship less open to ceremonial or Vestiarian criticism and encouraging a greater use of expository preaching. The term embraced embryonic Presbyterians—known as "Disciplinarians"—who regarded *John Knox's Genevan Service Book*[2] or its revisions prepared by English Puritans[3] as more truly patterned on God's word and the usage of the primitive church. Third, Puritanism included the semi-Separatists, like the Leyden minister of the Pilgrim fathers, John Robinson, and the proto-Independents such as the Brownists and Barrowists who desired to establish a more charismatic worship, with full opportunity for extempory, Spirit-inspired prayers. All three forms were used by Puritans.

Furthermore, the interest of Puritanism differed, even in liturgical matters, in three different decades. In the 1560s Puritanism centred on trivial sartorial and ceremonial issues. Could a sincerely Protestant minister wear the surplice, a garment worn by the sacrificing priests of the Roman Catholic church, without compromising his faith? Could he legitimately take part in an Anglican liturgy that required him to bow at the reception of the consecrated species in Holy Communion, and thus commit the idolatry of worshipping the created instead of the Creator, to say nothing of seem-

---

[1] My *The Worship of the English Puritans* (1948) is an extensive treatment of the theme from approximately 1550 to 1750, with full documentation. The present chapter reconsiders the issues only in the sixteenth century, and uses supplementary research materials.

[2] See William D. Maxwell, *John Knox's Genevan Service Book 1556*.

[3] The so-called Waldegrave and Middleburg editions. The former is entitled, "A booke of the forme of common prayers, administration of the Sacraments: &c. agreeable to Gods Worde and the use of the reformed Churches," and was probably printed in 1584 or 1585. There is a reprint of it in Peter Hall, *Fragmenta Liturgica*, I. The Middleburg edition was printed by Schilders of that city in 1586, 1587, and 1602. It is reprinted in Peter Hall, *Reliquiae Liturgicae*, I. The originals are in the British Museum. See British Museum Catalogue, vol. L 50, pp. 711-12.

ing to condone the rejected Roman doctrine of Transubstantiation? Could he decently make the sign of the cross in administering Baptism, since this was to introduce a sacramental sign, if not a new sacrament, without explicit sanction from the Bible? Could he insist on the wearing of rings in the service of Holy Matrimony when there was no authority for it in God's word? Small as the issues were, they raised the fundamental question of authority for worship—the Bible or tradition? As was seen before,[4] the bishops claimed that where the Scriptures were silent, the church could act, while the Puritans insisted on an explicitly Biblical sanction for every ordinance of worship and almost every detail of worship.

When, however, the battle for a more truly Reformed type of worship moved into the seventies, with the Admonition controversy, there was a far more radical criticism of the Book of Common Prayer, and a dissatisfaction with episcopacy (especially as prelatically involved in demanding obedience through the mechanisms of High Commission and Star Chamber) and with the ecclesiology of the Church of England since it lacked provision for disciplining and purifying the church membership. It was in this period, under the guidance of Cartwright, Travers, and Fenner, as well as Field and Wilcox, that the Genevan pattern in worship, church order, and government, and discipline was openly espoused.

In the 1580s, however, the difference between the patient and impatient Puritans widened. Robert Browne, a leader of the impatient Puritans, became a Puritan Separatist and established independent congregations in Norwich and elsewhere. After his release from prison, to which he was condemned for an act of schism, he and his congregation emigrated to Holland. In Middleburg in 1582 he published a call for creative impatience in *A Book which sheweth the Life and Manners of all true Christians*, and in *A Treatise of Reformation without tarrying for any and of the Wickedness of those Preachers which will not reform till the Magistrate command or compel them*. His conviction was that reformation "was not to be begun by whole parishes, but rather of the worthiest, were they never so few." He insisted that such "gathered Churches,"[5]

---

[4] See Chap. II.

[5] It is pertinent to cite Browne's definition of a "gathered" church: "The Church planted or gathered [from out of the world] is a company or number of Christians or believers, which, by a willing covenant made with their God, are under the government of God and Christ, and keep his laws in one holy communion: because Christ hath redeemed them unto holiness and happiness for ever, from which they were fallen by the sin of Adam." (From *A Booke which sheweth the Life and Manners of all true Christians* (Middleburg, 1582), C3 r.; also ed.

bound under God by covenant were to be independent of the state and to be self-governing. Barrow, Greenwood, and Penry were the martyrs of the new *ecclesiolae*, protesting in the name of the freedom of the gospel; but Browne recanted and received episcopal ordination in 1591, and was rector of Achurch, Northants until his death in 1633.[6] In 1588 and 1589 the *Marprelate Tracts* voiced the growing bitterness of the Puritans against the entrenched episcopal establishment.

More positively during the same period, Puritan ministers were giving an example of the godliness and learning that was at the heart of Puritanism. Richard Greenham fulfilled an exemplary pastorate at Dry Drayton, Cambridgeshire from 1570 to 1591, where he also taught pastors, including the brilliant Puritan preacher, Henry Smith, lecturer at St. Clement Danes from 1587, and his son-in-law, the long-lived John Dod, whose kindly witticisms were widely quoted.[7] William Perkins, fellow of Christ's College, Cambridge, had a great influence as a teacher and preacher. He was the most influential of all the early Puritan moral theologians;[8] his *Arte of Prophesying* was the major manual of Puritan preaching. Perkins died in 1602, and his mantle fell on his successors, Paul Baynes and the great theological casuist, William Ames.

It is necessary to look back at Puritanism with the eyes of John Geree, writing in 1646, in *The Character of an old English Puritane or Nonconformist* to find the essential spirit of Puritanism, especially as it relates to worship:

> The Old English Puritane was such an one that honoured God above all, and under God gave every one his due. His first care was to serve God, and therein he did not what was good in his own, but in God's sight, making the word of God the rule of his worship. He highly esteemed order in the house

---

A. Peel and L. H. Carlson, *The Writings of Robert Harrison and Robert Browne*, p. 253. The Church is to consist only of Christians, all members are to share the privileges and responsibilities of its government; and the Church is to be free of episcopal and royal interference.

6 Browne, however, died in Northampton gaol, where he was incarcerated for assaulting a policeman. It is understandable why Independents (later Congregationalists) disliked being called "Brownists."

7 He would, for example, inform friends who had come into money that although they had transferred from a boat to a ship, they were still at sea. Cf. Gordon S. Wakefield, *Puritan Devotion*, pp. 7-8.

8 In the seventeenth century he was considered the equal of Hooker and almost the equal of Calvin.

of God: but would not under colour of that submit to superstitious rites, which are superfluous and perish in their use.
. . . He made conscience of all God's ordinances, though some
he esteemed of more consequence. He was much in praier;
with it he began and closed the day. In it he was exercised in
his closet, family and publike assembly. He esteemed that
manner of praier best, where by the gift of God, expressions
were varied according to present wants and occasions; Yet he
did not account set forms unlawful. Therefore in that circumstance of the Church he did not wholly reject the liturgy but
the corruption of it. He esteemed reading an ordinance of God
both in private and publike; but he did not account reading to
be preaching. . . . The Sacrament of Baptism he received in
Infancy, which he looked back to in his age, to answer his
ingagements, and claim his priviledges. The Lord's Supper
he accounted part of his soul's food: to which he laboured to
keep an appetite. He esteemed it an ordinance of nearest communion with Christ, so requiring most exact preparation.[9]

Here was the absolute primacy of the Biblical criterion in Puritan
worship—"to serve God . . . making the word of God the rule of
his worship," the preference for extemporary prayer without denying the value of set forms, the importance of preaching as of meditation on God's word, and the centrality of the Lord's Supper, "an
ordinance of nearest communion with Christ." In all of it there was
the profound sincerity and inwardness, as well as intensity, of
Puritan worship. Geree's final words in the characterization of an
old Puritan are: "his whole life he counted a warfare, wherein
Christ was his captain, his arms, praiers, and tears. The Crosse
his Banner and his word *Vincit qui patitur.*"[10]

1. SOLA SCRIPTURA *as Liturgical Criterion*

The main principle of the absolute authority of God's word in
the Scriptures for faith, ethics, and worship was expressed by all
Puritans. To depart from this is the utmost human impertinence
and pretentiousness, for it implies that one knows God's will better
than He does, or that the inherent weakness of original sin does not
blind one's judgment through egocentricity. God's word is a full,
not a partial revelation of the divine purpose for humanity. Con-

[9] Cited by Wakefield, *Puritan Devotion*, p. xxii. Incidentally, the second chapter
is an excellent study of Puritan interpretation of Scripture.
[10] *Ibid.*

siderations of tradition or aesthetics are, therefore, strictly irrelevant once the Biblical criterion for worship has been accepted. Even when there are cases not mentioned in the Scriptures, St. Paul has provided four secondary criteria for testing for all orders and ceremonies of the church.[11]

On this basis of the binding authority of Scripture the Puritans established that the worship of the Apostolic Church was characterised by six ordinances—prayer, praise, the reading and preaching of the word, the administration of the sacraments of Baptism and the Lord's Supper, catechising, and the exercise of discipline. The *locus classicus* for Apostolic worship was Acts 2:41-42: "Then they that glady received his worde were baptized; and the same day there were added to the Church about thre thousand soules. And they continued in the Apostles doctrine and felowship, and breaking of bread, and prayers."[12] The types of prayer, supplications, intercessions, thanksgivings, were derived from I Timothy 2:1ff. and invocation or adoration and confession were derived from the example of the Lord's Prayer. For praise, they turned to Ephesians 5:19, with its reference to "psalmes, and hymnes and spiritual songs, singing and making melodie to the Lord in your hearts."[13] The proclamation of the Gospel became the central feature of Puritan worship, its importance attested to by the entire corpus of the Scriptures, which provided a saving knowledge of God, but especially in II Corinthians 1:12 and Romans 10:14-15. Their authority for the "Gospel Sacraments" was Matthew 28:19-20 and I Corinthians 11:23-26, among other texts. Puritans had to look hard to find a text to justify their using a set form of words for catechism when they tended to reject both set sermons (read homilies) and set forms of prayer (a liturgy). II Timothy 1:13 provided the required proof-text: "Kepe the true paterne of the wholsome wordes which ye has heard of me in faith and love which is in Christ Iesus."[14] Finally, the demand for purity in the church to be protected by the exercise of ecclesiastical censures was sanctioned by Matthew 18:15-18 and 18:2; I Corinthians 5:3-5, and the Third Epistle of John 10, which also provided the complete procedure for admonishment, excommunication, and readmission of penitent offenders.

---

[11] See Chap. II for the four Pauline epistolary references for the secondary criteria of worship.
[12] The Genevan version of the Bible, 1560.
[13] *Ibid.*   [14] *Ibid.*

The Scriptures also provided the Puritans with the authority for their occasional ordinances. Prophesyings, gatherings for expounding the Scriptures and for answering questions due to difficulties in comprehending their meaning, were sanctioned by I Corinthians 14:1 and 31. If a day of humiliation was held for the corporate expression of the people's penitence following a great natural, political, or military calamity, the invariable order was fasting, prayer, and sermon, as in Acts 13:1-3 and 14:23. When the exiled Puritans in Arnhem anointed their sick with oil and prayed for their recovery, it was strictly according to the model supplied by James 5:14-15.

Even such details as the frequency of divine services were sought in the Scriptures. The double burnt offering in Numbers 28:9 sanctioned two services on the Lord's Day; even the necessity for the Lord's Supper being preceded by a sermon was supplied by the precedent of Acts 20:27. Acts 4:36 and I Corinthians 16:2 enabled the Puritans to determine who should collect the offertory and at what point in the service it should be presented. Text scrutinizing went to ludicrous extremes on occasion, however. It was argued that marriage, not being a sacrament, did not require the services of a minister for its solemnisation, since it was not included in the list of pastoral duties in II Timothy 4:2ff. A text became a pretext when Matthew 11:28, "Come unto me, all ye that are wearie & laden, and I will ease you," was taken as proof that the only appropriate posture at the Lord's table was sitting, as witnessing to the rest that Christ promised to his disciples. Of equal perversity and ingenuity was Cartwright's argument for stability of position in worship (over against the mobility of the one presiding over Anglican worship who moves from prayer desk to lectern to pulpit to altar) drawn from Acts 1:15: "Peter stood up in the midst of the disciples." The most extreme example was the denial of the right to have responses in public prayers because the flesh of the repetitive cuckoo was forbidden in Leviticus.[15]

While the Puritans in their controversy with the Anglicans were not above straining the natural meaning of texts to yield significances unintended by the writers of the Biblical books, yet normally they were far from being legalists. It was they who disapproved of burdensome ceremonies, citing Acts 15:28 as their warrant for

[15] This example is given without bibliographical reference by the distinguished Scottish liturgiologist, the Rev. Dr. William McMillan, author of *The Worship of the Scottish Reformed Church 1550-1638*, in a book review in *The Dunfermline Press and West of Fife Advertiser* (24 July, 1948).

Christian freedom. Again it was not the Puritans but the Anglican bishops who were trying to impose vestments, which seems Aaronical rather than Christian, and became a stumbling block to the weaker brethren. It was in the strength of their conviction of the Scripture as the sole criterion for determining the ordinances and the details of worship that they went on to attack the Book of Common Prayer.

2. The Puritan Critique of the Prayer-book

Hooker provided a fair summary of the Puritan critique of the Book of Common Prayer, because it had

> too great affinity with the Form of the Church of Rome; it dif-fereth too much from that which Churches elsewhere reformed allow and observe; our attire disgraceth it; it is not orderly read, nor gestured as beseemeth; it requireth nothing to be done which a child may not lawfully do; it hath a number of short cuts or shreddings, which may be better called wishes than Prayers; it intermingleth prayings and readings in such manner, as if suppliants should use in proposing their suits unto mortal Princes, all the world would judge them mad; it is too long, and by that mean abridgeth Preaching; it appoint-eth the people to say after the Minister; it spendeth time in singing and in reading the Psalms by course, from side to side; it useth the Lord's Prayer too oft; the Songs of *Magnificat*, *Benedictus* and *Nunc Dimittis*, it might very well spare; it hath the Litany, the Creed of Athanasius, and *Gloria Patri*, which are superfluous; it craveth earthly things too much; for deliverance from those evils against which we pray it giveth no thanks; some things it asketh unseasonably, when they need not be prayed for, as deliverance from thunder and tempest, when no danger nigh; some in too abject and diffi-dent manner, as that God would give us that which we for our unworthiness dare not ask; some which ought not to be desired, as the deliverance from sudden death, riddance from all adversity, and the extent of saving mercy towards all men. These and such like are the imperfections whereby our form of Common Prayer is thought to swerve from the Word of God.[16]

16 *Ecclesiastical Polity*, Book v, chap. xxvii, sec. 1.

Within its brief scope this was a remarkable summary of the many incisive criticisms of the Prayer Book of the Church of England by the Puritan writers. While the Second Prayer Book of Edward VI required a priest or deacon only to wear a surplice, in 1559 Queen Elizabeth reinstated the ornaments rubric of the second year of King Edward VI's reign, requiring alb and vestment or cope. No amount of episcopal insistence that the vestments were to be worn for the sake of decency, comeliness, and good order blinded the Puritans to the fact that the vestments, theoretically *adiaphora* and of their nature indifferent, were made essential requisites of Anglican worship. Puritans argued that the requesting of the vestments was an infringement of the crown rights of Christ the Redeemer, who had freed Christians from the bondage of the ceremonial law of the Old Covenant which was now being reintroduced by Archbishop Matthew Parker in his *Advertisements*.[17] Other Puritan arguments used against the vestments were their association with the now discredited Roman Catholic faith, that they were garments of priests who believed in the sacrifice of the Mass and in Transubstantiation, and not at all suitable for a faith which asserted the doctrine of the priesthood of all believers, and that they were symbols of pomp and grandeur wholly unsuitable for the disciples of a humble Christ. Furthermore, they were out of step with the Continental Reformed churches in the matter of vestments.

The three "noxious" ceremonies were discarded as having no Biblical authority. *A Survey of the Booke of Common Prayer by way of 197 Queres grounded upon 58 places*[18] argued that kneeling at the Communion was un-Apostolic, unprimitive, unreformed, and utterly unsuitable for those who wished to avoid the appearance of evil. The same was true of the other two ceremonies, namely the crossing of the child in Baptism and the use of the ring in marriage, which were considered by Puritans to be unwarrantable human additions to the Dominical institutions as recorded in Scripture, and therefore arrogantly presumptuous. Query 95 of the *Survey* asked:

17 *A Parte of a Register* (c. 1590, MS in the Dr. Williams' Library, Gordon Square, London), p. 41, states: "if it be abolished and *Christ* bee come in steede, then a great iniurie is done to Christ for manie causes. The one is, that those ceremonies which Christ by his passion did abolishe should in contempte of him and his passion be taken agayne." The interpretation of the Ornaments Rubric and the relative authority of Parker's *Advertisements* are controversial issues. For a brief note, with suggested further reading, see the *Oxford Dictionary of Church History*, ed. F. L. Cross, p. 995b.

18 It is to be found in W. H. Frere and C. E. Douglas, *Puritan Manifestoes*, reissued with a preface by Norman Sykes, 1954.

"Whither the childe be not received againe by and with Crossing, and so may seeme to be a sacrament as well as Baptism for that cause . . . as if regeneration were by baptisme, and incorporation by crossing?"[19] Robert Johnson of Northampton asked where was the consistency, why the shaven crown was rejected but not the square cap, why the tippet was commanded and the stole forbidden, and "Wee would knowe why you do reiect *hallowed beades*, and yet receyve *hallowed ringes.* . . . ?"[20] His conclusion was: "And of these things I say we would have and heare some reason taken and gathered out of the word of God, which if we shall heare, we shall be gladde to learn. . . . If the conscience be perswaded, the hande shall straightway subscribe." If God did not ordain these ceremonies, how can they either please Him or build up his people in the faith?

Puritans were also greatly concerned that officially prescribed homilies, produced during a shortage of educated Protestant clergy, would oust sermons. This was why they berated the clergy of the Church of England as "dumb dogs" who could read the Book of Common Prayer and the homilies, but who could not or would not preach. The Puritans argued that while the word was all one, whether read or preached, nonetheless the preached word was more effective when it was applied to the minds and hearts of the congregation, whether for information, consolation, or rebuke. The point was, "As the fire stirred giveth more heat, so the Word, as it were, blown by preaching, flameth more in the hearers than when it is read."[21]

Archbishop Grindal, in his historic letter to the queen, stated the Puritan position strongly in his defence of "Prophesyings": "The Godly Preacher is termed in the Gospel a Faithful Servant, who knoweth how to give his Lord's family their appointed food in season; who can apply his speech to the diversity of times, places and hearers; which cannot be done in Homilies: exhortations, reprehensions, and persuasions, are uttered with more affection, to the moving of the Hearers, in Sermons than in Homilies."[22] Hooker himself grudgingly admitted the power of Puritan sermons as "the art which our adversaries use for the credit of their Sermons," and acknowledged the "especial advantages which Sermons naturally

19 *Ibid.*, p. 90.
20 *A Parte of a Register*, p. 104.
21 Hooker's *Ecclesiastical Polity*, ed. Hanbury, vol. ii, p. 76n.
22 See John Strype, *The History of the Life and Acts of Edmund Grindal*, p. 595.

have to procure attention, both in that they always come new, and because by the hearer it is still presumed, that if they be let slip for the present, what good soever they contain is lost, and that without all hope of recovery."[23]

The Puritans objected also to the lections in the Book of Common Prayer. They took exception to the number of readings from the Apocrypha as implying a slight on the sufficiency of the canonical Scriptures,[24] and to the fragmentation of Scripture in the epistles and gospels.[25] Chillingworth would later affirm that the Bible was the religion of Protestants. The Bible, the whole Bible, and nothing but the Bible, is the whole religion of Puritans.

The prayers of the prayer-book were also heavily criticised. The collects were, in their view, as Hooker suggested, "wishes" rather than prayers, mere "short-cuts" to prayer, as contrasted with the longer prayers of the primitive and Reformed churches.[26] Their brevity was distracting for the Puritans, but for Hooker they had the advantage of keeping the worshipper awake by a "piercing kind of brevity."[27] The accuracy of collects for celebrated anniversaries was also questioned[28] and the stylized endings were considered "vain repetitions." The Litany was thought to be something of an anachronism, since its petitions were unrealistic and inappropriate to the condition of most who used it. The people's responses were stigmatised as "vain repetitions" and a mere tossing to and fro of tennis balls. The latter was a common derisive metaphor employed by the Puritans for responses which they considered were forbidden by St. Paul's words in I Corinthians 14:6, requiring only one person to speak at once, and approving *Amen* as the only word of corporate response permitted.[29]

Not only the types of prayer were criticised, but also the content. It was asserted that the prayer-book craved too many material blessings. For example, the author of *The Second Admonition*

23 *Ecclesiastical Polity*, Book v, chap. xxii, sec. 20.
24 *A Survey of the Booke of Common Prayer* complains: "In place of 182 chapters canonical left unread, there be 132 out of the Apocrypha appointed by the Calendar to be read:" Frere and Douglas, *Puritan Manifestoes*, p. 26.
25 Query 44 of *A Survey* asks: "Whither the reading of these Epistles and Gospels (so called) be not the same fault which is blamed as unorderly in the preface of the Communion Book, *viz.* a breaking of one peece of Scripture from another?"
26 See *The Second Admonition*, i, 138; iii, 210ff.
27 *Ecclesiastical Polity*, Book v, chap. xxxiii.
28 For example, the Collect for the Sunday after Christmas reads: "Almighty God who hast given us thy only begotten Sonne to take our nature upon him and *this day* to be borne of a pure Virgine, &c." See Frere and Douglas, *Puritan Manifestoes*, p. 57.
29 *Ibid.*, p. 49.

believed that more than a third of the prayers, excluding psalms and Scriptural citations, were "spent in praying for and praying against the (commodities and) incommodities of this life, which is contrary to all the arguments or contents of the Prayers of the Church set down in Scripture, and especially of our Saviour Christ's Prayer, by the which ours ought to be directed."[30] Two petitions in the collects were criticised for smacking more of Popish fear than Christian confidence.[31] Finally, the criticisms of the prayers revealed a fundamental difference of attitude between the Anglicans and the Puritans on the repetition of the Lord's Prayer. For some Puritans the Lord's Prayer was a pattern for prayer to be imitated; for all Anglicans it was a set prayer to be repeated.

The Puritans also had criticisms to offer of the orders for the celebration of the sacraments in the prayer-book. In baptism the Puritans took exception to the ceremony of the signing with the cross, private Baptisms, Baptism by women, the interrogatories put to the child, and the custom of godparents acting as sponsors of the child. These practises were all contrary to the Reformed doctrine of Baptism as expounded, for example, in Calvin's *Institutio Christianae Religionis*.[32] The practices denied the fact that God's promises are sealed in the presence of the Christian community, not in the corner of a house. Women were forbidden by St. Paul to speak in the church; if they baptised they usurped the pastor's responsibility. Children inherited the promises made to the faithful and their seed in virtue of the election to grace of their parents, and it was parents who must stand as sponsors for their children, not godparents. The interrogations addressed to children were criticised as simply irrational.[33]

Puritan objections to the Anglican way of celebrating Holy Communion were many. The mode of individual delivery, *seriatim*, was thought to contradict the simultaneous corporate reception envisaged by Christ whose words were in the plural number.[34] They

[30] *Second Admonition*, i, 136.

[31] The Collect for the twelfth Sunday after Trinity contains the clause, "and giving unto us that that our prayer dare not ask" and the post-offertory prayer in the Communion Office with the clause, "and those things which for our unworthiness we dare not . . . ask." Frere and Douglas, *Puritan Manifestoes*, sec. vii, "The Humble Petition of 22 Preachers."

[32] *Ecclesiastical Polity*, Book iv, chap. xvi.

[33] The Puritan baptismal criticisms are detailed in the first *Admonition to Parliament* as found in Frere and Douglas, *Puritan Manifestoes*, p. 14; and in *A Parte of a Register*, pp. 55ff.

[34] Query 62 of *A Survey of the Booke of Common Prayer* asks: "Whither the delivery of the Communion into the hands of the communicants be according to

also disapproved of substituting for the Scriptural words of delivery others neither scriptural nor evidently grounded in Scripture.[35] Even greater was the Puritan deprecation of the cheapening of the Holy Communion by admitting the unworthy to it without any examination of the quality of their life.[36] Criticism was also voiced at the requirement of a thrice-a-year minimal attendance at the Lord's table.

An admirable summary of Puritan criticisms of the Anglican Communion was made by the authors of *An Admonition to Parliament*, by means of contrasting the primitive order of Communion with the prayer-book order:

> They had no introite . . . but we have borrowed a piece of one out of the masse booke. They read no fragments of the Epistle and Gospell: we use both. The Nicene Crede was not read in their Communion we have it in oures . . . (examination of communicants then, not now). Then they ministred the Sacrament with common and usual bread; now with wafer cakes. . . . They received it sitting; we kneeling, according to Honorius decree. Then it was delivered generally and indefinitely, Take ye and eat ye: we particularly and singulerly, Take thou and eat thou. They used no other wordes but such as Chryste left: we borrow from Papistes The body of our Lorde Iesus Chryste which was geven for thee,&c. They had no Gloria in excelsis in the ministerie of the Sacrament then, for it was put to afterward. We have now. They took it with conscience. We with custume. They thrust men by reason of their sinnes from the Lords Supper. We thrust them in their sinne to the Lordes Supper. They ministred the Sacrament plainely. We pompously, with singing, pyping, surplesse and cope wearyng. They simply as they receeved it from the Lorde. We, sinfullye, mixed with mannes inventions and devises.[37]

Here we may pause to recognize the strength of the characteristic Puritan pleas for simplicity, for fidelity to God's Word and the example of the primitive Church, for the sacredness of God's ordinances, and for a high seriousness and sincerity in worship.

Christ, his institution. Seeing he said, and in the plural number said, Take ye, eate ye, drink ye all, &c."

[35] This objection is found in the second part of the 62nd Query of *A Survey of the Booke of Common Prayer.*

[36] Frere and Douglas, *Puritan Manifestoes*, p. 13.

[37] *Ibid.*, p. 134.

The Puritans did not spare the other occasional orders of worship in the prayer-book from criticism. Marriages and burials were considered essentially civil, not religious. Exception in the marriage order was taken to the words, "With my body I thee worship," as a derogation of the recognition that worship is alone due to God and not to his creatures, however lovely or distracting. In the burial service exception was taken to the indiscriminate, euphoric, and universalistic words: "We commit his body to the ground in sure and certain hope of the resurrection to eternal life."[38] The order for "The Churching of Women" was approved as a thanksgiving for safe delivery of a mother from childbirth, but in its form it was criticised as resembling too much the Jewish service of purification, and confusing ritual impurity with ethical sin.[39] The Puritans objected to Confirmation as virtually the creation of a third sacrament, which since it required a bishop for its administration, seemed to reduce the significance of the other two sacraments which required no bishop for their administration. It was also felt that the Lord's Supper added the grace of confirmation of Baptism and that a separate service of Confirmation was therefore otiose. In the third place, the Puritans objected to the laying on of hands in a manner that was un-Apostolic while claiming to follow the model of the apostles. The Puritans, with their Genevan view that bishops or ministers (apart from apostles and prophets) were the sole order of the ministry, recognised in the Church of New Testament days, while elders and deacons, who were lay officials, found the Ordinal unacceptable. They also vigorously attacked the presumption of the presiding bishop being required to say, while laying hands on the ordinand, "Receive ye the Holy Ghost," interpreting, instead, the mood of the verb as imperative instead of optative. Therefore it seemed blasphemous to them to imply that the Holy Spirit was in episcopal captivity.

Finally, the Puritan, while rejoicing that so many saints' days had been removed from the calendar of the Church of England, still believed that there was only one festival of the church, and that was a weekly one—the Lord's Day, commemorating the Lord's resurrection from the dead, the new exodus of the New Covenant. On this day every week he rehearsed the mighty saga of God in

[38] *Ibid.*, p. 144. Query 173 of *A Survey*.

[39] Query 179 of *A Survey* asks: "Whither weake and superstitious womē may not be occasioned to thinke this Service rather a Purification (which were Iewish) than Thankes-giving . . . as if women were by childbirth uncleane, and therefore unfit to go about their business, until they are purified." Frere and Douglas, *Puritan Manifestoes*.

the creation, redemption, and sanctification of man through the celebration of the incarnation, life, atonement, and resurrection of Christ. Thus the entire drama of salvation was to be recalled each Lord's Day. There was, therefore, no need for separate festivals, which would only fragment God's self-revelation in the continuing deed of Christ. Saints' days were also regarded as a denial of the sole mediatorial role of Christ, as well as diminishing the glory due to God alone. Puritans had special Fast Days and Thanksgiving Days in which they reflected on particular manifestations, respectively, of God's judgment and mercy. However, Thanksgiving Days were occasional and rare, and were no more than the foothills of their corporate devotion. The Lord's Day services were the peaks of their life of public worship.

3. Prayer-books or Prayers without a Book?

One major issue for the Puritans in our period was whether corporate worship should be according to the book or the spirit. At issue was whether God was best worshipped through a liturgy (and one which was more fully based on the word of God than the Book of Common Prayer), or through extemporary prayer more fully responsive to the particular needs of a given congregation than a liturgy could be and which trusted the guidance of the Holy Spirit in the leading of worship. It seems, on the face of it, as if the Presbyterian Puritans preferred a liturgy like Calvin's Englished by John Knox, which also allowed a very restricted area of freedom for the minister in the words he might use in the prayer of illumination before the sermon, and that the non-Presbyterian Puritans considered a liturgy allowable but extemporary prayer by the minister as preferable. Both viewpoints were held by Puritans until the Westminster Assembly approved a compromise which allowed for order and liberty, by providing a manual with some set prayers and alternatives, as well as suggested topics for prayer and a theological structure and order of items. But that came a generation and a half after the period under study here.

The Puritans ransacked the Scriptures for light on the debate. Their search was inconclusive, however; for although the Old Testament yielded set forms of praise, as in the Psalms (many of which were equally prayers) and the Aaronic blessing (Numbers 6:24ff), these were given under the Old Dispensation and were not necessarily to be imitated by Christians. The New Testament was equally inconclusive, for Anglicans could argue that the Lord's

Prayer was intended to be repeated (as the Presbyterians did), while the semi-Separatist Puritans such as John Robinson insisted that it was a model of prayer on which the pastor was to create his own prayers. Furthermore, the non-Presbyterian Puritans, who might be denominated proto-Independents, insisted that responsive prayers were vetoed by the insistence in I Corinthians 14:14 and 16 that only the minister's voice was to be heard in prayer, except for the concurrence of the congregation in uttering "Amen," and that the use of read prayers would seem to be a "quenching of the Spirit," for "the Spirit helpeth, our infirmities for we know not to pray for as we ought."[40] Robinson's comment on this text ironically suggests that the Anglicans considered their prayer-book superior to the inspiration of the spirit: "Yes, Paul, with your leave, right well; for we have in our prayer-book what we ought to pray, word for word, whether the Spirit be present or not."[41]

The Brownists, the Barrowists, and the earliest Independents were decidedly against the repetition of the Lord's Prayer. The Brownists in their Confession of 1596 complained that "we are much slandered, if we denyed or misliked that forme of prayer commonly called the Lord's Prayer"[42] because they did not use it in public worship. John Penry, the Barrowist leader, denied that Christ's wish was, "that the disciples or others should be tyed to use these very wordes, but that in prayer and geving of thankes they should follow his direction and patterne which he had geven them, that they might know to whom, with what affeccion and to what end to pray. . . ."[43] John Robinson, Leyden pastor of the Pilgrim fathers, decided, among other considerations, that the different Matthaean and Lucan forms of the Lord's Prayer indicated that the Lord's Prayer was not intended to be repeated word for word; the absence of apostolic testimony to the use of this prayer demanded that it be regarded as a model for prayer rather than as a prescribed prayer.[44] Though the Puritans acknowledged Scripture as sole authority for doctrine and practice, they would for argument's sake turn to tradition occasionally as a support. For example, Robinson produced Tertullian as a witness from the early Church for the practice of *ex tempore* prayer, citing the *Adversus*

---

[40] This key Puritan proof text comes from Romans 8:26.
[41] *A just and necessary Apology* in *Works*, ed. Robert Ashton, 3 vols., III, 21ff.
[42] Williston Walker, *The Creeds and Platforms of Congregationalism*, p. 61.
[43] Champlin Burrage, *The Early English Dissenters in the light of recent Research*, 2 vols., I, 56.
[44] *Works*, vol. 1, p. 23.

Gentes: "We pray, saith he, without any to prompt us, because we pray from the heart."[45]

In making the case for free prayer as against liturgical forms the Puritans combined theological and practical objections. In the first place, prayers that were read implied the self-sufficiency of man and denied the effects of original sin in him as well as the conviction that God must be served as his word dictates, and his word dictates a dependence on the holy spirit. It was also felt increasingly that if ministers did not exercise the gift of free prayer the talent would atrophy in them. Another consideration was that set prayers had insufficient flexibility to meet the changes of providence in the life of the nation or the individuals making it up. Yet another reason for desiring to use free prayer was the conviction that it might be thought to be the only way to worship God if set prayers in a liturgy were made legally binding. The Puritans were careful in distinguishing the demands of God from the preferences of men; they considered the demands of God to be prayer; the preference of men in the Established church was for liturgy with unchanging formulas. Finally, the imposition of set liturgies brought persecution in its wake.

In the last analysis the two types of prayer, liturgical and free, represented two differing conceptions of the nature of the church. The Book of Common Prayer and all liturgies stressed the corporate nature of the church, while free prayer emphasized the need of individuals in a family church. If liturgical prayer reflects what is held in common in its creeds, confession, and abstract collects praying for the graces of the Christian life needed by all Christians, then free prayers made room for the distinctiveness of the individuals in the gathered church. Liturgical prayer did not require that the minister know all the members of his congregation. Free prayer, on the contrary, presupposed as its context a compact community, all of whom were known by the minister and who could voice their aspirations and needs in prayer. The two kinds of prayer seemed to bear out Ernst Troeltsch's demarcation between "Church" and "Sect" types of the Christian life. Liturgical prayer required a parish as a background, with the nation as the wider horizon; free prayer suggested not Israel but the holy remnant, a committed and closely knit family for whom the charismatic, intimate, and informal ways of free prayer were congenial.

---

[45] The relevant part of the Latin original is: *sine monitore, quia de pectore*, citation from Robinson, *Works*, III, 28.

Each type of worship had the defects of its own qualities. Precisely because the Book of Common Prayer envisioned the English nation on its knees before God, its prayers had to be general, "common" prayers, and therefore impersonal ones. This was where the author of the Second Admonition found the greatest weakness in the Book of Common Prayer: "there is no such praying as should touch the hearte."[46] Because free prayer thinks in terms of the "gathered" church its characteristic is a sincere and informal simplicity as between friends; its almost intrinsic defect is disorder.

Anglican critics, of course, stressed the weaknesses of free prayer as strongly as the Puritans did those of the Anglican liturgy. They claimed that free prayer might be no more than the result of mental laziness, could tend to ostentation or vulgarity, and wrongly presupposed that all ministers could express themselves fluently and felicitously in public, or that the people could give their assent on the first hearing of a prayer with deep meaning. The criticisms were put comprehensively by Hooker:

> To him that considereth the grievous and scandalous inconveniences whereunto they make themselves daily subject, with whom any blind and secret corner is judged a fit house of Common Prayer; the manifold confusions they fall into, where every man's private spirit and gift (as they term it) is the only Bishop that ordaineth him to this ministry; the irksome deformities whereby, through endless and senseless effusions of indigested Prayers, they oftentimes disgrace in most insufferable manner the worthiest part of Christian duty towards God, who herein are subject to no certain order, but pray both what and how they list; to him, I say, which weigheth duly all these things, the reasons cannot be obscure why God doth in public Prayer so much respect the solemnity of places where, the authority and calling of persons by whom, and the precise appointment even with what words or sentences his name should be called on amongst his people.[47]

In considering such a citation, one is inclined to wonder whether there was a different theology, as well as ecclesiology, differentiat-

---

[46] Frere and Douglas, *Puritan Manifestoes*, p. 115. The quotation continues: "And therfore another hath so little feeling of the common prayer that he bringeth a booke of his owne, and though he sitte when they sitte, stand when they stande, kneele when they kneele, he may pause sometime also, but most of all he intendeth his owne booke, is this praying?"

[47] *Ecclesiastical Polity*, Book v, chap. xxv, sec. 5.

271

ing Anglican and from Puritan worship. Hooker, at the end of the citation above, seems to imply that worship is court etiquette, an approach to the King of kings, where awe dominates over affection, whereas however strong their sense of holiness in approaching God the Puritans did not forget that it was holy *love* to which they responded when they cried, "*Abba*, Father" in their simple and often spontaneous confessions and petitions.

There was, I must again insist, a difference of opinion among the Puritans on the question of set and free prayers. There was agreement that the Book of Common Prayer was unacceptable as the permanent liturgy of Protestant England. Objection to this particular liturgy was before too many decades to lead to a radical rejection of all liturgies. Cartwright, the first leader of the English Presbyterians, who was well acquainted with the practice of the Continental Reformed churches in spells of involuntary exile, strongly favoured a set form of prayer as a medium for expressing the unity of Reformed churches, as a direction of ministers in sound doctrine and prayer, and as a means of assisting the weaker brethren—"and yet so, as the set forme make not men sluggish in stirring up the gift of prayer in themselves, according to divers occurents; it being incident to the children of God in some measure."[48] William Perkins fully allowed that a set form of prayer might be used, but only because, "to conceive a forme of prayer requires gifts of memorie, knowledge, utterance, and the gifts of grace," and not every child of God had these gifts, though he might have an honest heart. Hence, "in the want of them [such] may lawfully use a set forme of prayer, as a man that hath a weake backe, or a lame legge, may lean upon a crutch."[49] Perkins even met the objection that set forms of prayer "limit and binde the Holy Ghost" with the answer: "If we had a perfect measure of grace, it were somewhat, but the graces of God are weak and finall in us. This is no binding of the Holy Ghost, but a helping of the Holy Spirit, which is weake in us by a crutch to leane upon: therefore a man may with good conscience, upon defect of memory and utterance &c. use a set forme of prayer."[50] The implication was, of course, that the spiritual athlete will dispense with crutches.

Richard Rogers, on the other hand, denied that only *ex tempore* prayers were acceptable worship, because all Reformed churches

48 *A Treatise of Christian Religion or The Whole Bodie and substance of Divinitie* (posthumously published in 1616), p. 256.
49 *The Whole Treatise of the Cases of Conscience* (1608), Book II, p. 77.
50 *Ibid.*, pp. 77-78.

had a "prescript forme of prayer."[51] It would have been astonishing
if Puritans who acknowledged their Genevan ancestry had not
approved of a Reformed liturgy, bearing in mind that both in
Strassburg and Geneva Calvin had used one during his ministry,
even if he also recognised the right of the ministers to frame their
own prayers from time to time.

4. Puritan Prayer-books

The Puritans had two examples of Protestant liturgies in English to consider adopting or modifying. One was the most Protestant of the Anglican liturgies, that of 1552, which, as we have
seen, was unacceptable, despite the fact that it had won Knox's
approval[52] after the inclusion of the black rubric and the qualified
approval of John Calvin himself.[53]

The other liturgy was William Huycke's English translation of
Calvin's *La Forme des Prières* which appeared in 1550 and may
have been taken from Calvin's 1542 or 1547 edition.[54] For the
learned there was available the longer Strassburg form of Calvin's
liturgy which Poullain or Pollanus had translated from French
into Latin in 1551 and which was entitled, *Liturgia sacra, seu Ritus
ministerii in Ecclesia peregrinorum, profugorum propter Euangelium Christi Argentinae.* This form in particular owed much to the
influence of the Reformer Martin Bucer,[55] as Calvin acknowledged.

The most important of the prayer-books used by the Puritans
was almost certainly the Calvin-inspired volume, known popularly
as "John Knox's Genevan Service Book," the correct title of which
was *The Forme of Prayers and Ministrations of the Sacraments,
etc., used in the English Congregation at Geneva: and approved
by the famous and godly learned man, Iohn Calvyn. Imprinted at
Geneva by Iohn Crespin. MDLVI.* This was drawn up by the
Presbyterian-Puritan party of "Knoxians" at Frankfort-am-Main,
consisting of Knox himself, Whittingham, Gilby, Foxe and Cole,

[51] *Seven Treatises* (1603), Book II, chap. 4, p. 224.
[52] Laing (editor of Knox's *Works*, 1855), in vol. 4, pp. 43-44, states that
despite Knox's former "good opinion of the Book [of Common Prayer], he discovered in the English Book . . . things superstitious, impure, unclean, and
unperfect."
[53] Calvin's letter in the original Latin (with the deprecatory phrase, *multas
tolerabiles ineptias*) will be found in Laing's edition of Knox's *Works*, vol. 4, p.
51. An English translation is in *A Brief Discourse of the Troubles at Frankfort*
(1846 edn.), pp. xxxiv-xxxvi.
[54] See W. D. Maxwell, *John Knox's Genevan Service Book, 1556*, p. 71.
[55] See the statement by James Hastings Nichols, in *Corporate Worship in the
Reformed Tradition*, p. 57: "Consequently Martin Bucer must be given credit as
the chief architect of the Calvinist form of worship."

all Marian exiles. In addition to being originally written for Englishmen who wished to reform the Reformation, it came to be used widely in Scotland until the *Book of Common Order* (as it was retitled when it became, with some modifications and additions, the official service book of the Scottish Kirk) was ousted first by the "Laud's Liturgy" and shortly afterward by the *Westminster Directory* in 1644.

A Brief Discourse of the Troubles begun at Frankfort in the year 1554 about the Book of Common Prayer and Ceremonies is probably Whittingham's account of four attempts to meet the liturgical demands of the "Coxian" Anglican and "Knoxian" Presbyterian-Puritan parties, none of which was successful. When the "Knoxians" regathered in Geneva in 1556 they revived the work of the Calvinist committee which had begun its task in January 1555. Their Genevan Service Book was used with little adaptation by the English exiles for their worship in the church of Marie la Nove, which was in Maxwell's words, "the cradle of Puritanism."[56] Here, with a congregation of 186 members, tranquillity reigned, so much so that Knox wrote in December 1556 of the place as "the maist perfyt schoole of Chryst that ever was in the erth since the dayis of the Apostillis. In other places, I confess Chryst to be trewlie preachit; but maneris and religioun so sinceirlie reformat, I have not yit sene in any uther place."[57]

The Sunday morning service consisted of: a Confession of sins; a Prayer for pardon; a metrical Psalm; a Prayer for illumination; scripture reading; sermon; baptisms and publication of banns of marriage; long prayer (of intercessions and petitions) and Lord's Prayer; Apostles' Creed (recited by the Minister); a metrical Psalm; and the blessing (Aaronic or Apostolic). Three characteristics of the service were unmistakably Calvinist. It was Biblical, didactic, and congregational. Its Biblical basis is seen in the Confessional prayer based largely on the ninth chapter of Daniel, in the use of metrical Psalms, and in the preference for Biblical blessings compared with the Anglican blessing ("The peace of God"), which was used in the original interim order of Frankfort. It was didactic in that the climax of the service was approached by a prayer that the preacher may be illuminated by the Holy Spirit to declare God's word and the congregation to receive the seed of God's word in the sermon, which was an exposition of the lesson read; the climax of the worship and the summary of the historic

[56] *Ibid.*, p. 7.   [57] Knox, *Works*, vol. 4, p. 240.

faith, the Apostles' Creed, preceded the closing acts of worship. The congregational and participatory character of the service was in the provision of easily memorized metrical psalms. Perhaps the clearest indication of Calvinism was the statement of the doctrine of original sin so dominant in the opening prayers of confession, particularly in the alternative form.[58]

The order for the Lord's Supper was: The Words of Institution and Exhortation; The Eucharistic Prayer (including Adoration, Thanksgiving for Creation and Redemption, Anamnesis, Doxology); Fraction; Delivery; People's Communion while Scripture was read; Post-Communion Thanksgiving; Psalm 103 in metre; Blessing (Aaronic or Apostolic).

The Lord's Supper was to be celebrated monthly. The service was remarkable in four ways. It rightly gave a large element in the Eucharistic prayer to thanksgiving for creation as well as redemption. It warned against the cheapening of grace by "fencing the tables" against scandalous recipients, while at the same time insisting in the exhortation that "this sacrament is a singuler medicine for all poor sicke creatures, a comfortable helpe to weake soules, and that our lord requireth no other worthines on our parte, but that we unfaynedly acknowledge our noghtines, and imperfection."[59] It had an implicit *Sursum corda* in the end of the exhortation, "to lift up our mindes by fayth above all thinges worldlye and sensible, and therby to entre into heaven, that we may finde, and receive Christ, where he dwelleth undoubtedlye verie God, and verie man, in the incomprehensible glorie of his father, to whom be all praise, honor and glorye now and ever. Amen."[60] It lacked an *epiklesis*,[61] or explicit invocation of the Holy Spirit on the elements and the people of God, but this may have been thought unnecessary, since Calvin had a strong sense of the Holy Spirit as effective in all services in the reading and preaching of the word of God, as well as the sacraments, in taking the things of Christ and applying them to the faithful. Once again, there was the concern to be faithful to the Biblical authority in the recital of the Institution narrative as a warrant for the ordinance.

Strype, the ecclesiastical historian, reports that John Knox's

---

[58] ". . . we are miserable synners, conceyved and borne in synne and iniquitie, so that in us there is no goodnes." Maxwell, *Book of Common Prayer*, p. 87.
[59] *Ibid.*, p. 124.
[60] *Ibid.*
[61] This lacuna was filled by later Puritans, both in the Order for the Lord's Supper in the Westminster *Directory* and in Baxter's *Reformed Liturgy*.

Genevan Service Book was being used as early as 1567 in England. After mentioning the moderate Puritans who remained "in the Communion of the Church" of England, he continues: "But another sort there was, that disliked the whole constitution of the Church lately reformed; charging upon it many gross remainders of Popery; and that it was still full of corruptions not to be borne with, and Anti-Christian; and especially the habits the clergy were enjoined to use in their ministration and conversation. Insomuch that the latter separated themselves into private assemblies, meeting together not in Churches, but in private houses, where they had ministers of their prayer, they used a Book of Prayers framed at Geneva, for the congregation of English exiles lately sojourning there. Which book had been overseen and allowed by Calvin and the rest of his divines there, and indeed was for the most part taken out of the Genevan form."[62] There is confirmation of Strype's assertion in the reprint of "The examination of Certain Londoners before the Ecclesiastical Commissioners June 20, 1567."[63] In his answer to the Bishop of London (Grindal), the Puritan Smith admitted that the cause of attending private assemblies was the displacing of good preachers, and continued: "and then were we troubled and commanded to your courts from day to day, for not coming to our parish churches: —then we bethought us what were best to do; and we remembered that there was a congregation of us in this city in Queen Mary's days; and a congregation at Geneva, which used a book and order of preaching, ministering of the sacraments and discipline, most agreeable to the word of God; which book is allowed by that godly and well-learned man, Master Calvin, and the preachers there; which book and order we now hold." It is significant that the description of the *Forme of Prayers* and particularly the terms in which Calvin is referred to almost exactly correspond with the Crespin edition of 1556.

Strype has two more references to the use of a Genevan book by English Puritans. In 1571 the Puritans "did still in their own or other churches, or in private houses, read different from the established office of Common Prayer: using the Genevan form or mangling the English book, and preached without licenses."[64] He also reports that in 1584 there was a petition to Parliament for which the Puritans "had compiled and got in readiness a new platform of

---

[62] John Strype, *The History of the Life and Acts of Edmund Grindal*, pp. 168f.
[63] *The Remains of Edmund Grindal*, ed. W. Nicholson, pp. 203f.
[64] *The Life and Acts of Matthew Parker*, 3 vols., II, 65.

ecclesiastical government, agreeable to that of Geneva, and another form of prayer prescribed therein, in room of the old one, for the use of this Church."[65] These instances made it clear that the *Forme of Prayers* had an important clandestine existence in England to complement its open acceptance and use in Scotland, and that it served as a model for the worship of the proto-Presbyterian[66] Puritans in England.

It is also important to note that there were two adaptations of the John Knox's Genevan Service Book current among English Puritan congregations, the first of which was used in England, and the second by the Puritan exiles in Middleburg. The first was printed by Waldegrave and entitled, *A booke of the forme of common prayers, administration of the Sacraments, &c. agreeable to Gods Worde, and the use of the reformed Churches.*[67] It was probably printed in 1584 or 1585, almost certainly in the earlier year in support of the petition to Parliament for the Genevan discipline and Genevan prayer-book. Both Bancroft and Hooker referred to this or to one of the further editions printed for the English Puritans in 1586, 1587, and 1602. From Bancroft it is learned that prior to its introduction into the Low Countries in 1587, it was used possibly exclusively in Northamptonshire under the direction of a notable Puritan, Edmund Snape, an associate of Cartwright. Bancroft alluded to the formulary to substantiate his claim that the Puritans who claimed liberty to worship God according to their consciences and the Word of God, yet proposed, with Parliamentary sanction, to impose this book exclusively in corporate worship.[68]

Hooker also referred to the Waldegrave Liturgy in the fifth book of *The Laws of Ecclesiastical Polity* under the engaging title "Of them who allowing a set Form of Prayer allow not ours." He claimed that the Puritans changed from supporting free prayers to demanding a set formulary: "Now, albeit the Admonitioners did seem at the first to allow no prescript form of Prayer at all, but thought it best that their Minister should always be left at liberty to pray as his own discretion did serve; yet because this opinion on better advice they afterwards retracted, their defender and his

65 *The Life and Acts of John Whitgift*, 3 vols., I, 348.

66 A term coined to designate would-be Presbyterians who were unable in an Anglican state church to establish the full Presbyterian discipline and government with classes, synods, and a general assembly.

67 Reprinted in Peter Hall's *Fragmenta Liturgica*, 7 vols., I, 1f.

68 Richard Bancroft, *Dangerous Positions and Proceedings* . . . (1593), Book III, x.

associates have sithence proposed to the world a form such as themselves like."[69]

Bancroft's evidence, corroborated by Hooker, includes two interesting suppositions. The first is that the Puritans moved from free prayer to set forms; their alarm at the freedom of the prayers of more radical Puritans such as the Brownists may have encouraged them to follow Knox rather than Barrowe. Second, it was believed that the "Admonitioners" were behind the publication of the Waldegrave Liturgy, along with the Middleburg Liturgy. Cartwright had ministered to the English congregation there, and his friend Dudley Fenner was in Middleburg in 1586. Unquestionably Waldegrave was the printer for the Puritans. Since he had issued in 1584 the "Brief and Plain Declaration concerning the Desires of all those Faithful Ministers that have and do seek for the Discipline and Reformation of the Church of England," it would be a natural sequel to publish in the same or following year the formulary of worship appropriate for a Reformed Church of England. The Star Chamber in June 1586 restricted Puritan printing, at which time the Waldegrave Liturgy was taken to Middleburg to be republished by Richard Schilders in 1586.

The Waldegrave and Middleburg Liturgies were firmly based on Knox's Genevan service book, with a few significant variations. Their Puritan character is evident from the marginal Scriptural references and warrants for each liturgical action and for the content of the prayers. The first rubric of the Middleburg Liturgy prescribed a "Reader's Service,"[70] which was to take place before the entire congregation gathered to begin the liturgy proper. It was to be led by "one appointed by the Eldership [who] shall reade some Chapters of the Canonicall bookes of Scripture, singing Psalmes between at his discretion: and this reading to bee in order as the bookes and Chapters followe, that so from time to time the holy Scriptures may bee readde throughout."[71] This was the Puritan answer to Anglican shredding the Scripture in "pistling" and "gospelling."

When the minister arrived, the liturgy began with solemn confession of man's insufficiency and the affirmation of God's suffi-

[69] *Ecclesiastical Polity*, Book v, chap. xxvii, sec. 1.

[70] The Reader's Service was introduced into Scotland in 1560. See Maxwell, *Book of Common Prayer*, pp. 177-79, for a description.

[71] The title is *A Booke of the Forme of Common Prayers, Administration of the Sacraments, &c. agreeable to Gods Worde and the Use of the Reformed Churches* (Middleburg, 1586). It is reprinted in Hall's *Reliquiae Liturgicae*, vol. 1; and Bard Thompson's *Liturgies of the Western Church*.

ciency as Creator: "Our helpe be in thee name of the Lorde, who hath made both Heaven and Earth." So had Calvin's own service began, yet this opening sentence of Scripture was omitted in the Knox book.

There were other significant changes. The first alternative form of confession in Knox's book was omitted, leaving the Bucer-Calvin prayer of confession, to which Knox and his associates had added a petition for pardon. Three alternatives were provided for the prayer of intercession, instead of one in Knox; both Knox and the Puritan liturgies permitted the minister to substitute his own prayer of intercession. The third alternative prayer of intercession was heavily penitential in character. One has the impression that it may well have been used by Puritans for additional services on "days of humiliation."

The Waldegrave Liturgy added to the recitation of the Apostles' Creed required in the Knox book that of the Decalogue; while the Middleburg Liturgy omitted both (this, incidentally, is the only significant difference between the two Puritan liturgies of Waldegrave and Middleburg). In fact, with the exceptions already noted, there is a remarkable unity in word and action between Knox's "Genevan Service Book" and the Waldegrave and Middleburg Liturgies. All three agree in their orders for Baptism and the Lord's Supper and for the Marriage service. The Waldegrave and Middleburg orders make no provision for burials, while the Knox book reduces the burial service to its barest essentials.

The spirit of the three liturgies is well expressed in the appended comment in all three to the order for the Lord's Supper, to the effect that their aim had been, "that Christ might witness to our faith, as it were with his own mouth . . . so that, without his word and warrant, there is nothing in this holy action attempted."[72] In ethos, structure, and words the three liturgies agree. The few unimportant changes in wording made in the Waldegrave and Middleburg order were for clarity and solemnity. Knox's archaism, "the Sacrament is a *singuler* medicine for all poor sick creatures," was changed to "*excellent* medicine." The Knoxian order for the Lord's Supper in fencing the tables against persons of evil life says that if we are unworthy receivers "we kindle God's wrath against us," which the Puritan liturgies qualified as "*heavy* wrath." The list of those anathematized was lengthened in the Waldegrave and

---

[72] Maxwell, *Book of Common Prayer*, p. 128; Waldegrave: Hall, *Fragmenta Liturgica*, vol. i, p. 68; Middleburg: Hall, *Reliquiae Liturgicae*, vol. i, p. 61; and Thompson, *Liturgies*, p. 340.

Middleburg Liturgies to include the man who committed the chief liturgical sin for Puritans, "a mainteyner of Images, or mannes inventions in the service of GOD."[73] Once again we see in Puritan worship and life the vivid understanding of the obedience of faith and the solemn demand for sanctification. This worship may seem aesthetically bleak, but it was, above all, theologically vertebral and ethically vigorous. The Puritan liturgies also show that at least the Cartwrightian Puritans, while allowing a subordinate place for free prayer, never gave up the idea of a liturgy for the unity, discipline, and common instruction of congregations of the Reformed church.

5. The Gospel Sacraments[74]

According to Puritan thought the possibility of communion with God was established when God comes to his elect in the form of the "word," and the "word" is a term combining the meanings of first, Jesus Christ the Incarnate Word, second, the scriptures, and third, the proclamation of the gospel in preaching and in the sacraments. Perkins defined a sacrament as "A Signe to represent, a seale to confirme, an instrument to conveye Christ and all his benefits to them that doe beleeve in him."[75] The holy spirit used the senses to convey the benefits of Christ to men, "because wee are like *Thomas*, wee will not beleeve till wee feele them in some measure in our hearts."[76] There were only two sacraments of the gospel for the Puritans—Baptism, the sacrament of admission or initiation into the church of God, and the Lord's Supper, the sacrament of nourishment and preservation in the church. If we could have asked Perkins what is done in Baptism, he would have a ready, concise reply: "In the assemblie of the Church, the covenant of grace betweene God and the partie baptised is solemnly confirmed and sealed."[77] In the Lord's Supper the former covenant "solemnly ratified in Baptisme, is renewed in the Lord's Supper, between the Lord himselfe and the receiver."[78] What is the meaning of the eating of the bread and the drinking of the wine in the Lord's Supper? Perkins insisted that the outward actions "are a second seale, set by the Lords owne hand unto his covenant. And they do give every

---

[73] Middleburg: Thompson, *Liturgies*, p. 335.

[74] See also Chap. II for a comparison of Anglican and Puritan interpretations of the sacraments.

[75] *The Foundation of the Christian Religion gathered into sixe principles* (1595), sig. C4 v.

[76] *Ibid.*   [77] *Ibid.*   [78] *Ibid.*, D i recto.

receiver to understand, that as God doth blesse the bread and wine to preserve and strengthen the bodie of the receiver: so Christ apprehended and received by faith, shal nourish him, and preserve both bodie and soule unto eternall life."[79]

In the divine economy and ordering, sacraments fulfilled five important functions. They were instruments of signification, showing the cleansing and renewing power of God's pardoning and empowering grace. They were also confirmatory of faith; that is, they pointed to God's fulfillment of his promises, which encourages the wavering and unsteady of purpose. Third, the sacraments are not only witnesses to God's generous love but also witnesses to the believer's inner commitment to Christ. They are, as it were, badges of membership of the church, signs of the identification with the company of God's people. Finally, sacraments foster the brotherly love that there should be between members of Christ's church: they express and extend the reconciling love of Christ which is in the brethren to the neighbour.[80]

Puritans insisted that infant Baptism was appropriate because the sacrament does not declare our faith but testifies to the initiative of God's grace in Christ for man's salvation. Children, as the seed of covenanted members of the body of Christ, are entitled to the fruits of the divine promises.[81] God both seals his promise and also the means by which He fulfills His promise, that is, incorporation into Christ. Since this was clearly a sacrament of the church for admission into the church, it should be performed publicly before the gathered congregation and not privately in a home. Since it was an ecclesial act it should be performed by an ordained minister, without allowing laymen or women to perform it in exceptional circumstances.[82] Furthermore, no one could take away the responsibility of parents to bring up their children in the knowledge and love of God and in the ways of His ordaining; for this reason

[79] *Ibid.*
[80] The points made in this paragraph can be verified in Perkins, *Works*, I, 73; II, 92; Cartwright, *A Treatise of the Christian Religion* (1616), pp. 219-29; and Richard Rogers, *Seven Treatises* (1603), pp. 217f.
[81] Perkins, in *Works*, I, 129, argued that it is the faith of parents that gives their child "a title or interest in the covenant of grace."
[82] John Knox's *Forme of Prayers* begins the Order of Baptism with the following ominous rubric: "First note, that for asmooche as it is not permitted by Godswoord that wemen should preache or minister the sacramentes: and it is evident, that the sacramentes are not ordeined of God to be used in privat corners, as charmes or sorceries, but left to the congregation, and necessarily annexed to Gods woord, as seales of the same: therfore the enfant which is to be baptised shalbe broght to the churche on the day appointed to comen prayer and preaching. . . ." Maxwell, *Book of Common Prayer*, p. 105.

godmothers were unknown at Puritan baptismal ceremonies.[83] Nor, of course, was there any signing of the cross, for this was a human addition to God's appointment as described in the New Testament.

The order for Baptism in the Puritan liturgies (Waldegrave and Middleburg) has an identical structure consisting of six items: (1) The Interrogation (immediately following the Sermon); (2) A lengthy Exhortation and Explanation of the purpose of Baptism; (3) Recitation of the Apostles' Creed by the father (or by a surety in his unavoidable absence); (4) Prayer for grace and for the reception of the child into Christ's kingdom, completed by the Lord's Prayer; (5) The act of Baptism in the Triune Name; (6) Concluding Thanksgiving. The exhortation made it clear that all who received baptism need not "have the use of understanding and faythe, but chiefelye that they be conteyned under the name of gods people," so that children "wythout iniurie . . . can not be debarred from the common signe of Gods children." Still the lack of the outward action was not prejudicial to their salvation, should they die before being presented to the church.[84] Further, there was no magic in the water; it is the Holy Spirit that works regeneration in the heart of the elect.[85]

The demand made of the father of the child was to provide that his children shall be taught, "in all doctrine necessary for a true Christian: chiefely that they be taught to rest upon the justice of Christ Iesus alone, and to abhorre and flee all superstition, papistrie, and idolatrie."[86] The father, as proof of his consent to the performance of these duties, or in his absence the surety, was then required to recite the Apostles' Creed. Theological clarity and fidelity are not incompatible with human tenderness, as can be seen in the moving prayer preceding the act of Baptism: "ALMIGHTIE and everlasting God, which of thy infinite mercie and goodnes, hast promised unto us, that thow wilt not only be our God, but also the God and father of our children: we benseche thee that as thou hast vouchesaved to call us to be partakers of this thy great mercie in the felowshipe of faithe: so it may please thee to sanctifie with thy sprite, and to receive in to the number of thy children this infant, whom we shall baptise according to thy woord, to the end that he comming to perfite age, may confesse thee onely the true God and whom thow hast sent Iesus Christ: and so serve him, and be profit-

---

[83] Both Calvin and Pullain allowed godfathers as alternatives, but never godmothers; in this they were followed by the Presbyterian type of Puritans. The Independent Puritans, however, utterly rejected male and female godparents.
[84] *Ibid.*, p. 106.   [85] *Ibid.*, p. 107.   [86] *Ibid.*, p. 109.

able unto his churche, in the whole course of his lyfe: that after this life be ended, he may be broght as a lyvely member of his body unto the full fruition of thy ioyes in the heavens, where thy sonne our Christ raigneth world wyth out end."[87] The service is simple, solemn, sincere, and a clear testimony to the covenant conception that dominated Puritan thinking and living with God, with the family, and with the households of faith, and which made of the sacraments of Baptism and the Lord's Supper the seals of God's covenant—grace.[88]

Perkins was the clearest and most popular of the early Puritan theologians; his teaching on the theology of the Lord's Supper is typical. Applying his general definition of sacraments to the sacrament of the Eucharist, it should be noted that it is a sign representative of the reconciling sacrifice of Christ on the Cross anticipated at the Last Supper, a seal confirming God's holy covenant of love with His people, and an instrument conveying Christ and all his benefits, chiefly pardon and eternal life, to the faithful. Their meaning is summed up in his catechetical answer to the question, "What meaneth the bread and wine, the eating of the bread and drinking of the wine?" The answer he gave: "These outward actions are a second seale, set by the Lords owne hand unto his covenant. And they doe give every receiver to understand, that as God doth blesse the bread and wine, to preserve and strengthen the bodie of the receiver: so Christ apprehended and received by faith, shal nourish him, and preserve both bodie and soule unto eternall life."[89]

I have already referred to the earliest of the English Puritan liturgies, the Order for the Communion,[90] and to some of its leading characteristics, particularly the inclusion of thanksgiving for creation as well as redemption in the Eucharistic prayer, the implicit *Sursum corda* in the exhortation, and the lack of an explicit *epiklesis*. It is now appropriate to consider the rationale for this important service as contained in a concluding paragraph addressed

[87] *Ibid.*, p. 110.
[88] The significance of the sacraments as seals of the covenant of grace in its practical implications is well brought out by Walter Travers, in *A full and plaine declaration of Ecclesiasticall Discipline owt off the word off God and off the churche off England from the same* (published probably in Zurich, 1574), p. 24: "So that seing the lorde to have sete his sygnet to the confirming of our Salvacion and to have sealed yt up, we might the more quietlie rest and acquiete our selves in his faith and custodie."
[89] Perkins, *The Foundation of the Christian Religion gathered into sixe principles* (1595), sig. D i r.
[90] This was considered in the previous section of this chapter.

"To the reader." The primary intent was to renounce the error of the Papists by denying Transubstantiation, which in the Reformed view overthrew the nature of a sacrament since it confused the sign with the reality signified. Other aims in arranging the order were to "restore unto the sacramentes theyr owne substaunce, and to Christe his proper place." This place is thought to be usurped by the Catholic priests as though "the intent of the sacrificer should make the sacrament." The words of institution were read not as a consecratory formula but "to teache us how to behave in our selves in the action, and that Christ might witnes unto our faith as it were with his owne mowthe, that he hath ordeined these signes for our spiritual use and comforte."

After the warrant for the action had been read, "wee do first therefore examyne owr selves, accordyng to saint Pauls rule, and prepare our myndes that we may be worthie partakers of so high mysteries." Then followed in full fidelity to the Pauline record the fourfold action: "Then takyng bread wee geve thankes, breake, and distribute it, as Christ our Saviour hath taught us. Fynally, the ministration ended, we gyve thankes agayne accordyng to his example. So that without his woorde, and warrant, there is nothynge in this holy action attempted."[91]

Most impressive are the strong sense of obeying the divine commands in worship, the vivid dramatic power of the prophetic symbolism of breaking bread and pouring wine, as well as the strong sense of sharing in a banquet and holy supper with Christ and His friends. The memorial aspect, as those also of mystery and thanksgiving, are strong. The complementary dimensions of the eschatological banquet and the sense of the communion of saints, as well as the reinvigoration of the Christian hope through Christ's resurrection, are weaker in this rite. Little is said of the sacrifice of the church linked with Christ's sacrifice. The concept of sacrifice dominates the Roman Catholic Mass, and the sense of the rehearsal of the mystery of our redemption and of incorruption in eternal life controls the Orthodox Eucharist. The Anglican Holy Communion, as its name implies, is suffused with a sense of the great privilege of the meeting of God with men, and men with men, in their coinherence with Christ. In Puritanism it was the holy banquet of God's elect that was the uppermost thought. Such a concept drew greatly on the example of the Last Supper, and had an appeal as strong as it was simple.

[91] Maxwell, *Book of Common Prayer*, pp. 127-28.

6. *A Critique*

Puritanism was a new and vigorous movement of protest against the dead hand of tradition, against the idolatry that worships man's works instead of yielding first to the behests of the living God, and against the superstition that tries to dominate and domesticate deity. It was a movement of the "learned godly," the religious intellectuals of the day, a movement that found its strongest support in university circles, especially the University of Cambridge.[92]

What was the novelty of Puritanism? It was, in the first place, a new criterion of Scriptural authority that it offered, or at least the return to a criterion of the early church which cut away the massive tangle of encumbrances which had grown over the central oak of Scripture. With a rude hand it scraped away the parasitical ivy and mistletoe, however aesthetic, so that the Scripture might stand out in its splendid solitude. Or, to use a more apt metaphor, it allowed the voice of Scripture to speak the divine promises and commandments in the power of the Holy Spirit without the seductive sounds of the carousel of church tradition to interrupt it. This strong, simple Biblical authority seemed a liberation to those who were appalled by the somnolence, the superstition, and the sheer accommodation of the church to the worldliness of the Renaissance, though in its concern to return to the original texts of Christianity, as the Renaissance scholars returned to the original texts of the classical world of Greece and Rome, there was gain as well as loss. This digging down to the bedrock of Scripture, even when it led to bibliolatry, as we have seen, and even to deep disagreements between private interpretations, was an immense gain in simplicity for the learned and the unlettered. In worship it gave a sense of direct Biblical (and therefore divine) authority for every ordinance practised by the Puritans, from preaching to the celebration of the sacraments, from the types and order of the prayers to the form of praise (Psalms but not hymns). Hence the emphatic rejection within Puritanism of "men's inventions," or "will-worship," and the overpowering sense that God's elect were worshipping God in His own way as demanded and as exemplified in His word. No other conclusion could be drawn from the use of the lengthy readings from the Scriptures, entire chapters at a time, the baptismal formula used in the sacrament of Initiation, and the Warrant for

[92] This has been admirably demonstrated by Harry C. Porter in his *Reformation and Reaction in Tudor Cambridge.*

the Lord's Supper in the reading of the Pauline words of institution, and the use of the Biblical words at the delivery of the holy bread and wine to the people, the careful way the sermon elucidated the Scriptures, and the metrical versions of the Psalms used in praise. Each of these elements in Puritan worship contributed to the impression of the utmost fidelity on the part of minister, office-bearers, and congregation to God's holy word and law, the Bible.

The second impression Puritan worship made on those accustomed to the Roman Catholic or Anglican traditions was its extreme simplicity. It was decidedly a simple, unornate type of worship, which suitably met in white-walled churches with only tables of the commandments for instruction rather than adornment, and which despised, as distracting, "storied windows richly dight casting a dim, religious light" and allowed windows to cast the natural light of the sun as the worship gives place to the interior illumination of the Holy Spirit. So strong was the sense of the numinous in the approach of God in word and sacrament that any symbolism or decoration is felt to be utterly adventitious, not a gilding the lily, but a varnishing of sunlight. It was, to change the metaphor, a bare stage—like the Elizabethan theatre—so that the provocative images of Scripture could make their own impact on the *tabula rasa* of the sensitive imagination without tinsel trappings, or merely distracting stage decor. This is why the only ceremonies permitted were sacramental signs, such as the affusion of water in Baptism and the breaking of bread and pouring of wine in the Lord's Supper.

The third impression which Puritan worship made, especially in the Independent-Puritan rather than in the Presbyterian-Puritan circles, was of an affirmation of the dynamic role of the Holy Spirit. This was clearly allowed for in Presbyterianism in the proclamation of the word, as the preceding prayer of illumination makes clear. It was always theoretically possible for the Presbyterian minister to provide his own words in the prayers of intercession, but (watched by the eagle eye of the session and warned by the presbytery and the classis of the dangerous freedom of the semi-Separatists and the Anabaptists) he usually chose not to exercise that liberty. In fact, however, it was the proto-Independents who took with the utmost seriousness the Pauline conviction that the Holy Spirit would be the guide in extemporary prayers, and gave up prayer-books of every kind.

Puritanism was responsible for the insistence on the interiority of worship, for it is the Spirit of God that searches the heart. The same concern made Puritan fathers prepare for Sunday worship in the congregation by daily family worship in the home.[93] It was the Spirit that motivated Richard Rogers to assert that public worship without private help was cold: "for hearing of the word read and preached, doth little profit where it is not ioyned with preparation to heare reverently and attentively, and where it is not mused on after, yea and as occasion shall offer, conferred on also. . . ."[94] The very intensity of Puritan worship and life, with its deep seriousness so amusingly criticised by the Elizabethan and Jacobean dramatists, springs from the overriding concern for sincerity before God. In an emphasis that looked forward to the Quakers, Perkins gave priority to inward worship, consisting of the adoration of God and cleaving to Him, over outward worship. Adoration comprised four virtues—fear, obedience, patience, and thankfulness, while cleaving to God was possible through faith, hope, love, and inward invocation.[95]

The doctrine of the Holy Spirit in Puritanism is also seen in the demand for integrity of life. The fruits of righteousness, the harvest of the Holy Spirit, are the evidences of sanctification and proofs of the calling of the elect. This was manifested externally in the insistence that the holy table at Communion be fenced against all who lived scandalously. It was all the more honoured among Presbyterian Puritans because Farel and Calvin were expelled from Geneva for maintaining the purity of the Genevan church in precisely this manner. "Holy Discipline" was regarded as the third distinguishing mark of the true church by all Puritans, the other two being the preaching of the word of God and the true and due administration of the Gospel sacraments. So seriously was the duty of disciplining the faithful taken that they appointed officers whose task it was to rebuke the members who did not live according to their Christian profession, and to bring them to the parting of the ways at which they either confessed their sorrow before the congregation and produced fruits worthy of repentance and were restored to the communion of the local church, or, if they persisted in their course of scandalous conduct, they were solemnly excommunicated as "limbs of Satan."[96]

[93] See Chap. XII for Puritan spirituality.
[94] *Seven Treatises*, p. 225.   [95] *Works*, 3 vols. (1613), II, 62-63.
[96] See my *Worship of the English Puritans*, chap. 14, for a description of procedures for excommunication and restoration.

The overwhelming concern to prevent worship becoming stereotyped and stale is seen not only in creative prayers (instead of set prayers) and original sermons (instead of official homilies), but also in an innovation within Puritanism of days of thanksgiving and humiliation. This practise appears to have derived directly from the observation of Old Testament practise, and seems not to have been used elsewhere corporately and consistently in the history of Christianity. It was not so much that the Puritans rejected the Church Year or even that they telescoped it so that each Sunday celebrated the mighty acts of God from Creation to Incarnation, and from the first to the second Advent.[97] Rather it was the recognition that history is the acting out of the divine providence and that God continues to reveal Himself in judgment and mercy many times in the life of a nation and of a family. The appropriate response to these acts of judgment was a Day of Fasting and Humiliation. The appropriate response to the acts of mercy or deliverance was a Day of Thanksgiving. Each was kept by fasting and prayer, whether the act of judgment was national or family calamity and whether the act of mercy was the deliverance of the nation from the hands of its enemies or the deliverance of a mother from childbirth. But it was this recognition of the continuing guidance of God in special providence, as recorded in the private spiritual diaries of the Puritans or referred to in the sermons of their ministers, which gave a sense of the immediacy of the Holy Spirit in their lives. It is well known that the entire army of Cromwell gathered for fasting and prayer to seek the divine will. It is less well known that this was a feature of the earliest Elizabethan Puritanism to mark the understanding of the ways of divine providence by days of fasting and prayer, set apart for that purpose.

William Perkins said there were two just occasions for religious fasting: "The first is, when some iudgement of God hangs over our heads, whether it be publike, as Famine, Pestilence, the Sword, destruction, &c. or private. The second cause of fasting is, when wee are to sue, and seeke by prayer to God for some speciall blessing, or for the supply of some great want."[98] As an example of the latter, Perkins listed Christ fasting and spending the whole night in prayer before choosing His 12 disciples. Perkins insisted that

[97] A. G. Matthews, in *Christian Worship*, ed. Nathaniel Micklem, p. 173: "The Sabbath retained its lonely splendour as the sole red-letter day of the Puritan Calendar."
[98] *Works* (1613 edn.), II, *Cases of Conscience*, p. 102.

the true way to keep a fast was to abstain from meat, drink, and all delights that "may cheare and refresh nature"; by such abstinence as will afflict the body, and subdue the flesh to the "will and word of God," we stir our devotion to God, lead to contrition or inward sorrow, and admonish us of guiltiness before God. All this may seem masochistic to modern ears, though to the sixteenth century Catholic or Protestant it was recognised that the refractory and rebellious will of man is not easily brought into the captivity of Christ. It is all the more interesting to read of the rapture with which these exercises of fasting and prayer were entered into by the Puritans, as in the case of one John Lister (b. 1627) who testified in his autobiography how his spirituality was formed by memorising the sermons and lectures he heard at monthly preaching marathons, at funerals, and at Sunday worship. He vividly recalled the days of fasting: "O what fasting and praying, publicly and privately, what wrestling with God was there day and night? Many of those weeping, praying and wrestling seasons, both day and night, were kept in my dear mother's house, and the feasts were kept with great strictness and severity; not many of us, old or young, eating so much as a morsel of bread for twenty-four hours together; this was a great weariness for men. . . ."[99] For those who believed they were wresting a great blessing from God the experience would be one of exhilaration, not exhaustion. Puritans made few concessions to children, and this may account for Lister's memory of his own childhood boredom, but also of the intense interest that the "weeping, wrestling, and praying seasons" had for the adults. Here, again, is evidence of the intensity of the interior civil war of the Spirit in which all Puritans were engaged.

The fourth quality of Puritan worship was its deep concern that the people of God should not be spectators or listeners, but sharers in the worship, a feeling that arose from a profound conviction of the solidarity of God's chosen people. At first glance, it might seem that Anglican worship, with its responses in the versicles and the litany, gave a larger place to the laity than Puritanism did, for the latter required the minister's voice alone to be heard in prayer, except for the concluding amen, which indicated the congregation's concurrence with the minister. In fact, however, in gatherings of Puritans for worship, there were several points at which the congregation played a significant part. Even where they were theo-

---

[99] *The Autobiography of Joseph Lister*, ed. T. Wright, pp. 5-6, cited in Ronald A. Marchant, *The Puritans and the Church Courts in the Diocese of York, 1540-1642*, p. 115.

retically passive, as in listening to the preaching, a high degree of silent cooperation was expected, as they "mused" on God's word and its relevance to their lives. The metrical Psalms were popular in worship, not only because of the attractive tunes to which they were set, but also because, being so often in common metre and rhymed, they were easily memorised. They provided a more democratic vehicle for praise than the intricate chants that required trained choirs to sing them in the Roman Catholic and Anglican cathedrals and town churches of note. Furthermore, there was no division in Puritan meetinghouses between the sacred space of the sanctuary or chancel[100] and the secular space of the nave. Puritan elders (if Presbyterian) and deacons (if Independent) served the congregation with the sacred elements of bread and wine, which all members received. There was every incentive for members of the congregation to listen attentively to the sermon, not only because the exposition of the oracles of God was the climax of the service, but also because children and servants knew that it was the duty of the heads of Puritan households to test how carefully they had remembered the main "heads" of the sermon. Indeed, it was often noted that eager Puritan church members assisted their memories with notebooks and recalled the "doctrines" and "uses" of the sermons in their Sunday evening private devotions. Also, many unscrupulous printers published pirated sermons of famous preachers.[101] The Puritans never forgot, whether in church, home, or the marketplace, that there was never a holiday from God for His elect. Their worship was a shared commitment.

Was this solemn, simple, Biblical, shared worship without defects? It was only to be expected that Sunday services which aimed to make every gathering of worship a renewed Pentecost should occasionally become merely a "Low Sunday." Extemporaneous prayer could be sadly abused, by prolixity,[102] sensationalism, rambling thought or speech, and spiritual superficiality.

It can be, and was, argued that Puritan worship made too great demands on the spiritual and intellectual capacities of both minis-

---

100 Cartwright criticises the Anglican minister because "in saying morning and evening prayer [he] sitteth in the chancel with his back to the people as though he had some secret talk with God which the people might not hear." Cited in Whitgift, *Works*, II, pp. 487f.

101 See Alan F. Herr, *The Elizabethan Sermon, A Survey and Bibliography.*

102 Bishop Aylmer of London is accused by "Marprelate" (*The Marprelate Tracts*, ed. William Pierce, p. 187) of saying that long Puritan prayers before and after sermon are nothing else but "beeble-babble, beeble-babble." See also Benjamin Brook, *Lives of the Puritans*, 3 vols., I, 433.

ters and congregations. Hooker thought Puritan prayers presumed an attention span greater than it was reasonable to expect of congregations, and defended collects and responses on the ground that they would more easily retain the interest of congregations. He made a plea for an aspect almost forgotten in Puritan worship with the reminder that "in public Prayer we are not only to consider what is needful in respect of God, but there is also in men that which we must regard."[103] Concentrated Calvinism may overwind the human mechanism, as the cases of Jonathan Edwards' Uncle Hawley and of the gentle castaway from grace, the poet Cowper, remind us by their tragic suicides. Some who found Puritanism too demanding must have slipped with relief back into the comfortable compliance of the Church of England—understandably so, whether ministers or church members. It cannot have been easy for ministers to produce two new sermons for each Lord's Day, as well as a frequent weekday lecture, with the high level of godly learning their congregations had a right to expect. Every public prayer they offered was expected to be extemporaneous. They had responsibilities to take their part in "prophesyings" and in days of humiliation and thanksgiving. They had to visit their congregations so as to bear in mind the special needs of the poor, the sick, the infirm, the old, and especially those who were distressed in conscience. Ministers were expected to keep abreast of the latest casuistry and to keep up to date in Biblical knowledge by reading some of the vast commentaries on books of the Bible which proliferated in the sixteenth and seventeenth centuries. In addition to all this, they had to conduct their own household prayers and private meditations, and keep a spiritual diary. The Puritan ministry was a zestful, but highly strenuous calling, testing the powers of mind, heart, and will to the utmost.

Given this expectation, one can hardly wonder that Puritan sermons and the whole ethos of the services of the Puritans were excessively didactic. This was, of course, merely the defect of a good quality, the application of the mind to the understanding of faith. Yet, granting the primacy of the theological task in an era of profound change and transition such as the sixteenth century was, there came a time when criticism must yield to the "obedience of faith" and the critical intellect seemed unwilling to give way to devotional "musings" and meditation in worship. There was a serenity, at least quietness, of spirit in Roman Catholic and Angli-

[103] *Ecclesiastical Polity*, Book v, chap. xxxii, sec. 2.

can worship, as there was also in the worship of the Religious Society of Friends, but this is not to be found in the agitated worship of the Puritans and their successors. E. C. Ratcliff saw that the prophetic protest would have to stabilise and become sacerdotal: "In general, Independent worship has tended to place too great a burden on the minister, and to leave the congregation too much to the minister's direction, with the result that a ministry and worship designed to be prophetic have not infrequently developed features reminiscent of the 'sacerdotalism' against which the Independents made their earnest protest."[104]

The fear of Catholicism became itself a superstition among the Puritans. They did not give sufficient heed to the customs of the primitive Church and were even neglectful of the traditions of the Reformed Churches in Europe. It must, however, be acknowledged that very little was known at the time about the early families of Christian liturgies; the Puritans were not in any case likely to look to those quarters where liturgical lore was to be found, however meagre. The result was that Puritan worship, for all its strengths, was also controlled by a naive primitivism which relegated all centuries between the first and the sixteenth to the discard. Its determination that the ethical was a primary category had the consequence of suspecting and distrusting the aesthetic, except in the realm of music. Bunyan might recall in the images of his allegory, that "the famous town of Mansoul had five gates . . . Ear-gate, Eye-gate, Mouth-gate, Nose-gate, and Feel-gate." Puritans, for all practical purposes, were so conscious of the dangers of absolutising the finite or confusing an image with the reality which it reveals and also hides that they were blind in their worship, or at least colour blind. Their simple worship and dignified meetinghouses had the beauty of etchings, with the result that the painter or sculptor must find his spiritual home in the Catholic, Orthodox, or Anglican traditions, not in Puritanism. It also meant that Puritans are starved of the glory of the Creation in their worship.

Finally, authoritative and august as the Biblical criterion was in Puritan worship, demanding and eliciting a splendid loyalty, it was too rigidly applied. For every detail of worship a Biblical sanction, or at the very least, a Biblical silence, was required. This legalistic application of Scripture can be illustrated from Cartwright who argued from the single case of John the Baptist

104 *The Study of Theology*, ed. K. E. Kirk, p. 474.

preaching a sermon before his baptising that all Baptisms must be preceded by preaching.[105]

The worship of the Puritans was characterised by fidelity, simplicity, spontaneity and spirituality, as well as relevance. From the perspective of history it can be recognised for what it was— a singularly effective, prophetic protest against formalism. But institutional forms need to be revitalised, not jettisoned; even spontaneity runs into a stereotype, and supposedly aliturgical worship has its own awkward structure that makes it a liturgy. The spirit without the forms of worship was only a ghost whispering through the individual keyhole. The forms without the spirit were merely puppetry. The choice was no longer liturgical or charismatic worship, but a structure with possibilities for freedom, treasuring the past but open to the present and the future, in short, a liturgy with liberty. This the Puritans did not discover until 1644, and then not fully.

[105] Cited in D. J. McGinn, *The Admonition Controversy*, p. 101.

CHAPTER VIII

PURITAN PREACHING

THE PURITANS were not the only group in England who emphasised the central importance of preaching. But no others—not even such eminent exponents of the art of commending the Gospel as Richard Hooker or Lancelot Andrewes—elevated the proclamation of the Word of God as highly.[1] For the Puritan it was the climax of divine service; for the Anglican it was "so worthy a part of divine service." Hooker admitted it to be "the blessed ordinance of God, Sermons as the keys to the Kingdom of Heaven, as wings to the Soul, as spurs to the good affections of man, unto the sound and healthy as food, as physic unto diseased minds."[2] This evaluation is hardly to be distinguished from that of the "silver-tongued" Elizabethan Puritan preacher, Henry Smith, who had spoken of the Word of God as "the *Star* which shuld lead us to Christ, the *Ladder* which shuld mount us to heaven, the *Water* that shuld clense our leprosie, the *Manna* that shuld refresh our hunger, & the *booke* that we shuld meditate on day & night."[3]

1. The Primacy of Preaching

There was, however, a difference. For Smith the proclamation of the Gospel was "the one thing necessary," the listening to the words of Christ which the Master had commended in Mary as contrasted with the Martha who was "busied with many things." Whereas for Hooker, himself a preacher whose matter was as solid as Andrewes' sermons, but much less animatedly delivered, the Gospel was equally conveyed in the reading of the lessons from the Gospel and the Epistle in the liturgy, and in the dramatic reenactment of the same Gospel through the sacraments of Baptism and Holy Communion.[4] For the Puritan the proclamation of

---

[1] For the Puritans the Sacraments were "seals" of the Word (*sigilla Verbi*), that is, confirmations of God's promises in the Gospel. For the Anglicans, however, the sacraments were means of grace, at least as important as preaching.

[2] *Treatise on the Laws of Ecclesiastical Polity*, book v, sec. 22.

[3] *The Sermons of Maister Henrie Smith gathered into one volume* (Richard Field for Thomas Man, 1593), from "The Arte of Hearing in two Sermons," p. 646.

[4] In *Ecclesiastical Polity*, book v, sec. 21, Hooker argued that one can come to a saving knowledge of God through "conversation in the bosom of the Church," through "religious education," by "the reading of learned men's books," as well

the Gospel through preaching brought men to the existential cross-roads, where the way led either to life everlasting or to destruction, and the aim of his preaching was to convert his hearers from worldliness to godliness. He wished to transform, not merely to inform.

William Bradshaw rightly insisted that the Puritans gave the primacy in worship to preaching: "They hould that the highest and supreame office and authoritie of the Pastor, is to preach the gospell solemnly and publickly to the Congregation, by interpreting the written word of God, and applying the same by exhortation and reproof unto them. They hould that this was the greatest worke that Christ & His Apostles did."[5] No Puritan preacher ever forgot that St. Paul had spoken of preaching as "the power of God unto salvation."[6] Preaching was, in short, the declaration of the transforming revelation of the living God confirmed in the hearts of the believers by the interior witness of the Holy Spirit.

I have already indicated that the high evaluation of preaching by the Puritans was due to their conviction that this was God's primary way of winning men to His allegiance, hence it was the most important task of the first apostles and of all subsequent witnesses. In addition, there was a series of confirming considerations, all of which reinforced this evaluation.

Among the additional incentives was the recognition of the great dearth of preaching in Reformed England, itself a paradox for reformation was accomplished only "according to the Word of God." The "apostle of the North," Bernard Gilpin, had apprised the young King Edward VI of this desperate lack of preachers in 1552. He declared of Christ's Gospel: "But yet, it is not heard of all Your people, a thousand pulpits in *England* are covered with dust, some have not had foure Sermons these fifteene or sixteene

---

as by lections of the scriptures and by homilies. He excoriated the Puritan claims, "how Christ is by Sermons lifted up higher, and made more apparent to the eye of faith; how the Savour of the Word is made more sweet, being brayed, and more able to nourish, being divided, by Preaching." (Book v, sec. 22, p. 12.) Fuller said: "Mr. Hooker's voice was low, stature little, gesture none at all. . . ." *The Church History of Britain*, edn. of 1837, III, 128.

5 *English Puritanisme, containing the maine opinions of the rigidest sort of those that are called Puritanes in the realm of England* (1605), p. 17. See also Christopher Hill, *Puritanism and Revolution*, pp. 269-71; and Stuart Barton Babbage, *Puritanism and Richard Bancroft*, p. 372: "The Puritans magnified the office of preaching and it was through preaching that they exercised their greatest influence."

6 Romans 1:15-16. See also Romans 10:13-15, I Corinthians 1:21, and Ephesians 1:13. John Penry cited the last three references before the High Commission. Cf. *The Marprelate Tracts*, ed. William Pierce (1589), pp. 64-65.

years, since Friers left their limitations and a few of those were
worthy the name of Sermon."[7] The Catholic reign of Mary
worsened the situation for Protestants, and Elizabeth and her
Council clearly preferred to restrict the issuing of the episcopal
licenses to preachers who were compliant and not potential
troublemakers. Muzzled, if not mute, many of Elizabeth's ministers of religion became minions of state. Hungry parishioners
were fed the driest form of homilies. If they were so fortunate as
to have a preaching rector or vicar and lived near a fair-sized
town, they could count on good sermons four times a year.[8] As
Millar Maclure has characterised the situation, "The drone of the
homilies replaced the mutter of the Mass."[9] In these circumstances,
the Puritan preachers were indeed pastors, that is, shepherds,
sustaining the sheep with solid provender, high in theological
vitamins, often indigestibly so, but a great strengthening after
the starvation diet they were used to.

The second ground for Puritan insistence on preaching was the
complete inadequacy of the alternative—the reading of approved
homilies. It was not only that a sermon was unpredictable and
therefore provided an element of novelty in a prescribed service;
it was the applicability of a good sermon to the conditions of the
congregation. Cartwright, or the author of the *Second Replie* (in
the "Admonition Controversy") put his finger on the weak point
of homilies: "For where the preacher is able according to the
manifold windings and turnings of sin to wind and turn in with
it to the end he may strike it, the homilies are not able to turn
neither off the right nor off the left, but to what quarter soever
the enemies are retired it must keep the train wherein it was set
of the maker."[10] These printed homilies, while their doctrine was
admirable (often they had been prepared by the leading divines),
nevertheless were cold rations and no adequate substitute for the
"living oracles" of God spoken with conviction and sincerity, illus-

[7] *A Sermon preached in the Court at Greenwitch, before King Edward the
Sixth, the first Sonday after the Epiphany, Anno Domini, 1552* (1630), p. 25.
Gilpin observed on p. 23 that at Oxford and Cambridge "the decay of students is
so great, there are scarse left of every thousand, an hundred."
[8] From Bishop Cox's interpretation of the Royal Injunctions of 1560, the regulation went as follows: "If the person be able, he shall preach in his own person
every month; or else preach by another, so that his absence be approved by the
ordinary of the diocese in respect of sickness, service, or study at the universities.
Nevertheless, for the want of able preachers and parsons, to tolerate them without penalty, so they preach in their own persons, or by a learned substitute, once
in every three months of the year." Strype, *Annals of the Reformation*, I, 318.
[9] *Paul's Cross Sermons*, p. 54.
[10] D. J. McGinn, *The Admonition Controversy*, p. 395.

trated from the pursuits and interests of the parish, and applied to the needs of people who were if not the friends of the minister, at least well-known acquaintances, and, above all, his spiritual responsibility. The Puritans raised an outcry against both "dumbe dogges" (unpreaching parsons) and ersatz sermons such as these homilies were, which were spoiling the taste for real preaching. Henry Smith asserts, "And the droane never studieth to preach for hee saith, then an Homelie is better liked of than a Sermon."[11] Learned preachers often poked fun at the homilies, calling them "Homelies," and "humbles."

Furthermore, sermons could penetrate to the hearts of the congregation in a manner impossible for prescribed homilies. Cartwright made the point: "As the fire stirred giveth more heat, so the Word, as it were, blown by preaching, flameth more in the hearers, than when it is read."[12] Topsell indicated that a true listening to preaching is one that expects not only the light of information but the warmth of affection: "Let us therefore so sorte our selves in the congregation, where our eares may be beaten with an understanding sounde, and our hearts bee touched with a heavenly power, that the coales of zeale may be enflamed, and the light of knowledge may be kindeled."[13] Even Hooker grudgingly admitted that sermons were more popular than homilies, due, he suggested, to the exaggerated claims Puritans made for sermons, but also to more sensible considerations. The first was that men knew they could always read over a homily another time if they did not give it their full attention the first time, whereas the sermon was unrecoverable; the second was that sermons "come always new."[14]

A combination of the circumstances just mentioned, including the tight control of preachers by the bishops, the dearth of preaching, and the substitution of printed homilies for sermons, led the Puritans to try to make good the defect by the introduction of "Prophesyings," which constituted a kind of extramural theological course on homiletics organised by ministers with Puritan sympathies in the larger towns to train their younger colleagues in the art and craft of expounding and applying the Gospel. These meetings began as early as 1564, and flourished in the following

11 Ibid., pp. 309-10. A satirical name for "homilies" in a sermon of Thomas Lever is cited in The Cambridge History of English Literature, IV, 233.
12 Ecclesiastical Polity, vol. 2, p. 76, cites the Cartwright observation.
13 Times Lamentation (1599), p. 303.
14 Hooker's Ecclesiastical Polity, Book v, sec. 22, p. 20.

decade. Writing in 1577, Harrison gave a careful account of the procedure at a "prophesie or conference." The exercises took place weekly at some places, fortnightly at others, monthly in some places, and elsewhere twice a year. It was "a notable spurre unto all the ministers, thereby to applie their bookes, which otherwise (as in times past) would give themselves to hawking, hunting, tables, cards, dice, tipling at the alehouse, shooting and other like vanities, nothing commendable in such as should be godlie and zealous stewards of the good gifts of God, faithfull distributors of his word unto the people, and diligent pastors according to their calling."[15] According to Harrison, "such is the thirstie desire of the people in these daies to heare the word of God, that [the laity] also have as it were with zealous violence intruded themselves among" the ministers, though only as hearers.[16] The method was:

> Herein also (for the most part) two of the yoonger sort of ministers doo expound, ech after other, some peece of the scriptures ordinarilie appointed unto them in their courses (wherein they orderlie go through with some one of the evangelists, or of the epistles, as it pleaseth the whole assemblie to choose at the first in everie of these conferences); and when they have spent an houre or a little more betweene them, then commeth one of the better learned sort, who . . . supplieth the roome of a moderator, making first a brief rehearsall of their discourses, and then adding what him thinketh good of his owne knowledge, wherby two houres are thus commonlie spent at this most profitable meeting. When all is doone, if the first speakers have showed anie peece of diligence, they are commended for their travell, and incouraged to go forward. If they have beene found to be slacke, their negligence is openlie reprooved before all their brethren, who go aside of purpose from the laitie, after the exercise ended, to iudge of these matters, and consult of the next speakers and quantitie of the text to be handled in that place.[17]

The new scene of Puritan activity in England originated in the Swiss and Rhenish cities, where the Marian exiles had undoubt-

[15] *Harrison's Description of England in Shakespere's Youth*, ed. F. J. Furnivall, pp. 18-19.
[16] *Ibid.*, p. 18.
[17] It should be noted that Thomas Fuller's account of prophesying, in *The Church History of Britain*, ed. J. S. Brown, III, 6-7, is substantially the same, except that Fuller asserts that after the first young divine has preached "five or six more observing their seniority, successfully dilated on the same text."

edly come across them and been impressed by them, and introduced them into England.[18] It was an excellent device for improving the education of the clergy, for keeping them in a state of holy emulation, for raising the standard of Biblical studies (and the use of good commentaries) as well as preaching. Notably the young ministers were taught to expound successive passages of Scripture, not to take texts as mere pretexts to illustrate topics. They were, moreover, encouraged to read whole chapters of the Bible, which was to account in part for their dislike of Anglican shredding of the Scripture lessons as mere "pistling and gospling." The example set by the older men—with their knowledge of the Biblical languages in many cases, the disciplining of their minds to the Word of God, and the fervency of their preaching—must have been profoundly influential.

In time the exercises became not only miniature theological colleges for continuing ministerial education (and, as was typical of the Puritan movement, a means of further theological education for laymen), but they also became centres of Puritan propaganda for the "Discipline," and as such provided the framework of the later classes and presbyteries.[19]

Queen Elizabeth felt they were a threat to the established Church, and instructed Archbishop Grindal to request his brother bishops to forbid all "prophesyings" in their dioceses. On December 20, 1576 Grindal sent his famous letter to Queen Elizabeth in defence of prophesyings: "I am forced with all humility and yet plainly to profess that I cannot with safe conscience and without the offense of the Majesty of God, give my consent to the suppressing of the said exercises; much less can I send out my injunction for the utter and universal subversion of the same. . . ."[20] Ignoring her archbishop the queen on May 7, 1577 issued instructions directly to the bishops, commanding suppression. Grindal was sequestered by the royal orders and confined for six months to his palace in Lambeth. Because he remained obdurate his suspension was never fully removed, despite frequent appeals made on his behalf to the queen by his bishops and by the House of Clergy. This single event shows that even if the queen did not like to make windows in men's souls, she was not adverse to restricting their views. It also indicates how desperate the need

18 See Leonard J. Trinterud, "The Origins of Puritanism," in *Church History of Britain*, vol. 20, no. 1, pp. 46f. *Prophezei* were organized by Zwingli in Zurich in 1525 and in England in John à Lasco's Church of the Strangers.

19 P. M. Dawley, *John Whitgift and the English Reformation*, p. 146.

20 The letter is reproduced in Strype, *The History of the Life and Acts of Edmund Grindal*, pp. 578-79.

was for the Puritan protest for more preaching, that it was supported by the Primate of All England at the cost of his authority. Now the only way open to the Puritans to supply the lack of preachers was to get well-trained Puritan ministers appointed as lecturers in a given parish, their stipends being provided by men of substance and most often in London. A preeminent example was Henry Smith, an immensely popular lecturer at St. Clement Dane's Church in the Strand from 1587 to 1590, whom Bishop Aylmer tried to remove from this position but who was reinstated almost certainly by the influence of his stepmother's brother, Lord Burghley, to whom he dedicated the collected edition of his *Sermons* in 1593. Less conspicuously there were secret gatherings of Puritan ministers to continue these exercises.

Important as the role of preaching was in the Puritan discipline and culture of the spiritual life,[21] it must be remembered that it was reinforced by a cluster of ancillary practises, all dependent on the recognition of the primacy of the Bible as containing the will of God. Psalm-singing might be contemptuously termed "Genevan jigs" by Elizabeth, but these religious songs confirmed the lessons of obedience to God's law, the privileges of the people of God, and men's security in the providential rule of the Shepherd God. Fast-days and conferences were the Puritan substitutes for Catholic pilgrimages, as sermons on "cases of conscience" and Puritan diary-keeping was the substitute for the Catholic confessional.[22] Private prayer in the language of the heart was another exercise, along with private Bible-reading, that made men and women eager to respond to the marching orders of the living God. The sermons themselves were to be summarised carefully from notes taken during the service so that the heads of households could be sure their children had heeded the Word of God preached from the pulpit. As Marshall Knappen has observed of these mutually reinforcing spiritual techniques: "These called for more intelligence and more concentration than any of the Catholic techniques." Yet one may dissent from his conclusion, at least as applied to the great Catholic mystics, like St. Teresa of Avila or St. John of the Cross, that "doubtless, if used properly, they were capable of

---

[21] The culture of the spiritual life of Catholics and Anglicans, as well as Puritans, is treated in detail in a later chapter.

[22] The Puritan preachers provided a whole genre of sermons denominated "cases of conscience." To take only two examples among hundreds, Richard Greenham, the teacher of Henry Smith, produced *Godly instructions for the due examination and direction of all men* (1599) and William Perkins, *The whole treatise of the cases of conscience* (1604).

putting a finer edge on the spiritual life also."²³ Puritan spirituality
was certainly admirably adapted for the thoughtful laymen of the
age; it was a Bible-centred spirituality.

It may well have been Hooker's practise of reading his heavily
freighted sermons that blinded him to the importance of a face-to-
face confrontation with his congregation, and accounted for the
strange observation: "Men speak not with the instruments of writing, neither write with the instruments of speech; and yet things
recorded with the one, and uttered with the other, may be preached
well enough with both."²⁴ The point was that a man could address
other men directly from the pulpit, whereas the most eloquent
writing was indirect speech. Preaching was a striking on the soul,
as Miles Mosse, a preacher at Paul's Cross, well knew, when he
said: "*Vox est ictus animi*: passing through the eare, and braine,
and blood, it smiteth (as it were) and giveth a stroke upon the
verie soule, and so with a kind of Violence doth deeply affect
it. . . . But yet further, besides all the worke of Nature, there is
in Preaching a speciall gift of grace: which enableth a man to
speak with such evidence of the Spirit, & with such power to the
Conscience, as no pen of man by writing can expresse: whereof
Preaching is the most likely, and effectuall instrument of salvation and so to be respected."²⁵

For these reasons, then, the Puritans believed that preaching
was the means chosen by God for illuminating the minds, mollify-
ing the hearts, sensitising the consciences, strengthening the faith,
quelling the doubts, and saving the souls of mankind. To that
end the Puritan brethren dedicated their chief energies to preaching clearly, faithfully, sincerely, and movingly, trusting that the
Holy Spirit would take their human words and make them the
"lively oracles of God" to their congregations.

2. Biblical Interpretation

The Puritans were not the first Protestants. It is therefore not
surprising that in their Biblical interpretation they adopted many
of the early Protestant principles. Perhaps the most distinctive
Protestant principle for the interpretation of Scripture was the
determination to use as the primary meaning the literal or historical
sense of the texts, as contrasted with the medieval allegorical inter-

²³ *Tudor Puritanism; A Chapter in the History of Idealism*, p. 399.
²⁴ *Ecclesiastical Polity*, book v, sec. 21.
²⁵ *Justifying and Saving Faith* (1614), sig. gg2, cited in Maclure, *St. Paul's Cross Sermons*, p. 145.

pretation. Tyndale referred to the Catholic fourfold mode of scriptural interpretation, for contrast: "They divide Scripture into four senses the literal, tropological, allegorical, and anagogical. The literal sense is become nothing at all: for the Pope hath taken it clean away. . . . The tropological sense pertaineth to good manners (say they), and teacheth what we ought to do. The allegory is appropriate to faith; and the anagogical to hope, and things above."[26] This led to a great deal of ingenious subjectivity in the interpretation of Scripture. But it also had the advantage of providing intelligibility when there were contradictions or obscurities. It also had some practical purposes in mind, for the tropological meaning provided ethical guidance, the allegorical meaning gave spiritual assistance, and the anagogical sense kept the eschatological dimension of Scripture vivid. Protestantism was also interested in the same guidance, but it was content to try and find it by exegesis rather than by eisegesis, however brilliant the latter might be.

Tyndale, the great English translator, stands with all Protestants on the primacy of the literal sense. "Thou shalt understand," he writes, "that the scripture hath but one sense, which is the literal sense. And that literal sense is the root and ground of all, and the anchor that never faileth, whereunto if thou cleave, thou canst never err or go out of the way. Nevertheless, the scripture useth proverbs, similitudes, riddles or allegories, as all other speeches do: but that which the proverb, riddle, or allegory signifieth, is ever the literal sense, which thou must seek out diligently: as in English we borrow words and sentences of one thing, and apply them to another and give them new significations."[27]

It is clear that the older Henrician divines like St. John Fisher, an accomplished preacher of ornate sermons in the older mould, preferred the allegorical approach. Fisher would, for example, interpret the four rivers of paradise referred to in Genesis 2:10-14

[26] From "The Obedience of a Christian Man," *Doctrinal Treatises*, ed. H. Walter, p. 303.

[27] *Ibid.*, p. 304. William Whitaker, the Calvinist Master of St. John's College, Cambridge, was more guarded in his qualified rejection of the nonliteral senses of scripture: "These things we do not wholly reject: we concede such things as allegory, anagoge, and tropology in Scripture; but meanwhile we deny that there are many and various senses. We affirm that there is but one true, proper, and genuine sense of Scripture arising from the words rightly understood, which we call the literal: and we contend that allegories, tropologies, and anagoges are not various senses, but various collections from one sense, or various applications and accommodations of that one meaning." *A Disputation on Holy Scripture against the Papists, especially Bellarmine and Stapledon*, tr. and ed. W. Fitzgerald, p. 403.

as symbolising the four cardinal virtues,[28] and use Sinai as symbolising a Jewish synagogue and Sion the Christian church.[29] This was the very characteristic that the racy Latimer criticised in Catholic exegesis, so that the Catholics were inclined to see subtleties where none existed. For example, Latimer argued that when Christ chose the boat of Peter, it was misinterpreted by Catholics as teaching the primacy of Peter among the apostles.[30] While the Catholics were attracted to the poetical and prophetical books of the Bible for their texts and expositions, the Protestants were rediscovering the historical books of the Old Testament.[31] Latimer was a master at retelling the narrative portions of the Old Testament and applying them to current situations. He expounded the first two chapters of the first book of the Kings in order to apply the usurpation of Adonijah to the overweening ambition of Lord Seymour, at the time imprisoned in the Tower of London.[32] Equally, Latimer was glad to preach one of his "nipping, rough, and hard biting" sermons against bribery and unpreaching prelates as he related the story of the sons of Eli from the second and third chapters of the first book of Samuel. Similarly, the most popular of Elizabethan Puritan preachers, Henry Smith, had three sermons on King Nebuchadnessar, as a type of pride.

A second characteristic of Protestant Biblical interpretation, as contrasted with medieval Catholic exegesis, was a deep concern to discover the native meaning of Biblical expressions and derive them from the original languages. John Bradford, chaplain to Nicholas Ridley, was not unusual in emphasising that μετάνοια, the Greek word for "repentance," means a "forethinking" and has no reference to the developed Catholic sacrament of Penance. This emphasis on Biblical and Patristic learning reached its climax in

---

[28] *The English Works of John Fisher*, part I, ed. J.E.B. Mayor, 2nd edn., E.E.T.S., e.s. xxvii, 1935, pp. 24-25.

[29] *Ibid.*, p. 167.

[30] *Sermons by Hugh Latimer, sometime Bishop of Worcester, Martyr, 1555*, ed. G. E. Corrie, pp. 205-206.

[31] The contrast between Catholic and Protestant exegesis of the scriptures must not be too rigidly drawn, however, for several English Catholic preachers imbued with the "New Learning" emphasised the literal sense, such as Colet, Stafford, and Taverner. Moreover, even Latimer and Hooper were prepared to accommodate historical narrative to the current political situation in the search for monitory lessons for their contemporaries.

[32] Seymour had aspired to marry the widow of Henry VIII, Queen Catherine Parr, as Adonijah aimed to marry Queen Abishag on the death of King David. The names of the living aristocrats are not once mentioned, but the implications are clear. Cf. F. E. Hutchinson's "The English Pulpit from Fisher to Donne," vol. 4, *Cambridge History of English Literature.*

the Jacobean erudition (if not occasional pedantry) of the sermons of Bishop Lancelot Andrewes, but it was often used by Protestants in the interpretation of such disputed texts as the Vulgate, *Hoc est corpus meum.*

Since only the Bible, as the divinely inspired Word or message of God, was able to serve as the standard by which a corrupt church could be corrected, it was of crucial significance that it be expounded literally for Protestants. For them it was the standard of doctrine—of information, as well as of correction—of reformation. For the Puritans, however, it was not only a standard of faith and morals, but it served as the only authoritative criterion for the ordinances and the details of worship and of ecclesiastical government. There were Puritans (like Cartwright) who would have curbed the magistrates with a theological whip in a theocracy and applied the regulations of *Leviticus* to Leviathan, as Hobbes was to name the state. To the reflective mind of Hooker, this was bibliolatrous, but the very excess of respect given to the Bible rather than to the Christ, who was its climax and criterion, helps us to understand the energy and intellectual zest that was given to the study of the Scriptures and the application of its lessons for the people from the pulpits of the Puritan preachers.

3. The Structure of the Sermon

If the primary purpose of preaching is to present the claims of God on man through the criticism and comfort of the Gospel, the structure of the Puritan sermon had to be simple, memorable, and practical. As such, it aimed to produce light and heat, illumination of the mind and warming of the affections.

The structure of the Puritan sermon was anticipated by Bishop Hooper, who in this respect, as in his objections to ornate vestments,[33] deserves the title, "the father of Nonconformity." It was exemplified in his series of sermons on Jonah preached at the court of King Edward VI. The structure took the form of the exposition of a passage of Scripture *secundem ordinem textus*, by collecting lessons (or "doctrines") from each verse and adding the moral applications (or "uses") of them.[34] The outstanding Puritan

[33] See J. H. Primus, *The Vestments Controversy.* See also Geoffrey F. Nuttall and Owen Chadwick, *From Uniformity to Unity, 1662-1962*, p. 162; and W.M.S. West, "John Hooper and the Origins of Puritanism," *Baptist Quarterly* (October 1954-April 1955).

[34] See J. W. Blench's *Preaching in England in the late fifteenth and sixteenth centuries*, pp. 94, 101-102, to which I owe much in this chapter. Blench has

lecturers used the same method as Hooper, which had been made widely known through the Latin commentaries of Musculus.

It was outlined with characteristic clarity in the famous *Art of Prophesying* of William Perkins, lecturer at St. Andrew's Church, Cambridge, the most eminent Puritan scholar of Elizabethan times, and an admirable preacher. The preacher's task was: (1) "to reade the Text distinctly out of the Canonicall Scriptures"; (2) "to give the sense and understanding of it being read, by the Scripture it selfe"; (3) "to collect a few and profitable points of doctrine out of the naturall sense"; (4) "to apply (if he have the gifte) these doctrines rightly collected to the life and manners of men in a simple and plaine speech."[35]

What was perhaps most interesting about the structure of the Puritan sermon was that it was streamlined in the direction of changing man's mind with a view to improving his behaviour. There was little interest in speculative thought or even speculative divinity. Of paramount concern was that godliness which desires to know the will of God in order to follow it. Puritan theology was, as Perkins described it, "the science of living blessedly for ever."[36] If the form of the Puritan sermon was indeed so functional, was there a place for the imagination? The answer was that metaphors, similes, and *exempla* provided the illustrations that were the windows of the sermon, illuminating the doctrine while sustaining the interest of the auditors, and possibly even nerving their wills to action.

For an example of the typical structure of a Puritan sermon one could turn to the sermons of Thomas Cartwright, which were constructed in the approved Swiss manner, and which were dutiful and dull, or to the livelier sermons of Edward Topsell, the author of two sets of Old Testament sermons, the first on the book of Ruth, entitled *The Reward of Religion*,[37] and the second on the

---

shed almost as much light on Renaissance preaching in England as Owst did on medieval English preaching.

[35] *Works*, vol. 2, p. 673.

[36] "Perkins," *Dictionary of National Biography*, vol. 45. The exact source of the definition is Perkins' *A Golden Chaine* (1600), p. 1. It is interesting to note that Perkins, the teacher of William Ames, was said to have preached so that "his sermons were not so plain but that the piously learned did admire them, nor so learned but that the plain did understand them." Thomas Fuller, *The Holy State* (1648), p. 81.

[37] *The Reward of Religion. Delivered in sundry Lectures upon the Booke of Ruth, wherein the godly may see their dayly both inward & outward trials, with the presence of God to assist them, & his mercies to recompense them* (1597, first published 1596).

book of Joel, called *Times Lamentation*.[38] The more interesting alternative has been selected; the 16th of the 17 sermons on Ruth will be analysed.

The text is Ruth 4:16-17: "And Naomi took the child and laid it in her lappe, & became nurse unto it. And the women her neighbours gave it a name, saying, there is a child borne to Naomi and called the name thereof Obed; the same was the father of Ishai the father of David."

Topsell, after reading out the text, put it in its historical context and observed what comfort the birth of this son of Boaz brought to his grandmother Naomi, how she helped her daughter nurse him, and how the women of Bethlehem helped Naomi. Topsell derived four practical lessons from verse 16 and one from verse 17. His structure consisted of providing five doctrines and applying them.

(1) The first doctrine is that just as previously miseries were heaped on Naomi, now she knows the multiplied mercies of God in the birth of this boy. The application, or "use," is: "Thus we see it is a righteous thing with God, first to wound and then to heale, first to strike and then to stroke."[39] Then examples of a similar experience are given from the lives of Job, Joseph, and David. (2) The second doctrine from the same verse is that here is an example of a worthy grandmother who, out of gratitude to God, love of her grandson, and a desire to help her daughter, helped in the nursing of the child. The use of this doctrine is that it reminds aged parents who live to see their children's children that they have a duty to care for them. (3) Since Naomi is said here to be nurse of the child of Ruth, it is appropriate to raise a question with two parts. Is it lawful (A) to put children out to nurse from their own mother? Is it lawful (B) to commit their tuition to others? "Unto the first question I aunswere, that every woman being of health of body and minde, is bound by the worde of God to nurse her owne children. . . ."[40] This is confirmed by the examples of Sarah, Manoah's wife, Bathsheba, Elizabeth the wife of Zachariah, and the Virgin Mary who suckled the Saviour. To have the child nursed by some woman who is not its mother is a sign that its parents lack natural affection. Moreover, it is unnatural for a child to be nursed by a woman who did not bear

[38] The full title is *Times Lamentation, or an exposition of the prophet Joel in sundry sermons or meditations* (1599). It was reissued in 1601 and 1613.
[39] *The Reward of Religion*, pp. 285-86.
[40] *Ibid.*, p. 288.

him, "for the same body whereof he had his being is most fitte for his feeding."[41] The second part of the question must now be answered: is it right for parents to entrust their children to others for their upbringing? "I aunswere that every man and woman are bounde to see their childrens first instruction, that is, if it be possible to have them in their keeping at their first entrance to knowledge, and when they are first capable of any goodnesse."[42] Biblical examples are given in the instances of Isaac remaining with Abraham and little Benjamin with old Jacob. (4) After the important double question has been answered Topsell returns to the normal structure of the sermon. The fourth doctrine and use is the recognition of how important is the bringing up of children.

Topsell finally devotes his attention to the seventeenth verse. He reiterates that the child was named Obed, which signifies serving, or a servant, showing how he should serve for the comfort of Naomi, Boaz and Ruth. The women of Bethlehem proved to be good neighbours of Naomi. So the final doctrine (5) is to point out "the duty of the faithfull to be helpers one to another in the service of God, and admonition of their dueties."[43] The preacher concludes by giving the examples of the ruler of the temple who went to Jesus to seek help for his ailing child, the four friends who carried the palsied man to Christ for healing, and the friends of Dorcas who sought the aid of Peter to restore her to life.

The Puritan sermon structure was simple because it drew its lessons as the narrative proceeded. It was amply illustrated by godly examples drawn from other parts of the Bible to supplement the text. It was easily remembered, because to reread the texts at home was to recall the preacher's commentary and application of the passage. It did not provide a vehicle for the richer resources of rhetoric, with set pieces of sustained eloquence, grandiose comparisons and contrasts, tirades, apostrophe, and so forth, such as can be found in the Catholic sermons of St. John Fisher or of Hugh Weston, the Marian Dean of Westminster. Nor did the sermon form lend itself to the brilliant word analysis, patristic erudition, metaphysical and far-fetched conceits, and sheer sparkle of Lancelot Andrewes. Such elaborate rhetoric, "taffeta terms," and ornate diction, were suitable for sermons on state occasions, but it meant that the Word of God, which is sharper than any sword, and which could pierce to the quick of the conscience, remained sheathed in a

41 *Ibid.*, p. 290.   42 *Ibid.*, p. 291.   43 *Ibid.*, p. 293.

307

jewelled scabbard. The Puritan sermon was no ceremonial sword; least of all was it like the painted, wooden sword of a homily officially prescribed. It was a lithe, lean, sharp instrument, poised to strike the soul.

Thus a simple, straightforward form for the sermon was felt to be appropriate to the Gospel itself, a reminder that the preachers were after all clay vessels, yet containers of the Gospel, that the greatest preacher of Christ the world had seen, St. Paul, had insisted that the Gospel was to be preached not by words of eloquence but by the integrity of conviction.

Great ingenuity was required of the preacher, not so much in providing the doctrines of the scripture, nor even in the illustrations from literature and life, but chiefly in the application of the lessons to the variety of persons and conditions in his congregation, especially in a city church. Perkins defined application as "that, whereby the doctrine rightly collected, is diversely fitted according as time, place, and person doe require."[44] He even categorised seven types of application, referring specifically to their conditions in relation to salvation. (Other preachers would of course apply lessons to the age, sex, and callings of the persons represented in their congregations.) But Perkins' sevenfold application is interesting. It includes: "unbeleevers, who are both ignorant and unteachable"; some who "are teachable, but yet ignorant"; some who "have knowledge but are not as yet humbled"; those who are humbled; some who believe; some who "are fallen"; and, finally, "there is a mingled people."[45] The requirement that every doctrine have a use is the proof of the practical nature of Puritan preaching, being directed to the conversion of the will and the betterment of behaviour.

4. Sermon Style

While one can refer to a Puritan sermon structure, there was no single Puritan style of sermon. J. W. Blench, in his analysis of the style of Elizabethan preachers, maintains that there were three major styles—plain, colloquial, and fully ornate—but he finds it necessary to subdivide the first category, to which most Puritan sermons belong, into three parts: the first, bare and austere; the

---

[44] *The Arte of Prophesying, Or, a Treatise concerning the sacred and onely true Manner and Methode of Preaching*, included in *The Workes of that famous and worthy Minister of Christ in the University of Cambridge, M. William Perkins* (1613), 3 vols. This treatise was first issued in Latin, as the preface to the three-volume edition of his works indicates. The citation reference is to vol. 2, p. 664.

[45] *Ibid.*, pp. 665-68.

second, less colourless, employing tropes but not schemata; and the third, moderately decorated, using tropes but rarely using schemata.[46]

The silver-tongued Henry Smith used the fully ornate style, as did the Calvinist and future Archbishop George Abbott. Abbott defended this, a little self-consciously, in asserting that ministers of the Gospel have liberty "not only nakedly to lay open the truth, but to use helps of wit, of invention and art (which are the good gifts of God), so to remove away all disdain and loathing of the word from the full hearts of the auditory—similitudes and comparisons, allusions, applications, yea parables and proverbs which may tend to edification and illustrating the word; for they have to do with weak ones as well as with the strong—with some of queasy stomachs, with some who must be enticed and allured with a bait of industry and eloquence, of pretty and witty sentences."[47] Smith and Abbott were exceptions, for the vast majority of those with Puritan leanings used the first or second, or in rare cases, like Richard Greenham (tutor of Perkins), the third variation of the plain style.

Perkins, Laurence Chaderton (first Master of Emmanuel College, Cambridge), Fulke, Gifford and Thomas White, preferred the austerely plain style, considering the use of even the simplest rhetorical devices or of classical learning as a distraction and a drawing of attention to the preacher rather than his message. Perkins, for example, insisted that "*humane wisedome* must be concealed, whether it be in the matter of the sermon, or in the setting forth of the words," because the preaching is the proclamation of the wisdom of Christ and not of human skill, and because the hearers must ascribe their faith to divine, not human agency. At the same time, he would not allow the same argument to be made in justification of barbarism and Philistinism in the pulpits. The minister "may, yea and must privately use at his libertie the arts, Philosophy and variety of reading, whilest he is framing his sermon: but he ought in publike to conceal all these from the people, and not to make the least ostentation. *Artis etiam est celare artem: it is also a point of art to conceal art.*"[48] Perkins contended that this principle ruled out the use of Greek or Latin phrases, for they disturb the mind of the auditors, "that they cannot fit those

46 *Ibid.*, p. 163.
47 *An Exposition of the Prophet Jonah*, II, 331, cited in Blench, *Preaching in England*, p. 182.
48 Perkins, *Works*, II, 670.

things which went afore with those that follow," strange words prevent the understanding of what is spoken, thus there is great distraction.[49] Perkins also condemned telling stories and all profane and ridiculous speeches. The most appropriate style for a preacher, in Perkins' view, was "a speech both simple and perspicuous, fit both for the people's understanding, and to expresse the Maiestie of the Spirit."[50]

Unquestionably Bartimaeus Andrewes would have agreed with him, for this Andrewes believed that the use of rhetorical devices and citations from the Fathers and classical authors was a departure from the simplicity of the Gospel. With irony he said: ". . . there are some who thinke Christe too base to be preached simply in him selfe, and therfore mingle with him too much the wisdom of mans eloquence and thinke that Christ commeth nakedly, unlesse cloathed with vaine ostentation of wordes. Others esteem him too homely, simple and unlearned, unlesse he bee beautified and blazed over with store of Greeke or Latin sentences in the pulpits; some reckon of him as solitarie, or as a private person without honour and pompe, unlesse he bee brought foorth of them very solemnly, accompanied and countenaunced, with the auncient Garde of the fathers and Doctors of the Churches to speake for him: or els he must be glossed out and painted with the frooth of Philosophie, Poetry, or such like."[51] Yet while one sees the point of this diatribe, one must also recognise that if Lancelot Andrewes had taken the advice of Bartimaeus Andrewes, this great star would have been eclipsed in the homiletical firmament.

There were, however, eminent Puritan preachers who denied that preaching should be devoid of illustrations. They recognised

[49] Ibid., pp. 671-72. With this view, however, should be contrasted the words of Henry Smith which he states here (Sermons, collected edn. of 1593, pp. 311-12) and repeats (ibid., p. 662): "There is a kinde of Preachers risen up but of late, which shrowde and cover everie rusticall and unsaverie, and childish and absurd sermon, under the name of the simple kind of teaching, like the Popish Priestes, which make ignorance the mother of devotion, but indeede to preach simplie is not to preach rudely, nor unlearnedly, nor confusedly, but to preach plainly, and perspicuously, that the simplest man may understand what is taught as if he did heare his name." An early supporter of the plain Puritan preaching was Edward Dering, who challenged, "Let the sinner come forth, that hath been converted by hearing stories or fables of poets, I am sure there is none: for faith is onely by the worde of God: or let the preacher come forth that useth such things, and doth it not either to please men, or to boast of his learning." XXVII Lectures or Readings upon part of the Epistle written to the Hebrues (1576), no. 20.

[50] A Dilucidation or Exposition of the Apostle St. Paul to the Colossians, ed. A. B. Grosart, p. 6.

[51] Certaine verie godlie and profitable Sermons upon the fifth Chapiter of the Songe of Solomon (1583), p. 26.

310

that illustrations, and even divagations from the theme of too intense an exposition and application of Biblical doctrine, provided a welcome relief for the congregation and might even drive the lessons home. Such were preachers like Anthony Anderson or Thomas Cartwright, the father of Presbyterian Puritanism in England. Cartwright pointed out that a lie might seem more credible than the truth, when he employed the simile: "as the fruit that groweth now in Sodom hath a more excellent show than other fruit; and yet, come to feel it, it goeth to froth and wind, and that loathsome." Other preachers were more inventive. "In our daies," says Edward Topsell, "men quake in the congregation like steeples in the sea."⁵² The same Puritan preacher wished to illustrate the problem of King Claudius in *Hamlet*—namely the dryness of the spiritual life—and said of those with cold affections in worship, that "their voices in prayer are like unborne children, crie they cannot, much lesse speak anything, not so much as to say *Amen*."⁵³ Topsell could provide a series of metaphors that had a fine cumulative effect, as when he was describing the importance and effect of unity in worship: "Agreement in battle getteth victorie; content in a common weale maketh peace; unitie in musicke maketh harmonie; and the fellowship in praier conquereth the divell; getteth peace of conscience, and soundeth sweetely in the eares of God."⁵⁴ It may well have been, as Perry Miller has suggested,⁵⁵ that the modification of the bare and austere Puritan style of preaching to include tropes is due to the influence of the Ramist *Rhetorica* of Talon and to the feeling that clarity justified their use, whereas the schemata introduced ostentation and artificiality.

There was also a group of Puritan preachers that used not only tropes but some of the rhetorical schemata. Among them were Richard Greenham, as well as Arthur Dent, Stephen Gosson, and the most popular and skilled Puritan preacher of them all—Henry Smith. Topsell would also be in this category; that is, his style was relatively simple, and he used rhetorical devices moderately.

A lengthy excerpt will show the architectonic that lay behind the directness of Henry Smith, as he illustrated the different kinds of hearing which are given to a sermon:

As ye come with divers motiõs, so ye heare in divers manners: one is like an *Athenian* and he hearkeneth after newes:

⁵² *Times Lamentation* (1599), p. 212.
⁵³ *Ibid.*, p. 201.   ⁵⁴ *Ibid.*, p. 303.
⁵⁵ See *The New England Mind: The Seventeenth Century*, pp. 355-57.

if the Preacher say any thing of our Armies beyond sea, or Counsell at home, or matters of Court, that is his lure: another is like the Pharisie, and he watcheth if anything bee sayd that may bee wrested to be spoken against persons in high place, that he may play the divell in accusing of his brethren, let him write that in his tables too: another smackes of eloquence, and hee gapes for a phrase, that when he commeth to his ordinarie, he may have one figure more to grace and worship his tale: another is male-content, and he never pricketh up his eares till the Preacher come to gird agaynst some whom he spiteth, and when the Sermon is done, he remembreth nothing which was sayd to him, but that which was spoken agaynst other: another commeth to gaze about the Church, he hath an evill eye, which is still looking upon that from which Iob did avert his eye: another commeth to muze, so soone as hee is set, hee falleth into a browne studie, sometimes his minde runnes on his market, sometimes of his iourney, sometimes of his suite, sometimes of his dinner, sometimes of his sport after dinner, and Sermon is done before the man thinke where he is: another commeth to heare, but so soone as the Preacher hath sayd his prayer, he falles fast asleepe, as though he had been brought in for a corps, & the preacher should preach at his funerall.[56]

Only a master of the pulpit would have used so sinister a final simile and so ghoulish a reminder that the sleep of negligence is an image of the sleep of death, and yet done it so slyly. Similarly, we see a complex sentence structure yielding great clarity of meaning when Smith was arguing for simplicity in hearing the Word of God in the sermon: "As the little birds pirk up their heads when their damme comes with meat, and prepare their beaks to take it, striving who shall catch most (now this looks to be served, now that looks for a bit, and every mouth is open until it bee filled): so you are here like birds, and we the damme, the word the food; therefore you must prepare a mouth to take it."[57]

Another requirement that encouraged simplicity was the expectation that the sermon would be delivered without a manuscript. Perkins made this clear in his *Arte of Prophesying* by writing that something must be said about the training of the memory, "because

[56] "The Arte of Hearing, in two sermons," included in *The Sermons of Maister Henrie Smith gathered into one Volume* (1593), Richard Field for Thomas Man), pp. 642-43.
[57] Perkins, *Works*, II, p. 670.

it is the received custom for preachers to speake *by heart* before the people."[58] So excellent were the memories of the Puritan preachers, in fact, that it would not have been difficult for them to have preached extemporaneously, as one of the most famous, "Father Dod," was once required to do. The story goes that he had once preached in Cambridge against the drunkenness of the university students, which greatly enraged them. One day a group of them encountered Dod walking in a wood. They took him captive and held him in a hollow tree, promising to release him only on condition that he preach a sermon on a text they chose. They gave him the word "malt" for a text, and on this he expatiated, beginning: "Beloved, I am a little man, come at a short warning, to deliver a brief discourse, upon a small subject, to a thin congregation and from an unworthy pulpit," and taking each letter as a division of his sermon.[59]

Memorising of sermons gave the Puritan preachers a firm structure, a clear progression, and all the advantages of order, with the advantages of flexibility, so that as they watched the congregation they could explain further if there was incomprehension or add additional illustrations as needed. Also, the artificiality of reading from a manuscript was avoided.

It is interesting to consider the advice Perkins gave to would-be ministers in his influential treatise, particularly on matters of voice and gesture. The voice, he advised, should be moderate when inculcating the doctrine, but in the exhortation "more fervent and vehement."[60] The general rule for gestures was that they should be grave so that the body may grace the messenger of God. "It is fit therefore," Perkins wrote, "that the trunk or stalke of the bodie being erect and quiet, all the other parts, as the arme, the hand, the face and eyes, have such motions as may expresse and (as it were) utter the godly affections of the heart. The lifting up of the eye and the hand signifieth confidence. 2 Chron. 6. 13-14 . . . Acts, 7. 55 . . . the casting downe of the eyes signifieth sorrow and heaviness. Luk. 18.13."[61]

The impression has rightly been given that the Puritan preacher regarded his calling with the utmost seriousness and took great

---

[58] *Ibid.*, II, 672. Beside the words "*by heart*," there appears in the margin the word "memoriter."

[59] *Dictionary of National Biography*, vol. 15, p. 145b. This courageous minister, in a sermon preached in court, told Queen Elizabeth that the negligence of the general run of ministers was partly her fault, for she allowed it to go unchecked because it was of no particular political importance.

[60] Perkins, *Works*, II, 672.   [61] *Ibid.*

pains over it. It does not mean, however, that his preaching lacked humour. It almost entirely lacked the witty conceits of the metaphysical preachers like Lancelot Andrewes and John Donne, and it rarely emulated Latimer's raciness, but it did not disdain the whimsical or mocking strains. The former may be illustrated by the immensely popular Henry Smith in "The Affinitie of the Faithfull," where he remarked: "The Divell is afrayde that one Sermon will convert us, and we are not moved with twentie: so the divell thinketh better of us than we are."[62] The mocking mood was well expressed by Topsell:

> Surely I beleeve *Paul* was deceived when he said, *Faith came by hearing and hearing by the word of God.* What an impudent blasphemie were this, to say that Ladies and gentlewomen, on whose faces the sunne is not good inough to looke, whose legges must not walke on the ground, but either keep aloft in their bowers, or take the ayer in their coaches, whose hands must touch nothing but either chaines of pearle, cloathe of gold embrodered, and fine needle wrought garments; that these beautiful stars (I say) should come downe from their nicenes and learn faith at the mouth of preachers? Yet further, must our gallant youthes and proper servingmen, whose heads are hanged with haire as if they would fright away both Christ and his ministers from the place where they stand, come frō the taverns, from gaming houses, from the playhouses, frō the ale-houses, from the whoore-houses, and from all their disports, to be ratled up for their follies by preaching, & forsake the fashions of the world to be new fashioned in their minds, that in stead of infidelities (wherewith the most of that crew are infected) they may have faith engraffed in them by hearing the Gospell, least as they consume their purses, they condemne their soules, neither can robbe for more soules, as they do for more purses?[63]

The Puritans had one rare preacher, however, who was capable of producing conceits of the type that Andrewes and Donne delighted in. He was, of course, Henry Smith. Speaking of prayer, he said that it was "such a strong thing that it overcommeth God, which overcommeth all."[64] He was witty even in his prayers, for one petition went: "Teach us to remember our sinnes, that Thou

---

[62] Smith's *Sermons* (1593 edn.), p. 255.
[63] *Times Lamentation* (1599), p. 26.   [64] *Sermons* (1593), p. 878.

maiest forget them."[65] Referring to Satan, Smith says: "as a compasse hath no end, so he makes no end of compassing."[66] Though the Puritan sermons were very serious, they were not innocent of whimsicality, irony, or wit; these, like lightning flashes, lit up the terrain of preaching all the more dramatically because of their rarity.

How long was the Puritan sermon? Cartwright wished the unduly prolix preacher to be curbed: "Let there be, if it may be every sabbath-day, two sermons, and let them that preach always endeavour to keep themselves within one hour, especially on the week-days."[67] Knappen claims that there are records of two- and three-hour sermons.[68] It is difficult to tell from the written sermons how long the originals were, because they were so often greatly expanded in the writing out. Furthermore, speed of delivery varied. One incident, thought to be remarkable enough to record, would seem to indicate that a two-hour sermon was not a great rarity. Fuller tells that Laurence Chaderton, the first Master of Emmanuel College, Cambridge, who was brought up a Catholic under Vaux, the celebrated author of the *Catechisme*, and became a Puritan at that famous nursery of the "Brethren," Christ's College, was once visiting his native Lancashire. Probably because there was a dearth of Protestant preaching in this loyal Catholic shire, Chaderton's "plain but effectual way of preaching" was greatly appreciated. He was about to conclude "his sermon which was of two hours' continuance at least, with words to this effect, 'that he would no longer trespass upon their patience.' Whereupon all the auditory cried out (wonder not if hungry people craved more meat), 'For God's sake, sir, go on, go on.' Hereat Mr. Chaderton was surprised into a longer discourse, beyond his expectation, in satisfaction of their importunity. . . ."[69] Regular attenders at city congregations, who could count on getting two sermons each Sunday as well as a weekday lecture, would not require the gargantuan meal Chaderton gave while, as it were, on a "mission tour." One firm limitation was the impatience of his patron. Anderson referred to this so bitterly as to suggest that he or a friend had suffered such an indignity: "his service and Homilyes he must cut short, and measure them by the Cookes readynesse, and dynner dressing; the roste

[65] *Ibid.*, pp. 3-4.   [66] *Ibid.*, p. 1,066.
[67] D. Neal, *The History of the Puritans*, 5 vols., vol. 5, appendix 4, p. xv.
[68] M. M. Knappen, *Tudor Puritanism*, p. 390.
[69] Thomas Fuller, *The Worthies of England*, ed. J. Freeman, p. 301.

neare ready, the kitchin boye is sent to master Parson to bydde him make haste, the meate is readye, and hys mayster cals for dynner; he commeth at a becke, not daring to denye or make longer staye, least his delaye might cause the cook to burne the meate, and he be called of mayster and men, Syr John burne Goose. . . ."[70]

5. *Favourite Topics*

The Puritan preacher was never at a loss for a sermon topic. There were plenty in the Bible: history, prophecy, poetry, proverbs, parables, allegories, apocalypses. Puritan themes were as many and varied as those of the Bible itself. That this is hardly an exaggeration can be seen from the fact that Topsell can make an ingenious sermon from a genealogy.[71] But there were favourite themes peculiarly relevant to the Elizabethan situation to which the Puritans addressed themselves.

They were not particularly enamoured of Queen Elizabeth. For Spenser she might be "Gloriana," but to Knox she would only have been one other of the "monstrous regiment of women." The queen, of course, thought the Puritans a troublesome group, and frequently tried to tame them. Nor was another common topic of loyal Anglicans a favourite in Puritan pulpits, a eulogy of the Church of England, because the Puritan ministers wished to reform the Reformation of the Church.

In fact, criticism of the Church of England was frequently heard. It began with the controversies in which the Puritans crossed swords with the bishops—in the Vestiarian and Presbyterian or Admonitions controversies of the 1560s and 70s. It caused a furore in 1565 when Thomas Sampson, Laurence Humphrey, and William Fulke preached against the surplice, and eight years later at Paul's Cross, Richard Crick dared to plead for Cartwright's presbyterian principles. Throughout the reign the Puritan divines are highly critical of ignorant, unpreaching, negligent, and compliant parsons in the Church of England. Among many who employed this theme were Edward Topsell,[72] Edward Dering,[73]

[70] Anthony Anderson, *The Sheild of our Safetie set foorth* (1581), sig. T iv verso-V i recto.
[71] The reference is to the eighteenth and last sermon of his series on the Book of Ruth, in *The Reward of Religion* (1597).
[72] *Times Lamentation* (1599), p. 20.
[73] *A Sermon preached before the Quenes maiestie . . . the 25 day of February, Anno 1569*, sig. E iv verso-F i verso.

Bartimaeus Andrewes,[74] and Eusebius Paget.[75] In the criticisms of Roman Catholicism there was, of course, little to distinguish Anglican from Puritan preaching, so that Anthony Anderson, in contrasting the great uncertainties of a belief in purgatory with the assurance that justification by faith brings, was making a Protestant, rather than an Anglican or Puritan point.[76]

Many of the old secular topics were revived by the Elizabethan Puritans. Knewstub and Smith condemned covetousness in the usury of the day and the enclosing of common lands.[77] Henry Smith excoriated extravagant apparel. Adam and Eve may have covered themselves with leaves, but the women "[now] cover themselves with pride, like Satan which is fallen down before the like lightening: ruffe upon ruffe, lace upon lace, cut upon cut, four and twenty orders, until the woman be not so precious as her apparel."[78] Thomas White[79] criticised fastidious and delicate feeding, while Anthony Anderson ridiculed late rising.[80] Stockwood[81] and White[82] denounced the misuse of Sunday. Proof of the presence of continuity in preaching was the theme and the manner of Henry Smith's "set piece" on mutability and transience, which might have been preached as naturally by St. John Fisher: ". . . our life is but a shorte life: as many little sculs are in Golgotha as great sculs; for one apple that falleth from the tree, ten are pulled before they are ripe; and parents mourn for the death of their children, as often the children for the decease of their parents. This is our *April* and *May* wherein we flourish; our *June* and *July* are next when we shall be cut down. What a change is this, that within fourscore years not one of this assembly shall be

74 *Certaine verie godly and profitable Sermons upon the fifth Chapiter of the Songs of Solomon* (1595), p. 121.
75 *A godly Sermon preached at Detford* (1586), sig. A vi verso.
76 *The Sheild of our Safetie set foorth* (1581), sig. G i verso. One must observe, however, that the great space given in works of spiritual direction to such issues as, how one may be assured of election by God, would imply that in some Protestants at least this supposed certainty of election was a whistling in the dark. For two examples of many, see John Downame, *The Christian Warfare* (1604), book II, chap. 9, "Of the meanes whereby we may be assured of our election," and the first of Richard Rogers' *Seven Treatises* (1603), which demonstrated "who are the true children of God."
77 *The Lectures of John Knewstub on the twentieth chapter of Exodus and certaine other places of Scripture* (1579), p. 145.
78 From "Two Sermons on Usurie," Smith, *Sermons*, pp. 160-62.
79 *A Sermon preached at Pawles Crosse on Sunday the thirde of November, 1577*, p. 64.
80 *The Sheild of our Safetie set foorth*, sig. T iv v.
81 *A Sermon preached at Paules Crosse on Barthelmew Day* (1578), p. 50.
82 White, *Sermon at Pawles Crosse*.

left alive? but another Preacher, and other hearers shall fill these rooms, and tread upon us where our feet tread now."[83]

What then, were the new topics of the Puritan preachers? According to J. W. Blench,[84] the expert on preaching in the sixteenth century, the Puritans found only two new topics, a criticism of romances and stage plays and of the principle of *realpolitik* in the Elizabethan form of Machiavellianism, a very practical form of atheism. Stockwood considered the popular romances to be only pornography; there was an element of professional jealousy of the popularity of the plays when he asked: "Will not a fylthye playe wyth the blast of a Trumpette, sooner call thyther a thousands, than an houres tolling of a Bell, bring to the Sermon a hundred?" It would be two centuries before an English preacher would transform audiences into congregations by dramatic preaching that emptied the theatres. The playwrights would then retaliate by ridiculing the Rev. George Whitefield in their plays as "the Reverend Dr. Squintum."[85]

Interestingly enough, we have to wait for the advent of Puritanism before there was a pulpit celebration of the joys of marriage as companionship; this, from the mouth of Henry Smith, was the homiletical equivalent of Shakespeare's great sonnet: "Let me not to the marriage of true minds admit impediments." The novelty of this approach had to wait, one can only suppose, until divines needed no longer to be celibates. It is instructive to compare the attitudes of St. John Fisher and Henry Smith to marriage as exhibited in their sermons. In Fisher's exposition of the parable of the sower, which was allegorically interpreted, the ground which brings forth spiritual food a hundredfold, and therefore most abundantly, is virginity. The soil that produces sixtyfold is widowhood, where there are some seeds of carnality; but the soil which only produces thirtyfold is marriage, in which the weeds are more successful than the corn.[86] With this must be contrasted Smith's *epithalamium*. A good wife, he insisted, is "such a gift as we should account from God alone, accept it as if he should send us a present from heaven, with this name written on it, *The gift of God.*"[87] Recognising procreation of children and the avoidance of fornication as two of the causes of marriage, Smith stressed

---

[83] *Sermons*, p. 274. See also p. 348 for another typically medieval example.

[84] *Preaching in England in the late fifteenth and sixteenth centuries*, p. 305.

[85] See Horton Davies, *Worship and Theology in England*: vol. 3: *From Watts and Wesley to Maurice, 1690-1850*, pp. 176-79.

[86] *English Works of John Fisher*, ed. Mayor, pp. 235-36.

[87] Smith, *Sermons*, p. 8.

the importance of the third cause—"to avoyd the inconvenience of solitarinesse signified in these words, *It is not good for man to be alone*, as though he had said: this life would be miserable and irksome, and unpleasant to man, if the Lord had not given him a wife to companie his troubles."[88] Man, without a wife, is "like a Turtle, which hath lost his mate, like one legge when the other is cutte off, like one wing when the other is clipte, so had the man bene if the woman had not bene ioyned to him: therefore for mutuall societie God coupled two together, that the infinite troubles which lie uppon us in this world, might be eased with the comfort and help one of another, and that the poore in the worlde might have some comfort as well as the rich. . . ."[89] This tender preacher reminded every husband that his wife's "cheekes are made for thy lippes, and not for thy fists."[90] He also had a fascinating part of a sermon in which he advised what "signes of fitness" are to be looked for in choosing a wife or husband.[91]

The most distinctive and the profoundest of the Puritan sermons were those that offered spiritual direction. To a modern reader they would most readily be categorised as religious psychology, but to the Elizabethan world they were psychiatry—quite literally the healing of the soul. It is significant that Ezekiel Culverwell, in his introduction to the collection of Richard Rogers, *Seven Treatises . . . leading and guiding to true happiness both in this life and in the life to come* (1603), suggested that this work should in one respect be called "the Anatomie of the soule, wherein not only the great and principall parts are laid open, but every veine and little nerve are so discovered, that wee may as it were with the eye behold, as the right constitution of the whole and every part of a true Christian." It also deserved, Culverwell said, in another respect to be called "the physicke of the soule," because it supplies most of the "approoved remedies for the curing of all spirituall diseases, with like preservatives to maintaine our health, in such sorte as may be enjoyed in this contagious ayre. . . ." This was one common image for the application of the transforming Gospel to the troubled consciences of men.

Another equally common metaphor was taken from military life. It presupposed that the life of the Christian is like that of the knight in Dürer's etching, determinedly going his solitary way through the pitfalls prepared by hideous death and monstrous

---

[88] *Ibid.*, p. 15.
[90] *Ibid.*, p. 44.
[89] *Ibid.*, p. 16.
[91] *Ibid.*, p. 22.

devil, maintained by the courage of faith. The spiritual life was then interpreted on the analogy of continuing combat against the stratagems of Satan; considerable use was made of the sixth chapter of St. Paul's Epistle to the Ephesians, where the armour of salvation is described in detail. One of the most popular Puritan accounts of the spiritual life as an interior civil war was John Downame's *The Christian Warfare*, which first appeared in 1604.[92] Its lengthy subtitle provides an excellent summary of the author's intentions: *wherein is generally shewed the malice, power and politike stratagems of the spirituall enemies of our salvation, Satan and his assistants the world and the flesh; with the meanes also whereby the Christian may withstand and defeate them.*[93] It was a kind of Protestant counterpart to the famous *Spiritual Exercises* of St. Ignatius Loyola, the founder of the Jesuits, and aimed to train the will to the obedience of faith. It could not sustain the comparison, not only because Downame lacked Ignatius' power to grip the imagination, but also because it lacked rules within it for "thinking with the Church." However, the prevalence of such series of sermons, later made into handbooks for devotional direction, suggests that the Puritan ministers were filling a great gap left by the proscription of Catholicism in England. The third most common image for the Christian life, the pilgrimage, was medieval in origin. It would come into its own in Puritan thought in the next century with the sailing of the "Pilgrim Fathers" in 1620, and, supremely, in the creation of the classical Puritan prose epic, Bunyan's *Pilgrim's Progress*. A fourth image, the ship or the Ark of Salvation, was of course a popular medieval concept of the Christian life, but its stress on the corporate character of Christianity as the religion of *the Church*, made it generally unattractive to Puritans, for whom Christianity was the religion of the *elect*.

What is particularly interesting as well as impressive in this genre of Puritan sermons, was the almost scientific nature of the study of the soul, and the great practicality of the advice the Puritan brethren offered the members of their congregations in these

[92] By 1634 it had gone into a fourth edition. Another briefer example of the same approach was William Perkins, *The combat between Christ and the Divell displayed* (1606).

[93] The rest of the subtitle reads: "And afterwards more speciallie their particular temptations against the severall causes and meanes of our salvation, whereby on the one side they allure to securitie and presumption, and on the other side, drawing us to doubting and desperation, are expressed and answered. Written especially for their sakes who are exercised in the spirituall conflict of temptations and are afflicted in conscience in the sight and sense of their sinnes."

series of lectures. An elementary introduction to this "science of the soul" may be found in William Perkins' *The Foundation of the Christian Religion gathered into sixe Principles*,[94] which is in the form of a catechism. To the question: "How doth God bring men truely to beleeve in Christ?" the answer is: "First he prepareth their hearts, that they might be capable of faith: and then he worketh faith in them." But, the questioner persists, How does God prepare men's hearts? The answer is: "By bruising them, as if one would break an hard stone to powder: and this done by humbling them." But how does God humble a man? Answer: "By working in him a sight of his sinnes, and a sorrow for them."[95] Then Perkins shows that God proceeds to graft faith onto the heart, and he lists three "seeds" of faith. The first is a profound sense of humility when a man acknowledges the burden of his sins and feels that "he stands in great need of Christ."[96] The second is "an hungring desire and longing to be made partakers of Christ and all his merits," and the third is "a flying to the throne of grace from the sentence of the lawe, pricking the conscience."[97] What then follows? It is then that God, "according to his merciful promise, lets the poor sinner feele the assurance of his love wherewith he loveth him in Christ, which assurance is a lively faith."[98]

The practical question is then asked, as to what benefits a man receives from his faith in Christ. The summary answer, inevitable in a brief catechism, is justification and sanctification; other treatises give a fuller account of the stages in the economy of salvation. But how is a man to know that he is sanctified? What graces of the spirit will his life show? One undoubted sign will be repentance, "which is a setled purpose in the hart, with a carefull indevour to leave all his sinnes, and to live a Christian life according to all God's commandments."[99] A second sign of true repentance is a "continuall fighting and struggling against the assaults of a mans owne flesh, against the motion of the divell, and the enticements of the world."[100] So far so good. But what follows after a man wins victory over temptation or affliction? The answer is: "Experience of God's love in Christ, and so increase of peace of conscience and ioy in the holy Ghost."[101] What follows

---

[94] This first appeared in 1590. All references are, however, to the three-volume edition of the *Works* of 1608. The same stages are shown in chap. 4 of the first of *Seven Treatises* of Richard Rogers.

[95] Perkins, *Works*, sig. C i v.

[96] *Ibid.*, sig. C i r.

[97] *Ibid.*, sig. C ii v.

[98] *Ibid.*

[99] *Ibid.*, sig. C iii v.

[100] *Ibid.*

[101] *Ibid.*

if he should be defeated in a temptation? The answer is equally explicit: "After a while there will arise a godly sorrow, which is, when a man is grieved for no other cause in the world but this onely, that by his sinne he hath displeased God, who hath bene unto him a most mercifull and loving Father."[102] Every man had to reenact the drama of the fall and the restoration and redemption in his own soul; Perkins and other leading Puritan divines and preachers provided a prompter's copy of the part from the pulpit, or, to change the metaphor, the preacher provided a strategy for winning the interior civil war between the forces of light and darkness in his own soul. This was, however, elementary strategy that Perkins offered in this case.

The more detailed advice would be offered in a series of weekday lectures, offering a fuller anatomy and physic of the soul with Rogers, or more detailed plan of attack against the demonic strategy of the Evil One, in the case of Downame. A few citations from each will make clear the truth of William Haller's claim that "the formulas of Paul and Calvin, it seemed, offered the key to the problem of government not alone in church and state but also in man's inner life"[103] and of his other assertion of election, vocation, justification, sanctification, and glorification, that "here was the perfect formula explaining what happens to every human soul born to be saved."[104]

Rogers told his hearers what benefits they were to expect from hearing the Word preached; this was itself an indication of the high hopes of Puritan preachers evaluating their tasks. Rogers sees six firm benefits coming to attentive listeners: "Therefore, if he may by the preaching of the Word ordinarily *be led into all truth* necessarie for him to know, and be delivered from errour in religion and manners; *if he may be established and confirmed in the knowledge of the will of God*; if he may be reformed in his affections and life daily, more and more encreasing therin, and overcomming himselfe better thereby; if hee may both be brought to bestow some time of his life (as his calling will permit) in reading, and so as that he may profit thereby; and finally if he may *become an example* in time unto others, I may boldly affirme and conclude, that the ordinarie preaching of the word is a singular meanes whereby God hath provided that his people should grow and increase in a godly life."[105]

102 *Ibid.*
104 *Ibid.*, p. 90.
103 *The Rise of Puritanism*, p. 35.
105 *Seven Treatises* (1603), pp. 215-16.

Another example of the practicality and prudence of his spiritual direction can be seen in the fifth treatise, in which Rogers considers the "lets which hinder the sincere course of the Christian life" and which he interprets as diabolical tricks. One is to keep the believer in a wandering and unsettled course;[106] a second, by the leaving of one's first love;[107] a third, the lack of preaching.[108] Another group of Satan's stratagems are concerned with attacking the believer in his unmortified affections: first, the fear that he will not persevere and pride in his gifts;[109] second, of other unruly affections such as touchiness and peevishness;[110] third, worldly lusts.[111] In case the would-be Christian is discouraged by this time, the sixth treatise follows with the listing of the 10 privileges that belong to every true Christian.

A similar kind of book was *The whole treatise of the Cases of Conscience* . . . by William Perkins, which he gave as a series of weekday lectures which were collected and printed after his death. The aim of the book was "to discover the cure of the dangerousest sore that can be, *the wound of the Spirit*."[112] It covered the entire range of problems that would be raised by sensitive souls in matters of the spiritual or the ethical life, from "How a man being in distress of minde may be comforted"[113] to "Whether there be any difference between the trouble of Conscience and Melancholy?"[114] or, "How are we to use Recreations?"[115] We have here a complete spiritual and ethical directory for living the theocratic life impressive in its combination of Biblical guidance and sheer common sense. Again it should be noted that the Puritans had no interest in speculative theology or mysticism, but only in practical theology and ethics.

As a final example of the distinctive contribution of English Puritanism in the pulpit of this period, John Downame's *Christian Warfare* will be considered, especially as it enumerates the signs by which the elect soul may know that it has, in fact, been chosen and predestined to eternal salvation by God, so that it may con-

---

106 *Ibid.*, p. 425.   107 *Ibid.*, p. 432.   108 *Ibid.*, p. 437.
109 *Ibid.*, p. 441.   110 *Ibid.*, p. 447.   111 *Ibid.*, p. 451.
112 Perkins, *Works*, from prefatory letter of Thomas Pickering of Emmanuel College, Cambridge.
113 *Ibid.*, pp. 88f.   114 *Ibid.*, pp. 194f.
115 *Ibid.*, pp. 121f. The answer to the third question, in part, is: "Againe, the Bayting of the Beare, and Cockefights are no meete recreations. The baiting of the Bull hath his use and therefore it is commanded by civill authority, and so have not these . . . Games may be devided into three sorts. Games of wit and industry, games of hazzard, and a mixture of both." Athletic sports, chess, and draughts were approved, but games of "hazzard" were not.

fidently walk the razor edge between presumption and despair. Downame offered 10 "signes and infallible notes of our election." The first was "an earnest desire after the meanes of our salvation"; the second, the use of "the spirit of supplication"; the third, a mind on "heavenlie things" after it is weaned from the world; the fourth "the sight of sin and sorrow for it"; and the fifth, "an hungering desire after Christ's righteousnesse."[116] The sixth sign was the vigour of the "inward fight between the flesh and the spirit"; the seventh, a new obedience; the eighth, our "love of the brethren"; the ninth, the love of God's ministers; and the tenth and last, "an earnest desire of Christs comming to iudgement."[117]

With few exceptions Puritan preachers deliberately avoided rhetorical devices and stylistic graces that would have made their sermons more interesting. They were exceedingly sparing of humour; they did not show even in their hunt for similes and metaphor, which were permitted to illustrate their doctrines provided they did not distract the hearers, that imaginative fertility that distinguished Lancelot Andrewes and John Donne, who were the glories of the Jacobean age. Their qualities were Biblical fidelity and a profound conviction that will not be trivial about the majestic claims of God or toy with the plight of men. Their originality consisted in the assiduity and penetration with which they described the psychology of salvation and consoled distressed consciences.

[116] The "signs" will be found, respectively, on pp. 235, 236 *bis*, 237, and 239 of *Christian Warfare*.

[117] The sixth to tenth "signs" will be found, respectively, in *ibid.*, pp. 239, 240, 243, 245, 247.

CHAPTER IX
THE WORSHIP OF
THE SEPARATISTS

THE RADICAL DIFFERENCE between Puritans and Separatists was in their relation to the state church. The Puritans worked, however reluctantly, within the structure of the establishment, improving it and supplementing it whenever possible. The Separatists wished, in the words of Browne, for "Reformation without tarrying for any." Their patience with conformity and compromise was at an end. They claimed the text for their act of shaking the dust off their feet was the command of God, "Come ye out from among them and separate yourselves, saith the Lord."[1] The earliest Separatists may have believed that their self-exclusion from the national church would accelerate the process of reformation in the church which they had left. But as it hardened, Separatism became more rigid and regarded the parent religious community as a "false" church. Except for the difference in the attitude toward the established church, the Separatists were largely in agreement with the Puritans.

This was particularly the case in worship, because both groups took as their criterion the authority of the Scriptures as the Word of God. If the Independent Puritan, Henry Jacob, could say: "For as much as wee are in conscience throughly perswaded, that Gods most holy word in the New Testament is absolutely perfect, for delivering the whole manner of Gods worship . . . ,"[2] his thoughts were anticipated by Richard Fitz, the minister of the "Privye Churche" in London, a Separatist who wrote of his conventicle members, as "the myndes of them that . . . have set their hands and hartes, to the pure unmingled and sincere worshippinge of God, according to his blessed and glorious worde in al things, onely abolishing all tradicions and inventions of men. . . ."[3] Both parties insisted that only such ordinances as were warranted by the Word of God were to be permitted in divine worship. Every tradition, other than that of the primitive apostles, was written off as a "human devyce."

Apart from the intrinsic interest in watching a deeply dedicated

[1] II Corinthians 6:17.
[2] Champlin Burrage, *The Early English Dissenters*, 2 vols., II, 162.
[3] *The Trewe Markes of Christes Churches*, in *ibid.*, II, 13.

group of Christians experimenting in freedom, there was a historical significance in Separatist worship out of proportion to its small numbers. The Separatists provided the Puritans with practical illustrations of a worship modelled, as they believed, on the usage of the New Testament. While the Puritans denounced the idolatries and impurities of Anglican worship as then established, the Separatists provided the living remedy. There was, it is true, little community of sympathy between the Puritans and Separatists. Puritans considered Separatists schismatics, and Separatists thought Puritans timeservers. It was unlikely that there was much direct influence on the Puritans by the Separatists, so much were they at loggerheads. Yet there was some contact; this can be seen in persons who crossed the frontier between Puritans and Separatists. In the case of Francis Johnson, the conclusion might well be that Separatism was the left wing of Puritanism; for the successor of Travers and Cartwright as Puritan Minister to the Church of the Merchant Adventurers in Holland became an outstanding Barrowist minister. On the other hand, Puritanism seemed to the right wing of Separatism in the case of John Robinson, who began as a Brownist; and under the influence of Henry Jacob and William Ames became an Independent Puritan. Puritanism and Separatism aimed at the same end—a Church of England fully reformed in worship, discipline, and church government, as well as in doctrine. They differed only in their estimate of the capacities of the national church for improvement.

It will be assumed, with only a minimum of argument, that the term "Separatists" cannot legitimately be applied to the Independents, but that it is more properly reserved for the Barrowists, the Brownists, and the Anabaptists.[4] It is also assumed, since all parties, including the moderate Separatists, agreed in denouncing the obscure and eccentric sects such as the Family of Love, that their influence on Puritan or Separatist worship was negligible. The first assumption, that the Independents were not Separatists, can be shown to be reasonable by a brief examination of the writings of two of their earliest leaders, Henry Jacob and John Robinson. Jacob was the pastor of the first English congregation of Independents, founded in 1616. Among his papers was discovered a copy of "A third humble Supplication" of the Puritans addressed to the king in 1605 and corrected in Jacob's handwrit-

4 Nor is this statement intended to imply that the Independents had no antecedents among the Barrowists or that there was no connection between early Anabaptists and the later Baptists.

ing. He was representing the views of the Puritans who at that time had asked for toleration and permission "to Assemble togeather somwhere publikly to the Service & Worship of God, to use and enioye peacably among our selves alone the wholl exercyse of Gods worship and of Church Government . . . without any tradicion of men whatsoever, according only to the specification of Gods written word and no otherwise, which hitherto as yet in this our present State we could never enjoye."[5] Jacob and his fellow Puritans appended a guarantee that if this was allowed them, they would "before a Iustice of peace first take the oath of your Maiesties supremacy and royall authority as the Lawes of the Land at this present do set forth the same," and promised to keep communion with other churches, to pay all ecclesiastical dues, and to submit themselves, in any case of trespass, to the civil magistrate. Jacob was no Separatist; Robinson began as a Separatist and later became a Puritan.[6] Independents are, therefore, considered as Puritans, not Separatists.

1. *The Barrowists*

The worship of the Barrowist Separatists claims attention, first because it was established in England before that of the Brownists, with which it has affinities. The Barrowists flourished from 1587 to 1593. Their leaders, Barrow himself, Greenwood, and Penry, all died as martyrs to the cause of Separatism.

What were the reasons for their separation from the Church of England? Henry Barrow addressed himself to the question, providing four reasons related to worship, the character of church membership, the quality of the ministers, and type of church government. Anglicans worshipped God falsely, "their worship being

---

[5] Burrage, *English Dissenters*, I, 286.
[6] Robinson had penned *A Iustification of Separation from the Church of England* in 1610, but the following passage by the same hand shows his attitude after his conversion to charity by Jacob and his kindlier view of Puritanism: "To conclude, For my selve, thus I beleeve with my heart before God, and professe with my tongue, and have before the world, that I have one and the same faith, hope, spirit, baptism, and Lord which I had in the Church of England and none other: that I esteem so many in that Church, of what state or order, soever, as are truly partakers of that faith (as I account many thousand to be) for my Christian brethren: and my selfe a fellow-member with them of one misticall body of Christ. . . ." (*A Treatise of the Lawfulness of Hearing of Ministers* [1634], pp. 63f.) This is very like the spirit of the Puritans who sailed to Boston in 1629 and 1630 and of whom Cotton Mather wrote in the preface to his *Magnalia Christi Americana* (1702), to the effect that England's daughter New England was "by some of her *Angry* Brethren . . . forced to make a *Local Secession*, yet not a *Separation*, but hath always retained a Dutiful Respect unto the *Church of God* in England" (p. viii).

made of the invention of man, even of that man of sinne, erronious [sic], and imposed upon them."[7] Then, the "prophane multitude" receive all the privileges of membership of the Church of England without professing purity of doctrine or of life. Third, the ministry was "false" and "antichristian" and was imposed on the people as well as maintained by them. Fourth, "that their churches are ruled by and remaine in sujection unto an antichristian and ungodly government, cleane contrary to the institution of our Saviour Christ." These criticisms were curtly phrased, but not very illuminating because of the imprecise nature of their charges.

It is interesting that Barrow, when examined by Bishop Aylmer and Lord Burghley on March 18th, 1588/9, was specifically asked what he found idolatrous about the worship of the Church of England; Barrow replied that it was entirely idolatrous. Then he mentioned that the commemoration of the saints violated the first commandment, that the observations of fasts was idolatrous, and that the same was true of naming days after saints. Asked for his criticism of the Book of Common Prayer, he replied that prayers should not "be stereotyped and stinted, nor should they be tied to place manner, time, nor form."[8]

Barrow's criticism of set prayers or stinted forms was particularly severe. He had a superb definition of prayer, against which to test all read prayers. "Praier," he wrote, "I take to be a confident demanding which faith maketh thorow the Holy Ghost, according to the wil of God, for their present wantes, estate, *etc*."[9] He viewed read prayers as leading to the utter extinction of the Spirit of God and a denial of the sincere spontaneity that should characterise prayer: "How now? Can any read, prescript, stinted leitourgie, which was penned many yeares or daies before, be said a powring forth of the heart unto the Lord? . . . Is not this (if they wil have their written stuffe to to be held and used as praier) to bind the Holy Ghost to the froth and leaven of their lips as if it were to the holy word of God? Is it not utterly to quench and extinguish the Spirit of God, both in the ministerie and the people, while they tye both them and God to their stinted numbred praiers?"[10] Barrow further felt that the content of the Book of Common Prayer was borrowed too much from Roman Catholicism and that it tended to

---

[7] *The Writings of Henry Barrow, 1587-1590*, ed. Leland H. Carlson, p. 8.
[8] *Ibid.*, p. 16.
[9] *A Briefe Discoverie of the False Church* (1590/1, Dordrecht), pp. 64-65 and reprinted in Carlson, *Writings of Barrow*, p. 365.
[10] *Ibid.*, p. 65.

encourage a ministry that would not preach or pray in words of their own: "Is this the unitie and uniformitie that ought to be in al churches? And is amongst al Christe's servantes, to make them agree in a stinking patcherie devised apocrypha leiturgie, good for nothing but for cushsions and pillowes for the idle priestes, and profane carnal atheistes to rock them a sleepe and keep them in securitie, wherby the conscience is no way either touched, edified or bettered?"[11] What is perhaps most astonishing to learn is that the Barrowists rejected the repetition of the Lord's Prayer. Many Puritans repeated it as the Dominical warrant and model, but the Separatists regarded it as part of the Sermon on the Mount and as an illustration of the kind of prayer they should aim at, but not apishly imitate.[12]

The fullest statement of the radical position on the Lord's Prayer is expressed by John Penry, also a Barrowist martyr:

We answer first that the scrypture yt selfe sheweth his meaning herein to be, not that the disciples or others should be tyed to use these very wordes, but that in prayer and geving of thankes they should followe his direction and patterne which he had geven them, that they might know to whom, with what affeccion and to what end to pray as yt is expresslie sett downe in these wordes. After this manner therfore pray ye, and not as men will now have us: Say over these very wordes. Secondly we doubt not but that we may use anie of these wordes as others applying them to our severall necessyties as we see Christ himselfe did when he prayed. O my father yf this cupp cannot pass from me but that I must drinke yt, thy will be done where yt is plaine that Christ himselfe who gave the rule doth shew us how to use yt, to weet . . . in praying according as our specyall necessyties shalbe, whether we use any of these wordes or other, or pray with sighes & groanes that cannot be expressed.[13]

11 *Ibid.*, pp. 65-66.
12 See Leland Carlson's comprehensive summary of Barrow's objections to the Book of Common Prayer, in *The Writings of Henry Barrow, 1590-1591*, p. 6. Barrow, says Carlson, asserts that Anglican worship is too Roman, uses lections from the Apocrypha and others from the Bible that are arbitrarily selected, prayers are prescribed and are mechanical and second-hand, and are delivered in a superstitious and unscriptural manner. See also *ibid.*, pp. 100-101. It should also be noted that the Separatist Greenwood gave nine reasons against the use of set prayers in church, though he was willing for them to serve as the basis of meditation. See Leland H. Carlson, ed., *The Writings of John Greenwood, 1587-1590*, pp. 17-19.
13 The evidence for the Barrowist refusal to repeat the Lord's Prayer can be

Turning from theory to practice, there is a comprehensive and vivid account of the Barrowist gatherings for worship: "In the somer tyme they mett together in the feilds a mile or more about London, there they sitt downe uppon A Banke & divers of them expound out of the bible as long as they are there assembled."[14] It appears that they would arrange in advance where their next Sunday's meeting was to be held and there they would assemble at the selected rendezvous as early as five o'clock in the morning. There they might well remain for the entire day, kept there partly by the prolixity of their Biblical "exercises" and partly for fear of detection if they should stir while it was still daylight. There ". . . they contynewe in there kinde of praier and exposicion of Scriptures all that daie. They dyne together, After dinner make collection to paie for there diet & what mony is left some one of them carrieth it to the prisons where any of there sect be committed."[15] The kind of prayer they used was extemporaneous: simple, impassioned, sincere, and moving. It was the distinctive contribution of the Separatists to worship. The Independents and Baptists were to follow the example of the Separatists, and in time the Independents would persuade the English Presbyterians of the Westminster Assembly to adopt the expedient of a manual of worship, suggesting the order and the topics of the prayers but not the words themselves. It was, therefore, an influential precedent that the Separatists set in their insistence on "free prayer," for it had great strengths, as in simplicity, intimacy appropriate to those who cry "Abba, Father" in their approach to the august yet loving Creator, and directness, and great weaknesses, as when it was prolix, disordered, and a chain reaction of endless cliches.

At the beginning the Barrowists apparently used neither psalms nor hymns, unlike the Puritans, who were fond of psalm-singing at church and at home. The documents do not mention singing at all. Henoch Clapham, however, states that the Barrowists were first persuaded to sing in their assemblies by a man who urged its necessity, namely Francis Johnson, Barrowist pastor of the congregation in Kampen and Naarden: "*Franc-Iohnson* (being advised

---

found in Carlson, *Writings of John Greenwood*, pp. 261-62, where Greenwood tells Bishop Cooper that to repeat the Lord's Prayer is "popish doctrine, and such praier were superstitious babling." See also Burrage, *English Dissenters*, II, 56. It should be noted that the final phrase of the citation in Burrage, II, 74f., echoes Romans 8:17, which was the *locus classicus* of the authority for a pneumatic-charismatic type of prayer.

[14] Burrage, *English Dissenters*, I, 26.
[15] *Ibid.*

by one that talked with him thereabouts in the *Clinke* at London) did presse the use of our singing Psalmes (neglected before of his people for Apocrypha;) whereupon his Congregation publikely in their meetings used them, till they could have them translated into verse, by some of their Teachers. . . ."[16]

The Barrowists administered the sacraments of Baptism and the Lord's Supper. But, in the case of children, they held it to be unlawful to baptise children in the churches of the establishment, preferring to let them go unbaptised until a satisfactory Baptism could be administered by a true preacher of the Gospel; in this they resembled the Donatists of fourth-century Africa. The method of administering Baptism among them was described in the deposition of one Daniel Bucke. He reported that Johnson baptised seven children in 1592, as follows: "they [the congregation] had neither god fathers nor godmothers, and he tooke water and washed the faces of them that were baptised: the Children that were there baptised were the Children of Mr. Studley Mr. Lee with others beings of severall yeres of age, sayinge onely in the administracion of this sacrament I doe Baptise thee in the name of the father of the sonne and of the holy gost withoute usinge any other cerimony therein and is now usually observed accordinge to the booke of Common praier. . . ."[17] The same witness described the Barrowist administration of the Lord's Supper: "Beinge further demaunded the manner of the lordes supper administred amongst them he said that fyve whight loves [*loaves*] or more were sett uppon the table and that the Pastor did breake the bread and then delivered to the rest some of the congregacion sittinge and some standinge aboute the table, and that the Pastor delivered the Cupp unto one and he to an other, and soe from one to an other till they had all dronken usinge the words at the deliverye thereof accordinge as it is sett downe in the eleventh of the Corinthes the xxiiijth verse."[18] It is to be noted that the Lord's Supper as celebrated by the Barrowists was characterised by simplicity and fidelity to the New Testament narrative. In particular, the prophetic symbolism of Christ in breaking the bread and pouring out the wine, as representations of His body about to be broken and His blood about to be shed, was repeated with great solemnity. This fidelity to the words and

---

[16] *A Chronological Discourse* (1609), p. 36.

[17] British Museum, Harleian MS 6849, folio 216 v., reproduced in Burrage, *English Dissenters*, I, 142f.

[18] British Museum, folio 217 v., reproduced in Burrage, *English Dissenters*, I, 143.

actions of Christ in the institution of the sacrament was later to be a feature of both Presbyterian and Independent celebrations of Communion.

Another important ordinance of the Barrowists was the use of ecclesiastical "discipline" for the maintenance of the relative purity of the Church. It could in severe cases of dishonouring the name and fame of Christ lead to excommunication, or, to use the more sinister and also Biblical name, the "handing over to Satan." Where due regret had been expressed, and retribution made where necessary, the sinner was welcomed back into the fold publicly in the rite of "Restoration." The Barrowist excommunication against two offenders, Robert Stokes and George Collier, is described by Robert Aburne: "He saieth that they did use to excommunicate amongst them, and that one Robert Stokes, and one George Collier, and one or twoe more whose names he Remembreth not, wear excommunicated, and that the said Iohnson thelder did denounce thexcommunication against them, and concernynge the manner of proceedinges to excommunication he saieth, that they the said Stokes and the Rest beynge privatelye admonished their pretended errors, and not conforming themselves, and by Witnes produced to their congregacion, then the said Iohnson, with the Consent of the whole Congregacion, did denounce the excommunication, and that sithence they weare excommunicated which was a halfe yere and somewhat more sithence, they wear not admitted to their Churche."[19]

The Separatists provided the Puritans with an example of rigid discipline. Both parties insisted on the retention of the purity of the Church in doctrine and in life. Both Puritans and Separatists pleaded for a threefold reformation—Gospel-doctrine, Gospel-government, and Gospel-discipline. The doctrine of the "gathered church" and of the "visible saints," which played so large a part in Puritan theory and practice, was presupposed by the Barrowist discipline.

Since the Barrowists held that marriage was essentially a civil ceremony, there was no need for minister or church. Christopher Bowman, a deacon of the church of which Francis Johnson was the minister, declared: ". . . Mariage in a howse without a Mynister by Consent of the parties and frends is sufficient."[20] The Angli-

[19] British Museum, Harleian MS 6848, folio 41 v., cited by Burrage, *English Dissenters*, I, 143.
[20] British Museum, Harleian MS 6848, folio v., cited by Burrage, *English Dissenters*, I, 144.

can marriage ceremony was scrupled by the Separatists, not only because it required the use of the ring, a ceremony for which there was no warrant in the Word of God, but also because the New Testament was able to supply no example of a marriage service being performed by Christ or the apostles.

Burials were also considered to be civil rather than religious. In *A Briefe Discoverie of the False Church*, Barrow gave the subject more than cursory treatment. He found no authority "in the booke of God, that it belonged to the ministers office to burie the dead. It was a pollution to the Leviticall priesthood to touch a carcase or anything about it."[21] Barrow objected to the trappings of mourning gowns and the exorbitant costs of funerals for the families of poor men. He particularly excoriated encomia (laudatory sermons) for the dead because of their usual insincerity: "To conclude, after al their praiers, preachment, where (I trow) the priest bestoweth some figures in his commendations (though he be with the glutton in the gulfe of hell) to make him by his rhetorick a better Christian in his grave than he was ever in his life, or else he yerneth his money ill. After al is done in church, then are they all gathered together to a costly and sumptuous banquet. Is not this jolly Christian mourning?"[22]

Hitherto Barrow had criticised the insincerity of formal burial services. He must have been thinking of the common Puritan criticism of the Office in the Book of Common Prayer which assures the mourners that everyone is buried "in sure and certain hope of the Resurrection to eternal life" regardless of their faith or ethics. He went on to show the medieval fascination and fear of death and the Elizabethan's sense of being mocked by mutability, which finds the making of exquisite monuments a gilding of the skull. In the ensuing passage Barrow is pointing to a fact, namely that the Elizabethans developed a monumental art in their tombs which enabled the newly rich as the newly dead to dominate the parish church.[23] "Neither will I trouble them to shew warrant by the Word, for the exquisite sculpture and garnishing of their toombes, with engraving their armes and atcheavements, moulding their

21 Barrow, *Briefe Discoverie*, p. 126; cited in L. H. Carlson's edition of *The Writings of Henry Barrow, 1587-1590*, p. 459.
22 Barrow, *Briefe Discoverie*, p. 127.
23 A notable example of early seventeenth-century tombs in the parish church is that of the Fettiplace family at Swinbrook, near Burford in Oxfordshire. The phenomenal increase in Elizabethan funerary carving as status symbols for newly rich families will be considered more fully later in Chap. X.

images and pictures, and to set these up as monuments in their church: which church must also (upon the day of such burials) be solemnely arrayed and hanged with blacks, that even the verie stones may mourne also for companie. Is not this christian mourning, thinke you?"[24] The one recorded funeral of a Separatist known to me is that of Eaton. The account of it asserts that his many followers marched in procession behind the body to the graveside and without prayers or commendations or a sermon thrust it into the grave and stamped earth on it. This seemed more like good riddance than a good burial, but it was entirely in keeping with the Separatist conviction that it was a civil matter.

Separatists appear to have had no formal ordination to their church offices. Clapham informs us that Johnson and Greenwood were chosen as pastor and teacher, respectively, "Without any *Imposition of hands*." If they had already received the charismatic endowment of the Holy Spirit, there was apparently no need for the formal recognition of this on the part of a closely knit congregation. It seems, however, that Francis Johnson, when he came to Amsterdam a few years later, had a ceremony of the imposition of hands, and that this was performed by lay members of his congregation.[25] The same reporter gives us an amusing account of the way in which the offertory was taken at the Barrowist meetings:

And hereupon it was, that the *Separatists* did at first in their Conventicles, appoynt their Deacons to stand at the Chamber dore, at the people's outgate, with their Hats in hand (much like after the fashion of a Playhouse) into the which they put their voluntary. But comming beyonde seas, where a man might have seaven Doyts for a penny, it fell out, howsoever their voluntary (at the casting in) did make a great clangour, the *Summa totalis* overseene, the maisters of Play came to have but a few pence to their share. Whereupon, a broad Dish (reasonable flat) was placed in the middest of their convention, but when the voluntarie was cast in, others might observe the quantitie. But this way served not the turne, for a few doyts rushing in upon the soddaine, could not easily be observed, of what quantity it might be. Upon this, the Pastor gave out, that if (besides giftes from others abroad) they

24 Barrow, *Briefe Discoverie*, p. 127.
25 Henoch Clapham, *A Chronological Discourse* (1609), p. 31.

would not make him *Tenne pounds* yearely at least, he would leave them, as unworthy the Gospel.[26]

Barrowist sermons were all of an expository character; they must have been delivered in a homely, possibly rough-hewn manner. Clapham styled these preachers "Syncerians" and complained that they had spoiled preaching by "holding every howers talke, A Sermon: Insomuch as a number would not goe to meate (if a few were present of their faction) but there must be a kind of Sermon."[27] Barrow himself, again according to Clapham, disliked too casual and unpolished preaching, "saying of that, and of some Pinsellers and Pedlers that then were put to preach in their Thursedayes Prophecie, that it would bring the Scriptures into mightie contempt."[28] It is to be observed that the Separatists joined hands with the Puritans in regarding "Prophesying" as a valuable means of inculcating scriptural doctrine, and also as a way of training preachers.

Yet another parallel between Barrowists and Puritans was their invariable custom of founding a church on a covenant subscribed by officers and members. For instance, Abraham Pulberry in 1592 admitted upon examination that "hee hath made a promise to the Lord in the presence of his Congregacion when hee entred thereunto that hee would walke with them as they would walke with the Lorde."[29] In similar words, another declared, "as longe as they did walke in the lawes of God hee would forsake all other assemblies and onely followe them."[30]

As we look back on the worship of the Barrowists we can see that it was marked by a determined fidelity to worship God only as He had commanded in His holy Word, by a search for simplicity as well as sincerity, and by an intimacy characteristic of a small community. Above all, one recognises in these Separatists gatherings the intensity and dedication of a faith that would not only remove mountains but endure prisons and martyrdom. As ceremony was a vain show, so was the repetition of formal prayers like the collects of the Book of Common Prayer; mere "mumbling," and creeds that might be recited parrot fashion off the top of the mind were replaced with covenants which come from the warmth of the heart and were bound with the steel of a will converted and conformed to God. It was a narrow worship, but it ran deep. It

26 *Ibid.*, p. vi.   27 *Ibid.*, p. vii.
28 *Ibid.*   29 Barrow, *Briefe Discoverie*, II, 34.
30 *Ibid.*, p. 45; see also p. 60.

was critical of the Church of England, and risked Pharisaic contempt. But it was a way of life that was deeply compassionate to its own at the risk of martyrdom.

2. *The Brownists*

Browne's company achieved a complete separation from the Church of England in 1581 in Norwich. Though emerging a few years earlier than the Barrowists, their future was sought in Holland because Browne and his followers moved there in 1581 or 1582 to avoid persecution. The Barrowists made Separatism more widely known in practise than the Brownists, though Browne's writings gained a wide notoriety in England.[31] In fact, their ecclesiology and worship were very much like that of the Barrowists, and both were probably the parents of Independency.

The following description of the meetings of the Brownists for worship is given in *The Brownists Synagogue*:

> In that house where they intend to meet, there is one appointed to keepe the doore, for the intent, to give notice if there should be any insurrection, warning may be given them. They doe not flocke all together, but come 2 or 3 in a company, any man may be admitted thither, and all being gathered together, the man appointed to teach, stands in the midst of the Roome, and his audience gather about him. He prayeth about the space of halfe an houre, and part of his prayer is, that those which come thither to scoffe and laugh, God would be pleased to turne their hearts, by which meanes they thinke to escape undiscovered. His Sermon is about the space of an houre, and then doth another stand up to make the text more plaine, and at the latter end, he intreates them all to goe home severally, least the next meeting they be interrupted by those which are of the opinion of the wicked, they seeme very stedfast in their opinions, and say rather then they will turne, they will burne.[32]

Browne himself was the authority for an account of the service in Middleburg: "Likewise an order was agreed on ffor their meetinges together, ffor their exercises therein, as for praier, thanckes giving, reading of the scriptures, for exhortation and edifiing, ether by all men which had the guift, or by those which had a speciall

---

[31] Robert Browne's *A Booke which sheweth the life and manners of all true Christians*, published in 1582 in Middleburgh, was one of the more famous.
[32] (1641), pp. 5f.

charge before others. And for the lawefulness of putting forth questions, to learne the trueth, as iff anie thing seemed doubtful & hard, to require some to shewe it more plainly, or for anie to shewe it him selfe & to cause the rest to understand it."[33] This report reveals that "prophesying" occupied a central part in the Brownist gatherings, with opportunity to question the preacher. The distinction between "praier" and "thanckes giving" may mean no more than that prayers of petition and thanksgiving were offered. It might also mean that Browne was referring to special days of thanksgiving on which some signal mercy of Divine providence might be commemorated. Such days were not infrequently held, as were days of humiliation, during the Commonwealth and Protectorate. Since the Brownists had a set of officers named "receevers," it may be assumed that they took up collections from their church members.[34] Discipline was enforced along Puritan lines. This may be inferred from Browne's reference to "separating cleane from uncleane."[35] An autobiographical reference leads one to assume that Browne regarded marriage as a civil custom requiring no ecclesiastical solemnisation.[36]

For other ordinances practised by the Brownists we have no direct evidence. Since Browne and his friends, previously Puritans, aimed at setting up congregations as near as possible to the New Testament model, it is most probable that the sacrament of Baptism and the Lord's Supper were administered by them. There is, further, no reason to suppose that there was any liturgical difference between the Brownists and the Barrowists; the latter certainly celebrated both sacraments. Browne's innovations were in the sphere of church government and not, it would seem, in church worship. Like the Barrowists, the Brownists founded a church on a covenant. As Browne's writings appeared five years before the rise of the Barrowists, it is likely that they were indebted to him, as the Puritans were to both sets of Separatists.

3. The Anabaptists

As the relationship of the Brownists to the Independents is obscure, so is that of the Anabaptists to the Baptists. The Anabaptists seem to have had no organised existence in England before 1612. In the latter years of the short reign of Edward VI several

[33] *A True and Short Declaration* (1582), pp. 19f.
[34] *Ibid.*, p. 20.
[35] *A Booke which sheweth the life and manners of all true Christians* (1582), sig. K 2v.
[36] *Ibid.*

publications appeared, criticising the "pestilent" opinions of the continental Anabaptists. In 1560 there was issued "A Proclamation for the banishment of Anabaptistes that refused to be reconciled, 22 Septembris."[37] Throughout the country Anabaptists were encountered, not in organised societies, but as families or as individuals. After 1589 on there was a 20-year silence about them. It was broken only when William Sayer was imprisoned as an Anabaptist in the Norfolk county gaol in 1612. Hence Elizabethan Separatism was almost exclusively the contribution of the Barrowists and the Brownists. But the origin of English Anabaptism is to be found in Holland, that great asylum for religious refugees. Francis Johnson, the Barrowist minister, stated that there were "divers" Anabaptists in his congregation who were eventually excommunicated. The first Baptist congregation to be settled in England was that over which Thomas Helwys, who anticipated Roger Williams in his plea for religious toleration, presided, along with Thomas Murton. This group withdrew from John Smyth's congregation in Amsterdam and returned to England about 1612.

The characteristic difference between the Anabaptists and the other Separatists concerned the subjects and the mode of administering Baptism. Helwys made believer's Baptism, by sprinkling or pouring, a necessity for salvation. So dogmatic was he on this point that he insisted that the contrary practice of infant Baptism was sufficient to warrant the penalty of eternal damnation being inflicted on those who held it: ". . . if you had no other sin amongst you al, but this, you perish everie man off you from the highest to the lowest, iff you repent not."[38]

There is a comprehensive description of public worship in the Amsterdam congregation:

> The order of the worshippe and government of our church is 1. we begynne with a prayer, after reade some one or tow chapters of the bible gyve the sence thereof, and confer upon the same, that done we lay aside our bookes, and after a solemne prayer made by the 1. speaker, he propoundeth some text out of the Scripture, and prophecieth owt of the same, by the space of one hower, or thre Quarters of an hower. After him standeth up a 2. speaker and prophecieth owt of the said text the like time and space, some tyme more some tyme lesse. After him the 3. the 4. the 5. &c as the tyme will

[37] Burrage, *Briefe Discoverie*, I, 62.
[38] *Ibid.*, p. 253.

geve leave, Then the 1. speaker concludeth with prayer as he began wth prayer, wth an exhortation to contribution to the poore, wch collection being made is also concluded wth prayer. This morning exercise begynes at eight of the clock and continueth unto twelve of the clocke the like course of exercise is observed in the aftnwne from 2. of the clocke unto 5. or 6. of the Clocke. last of all the execution of the government of the church is handled.[39]

This description of Anabaptist worship in Amsterdam, where the Mennonites were prominent, demonstrates how opposed they were to the use of set forms in worship. They were logical enough in their attempt to gain a spirit-led worship to put away the Bible in their service, as a form of words. The second unusual feature of this worship was the special offertory prayer.

The minister of the Amsterdam Anabaptist congregation was John Smyth, who began as a Puritan within the Church of England. His famous treatise on the Lord's Prayer, *A Patterne of true Prayer*, is one of the fairest considerations of the claims of liturgical prayer ever made. His conclusion in 1605 about the legitimacy of repeating the Lord's Prayer was that ". . . Christ leaveth it arbitrie unto us, as a thing indifferent when we pray to say this prayer, or not to say it, so be that we say it in faith and feeling; or if wee say it not, yet to pray according unto it."[40] Three years later he decided to forego all set forms, as the initiator and defender of the most extreme form of pneumatic worship. In *The Differences of the Churches of the Separation* (1608) he declared: "Wee hould that the worship of the new testament properly so called is spirituall proceeding originally from the hart: & that reading out of a booke (though a lawfull ecclesiastical action) is no part of spirituall worship, but rather the invention of the man of synne it being substituted for a part of spirituall worship; therefore in time of prophesjng it is unlawfull to have the booke as a helpe before the eye wee hould that seeing singinging [sic] a psalme is a parte of spirituall worship therefore it is unlawfull to have the booke before the eye in time of singinge a psalme."[41]

The rejection of forms in worship was so complete that it almost approached Quakerism 40 or 50 years before it originated. This

---

[39] British Museum, Harleian MS 360, folio 71; Letter of the Bromheeds to Sir William Hammerton.
[40] *The Works of John Smyth*, ed. W. T. Whitley, i, 81.
[41] *Ibid.*, p. v.

"spiritual worship" while it does not altogether do away with books, regards the part of worship in which they are used as the mere preparation for the pure worship which proceeds without them. This interpretation of radical spiritual worship is confirmed by the words of Helwys, when contrasting the worship of his community with that of the Johnsonians, also in the city of Amsterdam. Helwys explained: "They as partes or meanes of worship read Chapters, Textes to preache on & Psalmes out of the translacion, we alreddy as in prayinge, so in prophesiinge & singinge Psalmes lay aside the translacion, & we suppose yt will prove the truth, that All bookes even the originalles themselves must be layed aside in the tyme of spirituall worshipp, yet still retayninge the readinge & interpretinge of the Scriptures in the Churche for the preparinge to worshipp, Iudginge of doctrine, decidinge of Controversies as the grounde of or faithe & of or whole profession."[42]

Singing in unison was not a feature of early Anabaptist or Baptist praise. The precedent of solo singing was so rigidly followed by succeeding Baptist congregations that as late as 1690 that great innovator, Benjamin Keach, had difficulty in persuading his own congregation to sing in unison.

The sermons, it can be surmised from the importance of "prophesying" in worship, were long expositions of Scripture, rarely lasting less than an hour.

Little is known about the administration of the two "Gospel Sacraments," as they were called in this period in these circles. Baptism was, as we have seen, dispensed only to believers, to those "which heere, beleeve and with penitent hartes receave ye doctrine of ye holy gospell: for such hath ye Lord Iesus commaunded to be baptized, and no un-speaking children."[43] Helwys did not insist on sprinkling or dipping as the mode of Baptism; immersion was insisted on by the London Baptists of 1633. This church, under the leadership of Jessey and Blunt, used immersion as the only legitimate mode of administering the scrament, "being convinced of Baptism, yt also it ought to be by diping ye Body into ye Water, resembling Burial & riseing again."[44]

There are no accounts of early Baptist celebrations of the Lord's Supper. Yet since their sacramental doctrine was so clearly Zwinglian or Calvinist, it is probable that their service resembled that of the other Separatists in its simplicity, its fidelity to the

[42] Letter of Sept. 20, 1608, cited by Burrage, *Briefe Discoverie*, II, 166.
[43] *Ibid.*, p. 196. [44] Cf. *ibid.*, p. 303.

account of the Dominical institution in the First Epistle of Paul to the Corinthians, Chapter XI, verses 23ff. Most likely it repeated not only Christ's words of delivery but also his manual actions of fraction and libation.[45] The ordinance of the Lord's Supper was almost certainly kept every Lord's day. In this respect the Barrowists in Amsterdam were in agreement with the English Anabaptist congregation there. Johnson asked, "Whether it be not best to celebrate the Lords supper where it can be every Lords day; this the Apostles used to do; by so doing we shall return to the intire practise of the Churches in former ages."[46] The Anabaptist Confession of faith insisted that they dare not omit it on the Lord's day: "Oblata iusta occasione impedimenti, affirmamus coenam dominicam omitti posse donec tollantur impedimenta: aut aliter non audemus omittere quoque die sabbati quum convenimus and praestandum caetera Dei publici Ministerii. . . ."[47] The same Confession forbade laymen to celebrate the sacraments, because the privilege was in inseparably linked to the office of the minister of the Word of God.[48] In this matter they concurred with the other Separatists. John Robinson, for instance, was unwilling to allow his godly ruling elder, Brewster, to celebrate the Lord's Supper aboard the *Mayflower*, though no clergyman would accompany them on the perilous voyage to Plimouth Plantation.

Marriage was sanctioned by the Anabaptists only if both contracting parties belonged to their own communion, a most restrictive device in the early days of a new movement. If this rule was broken it was often accompanied by ecclesiastical censure, sometimes leading to excommunication. The *Short Confession of Faith* declared: "Wee permitt none of our communion to marry godles, unbeleeving, fleshly persons out of ye church, but wee censure such (as other synnes) according to the disposition & desert of ye cause."[49] The practice was enforced in Baptist and Independent congregations until the end of the seventeenth century.[50]

It is probable that the Anabaptists held love-feasts. This was certainly a feature of the church life of the early Separatists. In 1568, we are informed, "About a week ago they discovered here

---

[45] *Ibid.*, p. 106.   [46] *Ibid.*, p. 293.   [47] *Ibid.*, p. 235.

[48] Ministrationem sanctorum sacramentorum inseparatim cum ministerio verbi, coniunctam esse agnoscimus et cuique membro corporis administrare sacramenta non licere. *Ibid.*, p. 235.

[49] *Ibid.*, pp. 198f.

[50] For the Baptist practice see the *Fenstanton, Warboys and Hexham Records*, ed. E. B. Underhill; for the Congregational practice see Norman Glass, *The Early History of the Independent Church at Rothwell*, pp. 75f.

a newly invented sect, called by those who belong to it 'the pure or stainless religion'; they met to the number of 150 in a house where their preacher used half of a tub for a pulpit, and was girded with a white cloth. Each one brought with him whatever food he had to eat, and the leaders divided money amongst those who were poorer, saying that they imitated the life of the apostles and refused to enter the temples to partake of the Lord's supper, as it was a papistical ceremony.[51] It is unlikely that they were Baptists, because the latter did practise the Lord's Supper. These sectarians may have been the "Family of Love" about whom little is known. On the other hand, there is evidence that love-feasts were held by the English Baptists as a general feature of their church life, but never as a substitute for the Lord's Supper. The church record of the Warboys congregation has this entry for 1655: "The order of love-feast agreed upon, to be before the Lord's Supper; because the ancient churches did practise it, and for unity with other churches near to us."[52] Another ordinance that appears to have been used exclusively in England by the Baptists was feet-washing. Their warrant for it was Christ's humility in performing this menial service for His disciples.[53] The Assembly of the General Baptists "had long agreed that the practice of washing the feet of the saints, urged in Lincolnshire by Robert Wright in 1653, and in Kent by William Jeffrey in 1659, should be left optional as not specified in Hebrews VI."[54]

Three distinctive contributions were made by the Anabaptists or Baptists to English Separatist worship. Their most distinctive custom was, of course, that they practised believers' Baptism by immersion. Second, in their opposition to set forms of worship they went further than the other Separatists. This position was radically expressed by John Smyth: "That the reading out of a Book is no part of spiritual worship, but the invention of the man of sin; that Books and writings are in the nature of Pictures and Images; that it is unlawful to have the Book before the eyes in singing of a Psalm."[55] The third influence exerted by the Baptists was the method of running exposition of the Bible or interpolated com-

---

[51] *A Calendar of Letters and State Papers relating to English Affairs, preserved principally in the Archives of Simancos*, II, 7.

[52] *Fenstanton*, ed. Underhill, p. 272.

[53] John 13:5.

[54] "Original Sin, Feet-Washing and the New Connexion," *Transactions of the Baptist Historical Society*, vol. 1 (1908-1909), 129f.

[55] This is a summary of Smyth's views as abstracted from his *The Differences of the Churches of the Separation* by Robert Baillie, in *A Dissuasive from the Errours of the Time* (1645), p. 29n.

ment during the public reading of the Scriptures. This was the practice of the General Baptists, as is demonstrated by Grantham's survey of their church life made in 1678; it was the method used by the New England Puritans. John Smyth seems to have been the initiator of this trend of teaching.[56]

4. Other Underground Congregations

It now remains to consider other influences on Separatist congregations, which derive from certain obscure congregations in England and on the Continent. As their relation to the established Church was questionable, they were not considered as Separatist. Although their worship was like that of the Separatists, their form of church government was not. The first in point of time and in importance as a precedent was the "Privye Churche" that met in London during the reign of Queen Mary.[57] Its importance lay in the fact that it was organised for worship under persecution, providing later Separatists with a precedent for gathering secretly to worship, in defiance of the laws of the land, according to their own conscience.

A similar underground church was the Plumber's Hall congregation. They were of the lineage of the Marian secret church, as they proudly claimed, "we bethought us what were best to doe, and we remembered that there was a congregation of us in this Citie in Queene Maries dayes: And a congregation at *Geneva*, which used a booke and order of preaching, ministring of the Sacraments and Discipline, most agreeable to the worde of God: which booke is alowed by that godly & well learned man, Maister *Calvin*, and the preachers there, which booke and order we now holde."[58] This congregation was Puritan rather than Separatist, and had the clearest affinities with Presbyterianism, except that it was pre-Presbyterian and pre-Puritan. The congregation was discovered in 1567.

A congregation similar to the Plumber's Hall was known as the "Privye Churche" of Richard Fitz, which the London authorities also discovered in 1567. Their minister made the three Puritan demands: "trewe markes of Christs churche" consist in "Fyrste and formoste, the Glorious worde and Evangell preached, not in bondage and subjection, but freely, and purely. Secondly to have the Sacraments mynistred purely, onely and all together accord-

---

[56] *Works*, I, lxxxvii f.
[57] For the details see Foxe's *Acts and Monuments*, VIII, 485f., 558f.
[58] Cited in Albert Peel, *The First Congregational Churches*, p. 7.

inge to the institution and good worde of the Lord Jesus, without any tradicion or invention of man. And laste of all, to have, not the fylthye Cannon lawe, but dissiplyne onelye, and all together agreable to the same heavenlye and almighty worde of oure good Lorde, Jesus Chryste."[59] The discipline so strongly insisted upon was practised, not after the second diet of worship on the Lord's day, as was the Baptist custom, but "on the fourth day in the weke we meet and cum together weekely to use prayer & exercyse disciplyne on them whiche do deserve it, by the strength and sure warrant of the lordes good word as in Mt. xviii. 15-18 and I Cor.v."[60]

The close affinities of this congregation with the later phenomenon of Independency are seen, not only in modes of worship, but in the adoption of a covenant. On this account Albert Peel argued that "there is no valid reason for moderns to deny to Fitz's congregation, and probably to others contemporary with it, the title of 'the first Congregational Churches.' " In my view, however, since Robert Browne had not yet written his justification for the autonomy of each local congregation, so that it might the better serve Christ, without interference from bishops or magistrates, it is premature to title this congregation "Congregational." Furthermore, by its attachment to Geneva it was proto-Presbyterian rather than proto-Congregational.

Two conclusions follow from this study of Separatist worship. The first is that the Separatists put into practice a "Reformation without tarrying for any," while the Puritans resolved to attain the liturgical achievements of the Separatists within the established Church. Some were immediately attainable as supplements, such as the "prophecyings," additional free prayers which could be added to the forms of prayer in the Book of Common Prayer, and the singing of psalms. It was particularly in the field of extemporaneous prayer that the Separatists were innovators. Other features of Separatist worship such as the founding of churches on covenants, if imitated, could have led to splits in the parish churches between the covenanted and the uncovenanted. Nor was it possible to celebrate worship in the parish churches without using the three "noxious ceremonies," the ring in marriage, the signing of the cross in Baptism, and kneeling to receive Holy Communion. It was also necessary to wear the Anglican vestments. But all these matters being of the *esse*, not the *bene esse* of true

[59] *Ibid.*, p. 32.   [60] *Ibid.*, p. 33.

churchmanship, could wait for a monarch or parliament more propitious to the Puritan demands in the future.

The truly decisive influence of the Separatists was in their radical opposition to any set forms of prayer, including the Lord's Prayer. It is worth recalling the irony of Barrow's attack on set prayers (as in the Prayer Book), to make his plea for free prayer ring with conviction in our ears: "Shall we think that God hath any time left these his servants so singularly furnished and destitute of his grace, that they cannot find words according to their necessities and faith, to expresse their wantes and desires, but need thus to be taught line unto line, as children new weaned from the brestes, what and when to say, how much to say, and when to make an end; to say this collect at the beginning, that at the end, that before the tother after, this in the morning, that at after noone, etc."[61] This fine conception of liberty in prayer, with great possibilities of achievement, as also of bathos, was the heritage that Separatism transmitted to Independency and through Independency to the English Presbyterians at the Westminster Assembly of Divines, and thence through the Pilgrims and Puritans to the new world. It was a noble heritage; it was this influence of Separatism which probably accounts for the departure of Puritanism from some of the liturgical customs of the European Reformed churches. Unquestionably it brought fervour, freedom, intimacy, and sincerity into worship, though sometimes at the cost of order and dignity. Often its freshness was a tonic for those who had developed the familiarity that breeds contempt for set forms, even for the richest and most felicitously phrased of literary forms such as the Book of Common Prayer was and is.

[61] *Briefe Discoverie*, p. 74.

PART THREE

LITURGICAL ARTS AND AIDS

1. Thomas Cranmer, 1489-1556, Archbishop of Canterbury, Anglican Martyr

2. Thomas More, 1478-1535, Lord Chancellor of England, Catholic Saint

3. John Jewel, 1522-1571, Bishop of Salisbury, Anglican Apologist

GVILIELMVS PERKINSVS S.THEO. Prof.
PERINSVS Christi defendens dogmata talis
Vultu erat, ingenium scripta facunda probant AB

4. William Perkins, 1558-1602, Puritan Divine

5. Edmund Campion, 1540-1581, Jesuit Martyr

SVNT MELIORA MIHI.

RICHARDVS HOOKER Exoniensis scholaris sociusq̃ Collegij Corp: Christi Oxon: deinde Londi: Templi interioris in Sacris magister, Rectorq̃ huius Ecclesiæ, scripsit octo libros Politiæ Ecclesiasticæ Anglicanæ, quorum tres desiderantur. Obijt Añ: Dõ: M.DC.III. Ætat: suæ L.
Posuit hoc pijssimo viro monumentum Anº Dõ: M.DC.XXX.V. Guli: Cowper Armiger, in Christo Iesu quem genuit per Evangelium. 1 Corinth: 4. 15.

Guil: Faithorne sculp:

6. Richard Hooker, 1554-1600, Anglican Apologist

The Ship of the
Romifh Church.

Burning of
Images.

¶ Ship ouer your
trinkets & be pac-
king you Papiftes.

The Temple
well purged.

The Papiftes packing away
theyr Paltry.

The cõmu-
nion Table

7. The Protestant Significance of the Reign of Edward VI

A table describing the burning of Bishop Ridley and Father Latimer at Oxford, D. Smith there preaching at the time of their martirdome.

Cranmar.

O Lord ſtrengthen them,

Si corpꝰ meū tradam igni caritatem auté non habeā nihil vtilitatis,&c.

Smith.

M. Ridley I will remember your ſute,

In manꝰ tuas domine,

Ridley

Latimer

Father of heauē receaue my ſoule.

9. The Martyrdom of Archbishop Thomas Cranmer, March 21, 1556

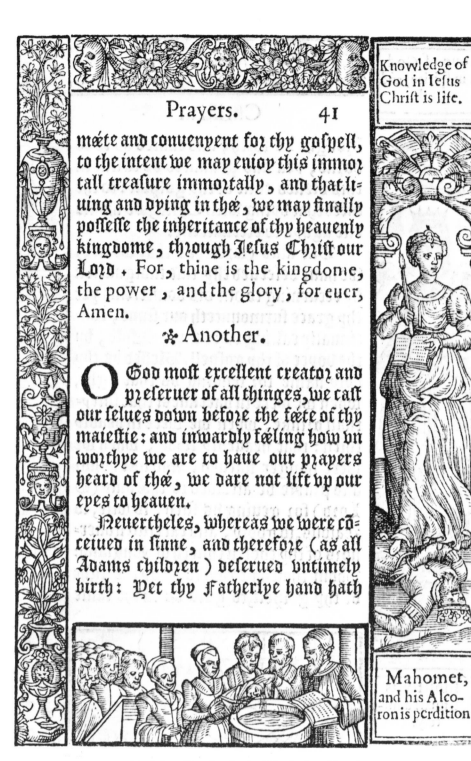

Knowledge of God in Iesus Chrift is life.

mǽte and conuenyent for thy gofpell, to the intent we may enioy this immor tall treafure immortally, and that lt= uing and dying in thǽ, we may finally poffeffe the inheritance of thy heauenly kingdome, through Iefus Chrift our Lord. For, thine is the kingdome, the power, and the glory, for euer, Amen.

�֍ Another.

O God moft excellent creator and preferuer of all thinges, we caft our felues down before the fǽte of thy maieftie: and inwardly fǽling how vn worthye we are to haue our prayers heard of thǽ, we dare not lift vp our eyes to heauen.

Neuertheles, whereas we were cõ= ceiued in finne, and therefore (as all Adams childrē) deferued vntimely birth: Yet thy Fatherlye hand hath

Mahomet, and his Alco- ron is perdition

10. Infant Baptism in the Elizabethan Church

oue of God
s in spirite, and
truth.

Idolatry, is
Spirituall adul-
tery.

Chriſtian

bꞁought vs foꝛth, and cauſed vs to be
boꝛne, yea and graunted vs to liue in
this bleſſed time of thy goſpell: to the
intēt that dying with Chꝛiſt, we ſhould
riſe agayn to eternall life.

But alas (wꝛetches that we are)
we haue deſerued thy iuſt indignation
by returning to our old vomit. And yet
thy grace ſurmounteth our ſinne, con=
tinually calling vs to thy ſhæpfold, by
the voyce of thy goſpell. Bleſſed be thy
holy name foꝛ ſending vs that light,
when we were in darcknes: that ſpiry=
tuall dꞁinke, when we were in deadly
thirſt: that heauenly fœde, when we
were hunger ſtarued. And like honoꝛ
and pꞁayſe be aſcribed to thæ only (O
Loꝛd) foꝛ geuing vs ſuch, ſo wiſe, ſo
zealous, ſo godly, and carefull gouer=
noꝛs of thy choſen church of England:
whom thou haſt rayſed vp by the light
of thy goſpell, to guid vs in the ſame

11. Holy Communion in the Elizabethan Church

CHAPTER X

RELIGIOUS ARCHITECTURE
AND ART

D ID PROTESTANTISM kill, secularise, or simplify religious architecture and art in England? This is almost as contro- versial and intriguing an issue as the much debated ques- tion whether Protestantism originated or abetted capitalism.

It has been fashionable for historians to deplore the aesthetic losses caused by the Reformation. They can be seen plainly in the contrast between the richness of the architectural setting of Catho- lic worship and the austerity of the context within which Protes- tant worship is celebrated. With typical vigour and a preference for the inanimate over the living, A. L. Rowse insists that "the sad- dest thing about the Reformation, sadder perhaps than the loss of lives, was the enormous loss in things of beauty, for they were in no way responsible for the fate that befell them."[1] G. G. Coulton,[2] that stern mentor among the medievalists, challenged the Pugi- nesque contrast between the "ages of faith" with all their variegated splendour of stone, wood, metal, and glass in cathedrals and mon- asteries, and the industrial ugliness of Victorianism, as a piece of romantic sentimentality. The balance seems to have been struck by A. G. Dickens.[3] He acknowledges that one of the most impres- sive features of medieval religion was its ability to inspire "lively art and craftsmanship" and observes that some of the most splendid Gothic edifices (churches and palaces) continued to be built in the reigns of Henry VII and Henry VIII. Yet he also remarks that much glass-making had become a wearying stereotype and that sacred statues were crudely mass-produced. Furthermore, he even raises again some of the ethical questions that Coulton had about the much vaunted "religious" achievements of the medieval artists. Was there no fear in the faith that built the cathedrals? Were the pious donors never expiating their sins with conscience money? Did not the Church itself use indulgences to raise its fanes? Did not its theologians and architects from the Abbot Suger on allow the central events of Christian history to be overlaid with super-

1 *The England of Elizabeth*, p. 465.
2 See G. G. Coulton's *Art and the Reformation*, chaps. 19, 20.
3 *The English Reformation*, pp. 9-10.

stition and almost incredible legends? Also, what of the vaunted anonymity of the medieval artist who is supposed not to have wanted his right hand to know what the left hand was doing? There is, of course, no question about the splendour of medieval art, even if there is a questioning of some of the naïveté in the rose-window-tinted view of medieval piety. Moreover, one is not likely to aggrandise the lowly status of Reformed architecture and art by the process of minimising the greatness of medieval Catholic art. The question remains: did Protestantism produce an authentic religious architecture and art in sixteenth-century England?[4]

1. *Iconoclasts and Iconophiles*

Unquestionably there was a strongly iconoclastic element in English Protestantism, which was actively fought by the English Catholics of the time. The views of Bishop John Hooper of Gloucester, in Edward's reign, and of William Burton, a Bristol preacher in Elizabeth's reign, were typical of the iconoclasts who desired to do away with all images of the Holy Trinity, the Incarnate Son of God, and the saints, as a breaking of the Second Commandment against "graven images."

Hooper had fled England in 1539 to avoid persecution for his extreme Protestant views, and had lived from 1547 to 1549 in Zurich, perhaps the most radical of the Reformed centres, where he was greatly influenced by John à Lasco. He returned to England in 1549 and became a prominent opponent of Bonner, a Henrician Catholic who was Bishop of London. A critic of priestly vestments, transubstantiatory doctrine, and devotion to the saints, he inevitably desired to remove all traces of "idolatry," which he interpreted images of God and the saints to be. These views were reflected in his injunctions for the Dioceses of Gloucester and Worcester, especially referring to images in windows:

[4] If the question had not been limited to England, it might in one way have been easier to answer because the impact of Lutheranism, especially in the Scandinavian countries, was not as iconoclastic as Calvinism; yet Calvinism produced some interesting variations in church shapes among the Huguenots (see Paul de Félice, *Les Protestants D'Autrefois*, 4 vols., I, chap. 1). It is, however, significant that Karl Holl, in *Die Kultur bedeuting der Reformation*, tr. as *The Cultural Significance of the Reformation*, claims that Protestantism contributed significantly to politics, economic life, education, and the understanding of history, philosophy, poetry, and music, yet the only first-rate painter to be claimed for Protestantism is Rembrandt, who was active a century after the Reformation and whose church affiliation (Mennonite or Dutch Reformed?) is uncertain, even if his work is thoroughly Biblical in its inspiration. See the fascinating work of W. A. Visser T'Hooft, *Rembrandt and the Gospel*.

Item, that when any glass windows within any of the churches shall from henceforth be repaired, or new made, that you do not permit to be painted or portrayed therein the image or picture of any saints; but if they will have anything painted, that it be either branches, flowers or posies [mottoes], taken out of Holy Scripture. And that ye cause to be defaced all such images as yet do remain painted upon the walls of your churches, and that from henceforth there be no more such.[5]

William Burton, a Puritan preacher of the Church of England, denounced the stained glass at St. Thomas's Church in Bristol:

It likewise appeareth what dishonour and disgrace have been offered by gross idolaters which would take upon them to paint and picture out the invisible and incomprehensible majesty of the Almighty like a man whose breath is in his nostrils, whose being is not of himself, whose years are but a span long, and in his best estate is altogether vanity. Whatsoever ye do, saith the Scripture, do all to the glory of God. That is, strive to do it so, as God may get the most glory by it. Are such representations of God to the advancing of his glory? What do they show and teach us that we might give Him everlasting praise for? He is painted as a man as you see in yonder story of creation in yonder window in a dozen places together. What may we learn by them? A man hath his being from another. If God be as he is painted forth, he hath so too, which to say is blasphemy.[6]

The two charges of Protestantism against the representation of God and the saints thus were blasphemy and idolatry. It allegedly was blasphemy to represent God anthropomorphically, though it could very well be argued that since in the Incarnation God had become man here was the true icon of the invisible God[7] and a perfect justification for the making of copies of a true image. This, in fact, was the heart of the Catholic contention. The charge of idolatry seemed more appropriate to the reverence for the saints,

___

[5] *Visitation Articles and Injunctions*, eds. W. H. Frere and W. M. Kennedy, ɪɪ, 289.

[6] *David's Evidence* (1596), pp. 140-41. A marginal note indicates that he is pointing to windows in St. Thomas's Church, Bristol. The D.N.B. says he preached at Bristol in 1590.

[7] Colossians 1:15 is usually translated in the West as "the image of the invisible God." Yet the Greek original, rendered "image," is actually ἐικὼν.

who are, after all, created beings, lacking aseity, which was Hooper's main point. The Catholic defence of reproductions of the saints was that these imitators of God were reminders of the potentialities of men and women responsive to the endowments of divine grace. It is of course arguable that poor representations of divine beings are worse than none.

The Catholic view of images may be illustrated from the defence of them in "The King's Book," and in Harding's controversy with Jewel. *A Necessary Doctrine* of 1534 provides an exposition of the Second Commandment, which interpreted the veto against graven images as applying not to their creation but to their extreme veneration. Their positive purpose was "that we (in beholding and loking upon them . . .) may call to remembraunce the manifolde examples of vertues, which were in the sainctes, whom they do represent. And so maye they rather be provoked, kendled, and styred, to yelde thankes to our lorde and to praise him and his sayde saintes, and to remembre and lamente our synnes and offences, and to praye God, that we may have grace to folowe their goodnes & holy lyvyng."[8] As an illustration, it was said that the image of Christ on the rood, or the depiction of the crucifixion on walls or windows, was made so that the beholder might learn Christ's virtues, remember his "paynefull and cruel passion," and "condemne and abhorre our synne, which was the cause of his so cruell death."[9] There were two caveats, however, against a misuse of images. Those who would trust more in one image than in another were in error, as were those who spent money on bedecking images rather than helping "poor christen people, the quicke and lyvinge images of god, which is the necessary worke of charitie commanded by god."[10]

Harding adduced three arguments for making and using images. The first was that images are the theological instructors of "the

[8] *A Necessary Doctrine and erudition for any chrysten man, set furth by the Kynges maiesty of Englande &c.* (1534), commonly known as the "King's Book," sigs. L recto and verso.

[9] *Ibid.,* L verso.

[10] *Ibid.,* L iii recto. It is also worth citing Sir Thomas More's view of the value of images: "Albeit that every good Christian hath a remembrance of Christ's passion in mind, and conceiveth by devout meditation a form and fashion thereof in his heart, yet is there no man I ween so good nor so well learned, nor in meditation so well accustomed, but that he findeth himself more moved to pity and compassion, upon beholding of the holy crucifix, than when he lacketh it." (*Dialogue against Tyndale* in *The English Works,* reproduced in facsimile from William Rastell's edition of 1557 and edited with a modern version by W. E. Campbell, II, 15-16.) See also Johan Huizinga, *The Waning of the Middle Ages,* chap. 12.

simple and unlearned people, which be utterly ignorant of letters [who] in pictures doo as it were reade and see nolesse then others doo in bookes, the mysteries of christen Religion, the actes and worthy dedes of Christ and of his Sainctes."[11] The second argument was that images are vivid reminders to sluggish souls of their indebtedness to Christ and the saints, and are encouragements "to the like will of doing and suffering, and to all endeavour of holy and vertuose life."[12] The third argument was in fact an extension of the second, to the effect that images are reminders of all things necessary to salvation. In recalling the benefits and merits of Christ and the virtuous examples of the saints, Harding said: "that if we bee such as they were, we may by Gods grace through Christ atteine the blysse they be in, and with them enioye lyfe everlasting."[13]

The Anglican rebuttal of the chief arguments for the use of images in worship was contained in the homily, "Against Peril of Idolatry." It was addressed particularly to the claim that arguments against idolatry do not apply to the images of the true God, Christ and the saints, that images of Christ may be made although He was God, because He took human flesh and became a man. The reply was: "For Christ is God and man: seeing therefore that of the Godhead, which is the most excellent part, no image can be made, it is falsely called the image of Christ. . . . Furthermore, no true image can be made of Christ's body, for it is unknown now of what form and countenance he was."[14] Two other arguments for images were rebutted. One was that it is not the image that is honoured but the saint whose image it is; the other was that tradition approves of creating saints' images. The reply was that honouring of saints by such was to dishonour them by despoiling God of honour; the true honouring of the saints is to live in charity, giving generously and living simply as they did, since the true service of God is in helping the poor, the living images of God, not in honouring dead simulacre of the saints.[15]

The importance of the arguments is that they had some reason

11 Thomas Harding, *An Answer to Maister Iuelles Chalenge* (1565), p. 189 verso.
12 *Ibid.*, p. 190 recto. 13 *Ibid.*, p. 191 recto.
14 *Certain Sermons* (1574 version, reprinted and edited by G. E. Corrie), p. 217. It should be noted that the outline of this homily is contained in Bishop Ridley's "Treatise on the Worship of Images," yet the contents are almost entirely a literal translation of Bullinger's treatise, *De origine cultus Divorum et simulacrorum erronea.*
15 *Certain Sermons*, ed. Corrie, p. 269.

on their side in their desire to protect men from false notions of God. However psychologically weak the understanding of iconoclasts was, they were not colour-blind Philistines utterly insensitive to aesthetic values.

In some measure in Anglican worship, and altogether in Puritan worship, the iconoclasts won the arguments with the iconophiles. The result was the almost complete smashing of the figures on the roodscreen and the ending of the making of religious sculptures in wood, as well as the almost total elimination of painting on walls or windows of churches. The monastic churches, except when they were kept as parish churches, must have suffered the greatest despoiling, for their windows no longer needed to be glazed. Many must have soon been emptied of windows for the lead.[16]

The melancholy story is told in part in the 28th Article of the Royal Injunctions of 1547: "Also, that they shall take away utterly extinct and destroy all shrines, covering of shrines, all tables, candlesticks, trindles or rolls of wax, pictures, paintings, and all other monuments of feigned miracles, pilgrimages, idolatry, and superstition: so that there remain no memory of the same in walls, glass-windows, or elsewhere within their churches or houses. And they shall exhort all their parishioners to do the like within their several houses."[17] Thirty years later William Harrison wrote of the imageless Anglican churches: "Churches themselves, belles and times of morning & evening praier remain as in time past, saving that all images, shrines, tabernacles, rood loftes & monuments of idolatrie are removed, taken down & defaced: Onlie the stories in glasse windowes excepted, which, for want of sufficient store of new stuffe, & by reason of extreame charge that should grow by the alteration of the same into white panes, throughoute the realme, are not altogether abolished in most places at once, but by little and little suffered to decaie that white glass may be set up in their roomes."[18]

Goldsmiths, silversmiths, glaziers, painters, and carvers must immediately have turned from ecclesiastical to secular work. Painters of glass turned to heraldry.[19] It is equally meaningful that

[16] Christopher Woodforde, *English Stained and Painted Glass*, p. 38.
[17] Eds. W. H. Frere and W. M. Kennedy, *Visitation Articles and Injunctions*, II, 126.
[18] *Harrison's Description of England in Shakspere's Youth*, 2 vols., ed. F. J. Furnivall, I, 18f.
[19] Woodforde, *English Glass*, p. 33.

painters in oils turned away from religious paintings for churches
(which were forbidden them) and found secular inspiration in
portraits and miniatures,[20] thus exhibiting that growing individualism which was the legacy of the Renaissance and Reformation.
Eric Mercer may indeed be right in distinguishing Renaissance
from Reformation individualism by regarding the former as celebrating the exceptional and the latter as recognising the importance of the common man.[21] In this sense the Reformation may
have assisted in the process of the secularisation of art and in
moving it from the church to the home, although this does not
necessarily mean that all its sacred quality has been lost, since
man is made in the image of God.

Did the advent of English Protestantism bring any relief to a
cluttered, complicated, and overdecorated worship? The various
uprisings, from the Pilgrimage of Grace to the Western Rebellion
in Cornwall and Devon, showed that the Catholics were conscious
of an impoverishment in worship. Yet Protestants rejoiced in a
simplification of worship and a functionalism which austerity in
aesthetics aided. This is plainly to be seen in the significant illustration in Foxe's *Acts and Monuments* in the folio edition of
1573,[22] which symbolises the religious meaning of the reign of
the young Protestant, Edward VI. It depicts a simple Renaissance
temple in the foreground, at which an attentive congregation is
listening to the preaching of the Gospel, while on a platform a
table replaces an altar and on it is a loaf and a large communion
cup from which all will partake, while the sacrament of Baptism
is being celebrated by a minister and an eager family gathered
around the font at the entrance of the church. In the upper left
of the woodcut, there is the porch of a Gothic church from which
Roman Catholics are emerging in haste, each with a sack or lug-
gage on his shoulder, as he rushes toward the "Ship of the Roman
Church" bearing the emblem of the five wounds of Christ. The
legend says: "The Papistes packing away their Paltry." What they
are removing is heavy service books (such as missals, sacramenta-
ries, pontificals, and breviaries), and with them vestments, tabernacles, croziers, candlesticks, censers, bells, monstrances, and

20 See Ellis Waterhouse's *Painting in Britain, 1530-1790*, p. 19, for the judgment: "Alone of painters in Elizabeth's reign, Hilliard and his rival, Isaac Oliver,
are worthy to be named as contributors to the age of Shakespeare." Both were
"limners," or miniaturists in portraits.
21 *English Art, 1553-1625*, p. 6.
22 II, 1,294, which is included in the illustrations to the present volume.

crucifixes, while behind there is a conflagration of images. Another legend reads: "Ship over your trinkets and be packing you Papistes." This is all very crude, but also telling. There is no mistaking the fact that for Protestants these pieces of liturgical paraphernalia were "paltry," mere trinkets.

The relief Protestants felt in the newly found simplicity of their worship is admirably expressed by Bernard Gilpin in a sermon preached before the court in 1552. He contrasted Catholic with Protestant worship: "They [Catholics] come to the Church to feede their eyes, and not their soules; they are not taught, that no visible thing is to be worshipped. And for because they see not in the church the shining pompe and pleasant variety (as they thought it) of painted clothes, candlesticks, images, altars, lampes, tapers, they say, as good to goe into a Barne; nothing esteeming Christ which speaketh to them in his holy Word, neither his holy Sacrament reduced to the first institution."[23] That kind of streamlining of worship, and simplicity of style, seemed to Protestants to comport more naturally with the worship of the New Testament as it had been recovered in the vernacular Bibles of the era for the benefit of the common man.

I have already mentioned the iconoclasm that destroyed so much medieval art, and the simplification of Anglican worship through the removal of so much priestly paraphernalia from the celebration of the liturgy, along with the secularization of painting and much architecture. Are we then to assume that Anglicans and Puritans merely adapted Catholic churches for their own worship, and that they made no architectural nor artistic contributions of their own? Such would be an oversimplification of an intricate question.

2. Anglican Architecture

The style of ecclesiastical architecture does not usually differ markedly from the leading domestic architecture of the age, particularly in a time of increasing secularity. It is therefore important at least to glimpse the character of Tudor architecture in general, as a preliminary to studying Anglican ecclesiastical architecture.

There appear to be three styles of Tudor architecture, corresponding roughly to three successive periods.[24] First, there was

[23] *A Sermon preached in the Court at Greenwitch, before King Edward the Sixth, the first Sonday after the Epiphany, Anno Domini 1552* (1630), sig. C3 recto and verso; also pp. 21-22.
[24] John Harvey, *Tudor Architecture*, pp. 12-13.

the rich, ornate work inspired by the ostentatious Edward IV, of which the most notable examples are St. George's Chapel, Windsor (1474-1500); Magdalen College, Oxford (1474-1490); and the central tower of Canterbury Cathedral (1490-1497). The succeeding style is simpler and more austere, and is expressed in expanses of plain brick walls, with cuspless arches and tracery. Typical of this style are Corpus Christi College, Oxford (1512-1518), and Wolsey's Hampton Court (1515-1525). The third mongrel style, mixing Gothic and Classical, can be seen in Henry VIII's Hampton Court (1531-1536), Titchfield Place (1537-1540), and Cowdray (1537-1548). The native Gothic, whether elaborate or simple, was mingled unnaturally with the new French and Italian influences. Gothic art was inspired by the horizontal and dynamic line, while Roman art stressed the horizontal and static. The demand for classical details and even for classical elevations ended in almost total dislocation.[25]

But before the Gothic died it had a glorious swansong in the designs of Robert and William Vertue in Henry VII's Chapel in Westminster Abbey. Begun in 1503 and practically completed in 1519 by William Vertue after Robert's death in 1506, this is the most opulent and splendid of all English royal chapels. The richness of the detail is prevented from being fussy or cloying because of the sureness of the structure and the lines of the design. This is a most attractive example of fan-vaulting, with pendants and eastern corona. As Harvey says, "Here all is airiness and grace, and it is as though the stone pendants were themselves borne up by a floating lacework of gossamer traceries."[26]

Perhaps the most remarkable fact about Anglican Church architecture is not the Gothic revival which flourished in the nineteenth century, but the Gothic *survival* which continued from 1558 to 1662.[27] It is most significant that Gothic survived longer in England, and was less altered from its medieval precedents, than

---

[25] *Ibid.*, pp. 38-39.   [26] *Ibid.*, p. 27.

[27] This thesis has been argued with a plethora of supporting evidence in John M. Schnorrenberg's "Early Anglican Architecture, 1558-1662: Its Theological Implications and Its Relation to the Continental Background," unpub. Ph.D. thesis, Princeton University, 1964. It is a view also put forward by Eric Mercer in *English Art, 1553-1625*, Oxford History of English Art, pp. 85-87, who claims that there were three reasons for continuing to use Gothic styles in architecture, of which the first was the strongest. "This appreciation of the aesthetic appeal of Gothic was the basis of its deliberate use, and the intention was either to advertise the religious functions of the building, or to give it an archaic air, or to emphasize a particular feature by the contrast of its ornament with the classical character of all the rest" (p. 85).

among the Lutherans of Germany and Sweden. The use of this style fully or partly Gothic in England seems increasingly a deliberate attempt to maintain the continuity of the medieval with the reformed Church of England; in this respect it is of a piece with English life in general. Outward forms are maintained even if their contents are transformed. A ruling monarchy becomes in time a mere figurehead of a sovereign. A religious primer looks like the old primers, but it has increasingly Protestant contents. So it was with ecclesiastical architecture: the old forms were maintained and repaired, and even the additions are not often unGothic, despite the fact that the worship was no longer a dramatic spectacle, but equally intended to be heard until Christopher Wren, a century after the beginning of Elizabeth's reign, designed definitely "auditory" and, therefore, Protestant churches. Even when this Gothic survival became a deliberate revival, as during the Commonwealth period, it was still used as an expression of loyalty to a persecuted Church.

It could, of course, be argued that in the sixteenth and early seventeenth centuries in England, no style other than Gothic was possible for the buildings of the Church of England. Had the Church of England so wished, however, it could have followed the Renaissance styles already being used in secular architecture in order to indicate its break with the older religion of Roman Catholicism. Anglican church-building could even have followed the example of the French Calvinists, the Huguenots, with their plans for single-roomed structures shaped as squares, circles, polygons, and ovals, and built in the classical style.[28] That this possibility was not taken seems a proof that the national church sought to affirm continuity rather than discontinuity in its architecture. The taste continued to be firmly Gothic, especially when the theology was conservatively orthodox, as, for example, that of Archbishop Laud. He favoured Gothic buildings, with occasional classical details, even though the most outstanding architect and designer of his time was the classicist, Inigo Jones. The classical style was not to triumph in Anglicanism until 1663, when Wren designed the first of his churches, Pembroke College Chapel, Oxford, in a classic style. Until then Gothic reigned supreme.

Nineteen churches or chapels were built or rebuilt during the reign of Elizabeth. Several of them were enlarged by the adding of

[28] de Félice, *Les Protestants D'Autrefois*, I, 11.

a tower, an aisle, a porch, a chapel, or combination of some of these features. Five have been destroyed, three remodelled, and the rest enlarged and altered.[29] In these circumstances generalisations have to be more tentative than usual. The number of churches involved seems small, which was due to lack of funds in the impoverished churches. The reforms of Henry VIII and Edward VI gave the sovereign as supreme "head," or "governor," of the Church of England a vast patronage of "livings," the tax on their first fruits, and on the presentation to bishoprics. Also, several sees were left vacant for long periods, enabling the queen to pocket the revenues.[30]

The few new churches were built in the traditional Gothic style —St. Peter's, Brooke, Rutlandshire in 1579, and St. Nicholas, Bedfordshire about the same year. The most elaborate, and largest, Elizabethan church surviving is St. Wilfred, Standish, Lancashire, which was rebuilt probably between 1582 and 1584, the largest donor being the rector, Richard Moody. It is of the standard Gothic plan and elevation, with an arch inside and two exterior turrets to mark the distinction between chancel and nave. The church of St. Michael, Woodham Walter of 1563-1564, is an early example of Gothic survival in Essex, and interesting because it is built of brick, and has stepped gables, and square-headed windows filled with tracery. Interestingly the donor, Thomas Radcliffe, the third Earl of Sussex, was thought to have strong Protestant leanings, yet he had the church consecrated and at the direction of the Calvinist Bishop Grindal.

With one exception the additions made to existing churches were Gothic added to Gothic. The exception was the unique seven-sided porch built by Bishop Jewel for St. Leonard's Church, Sunningwell, Berkshire, about 1562. Although it has flat-headed Tudor windows, the porch has Ionic pillars on pedestals with entablature *en ressant*.[31] It is possible that Jewel, as Bishop of Salisbury and a decided Protestant, recalled the Renaissance neoclassical architecture he had seen in his years of exile in Frankfort, Strassburg, or Zurich.

[29] Schnorrenberg, "Early Anglican Architecture," pp. 46f. In the appendix (p. 279) it is pointed out that about 152 Anglican churches were built or rebuilt between 1558 and 1662.

[30] Ely was without a bishop from 1581 to 1600, Norwich from 1578 to 1585, and Oxford from 1558 to 1567, 1568 to 1589, and 1590 to 1603.

[31] See Basil Clarke and John Betjeman, *English Churches*, illustration no. 129.

3. *The Elizabethan Conception of the Church*

Catholicism, with its realistic doctrine of the Eucharist, conceived of the church in the most literal sense as the house of God, since the tabernacle housed God and the church housed the tabernacle. Calvinists, however, believed that the church was the house of prayer and preaching—the assembling place of the people of God—and Puritans believed that the saints, that is, God's people in the process of being sanctified, were the temples of the Holy Spirit and that churches were meetinghouses. The Elizabethan Anglican view of the church seems to have been midway between the Roman Catholic and Calvinist concepts, insofar as it can be determined from the pertinent homilies dealing with this theme (Nos. 1, 2, 3, and 8 in the edition of 1563), from the writings of Hooker, and in comparison with the Lutheran and Calvinist views of the function and purpose of a church edifice.

The homilies present two views of the Church. In one, it was viewed as the earthly shadow of the heavenly mansion;[32] the building mediated numinosity. In the other, it was affirmed that its holiness was not of itself, "but because God's people resorting thereunto are holy, and exercise themselves in holy and heavenly things."[33] The former view was Catholic and the latter, Calvinist. The Anglicans combined both numinous and functional concepts of the religious sanctuary.

Richard Hooker typified these views in his *Ecclesiastical Polity*, for the edifice as well as the people are holy, dedicated to God. In a famous citation Hooker claimed that although worship is acceptable to God wherever it is offered, "the very majesty and holiness of the place where God is worshipped, hath *in regard of us* great virtue, force, and efficacy, for that it serveth as a sensible help to stir up devotion, and in that respect no doubt bettereth even our holiest and best actions in this kind."[34] Furthermore, he affirmed that the gravest of the Fathers had claimed that "the house of prayer is a Court beautified with the presence of the celestial powers; that there we stand, we pray, we sound forth hymns to God, having his Angels intermingled as our associates," hence, "how can we come to the house of prayer and not be moved with the very glory of the place itself so to frame our affections praying, as doth beseem them, whose suits the Almighty doth

---

[32] *The Two Books of Homilies*, ed. J. Griffiths, p. 166.
[33] *Ibid.*, p. 275.
[34] *Ecclesiastical Polity*, Book v, chap. 16, para. 2.

there sit to hear, and his Angels attend to further?"[35] We have only to substitute "saints" for "angels" to realize how medieval this appeal was. Yet Hooker could also insist on the functional importance of the church, without an addiction to "stiffness": "It cannot be laid to any man's charge at this day living, either that they have been so curious as to trouble bishops with placing the first stone in the churches they built, or so scrupulous, as after the erection of them to make any great ado for their dedication. In which kind notwithstanding as we do neither allow unmeet nor purpose the stiff defence of any unnecessary custom heretofore received; so we know no reason wherefore churches should be the worse, if at the first erecting of them; at the making of them public, at the time when they are delivered as it were into God's own possession, and the use whereunto they shall ever serve is established, ceremonies fit to betoken such interests and to accompany such actions be usual, as in the purest times they have been."[36]

Hooker's views were not exclusively Calvinist, for they did not heed the caveat of Calvin: ". . . there is great need of caution, lest we either consider them as the proper habitations of the Deity, where he may be nearer to us to hear our prayers—an idea which has begun to be prevalent for several ages—or ascribe to them I know not what mysterious sanctity, which might be supposed to render our devotions more holy in the Divine view."[37] The first Hooker citation above directly contradicts the second part of the Calvin citation. Nor did Luther support to any extent the views of Hooker, since he had an intense dislike for all ecclesiastical triumphalism—"all the high, big and beautiful churches, towers and bells in existence." Luther wrote: "For indeed, the Christian Church on earth has no greater power or work than common prayer against everything that may oppose it. This the evil spirit knows well, and therefore he does all he can to prevent such prayer. Gleefully he lets us go on building monastic houses, making music, reading, singing, observing many masses, and multiplying ceremonies beyond measure. . . . But if he noticed that we wished to practise this prayer, even if it were under a straw roof or in a pig-sty, he would indeed not endure it, but would fear such a pig-sty far more than all the high, big and beautiful churches, towers and bells in existence, if such prayer be not in them. It is indeed not a question of the places and buildings in which we assemble, but

[35] *Ibid.*, chap. 25, para. 2. [36] *Ibid.*, chap. 12, para. 1.
[37] *Institutes* III: xxx.

only of this unconquerable prayer, that we pray it, and bring it before God as a truly common prayer."[38]

The conclusion must be that while Edwardian Anglicanism was strongly iconoclastic and functional in its conception of the church building, Elizabethan Anglicanism combined the functional approach of the house of prayer with a recognition that churches themselves were capable of stirring up feelings of holiness and numinosity, a view closer to Catholicism than to Protestantism, as expressed at any rate by Luther or Calvin.

4. The Adaptation of Churches for Anglican Worship

There is no need to deal further with the iconoclasm that changed the churches from "tuppenny coloured" to "penny plain" in the days of Edward VI, nor even with the positive sense of the simplification and uncluttering of worship which ensued. My present concern is rather with first the Edwardian and then the Elizabethan rearrangements that were made for participation in a new vernacular liturgy and to make the edification of the people more feasible.

Hooper, Bucer, and Ridley were convinced that the medieval two-roomed church (with nave separated from chancel by rood-screen and arch, as well as by steps) was fundamentally unsuited to Reformed worship. The Church of England had inherited such buildings, yet it was convinced that the true liturgical unit was not the clergy, with the laity as onlookers, but the whole body of the faithful gathered at worship. Even the larger parish churches had insufficient space for vast gatherings of the people, since much of the space was taken up by chantry chapels, and the choir intervened between the nave and the altar; where there was a tympanum over the roodloft there was little chance of glimpsing the altar, and only the elevation of the Host in the Eucharist was visible.

The most radical solution to the problem was Hooper's suggestion that all chancels be closed off and only the naves used. The advantages were that the people would better understand what the minister was reading and the communicants would be able plainly to see and understand what was happening at the altar in Holy Communion. Moreover, the minister, being close to the people, would then be able to tell whether his reading was being under-

[38] "The Treatise of Good Works" (1520), tr. W. A. Lambert, in Luther's *Works*, I, 235.

stood or not.[39] Hooper was unsuccessful in his advocacy, but in his own dioceses of Gloucester and Worcester, as his injunctions of 1551-1552 show, he was able to enforce his own intense dislike of the screens that separated clergy from the people and which symbolised for him the veil of the Temple in the Old Dispensation which had been done away in Christ, thus making screens unnecessary in a Christian church.

Bucer, in his *Censura* of 1551, was strongly critical of the rubric before Matins in the First Prayer Book of Edward VI, which read, "The priest being in the quire shall begin with a loud voice . . ." because a minister in a returned stall in the chapel with his back to the congregation could not be heard clearly in a large church even if he raised his voice. Furthermore, Bucer objected that a special place for the clergy, such as the chancel was, gave the laity the impression that the ministry were nearer God than the laity and that the liturgy was the special prerogative of the clergy. He believed that the round, single-room churches of antiquity, with the priest surrounded by the congregation so that they might easily hear and see, was the most desirable church architecture. In the Reformed church at Strassburg, Bucer had put his liturgical ideas into practice. For the vernacular Communion on Sunday mornings the altar was moved to the western side of the chancel. The minister probably stood behind it, facing the people. Because of acoustic difficulties in huge medieval churches it became customary to pull the altars into the nave near the pulpit, and the service, apart from the actual Communion, was conducted from the pulpit. The baptismal font was placed near the altar and pulpit so Baptism could be administered in front of the entire congregation.[40]

The most striking event in Edward VI's reign from a liturgical standpoint was the movement officially led by Ridley, and suggested originally by Hooper, to abolish high altars of stone and replace them with wooden communion tables. Altars in chantry chapels had already been removed by the order for the dissolution of the chantries in 1548. Individual iconoclasts, without legal authority, had virtually been incited to take the law into their

---

[39] Hooper, *The Early Writings of John Hooper*, ed. S. Carr, pp. 491-92.
[40] This paragraph conflates material drawn from G.W.O. Addleshaw and Frederick Etchells, *The Architectural Setting of Anglican Worship*, pp. 23-24; and W. D. Maxwell, *John Knox's Genevan Service Book, 1556*, pp. 36-39, 46-47, 115-16.

own hands following Hooper's Lenten sermon before the court in 1550, during which he had hoped that the magistrates would be pleased "to turn the altars into tables, according to the first institution of Christ."[41] Bishop Ridley, in the injunctions for the diocese of London put out in the summer of 1550, exhorted curates, churchwardens, and others in authority "to erect and set up the Lord's Board, after the form of an honest table, decently covered."[42] Ridley argued that altars were appropriate for the Jewish sacrifices of the Old Testament, but as they had been brought to an end by Christ's one all-sufficient sacrifice on the cross, so also should altars cease to be used; in their places should be tables suitable for celebrating the Lord's Supper in which Christians spiritually feed on Christ. Ridley believed that the table agreed with Christ's institution of the sacrament and the practice of the apostles and the primitive Church.[43]

Addleshaw and Etchells, however, pointed out that the communion table had one great practical advantage over an altar: its mobility.[44] It could be placed in that part of the church where the communicants could best see and hear, and the most "convenient" place from 1550 to 1553 for the position of the communion table was the choir or chancel. Furthermore, Addleshaw and Etchells observe that the rubric in the First Prayer Book, which ordered the communicants to move into the chancel at the Offertory and to stay there until the rest of the service was still obeyed, but "instead of kneeling with their faces toward the Lord's board fixed in the form of an altar against the east wall, they now knelt all around it set in the form of a table in the middle of the chancel."[45]

The more radical Communion arrangements of the Edwardian reformers were not retained in the relatively conservative Elizabethan settlement of religion. The two-room medieval plan of architecture and worship was retained. Although at the beginning of Elizabeth's reign the great rood and its figures were taken down and the doom painted on the tympanum was whitewashed over, where they had survived or been restored, the chancel screens were not destroyed, thus indicating that there was to be a partition between chancel and nave. The communion tables were to remain as substitutes for stone altars and were to be covered with "a fair linen cloth" reaching to the ground on all four sides. The Royal

41 Hooper, *Early Writings*, p. 488.
42 *Visitation Articles*, ed. Frere and Kennedy, II, 242-44.
43 Nicholas Ridley, *Works*, ed. H. Christmas, pp. 322-23.
44 *Ibid.*, p. 27.   45 *Ibid.*, p. 28.

Injunctions of 1559 indicate that the table was to remain where the altar had hitherto stood, "saving when the Communion of the Sacrament is to be distributed," and on that occasion was to be put "in good sort within the chancel, as whereby the minister may be more conveniently heard of the communicants in his prayer and ministration, and the communicants also more conveniently, and in more number communicate with the said minister."[46] The Interpretations of the Bishops of 1560-1561 specified that the communion table was to be moved into the nave in front of the chancel door, "where either the choir seemeth to be too little or at great feasts of receivings."[47]

The whole process of adapting medieval churches for Anglican worship is admirably summarised in the words of Addleshaw and Etchells: "The process by which medieval churches were adapted for Prayer Book worship might be summed up as one of taking the communicants into the chancel for the Eucharist, so that they can be within sight and hearing of the priest at the altar; and of bringing down the priest from the chancel into the nave so that he could be amongst his people for Morning and Evening Prayer."[48] Though the one-room plan of an "auditory" church, which Wren was to use for his churches a century later, was more logical for Anglican worship, the Elizabethan arrangement had two advantages, quite apart from the continuity of using a Gothic building for Anglican worship. The first was that moving to the chancel for the Communion service seemd to give the Sacrament a special sacredness, which has been strongly emphasised through most of Anglican history; the chancel screen helped to separate the liturgy of the catechumens from the liturgy of the faithful, thus imparting to the climax of worship a sense of deep mystery.

5. Communion Plate

With the disappearance of shrines, crucifixes, reliquaries, ornate candlesticks, and elaborately chased chalices, the goldsmiths and silversmiths could hardly have made a living if they had depended solely on ecclesiastical patrons. The silversmiths, at least, were to spend a considerable part of their time in changing Roman Catholic chalices into ampler Anglican communion cups. Though there was no prescribed pattern for Edwardian commun-

[46] Gee and Hardy, *Documents illustrative of English Church History*, pp. 439-41.
[47] *Visitation Articles*, III, 62.
[48] Addleshaw and Etchells, *Architectural Setting*, p. 45.

ion cups, it is interesting that seven examples that survive are virtually identical in shape, having a bell-shaped bowl, a spool-shaped stem with a moulded rib round the waist, and a moulded base.[49] An excellent example of this type of communion cup is to be found in St. Mary's Church, Beddington.[50] Of even greater importance is the 1549 St. Mary Aldermary communion cup on loan to the Victoria and Albert Museum in Kensington, which originally was probably made for the Royal Chapel, since it bears Edward VI's royal arms.[51] Its great importance is that it has a cover which, when reversed, becomes a paten. The paten cover was to become a leading characteristic of Anglican communion plate for a century and a half after the creation of this the first surviving example.

The typical Elizabethan communion cup does not, however, have the shape of an inverted bell for its bowl, but the shape of a beaker. Its stem is spool-shaped, usually with a moulding round the middle, and it has a stepped and moulded base. The slightly domed cover that usually accompanied it had a flat knob which served as a foot when used as a paten.

The most significant fact about both Edwardian and Elizabethan communion cups, however, is their size as compared with the Roman Catholic chalices which they replaced and which often were smelted down to help make communion cups. The reason for this is that communicants at the Catholic Mass received the Host in the form of consecrated wafers, but not the consecrated wine, with the exception of the celebrating priest, for whom a small chalice was sufficient. Anglican Communions were, however, in both kinds,[52] and it was necessary to have a communion cup large enough to enable all communicants to partake, as well as a paten on which the consecrated wafers would stand. But when bread replaced wafers, larger patens became necessary.[53]

The decision that every English church had to have its chalice converted into a communion cup appears to have been Archbishop

---

[49] See Charles Oman, *English Church Plate, 1597-1830*, p. 192.

[50] The Beddington communion cup and paten of 1551 are illustrated in plate 49 of Oman's volume.

[51] Oman, *English Church Plate*, plate 50, illustrates the cup and paten-cover. The earliest surviving Edwardian cup is that of the church of St. Lawrence Jewry of 1548, which is shown on plate 52A in Oman's book.

[52] As from May 1, 1548.

[53] The Puritans insisted on household bread instead of wafer bread, and eventually their view prevailed. For the controversy see Archbishop Matthew Parker's *Correspondence*, eds. J. Bruce, T. T. Perowne, p. 478; and *Zurich Letters*, ed. Hastings Robinson, I, 248.

Parker's, with the probable assistance of Grindal. It seems rather strange that so tolerant a Primate should have taken so decidedly Protestant a step. Charles Oman makes the interesting suggestion that this decision was a means and not an end, a means to force ambivalent clergy to decide between Catholicism and Anglicanism. "It was," he says, "notorious that in the early years of this reign there were many clergy who were prepared not only to celebrate the Communion at the parish church but also to say Mass up at the manor for those who preferred the old service. The conversion of the parish chalice into a communion cup would help to curb the activities of these unreliable individuals."[54] In spite of this ingenious hypothesis it should be pointed out that the number of ambivalent priests was never large. Besides, there were other ways of curbing them; and after the Second Prayer Book of 1552 there was little possibility of mistaking Anglican Communion doctrine for Transubstantiation. But it is quite clear that the use of cups instead of chalices, combined with the vernacular liturgy and the Communion in both kinds, meant that Holy Communion was plainly distinguishable from the Mass.

Edmund Guest, the Bishop of Rochester, provided the earliest diocesan injunctions for the conversion of chalices into cups in 1565:

> Item, that the chalice of every parish church be altered into a decent cup therewith to minister the Holy Communion, taking away no more thereof but only so much as shall pay for the altering of the same into a cup. And the said cup to be provided in every parish within my said diocese by or on this side of the feast of Saint Michael the Archangel next coming after the date thereof.[55]

The symbolism on the silver communion cups was of the simplest character, especially when compared with the rich variety of medieval designs on chalices. On Edwardian and Elizabethan church plate we find only the sacred monogram, usually depicted with the letters I H S with a cross above and three nails below. On Recusant plate there was a growing tendency to add a heart beneath the nails.[56]

With this Anglican simplicity should be compared the splendid

[54] *English Church Plate*, p. 135.
[55] Frere and Kennedy, *Visitation Articles*, III, 162.
[56] Oman, *English Church Plate*, p. 225.

liturgical equipment used by the Catholic Recusants, as reported in John Gerard's autobiography. After listing vestments, Gerard continues: "Six massive silver candlesticks stood on the altar, and two smaller ones at the side for the elevation. The cruets, the lavabo bowl, the bell and thurible were all of silver work; the lamps hung from silver chains, and a silver crucifix stood on the altar. For the great feasts we had a golden crucifix a foot high. It had a carved pelican at the top, and on the right arm an eagle with outstretched wings, carrying on its back its little ones, who were learning to fly; and on the left arm a phoenix expiring in flames so that it might leave behind an offspring; and at the foot was a hen gathering her chickens under her wings. The whole was worked in gold by a skilled artist."[57] The Anglican church would not know such splendour until the influence of Bishop Lancelot Andrewes was felt through Archbishop Laud and others in the nascent High Church tradition early in the next century.

6. *Church Furniture*

Anglicans would probably have made do with the medieval arrangements of their churches had they not been provoked to consider change by the Puritans. We have seen that larger patens became necessary parts of communion plate because the Puritans insisted on the use of household bread instead of the wafers which could conveniently be contained in a cover-paten. Anglican priests might have been content with their fonts at the west end of their churches, fittingly symbolising the entry by Baptism into the Christian church.[58] The Puritans, however, objected because the congregation could not as a whole see and hear the service of Baptism, which was particularly important to them as signalising the covenant of God with His people and their children; they did not allow godparents to take away from the congregation its rights as cosponsors of the child with the parents. Even on grounds of convenience the Puritans had a point, especially when the Baptism took place in a crowded nave after the second lesson at matins or evensong, for only those in the immediate vicinity of the font could see and hear the Sacrament. The Puritan demand was not,

[57] John Gerard, *The Autobiography of an Elizabethan*, tr. from the Latin by Philip Caraman, pp. 195f.

[58] Bishop Richard Montagu, appointed to Norwich in 1638, is credited with the statement that the font should be placed in the West end of the church, "in order to signify our entrance into God's church by baptism" (*First Report of the Royal Commission on Ritual* (1867), p. 580b, cited by Addleshaw and Etchells, *Architectural Setting*, p. 65).

however, met by the Church of England; it was only in their own meetinghouses that they were able to use their portable fonts, which were attached to pulpits when needed or set permanently near the pulpit or communion table in accordance with the customs of the Reformed churches in Scotland and on the Continent.

The reading desk became an important article of furniture in the Church of England, primarily because on most Sunday mornings the entire service was conducted from it and the pulpit, except on Communion Sundays. It therefore had to be in a central position in the church where the priest could be heard best. Sometimes the desk stood alone. Sometimes it had the clerk's seat in front of it. Sometimes it was combined with the pulpit and clerk's seat to make a single piece of furniture. A combination of reading pew and pulpit became a pulpit with two storeys, of which there are extant examples in Halston Chapel, Shropshire and St. Ninian's, Brougham, Westmorland. When the clerk's seat, the reading desk, and the pulpit were combined, the "three-decker" was produced, of which there are extant examples in the churches of Parracombe, Devon; Kedington, Suffolk; and Coxwold, Yorkshire.[59] When the reading pew was a separate piece of furniture its position was on one side of the screen or chancel arch, or in a vast nave, where it might be placed against the first pillar; opposite it was the pulpit.

The desks had candlesticks attached to them. Pulpits often had an hourglass and at the back a wooden peg on which hung the parson's black gown which he wore while preaching. The clerk's seat occasionally had a music stand, since in parishes without any official choir he led the singing himself with a pitch pipe. There might also be a psalm-board attached to the pulpit.

The pews, of two types, were arranged about the pulpit and reading desk. The most common were the low-backed straight pews, or benches, many of which, especially in the west country, had carved ends. There are particularly fine examples of Renaissance heads in the carved benchends in Talland Church, Cornwall (c. 1537-1547),[60] and of Elizabethan roses and profiles in Milverton church, Somerset. The other type of pew developed in and popularised in the seventeenth century was the "box pew." It contained a small rectangular enclosure, with a seat on three sides and an opening on the fourth, with high sides and backs which permitted only the heads of the occupants to be seen. It has been

[59] *Ibid.*, p. 75n.
[60] Basil Clarke and John Betjeman, *English Churches*, plate 107.

said that their introduction was due to the desire of the Puritans to avoid such "high" liturgical gestures as bowing at the *Gloria* and at the name of Jesus, or turning to the east for the Gospel and the Apostles' Creed, without being detected by Laudian spies. The real reason was probably the purely functional one of enabling the pew's occupants to avoid draughts. Such distinctive and comfortable pews may also have carried a certain social cachet for their occupants, especially when the affluent were able to afford adding rails with curtains along the tops of these boxed pews for greater protection against draughts.

7. *Anglican Adornment of Churches*

It will come as no surprise that the adornment of Anglican churches was simple, even austere. Such one would expect from the iconoclasm of Edwardian days, but one might not have anticipated the erosion of neglect under Elizabeth. The neglect may even have been encouraged by the Puritan dislike of all aestheticism, which was guilty of association with Roman Catholicism. That, at least, was the implication of John Howson, chaplain to Queen Elizabeth and future Bishop of Oxford, in a sermon from Paul's Cross in 1598. Howson charged that covetous country gentry conveniently object to "idolatry" as a mask for their depredations. The result was, he complained, that "in Countrey Villages . . . the Churches are almost become . . . little better than hogstyes; for the best preparation at any high feast is a little fresh straw under their feete, the ordinary allowance for swine in their stye . . . and in cities and boroughes they are not like the Palaces of Princes as they were in the primitive church, but like a countrey hall, faire white limed, or a citizens parlour, at the best wainscotted; as though we were rather Platonists then Christians, who would neither have gold nor silver in their churches because it was *individiosa res*, and gave occasion to sacriledge."[61] These "irreligious Julianists" make a slender allowance to God, "like a Stoicks dinner, or Philosophers breakfast," consisting only of bare walls and roof to keep out the rain. Ironically, he observed, "Neither is it lawful to add any ornament . . . except perchance a cushion and a wainscot seate, for ones owne ease and credite."

Whatever the reasons—including iconoclasm, neglect, and con-

---

[61] *A Second Sermon preached at Pauls Crosse the 21. of May, 1598* (1598), sig. D3, cited in Millar Maclure, *The St. Paul's Cross Sermons, 1534-1642*, p. 133.

venience—the fact remains that Elizabethan churches were simple and bare compared with their appearance in the Middle Ages. The stained glass windows had not entirely disappeared, although the representations of the Deity and the saints had been removed, and increasingly clear glass was replacing coloured panes. The rood-lofts, with their carved and painted figures of the Christ, the Virgin and St. John, had been pulled down, and the painting of the doom on the tympanum had been whitewashed over. So also were all of the wall paintings of scenes in the life of the Virgin, or of St. Christopher bearing the Christ child. The pillars of the nave had had wooden or stone brackets with saints standing in the niches; all were torn out.

What was to replace all this aesthetic richness? Undoubtedly the two Gospel sacraments, with the affusion of water and the breaking of the bread and the libation of the wine, would stand out in solitary glory and not be lost in the fussy sacramentalia of medievalism. But were there no additions to the equipment of churches, apart from the utilitarian cover-patens and large communion cups, by the Tudor Church of England?

There were two: painted wooden boards bearing the Royal Arms, the Lord's Prayer, the Creed, and the Tables of the Ten Commandments, affixed by royal commandment, and the sculptured tombs placed there by rising families. The royal arms were generally set up on the tympanum where the rood had formerly stood, a most unfortunate position, since it implied that the worship of the sovereign had replaced the worship of Christ, a point not lost to Catholic critics of the Church of England such as Harding, and used in his controversy with Jewel. It was believed that the painted wooden boards gave a good indication of the ways of the Church of England, whose supreme governor was the sovereign, and which regarded the Lord's Prayer as a summary of devotion, the Decalogue as a summary of behaviour, and the Apostles' Creed as a summary of belief.

The Royal Arms, with their exotic animal supporters, the golden lion and silver greyhound, added a greatly needed touch of colour to the nave. In fact, although it accented the Erastianism of the Church of England, it was quite common in other Reformed and un-Reformed churches on the Continent to include the arms of monarchs or of local nobles in the churches.

The painted boards served a triple function. They demonstrated that the building was a church and was to be used for sacred purposes; they informed the worshippers about the essentials of the Christian way of life; and they provided attractive decoration.[62]

The provision of sculptured tombs in churches had a much more secular intention, and was a particularly interesting Elizabethan phenomenon. Sculptors found a livelihood in fashioning the representations of the saints in medieval churches, and even of the tombs of monarchs, queens, prelates, and nobles and their ladies. With the advent of the Reformation, however, the wealth and iconographical possibilities of the old faith largely disappeared from England. The newly wealthy classes could not express their religious feeling in the old manner by the erecting or adornment of churches; they expressed it in terms of ancestor worship, itself an index of their own wealth and worth. They had to be large to be impressive, for the tomb-makers of the period were mainly uninspired. It was only in size, material, and richness of decoration that tombs could be distinctive, not in design or quality of carving. "By about 1600," writes Eric Mercer, "it was almost *de rigueur* for a landed family to have a series of magnificent tombs in the local church."[63]

So important was the concept of tombs as family properties by the middle of the sixteenth century, that the government could not enforce their desire for religious uniformity in reference to mortuary sculpture and its type of decoration. In Sussex there is a series of tombs with representations of religious scenes and themes abhorrent to the Reformers; they extend in date from the dissolution of the monasteries to the accession of Elizabeth. In old Selsey church there are representations of St. George and the Dragon and of the martyrdom of St. Agatha on the tomb of John Lews, who died in 1537, and his wife, Agas. Richard Sackville's monument, c. 1540, in West Hampnett, depicts the Trinity with the naked Son reclining lifeless in the arms of the Father, while John Gounter's monument of about 1557 at Racton shows the risen Christ.[64]

Medieval tombs often showed cadavers in shrouds, to remind the beholder that the body is only the fragile and corruptible prison of the soul. The medieval tomb with its plea for a prayer for the soul of the departed was much humbler than the inscription on the Elizabethan tombs, reciting the deceased's ancestry and achieve-

---

[62] See Edward Cardwell, *Documentary Annals of the Church of England*, 2 vols., I, 262; a royal letter of 1560 states the triple purpose of the required tables of the commandments to be set up in the east end of churches.

[63] *English Art, 1553-1625*, p. 218.

[64] These examples are provided by Mercer, *English Art*, pp. 219f.

ments. The medieval *Orate pro anima* was changed to *Memoriae sacrum*.[65] Elizabethans had dismissed the Catholic doctrine of purgatory as unscriptural, and prayers for the dead were thought to be superstitious; so the spectator was not to think of the deceased as he now is, but as he was in the prime of life. This the effigy portrait helped him to do. These Renaissance men and women are shown at the crest of the wave of life. The man is depicted in his robes of office and the woman in her matronly dignity. As the fashions changed they were to be shown, not lying down with their hands folded in prayer and pious resignation, but very much alive —kneeling with their progeny in prayer or even sitting as if in conversation in parallel niches. They were painted richly and gilded, to give them verisimilitude. Furthermore, by heraldry and inscriptions there was an endeavour to stress less the death of an individual than the survival of the family.[66]

It is also interesting that although prayers were no longer to be offered for the dead in the Elizabethan Church of England, the tombs of the dead continued to bear the shape of altars at which such prayers could be offered. The most popular design showed husband and wife facing each other across a prayer desk on their knees, with their sons and daughters behind or below them. Other figures on the tombs turned toward the visitor in the church, which seemed, at least to the dramatist Webster, unseemly and irreverent, as he records in *The Duchess of Malfi*:

> Princes' images on their tombs
> Do not lie, as they were wont, seeming to pray
> Up to heaven, but with their hands under their cheeks
> As if they died of the toothache; they are not carved
> With their eyes fixed upon the stars, but as
> Their minds were wholly bent upon the world,
> The self-same way they seem to turn their faces.[67]

Here, clearly, Elizabethan sacred art had become secular.

8. *Religious Influences on Domestic Art*

In the change in the character and inscriptions of tombs we have seen one example of the influence of religion—the negative influence of Protestantism causing disbelief in purgatory and of the utility of praying for the dead. Are there others, comparable

[65] John Buxton, *Elizabethan Taste*, p. 136.
[66] *Ibid.*, p. 142.   [67] Act IV, scene ii, ll. 165-70.

to the restrictions imposed by Byzantine religion on art, forbidding sculpture because of its three dimensionality, banning all drama except that of the Liturgy itself, and prohibiting instrumental music, yet inspiring icons, mosaics, murals, and miniature ivory carvings?[68]

Certainly another negative influence was that of forbidding the painting on walls or windows of churches of representations of the Trinity and the saints, and of any incidents involving them. It helped to produce two positive effects, insofar as the Renaissance and the Reformation went together a considerable way. One positive effect was to drive painters to concentrate on portraits of the living with all the distinctive traits of an unique personality. While the Renaissance celebrated the man of distinction, the Reformation stressed the calling, significance, and dignity of every man as a cooperator with God. It is probably not without significance that the most private form of portrait painting, aiming at catching the secret, intimate, and personal life of the sitter so disregarded in the official oil paintings, was the miniature, or that Nicholas Hilliard, England's greatest miniaturist was a Protestant of the Protestants who had been a Marian exile.

A second influence Protestantism had on the development of art in England was to give a preference in domestic paintings, frescoes and tapestries, less for the pagan themes of a neoclassical art than for Biblical subjects. Moreover, with the growing ethical impact of Protestantism, and its delivery of individuals from a dependence on the mediation of the saints or on the priesthood in their approach to God, art turned from portraying the Madonna and the saints to the Biblical exemplars of faith.

Mural paintings are rare, because frescoes did not fare well in the damp English climate, and because there was probably a preference for either the cheaper painted cloths or the warmer tapestries on the walls. There are, however, in Queen Hoo Hall in Hertfordshire wall paintings of King Solomon worshipping false gods, and at Harvington Hall in Worcestershire can be detected the faint remains of battle scenes from the Old Testament.[69]

At the turn of the century and from 1600 to 1620, interior decoration of halls and houses was concentrated on the classics, on scientific interests, or on religious themes.[70] There were several

---

[68] Ernst Benz, *The Eastern Orthodox Church, Its thought and life*, tr. Richard and Clara Winston, from *Geist und Leben der Ostkirche*, p. 1.
[69] Buxton, *Elizabethan Taste*, p. 93.   [70] Mercer, *English Art*, p. 124.

causes, including the continuity of religious themes from the Middle Ages to the end of the Tudor age. Such are to be seen in a plaster ceiling with a tree of Jesse found in a house in the Butterwalk at Dartmouth, or a scene reminiscent of the Last Judgment on the chimney piece in the Hall at Burton Agnes.[71] Protestantism brought the vernacular Bible into wide use, which was another important cause for the increasing fondness for Biblical decoration. There were favourite scenes from the Old and New Testaments incorporated in domestic decoration, such as Daniel in the den of lions at Stockton House, Wiltshire, or Samson and the gates of Gaza at the Old House in Sandwich. At this time, although politics were no longer shaped by religious criteria, ethics were still based on the appeal to Christianity, another reason for the popularity of religious themes in decor. New Testament themes were late on the scene. When they did come, they had a strong moral didacticism inherent in the selection of the theme. For example, the most popular themes were Dives and Lazarus, the Return of the Prodigal Son, the Good and the Bad Steward. It is the same ethical impulse that was responsible for the innumerable emblematic figures, especially of the virtues, which were displayed on ceilings, friezes, and overmantels. It is interesting that halls and mansions of any pretension which had mural decoration were almost always classical in theme, but that such small houses as Knightsland Farm at South Mimms (c. 1590), and the Swan Inn at Stratford (c. 1580) used Biblical and moral subjects.[72] The implication is that the gentry and merchants preferred Biblical and moral subjects for the decoration of their homes, while the courtiers were equally interested in the classics and the Bible.[73] One may suppose that some of the gentry and merchants were Puritans, who might find such representation distracting in church but didactic at home.

Such a supposition seems less wide of the mark when it is recalled that the leader of the Elizabethan Puritan party, the Earl of Leicester, had a notable collection of paintings at Leicester House in London, including about 50 portraits, a group of historical paintings, and several religious paintings. The later group included a *Nativity* ("Christ how he was born in an ox-stall"), a *St. John the Baptist preaching in the Wilderness*, a *Persecution of Saul* [is a *Conversion* of Saul intended, at the time of which the risen Christ is reported to have said, Saul, Saul, why persecutest thou me?], a *Mary Magdalene*, and such Old Testament scenes as

71 *Ibid.*   72 *Ibid.*, p. 128.   73 *Ibid.*, p. 126.

Elijah taken up in the Fiery Chariot and *Noah and the Flood*. In other houses Leicester also had a *St. Jerome*, a *Deposition* from the cross, and a *St. John the Baptist* beheaded.[74] It is tantalising that we do not known who the Italian and Dutch painters of the works were. It is even more tantalising that we know so little of what other art collectors had gathered. But it is enough to show that although England had no artists of the highest quality in the visual arts, with the possible exception of Hilliard, her poets and dramatists were able to show that her imagination was not fettered by Protestantism, and, indeed, found the struggle between the old and the new faith, as between faith and science, issues worthy of the plays of Shakespeare and of the poems of Donne in which agape and eros fight over again the drama of religions and the Renaissance.

The chapter opened with a question which can now be answered, however tentatively. In matters of religious architecture English Protestantism was exceedingly conservative, building, rebuilding and adding exclusively in the Gothic style, but also adapting the Gothic interiors to the vernacular worship and without radical transformations. In liturgical furnishings there was a new functionalism and simplicity, and even a utilitarianism. In ecclesiastical decoration there was iconoclasm and almost total colour blindness, the carving of Tudor benches and choir stalls being the only general exception. In the matter of tombs secularity prevailed. The total effect was of the simplification of religious worship and instruction in an appropriately simplified church interior. The austerity was partly a conviction and partly the result of neglect. Surprisingly, while religious art was banished from the church, it made its way into the homes of the people who had the taste for it and who could afford it.

In the succeeding age High Anglicans would be the perceptive patrons of a richer ecclesiastical architecture and art, which is proof that Elizabethan religious art was not dead, but, like the daughter of Jairus, only sleeping.

74 Buxton, *Elizabethan Taste*, p. 100.

CHAPTER XI

CHURCH MUSIC

TWO STATEMENTS about sixteenth-century church music, itself a controversial subject, can be made. The first is that there was a great need for choral reform because of the chaos in musical practice. This was as readily agreed to by Roman Catholics as by Anglicans, Lutherans, and Calvinists. Second, there was disagreement about the best way to reform musical practice. Lutheran hymns and chorales, Anglican anthems and chants, and Calvinist psalmody were three different ways of meeting the need for change.

The early sixteenth-century situation in choral composition and singing is described by Stanford and Forsyth in writing of the time when unrelated themes were intermixed in the counterpoint in church music: "When the world turned topsy-turvy, and people first realized that music was not carpentering in 3-inch lengths, a sort of licentious orgy of music set in. It is difficult to explain with reverence just what happened. And if you want a modern analogy with the state of church music at that time, you may imagine one of our composers taking as his base an Anglican chant, and spreading it out so that each note occupied three or four bars; then for his treble using 'Take a pair of sparkling eyes' (*allegro con molto*), and for his alto part fitting in as much as he could of 'Tipperary' or 'Onward Christian Soldiers,' or both. . . . It has been described by contemporary sufferers, and if half what they say is true, it must have been like rag-time gone mad."[1] Further evidence of the seriousness of the situation is seen in the fact that the Council of Trent, in its meeting between 1545 and 1547, was disposed for a time to exclude music altogether from the service of the Roman Catholic church.[2] This drastic step was not taken, partly because of the worshipful and aesthetic music being produced for the Church by such Catholic musicians as Palestrina (himself one of the great figures of the Counter-Reformation), Orlando di Lasso, and Vittoria, and partly also because the Roman church refused to employ the Protestant remedy of vernacular hymns. In regard to the latter, however, it is interesting to recall that at the Council

---

[1] Charles V. Stanford and Cecil Forsyth, *The History of Music*, p. 138.
[2] Millar Patrick, *Four Centuries of Scottish Psalmody*, p. xxi.

of Trent the Emperor Ferdinand sought permission for the use of German hymns, as Cardinal Lorraine did for French hymns, but the Papacy used all its authority to suppress vernacular hymnody.

1. *Reformation Remedies*

It is, of course, impossible to understand Reformation remedies for the people's praise without first considering the diagnoses of the ills put forward by the leading Reformers, Luther[3] and Calvin.[4]

Luther had a great love of music, believing it to be a direct gift of God. He defended it against the iconoclasts, whose hatred of the Roman Catholic Mass was so unreasonable that they would have suppressed hymn-singing and organ-playing in religious services. Luther wrote in the preface to Walther's first collection of hymns in 1524: "I am not of the opinion that all the arts should be stricken down by the Gospel and disappear, as certain zealots would have it; on the contrary, I would see all the arts, and particularly music, at the service of Him who created them and gave them to us."[5] Luther's aim was not to suppress the Mass but to bring it within the spirit and scope of the gospel as he understood it. He wished it to be translated from Latin into German so the entire congregation would understand the words that priests and choristers were singing. As early as 1523 Luther wrote in the preface of the *Formula Missae et Communionis pro Ecclesia Wittembergensis*: "I also wish as many of the songs as possible to be in the vernacular, which the people should sing during Mass either immediately after the *Gradual*, and immediately after the *Sanctus* and *Agnus Dei*. For who doubts that once upon a time all the people sang these, which now only the choir sings or responds when the bishop is consecrating? But these songs may be ordered by the bishops in this manner, they may be sung either right after the Latin songs, or on alternate days, now Latin, now the vernacular, until the entire

---

[3] The following works are selected as introductions to the subject of Lutheran music: Christhard Mahrenholz, *Das evangelische Kirchengesangbuch. Ein Bericht uber seine Vorgeschichte*; Mahrenholz's *Luther und die Kirchenmusik*; Paul Nettl, *Luther and Music*; and Basil Smallman, *The Background of Passion Music*. See particularly the introduction to Luther's hymnody in vol. 35 of the Weimar *Ausgabe*.

[4] For Calvinistic music, which had a greater influence on English congregational music, see the following sources: Pierre Pidoux, *Le Psautier huguenot du 16e siècle*, 2 vols.; R. R. Terry, *Calvin's First Psalter* (1539); interpretations: O. Douen, *Clément Marot et le psautier huguenot*, 2 vols.; Charles Garside, Jr., *Zwingli and the Arts*; and Percy A. Scholes, *The Puritans and Music in England and New England*.

[5] Cited in Théodore Gérold's "Protestant Music on the Continent," *The New Oxford History of Music*, vol. 4: *The Age of Humanism 1540-1630*, ed. Gerald Abraham, p. 419.

Mass shall be made vernacular. But poets are wanting among us, — or they are not known as yet, — who can put together pleasingly pious and spiritual songs, as Paul calls them, which are worthy to be used by all the people in the Church of God."[6]

Luther made three striking contributions to the reform of worship: the recovery of the sermon (his own were marked by Biblical fidelity, a profound understanding of faith, vivid illustrations drawn from the observation of common life, and a marvellously racy and idiomatic German speech); the restoration of Communion to the people; and the introduction of German hymns, some of them written by him and set to Luther's own music.[7] An interesting development of the hymn was the chorale, in which portions of Scripture (especially narratives of the Passion of Christ) were sung with different voices representing the narrator (or evangelist), the principal characters, and the people's response. This particular development is to be found in embryo in Luther's *Deudsche Messe und Ordnung Gottis Diensts*[8] of 1526.

Luther's reforms were based on four principles. Worship was to be ruled by the shape and spirit of the gospel of Christ; its primary purpose is to create faith; it must be intelligible to produce the maximum participation of the people; and no liturgy was to be a law and hindrance to Christian freedom.[9]

Calvin, also, saw the need to deliver the common people from the two languages unintelligible to them—Latin and polyphony—and the need for two media which had to be found or created by those insistent on reform. What was necessary, in short, was "words which the people could understand, cast in a form in which they could without undue difficulty read or memorise them; and . . . music, of a type which they would be able to sing."[10] The major Protestant churches solved the double problem in the same way; they used versified texts set to simple melodies. The difference between Luther and Reformed church music depended, however, on two factors. The first was that the models for spiritual songs were different in each case. Luther used the customary verse forms,

6 See *Liturgies of the Western Church*, selected and ed. Bard Thompson, p. 119.
7 Gérold, "Protestant Music," pp. 422-23, lists five hymns (including the great "Ein feste burg is unser Gott"), the melodies and words of which were by Luther.
8 *Liturgies*, p. 132.
9 The dimension of Christian freedom in Luther, expressed in his wishing to prevent any Wittemburg liturgy from becoming a new liturgical law, and the primacy of faith, are finely brought out in Vilmos Vajta's interpretation, *Luther on Worship*, pp. 174f.
10 Millar Patrick, *Four Centuries of Scottish Psalmody*, p. 3.

stanzaic in form, which the German people were familiar with in the Latin hymns and in the vernacular songs of the day, both secular and sacred, permitting a succession of verses to be sung to the same melody. Second, while Calvin restricted the subject matter to the Psalms, Luther also used New Testament material and started the churches following his leadership on the path to Christian hymnody.

In 1537 Calvin and Farel, the reformers of the Genevan church, designed a scheme for introducing singing into worship, but before the plan took effect they were expelled from Geneva. While spending four years as pastor to the French Protestant church in Strassburg, Calvin's convictions about the importance of the people's praise were reinforced by the admirable congregational singing in the Lutheran churches in Strassburg, where the practice was then a decade old. Consequently he published a small volume of four sheets of 16 pages each, with the title *Aulcuns Pseaulmes et Cantiques / mys en chant. A Strasburg 1539.* This historic volume contained 17 Psalms in metre, five of which Calvin is believed to have composed;[11] the remaining 12 were the work of Clément Marot. Calvin persuaded Théodore Beza to complete the work, and the entire Huguenot Psalter was published in 1562. Its brilliance is not only in the literary quality, especially of Marot's religious verse, but also in its unique "ingenuity in using every kind of structural device to render impossible the monotony so characteristic of the Psalters used in England, Scotland, and America."[12] Calvin argued that the Psalter was also the book of Christian, as well as Jewish, praise, inspired by the Holy Spirit; to depart from it in themes as well as words, except insofar as the exigencies of rhyme and metre dictated, was presumptuous. Thus the Calvinist influence restricted both style and content in Christian praise much more than the Catholic or Lutheran influences and examples did. On the other hand, the solemnity of some English versions of the Psalms, sung to the melodies of Louis Bourgeois, were incomparable in their devotional depth and expression of the majesty of God and the humility of man *coram Dei.* Others were merely jogging doggerel.

[11] Psalms 25, 36, 46, 91, 138 and the additional versions of the Song of Simeon and the versified Decalogue, according to Douen, *Clément Marot.*
[12] Patrick, *Scottish Psalmody,* p. 19; see also Waldo S. Pratt, *The Music of the French Psalter of 1562: A Historical Survey and Analysis.*

2. *Luther's Influence on English Worship and Praise*

Luther's chief influence was in the gradual acceptance of the vernacular in English worship. The acceptance was so gradual that the English Litany of 1544 used for processions was the only form of service in the vernacular to be approved during the 38 years of Henry VIII's reign. Henry wished to use it to inspire patriotism in his wars with France and Scotland. This was not possible if the people did not understand and share in the worship, "forasmuch as heretofore the people, partly for lack of good instruction and calling, partly for that they understood no part of such prayers or suffrages as were used to be sung and said, have used to come very slackly to the Procession."[13] When the English Protestants got the opportunity to introduce changes in worship during the reign of the young Edward VI, they set to work immediately to produce a vernacular liturgy, which appeared two years after Edward VI succeeded. The Protector, Somerset, however, sent visitors within six months of the coronation to end all Popish practices and ceremonies in the cathedrals and churches. Their instructions reveal a great concern for intelligibility in worship, as also for simplicity, as they insisted on informed teaching and preaching, the reduction of candles and bell-ringings at Mass, and the use of English processions and Litany instead of the Latin.

The specific directions given at particular cathedrals further reinforced the desire for a comprehensible and uncomplicated liturgical style. At Winchester the singing of sequences was forbidden, and before Mass and evensong each day chapters from the Old and New Testaments were to be read to the choristers. At Canterbury sung Lady Mass was abolished on holy days so that a sermon or reading from the homilies could be substituted. At York the choir was forbidden to sing more than one Mass per day or to sing responds. It was instructed to replace Latin anthems or responds with English substitutes; it was also recommended that the services of the lesser hours, dirges, and "commendations" be ended.

The more iconoclastic Anglican Edwardian clergy such as Cranmer's chaplain, Thomas Becon, criticised at once the unintelligibility, oversubtlety, and expensiveness of Catholic music: "There

13 Cited in F. E. Brightman, *The English Rite*, 2 vols., I, lix. During the outgoing procession the Litany was sung, as also the Our Father, versicle, response and collect; during the returning procession there was sung the Anthem ("O Lord, arise . . ."), and before the choir steps the versicle, response, collects, which were followed by the Mass.

have been . . . which have not spared to spend much riches in nourishing many idle singing men to bleat in their chapels, thinking so to do God on high sacrifice . . . but they have not spent any part of their substance to find a learned man in their houses, to preach the word of God, to haste them to virtue, and to dissuade them from vice. . . . A Christian man's melody, after St. Paul's mind, consisteth in heart, while we recite psalms, hymns and spiritual songs, and sing to the Lord in our hearts. All other outward melody is vain and transitory, and passeth away and cometh to nought. . . . So ye perceive that music is not so excellent a thing, that a Christian ought earnestly to rejoice in it."[14] In fact, the Edwardian order of the day was the requirement for virtue rather than musical competence, as the 30th Article of the Royal Injunctions for St. George's Chapel, Windsor (February 8, 1550) clearly demonstrated:

> *Also*, whereas heretofore when descant, prick-song, and organs were too much used . . . in the church, great search was made for cunning men in that faculty, among whom there were many that had joined with such cunning evil conditions, as pride, contention, railing, drunkenness, contempt of their superiors, or such-like vice, we intending to have Almighty God praised with gentle and sober quiet minds, and with honest hearts, and also the commonwealth served with convenient ministers, do enjoin from henceforth, when the room of any of the clerks shall be void, the Dean and prebendaries of this church shall make search for honest and quiet men, learned in the Latin tongue, which have competent voices and can sing, apt to study, and willing to increase in learning: so that they may be first deacons and afterwards admitted priests; having always more regard to their virtue and learning than to excellency in music. . . .[15]

The great simplification in church music that had come about in Edward VI's reign was indirectly an expression of Luther's concerns in worship. It is most conveniently seen in the difference between the 1549 and the 1552 prayer-books. The first Book of

---

[14] In Becon's *Jewel of Joy*, in *The seconde parte of the bokes which Thomas Beacon hath made* (1560), vol. 2, part 2, folio xiii. In the *Catechism of Thomas Becon*, ed. J. Ayre, p. 429, Becon stated: "To say the trueth, musycke is a more vayne and triefelynge science than it becommeth a man borne and appoynted to matters of gravitye, to spēde muche tyme about it."

[15] Frere and Kennedy, *Visitation Articles*, ii, 775-76.

Common Prayer required in the Liturgy the following eight items to be sung: the Introit, the Kyrie (optional), Gloria in Excelsis, the Creed, the Offertory, the Sanctus, O Lamb of God, and Post-Communion. The second prayer-book required only a sung Gloria in Excelsis. The intensification of anti-Marian feeling during the reign of King Edward meant also that the Tudor composers lost "one of their most characteristically intimate forms of devotion—the motet in honour of the Blessed Virgin."[16] Even John Marbeck's *Book of Common Praier Noted* (1550) was unable to rouse any great enthusiasm, although its "quasi-mensural notion of unison chant should have satisfied the most ardent among the reformers."[17] It certainly attempted to put into practice Cranmer's desire for a simplicity that would take the form of a note to a syllable. Moreover, even in the reign of Elizabeth the general church music seemed to be characterised by "drabness,"[18] compared to the brilliant poetry of the age and the rich quality of the secular madrigals.

Clearly the chief influence of Luther was early and indirect, since it came through Martin Bucer and Peter Martyr, Regius Professors respectively in Cambridge and Oxford in Edward's time, in their help in the provision of the first English vernacular prayer-book. Neither they nor their sponsors in England followed Luther's explicit recommendation that choral music, together with the new congregational music, should be cultivated in the reformed churches and schools. Consequently (apart from the late Elizabethan cultivation of anthems and services, both developed in cathedrals and collegiate chapels, and not in parish churches), the congregational music of the English Reformed church followed the Calvinist rather than the Lutheran model.[19] The Elizabethan development of verse anthems can hardly be regarded as due to Lutheran influence, since it came so late (some 40 years after Luther's death. In any case they were limited to cathedral and collegiate churches; there was no provision for congregational participation. For the praise in parish churches the Calvinist impact was paramount.

16 Denis W. Stevens, *Tudor Church Music*, p. 19.
17 *Ibid.*, p. 55.
18 This term is used by Frank L. Harrison in "Church Music in England," The New Oxford History of Music, vol. 4: *The Age of Humanism, 1540-1630*, ed. Gerald Abraham, p. 470, to characterise all Elizabethan parish church music, except the music in cathedrals, collegiate churches, and a few schools such as Eton.
19 *Ibid.*, p. 465.

3. Calvin's Influence on English Praise

Although the decisive influence leading the directors of parish praise to accept metrical psalmody was Genevan, the initial impact in England was Lutheran. About 1539 there first appeared Myles Coverdale's *The Goostly psalmes and spirituall songs drawen out of the holy Scripture, for the conforte and consolacyon of soch as love to rejoyse in God and his worde*. It had a Lutheran character, not only because Coverdale and Luther had been Augustinian monks, but because it drew heavily on the Lutheran originals for title, introduction, words, and music. It contained 13 metrical psalms, together with metrical versions of the Magnificat, the Nunc Dimittis, Lord's Prayer, Creed, Decalogue, and over a dozen German and Latin hymns. None of the settings was harmonised. These metrical psalms were intended to be sung by the "lovers of Gods worde" at their homes or at their work, instead of "balettes of fylthynenes," and by children "in godly sports to passe theyr tyme." Coverdale wrote: "Yee [a] wolde God that our mynstrels had none other thynge to playe upon, neither our carters and plowmen other thynge to whistle upon, save Psalmes, hymnes, and soch godly songes as David is occupied with all. And yf women syttinge at theyr rockes [distaffs] or spynninge at the wheles, had none other songes to passe theyr tyme withall, than soch as Moses sister, Elchanas wife, Debbora, and Mary the mother of Christ have song before them, they shulde be better occupied, than with hey nony nony, hey troly loly, & soch lyke fantasies."[20]

The origin of the influential Huguenot Psalter, to which English and Continental congregational praise owed so much, is paradoxical. Sir Richard Terry observed that there was no more ironic historical development than "the chain of circumstances which led to metrical psalmody beginning as the favourite recreation of a gay Catholic court and ending as the exclusive 'hall-mark' of the severest form of Protestantism."[21]

It was the great and growing popularity of metrical psalmody that matters for our purposes, rather than its secular origins. Still, as R. E. Prothero recounts the story,[22] when Marot's *Psalms* first appeared, they were sung to the popular airs of the day by Catholics and Calvinists. The Dauphin (the future Henry II) loved the

---

[20] Cited by Morrison Comegys Boyd, *Elizabethan Music and Musical Criticism*, pp. 39-40.
[21] *Calvin's First Psalter* (1539), ed. R. R. Terry, p. iii.
[22] *The Psalms in Human Life*, p. 137.

sanctes chansonnettes, singing them, setting them to music, encouraging the adoption of a particular psalm as a motto for each leading courtier. Henry sang Psalm 42 (Like as the hart desireth the water-brooks; *Sicut cervus* . . .) as he hunted the stag in the forest of Fontainebleau, riding by the side of his favourite, Diane of Poitiers, for the motto of whose portrait as a huntress he selected the first verse of his favourite psalm. In France, says Prothero, "the metrical version of the Psalter, in the vulgar tongue, was one of the principal instruments in the success of the Reformed church."[23] So much was this so, where the children were taught to learn the metrical psalms by heart, that to chant psalms meant in popular parlance to turn Protestant. They fired Huguenot soldiers, as later they were to encourage Cromwellian troops in battle; they were equally potent sources of spiritual strength in defeat or in humiliation. It is to Calvin's credit that he edited the first printed edition of metrical psalms for church worship. Through this influence metrical psalmody became the people's praise par excellence, not only in the parish churches of England, but also in their open-air demonstrations and in their homes.

Calvin's attitude toward the metrical psalms was best expressed in his preface to the Genevan edition of Marot's Fifty Psalms, together with a liturgy and catechism, in June 10, 1543, where he said: "Nous ne trouvons meilleurs chansons ne plus propres pur ce faire, que les pseaumes de David, lesquels le sainct Esprit luy a dictez et faits."[24] The French Psalter contained great poetry, sublime music by Bourgeois, Greiter, and Franc, for which Goudimel made four-part settings.[25] Bourgeois was also fond of singing in parts, which Calvin resisted, partly because of the printing expenses involved, and probably also because harmonic practice was then in a state of flux.[26] It may be guessed that it was the music that carried the Calvinistic theology into the hearts of the people through the powerful sense of faith and utter trust in God as expressed so personally in the metrical psalmody. This view led Robert Bridges to write: "Historians who wish to give a true philosophical account of Calvin's influence at Geneva ought probably to refer a great part of it to the enthusiasm attendant on the singing of Bourgeois's melodies."[27]

The English success of the metrical psalms, both in the eager-

23 *Ibid.*, p. 138.
24 *Ibid.*, pp. 140-41.
25 Gérold, "Protestant Music," p. 443.
26 Patrick, *Scottish Psalmody*, p. 19.
27 Cited in Percy Dearmer, *Songs of Praise Discussed*, p. 391.

ness with which versions poured out of the presses and in the use made of them, was comparable to the Genevan. John Jewel's testimony to their popularity in a letter to Peter Martyr sent from London on March 5, 1560 is impressive: "Religion is now somewhat more established than it was. The people are everywhere exceedingly inclined to the better part. The practice of joining in church music [the psalms] has very much helped us. For as soon as they had once begun singing in public, in only one little church in London, immediately not only the churches in the neighbourhood, but even in the towns far distant began to vie with each other in the same practice. You may sometimes see at St. Paul's Cross, after the service, six thousand persons, old and young, of both sexes, all singing together and praising God. This sadly annoys the Mass priests and the devil. For they perceive by these means the sacred discourses sink more deeply into the minds of men, and that their kingdom is weakened and shaken up at almost every note. . . ."[28]

Further evidence of the popularity of the metrical psalms was the many editions published during the Elizabethan period. Between 1560 and 1579 at least 20 different editions of Sternhold and Hopkins came out, while between 1580 and 1599 there were 45.[29]

4. Sternhold and Hopkins—The "Old Version"

Pre-Sternholdian psalmody is of little significance. Coverdale was too dependent on German sources and too Lutheran in doctrine to suit Henrician church needs, while the more academic experiments of the poets Surrey and Wyatt were no more suitable for common worship than the later versions of Spenser. It was Sternhold's version, completed by Hopkins, that best supplied the need for church and domestic psalmody in metre.

Sternhold produced his first edition, with 19 metrical psalms, probably in 1548, as a minor courtier. The third edition of 1551 contained 37 psalms of his and seven by Hopkins, the schoolmaster who supplemented his work. It was not until 1562, however, that John Daye published *The Whole Booke of Psalmes collected into Englysh metre, by T. Starnhold, I. Hopkins, and others.* Thereafter the ensuing editions of the Old Version were the composite

---

[28] *Zurich Letters*, II, 71. See a slightly different version of the letter in Jewel's *Works*, ed. Ayre, IV, 1,231.
[29] The figures are taken from E. B. Schnapper, *The British Union Catalogue of Early Music printed before 1801.*

work of Sternhold, Hopkins, Whittingham, Wisedome, William Kethe, and John Craig. They remained in vogue until the "New Version" of Tate and Brady supplanted them in 1698.

Influential as the compilation was, it was, alas, often no more than divinity couched in the sorriest doggerel. The droll Fuller rightly estimated that the versifiers had drunk more of Jordan than of Helicon, adding that two hammerers on a smith's anvil would have made better music.[30] Queen Elizabeth disliked the new "Genevan jigs," as did the poets. Edward Phillips, Milton's nephew and himself a minor poet and Cavalier, described someone as singing with a "wofull noise":

> Like a crack'd saints' bell jarring in the steeple
> Tom Sternhold's wretched prick-song for the people.

The witty and dissolute Earl of Rochester had little time for the Puritans or their psalmody, as the following squib suggests, when he heard a clerk lining out a psalm:

> Sternhold and Hopkins had great Qualms,
> When they translated David's Psalms,
>   To make the heart full glad:
> But had it been poor David's fate
> To hear thee sing and them translate,
>   By God 'twould have made him mad![31]

How poor, in fact, was the work of Sternhold and Hopkins? It was awkward and obvious, but at times capable of a sustained dignity, as in Psalm I, verses 3 and 4, which celebrate the righteous man:

> He shall be lyke the tree that groweth
>   fast by the river side:
> Which bringeth forth most pleasant frute
>   in her due tyme and tyde.
> Whose leaf shall never fade nor fall,
>   but florish still and stand:
> Even so all thinges shall prosper well
>   that this man takes in hand.

The inspired Pegasus is often hobbled, as in Psalm XVI, in a complaint against idolaters:

---

[30] Thomas Fuller, *The Church History of Britain* (1655), IV, 73.
[31] Both versified comments on Sternhold and Hopkins were derived from R. E. Prothero's *The Psalms in Human Life*, p. 113.

They shall heape sorrows on their heads,
    which runne as they were mad:
To offer to the Idols gods
    alas it is too bad.

Perhaps the greatest achievement was Sternhold's version of the "Old Hundredth," which begins: "All people that on earth do dwell." Hopkins reaches the greatest heights in Psalm XVIII:

The Lord descended from above
    and bowd the heavens hie;
And underneath his feete hee cast
    the darkness of the skie.
On Cherubs and on Cherubins
    full royally he rode
And on the winges of all the windes
    came flying all abrode.

These are lines worthy of being illustrated by William Blake.

Despite the frequency of Double Common Metre, the inversions, the very obviousness of the plain rhymes, they were easily memorised and fitted to simple tunes for ordinary worshipers. They were, in fact, used with great delight in church and at home and entirely fulfilled the hopes of the subtitle of the 1566 edition: "allowed to be soong of the people together, in churches, before and after morning and evening prayer: as also before and after the Sermon, and moreover in private houses, for their godly solace and comfort, laying apart all ungodly songes and ballades, which tend only to the nourishing of vice, and corrupting of youth."

Unfortunately the old modal tunes and most of those originating from Geneva, which had been conserved in the Anglo-Genevan Psalter of 1566, 1568, and 1560, were increasingly dropped from the successive editions of Sternhold and Hopkins. The return to a heavy style was due in all probability to the demand of the Reformers for a style of music encouraging distinct articulation of the words, with one note to a syllable, and also to Cranmer's requirement that the harmonising should be note against note—that is, in plain chords.[32] The greatest criticism that can be made of the English metrical psalmody, both words and music, was the monotony it induced. As W. H. Frere remarked: "The English tradition

---

[32] Patrick, *Scottish Psalmody*, pp. 39-40.

hardly ever got away from the jog-trot of D. C. M."[33] Its chief
advantage was that it provided for the people of the average parish
church an easily memorised set of rhymes and tunes, thus returning to the common people the privileges snatched from them by professional choirs singing complex polyphonic motets and anthems.

Sternhold and Hopkins did not have the field of metrical psalmody to themselves by any means. But no other compilation
matched theirs in popularity. Robert Crowley's *Psalter of David
newly translated into Englysh metre* (1549) was the first complete English metrical Psalter, and the first to contain harmonised
music, although it was restricted to a single chant-like tune. The
first Sternhold and Hopkins with music was published in Geneva
in 1556 as *One and Fiftie Psalmes of David*. In later editions of
this Anglo-Genevan Psalter (so-called because it was prepared for
the use of English Protestant refugees in Geneva under the joint
ministry of Goodman and Whittingham, also the editor of the
Geneva Bible), Whittingham and Kethe broke away from standard common and double metre and thus broke some of the monotony. In 1567 Archbishop Parker issued his psalter privately. In
the preface he explained that friends had persuaded him to publish
it along with the tunes Thomas Tallis had composed to go with it.
Parker urged that a sad tune accompany a sad psalm and a joyful
tune a joyful psalm. He then provided a versified reaction to Tallis's
tunes:

> The first is meek, devout to see,
> The second, sad, in majesty,
> The third doth rage, and roughly brayeth,
> The fourth doth fawn, and flattery playeth,
> The fifth delighteth, and laugheth the more,
> The sixth bewaileth, it weepeth full sore,
> The seventh tradeth stout, in forward race,
> The eighth goeth mild, in modest pace.

Tallis's brilliant and imaginative pupil, William Byrd, also produced several psalm tunes of great delight. In 1588 he published
Psalmes, Sonets and songs of sadness and pietie, which included
10 metrical psalms, 7 devotional songs, and 16 secular compositions. A year later there appeared his *Songs of sundrie natures*,
including three-part settings of the penitential psalms, and 8

[33] In the introduction to the historical edition of *Hymns Ancient and Modern.*
"D.C.M." is Double Common Metre.

anthems (3 of which were in verse form). The psalms were appropriately set in the minor modes. In 1611 Byrd published *Psalms Songs and Sonnets*, which included 2 fine verse anthems, "O God that guides" and the penitential prayer, "Have mercy upon me," as well as further psalm tunes of great richness and solemnity.[34]

The psalm tunes were the work not only of the masters such as Tallis and Byrd, but also of many minor composers of the day, who also paid tribute to the popularity of this simple musical genre. Among those who produced collections of psalm tunes were Cosyn (1585), Daman (1591), East (1592, which contained settings by Farmer, Kirkby, Allison, Farnaby, Dowland, Cobbold, Johnson, and Cavendish), and Barley (1599).[35] There was no lack of metrical psalmody or tunes for psalms.

One question remains: why was popular praise limited to metrical psalmody? Why were there so few hymns produced in the Tudor age in England and those not of a popular character? Certainly it was not for lack of archiepiscopal support. For a letter of Cranmer's, written on October 7, 1544 or 1545, to King Henry VIII shows that he had translated the Latin hymn *Salva festa dies* into English and expected the king to appoint a better translator.[36] Yet in the successive editions of the Book of Common Prayer the only hymns included were translations of the *Veni Creator* in the Ordinal. Translations of ancient hymns were no new phenomena, either. Rough translations of Latin hymns were frequent in the pre-Reformation Sarum primers; they were to be found in the primers of Henry VIII, which were scrutinised by Cranmer.

Did the growing influence of Calvinism repress the production of hymns?[37] This is a strong possibility, considering the influence of the returning exiles from Geneva and other Reformed centres at the crucial beginning of Elizabeth's reign, or the importance of Puritanism during the last three decades of her reign. What is indubitable is that the Sternhold and Hopkins Old Version quickly acquired an almost canonical authority, and dominated the field to the virtual exclusion of experiments on the Lutheran or Latin models, which had seemed promising only two decades before. Not only did the metrical psalms lie to hand, but it was only too

[34] Peter Le Huray, *Music and the Reformation in England, 1549-1660*, pp. 371ff.

[35] Percy Young, *A History of British Music*, p. 152.

[36] Cranmer, *Works*, ed. Jenkyns, II, 412.

[37] This is the view of C. S. Phillips, in *Hymnody Past and Present*, p. 153; and H.A.L. Jefferson, in *Hymns in Christian Worship*, p. 32.

easy to find tunes for common metre verses. Perhaps the decisive reason for the failure of hymns to develop was the strong sixteenth-century sense in nascent Protestantism of "the Bible and the Bible only" as the liturgical, doctrinal and moral criterion. To develop original hymns of Christian experience would have seemed a human impertinence when God had already provided approved forms of praise in the Psalms. If it were argued that the New Testament Canticles were also approved sources of hymnody, some Protestants would have answered that our circumstances do not match those of the Blessed Virgin in the *Magnificat* nor of Simeon in the *Nunc Dimittis*, so these songs are not suitable for repetition.[38] Even if the point were conceded, it could yet be argued that it gave Elizabethans no authority for making their own transcripts of Christian experience into parts of public worship.

The only traces of Tudor hymnody left in modern English hymnbooks are poems recently made to serve the purpose of common praise for which their high individual lyrical or meditative quality made them unsuitable. "Most glorious Lord of life" is a Spenserian sonnet which now appears as No. 283 in the *English Hymnal*, while Shakespeare's "Poor soul, the centre of my sinful earth" appears inappropriately in *Songs of Praise* (No. 622). Edmund Campion's moving song, "Never weather-beaten sail," is also included in *Songs of Praise* (No. 587). Three other poems, however, are included in *Songs of Praise* which seem more suitable as hymns—Campion's "Sing a song of joy" (No. 639); Thomas Gascoigne's "You that have spent the silent night" (No. 38); and Sir Henry Wotton's "How happy is he born and taught" (No. 524).[39]

For the best English Church music in Elizabethan times one had to go to the cathedrals, the collegiate churches, and the Queen's Chapel. There one could find music fit to match the brilliance of the poetry of the period.

5. *Music in the Great Churches*

Although great music was composed for the Church of England in Elizabethan and Stuart times, the Church played only a minor role in the general musical life of the country. Madrigals and masques were greatly preferred to motets. The choral music of

---

[38] This was, in fact, a point of the polemic of the Puritans against the use of the Canticles of the New Testament.

[39] Sir Philip Sidney's exquisite "Lord in me there lieth naught" is a paraphrase of Psalm 139 rather than a hymn. It appears in *Songs of Praise* (no. 605).

Elizabethan England was cultivated mainly by professional musicians in cathedrals and in a few other endowed churches.[40] As Walter Woodfill observed, "However much psalms were sung in the parish churches, the fine religious choral music of the age was virtually the monopoly of some two dozen cathedrals, royal peculiars such as Westminster, the principal colleges of Oxford and Cambridge, and schools such as Eton."[41]

It was in the cathedrals and abbeys of the old faith that the most glorious choral music was to be heard. Oddly enough, it was a loyal Catholic, William Byrd, who composed the best church music in Elizabethan days, who had protested in his will that he hoped "to live and dye a true and p[er]fect member of his holy Catholycke Church wthout wch I beeleve theire is no Salvation."[42] For a time it looked as if the rich and ancient tradition of choral singing might die out. Peter Le Huray, for example, thought the Protestant iconoclasts of Edward VI's reign might have swept it away, had not the Catholic Mary rescued it by her accession. "Is it," he asked, "too fanciful to see in the accession of a Catholic monarch, its ultimate salvation?"[43]

Several divines were anxious to revise the Book of Common Prayer in a more Protestant direction, especially Archbishop Holgate of York, Bishop Ridley of London, and Bishop Hooper of Gloucester. Their view was that choral and instrumental music was of little value in worship, and they reduced it where they could. The Royal Injunctions for St. George's Chapel, Windsor (February 8, 1550), were straws in the wind of iconoclasm:

> Article 27: Also, because the great number of ceremonies in the church are now put away by the King's Majesty's authority and act of Parliament, so that fewer choristers be requisite, and the College is now otherwise more changed than it hath been, we enjoin from henceforth there shall be found in this College only ten choristers. . . .[44]

This was, in fact, extending the process of reducing the choral establishments which had begun with the spoliation of the monasteries and continued with the abolition of the chantries.

---

[40] Woodfill, *Musicians in English Society from Elizabeth to Charles I*, p. 153. See also the authoritative monograph by Edmund H. Fellowes, *English Cathedral Music from Edward VI to Edward VII*.
[41] Woodfill, *Musicians*, p. 156.
[42] Cited in Young, *British Music*, p. 144.
[43] Le Huray, *Music and the Reformation*, p. 29.
[44] Frere and Kennedy, *Visitation Articles*, II, 224.

Unquestionably the Catholic services had given the composer greater and more frequent opportunities for choral compositions than the Protestant composer ever had. The imposing ceremonial of High Mass in the Sarum Use demanded music of great dignity and rich complexity. Every third day was a festal occasion, and every day the Daily Office (especially matins and vespers) generated a need for solemn settings for antiphons, hymns, responsories, and canticles. Choirs were now smaller than in pre-Reformation days, partly because of the royal confiscation of ecclesiastical lands and revenues and partly because of inflation. Even so, the great churches often retained a choir of 20 voices; where they were thinner, they were helped by organs, lutes, and hautboys. In a pinch the choirmaster could manage with 16 voices, because the norm of later Tudor texture was to require five-voice parts.[45]

The basic question is, how much of the venerable choral tradition inherited from Catholicism, which had been repudiated so decisively in the parish churches by the satisfaction with psalmody of a metrical kind, survived in the great churches in the reigns of Edward and Elizabeth? It must be made clear at the outset that there were large problems to be faced, even supposing there had been the financial and musical resources necessary to maintain the tradition. It was difficult, though not impossible, to fit English texts to music previously composed for Latin texts, for English is a stress language and Latin a quantitative language. The difficulty had, indeed, been surmounted in paraliturgical ways by the provision of vernacular carols for Christmas and Easter and other festivals.[46] It was even popular to mix Latin and the vernacular since the days of the macaronic songs of the Goliards.

A second problem was to fit English choral music into Cranmer's "contortions," that is, his demand that each syllable of a word should be represented by a musical note. According to Denis Stevens, Cranmer confused the musico-liturgical functions of homophony and polyphony.[47] Cranmer had prescribed that the new

45 Stevens, *Tudor Church Music*, p. 13. Fayrfax, the famous composer who died in 1521, and had been associated with the Abbey of St. Albans, wrote music in five parts. On the other hand, a Scottish priest and contemporary, Robert Carver, composed a motet, *O bone Jesu*, with 19 voice parts in the acclamations.

46 This problem was not, however, an entirely new one, nor insuperable, for the paraliturgical forms such as carols for Christmas and Easter had been closely bound up with the vernacular since the mid-fifteenth century; it was fashionable to mix Latin and vernacular texts as in the macaronic songs.

47 Stevens, *Tudor Church Music*, p. 18.

style of church music was not to be "full of notes [i.e. melismatic] but, as near as may be, for every syllable a note."[48] In making this suggestion Cranmer was proposing music for a processional litany, for which it was appropriate. But for nonprocessional choral pieces it was wholly unsuitable. Furthermore, since processions in Edward's time were thought to be either superstitious or unnecessary by the reforming party, the projected English version of the Sarum *Processionale* came to nothing.[49]

Yet Cranmer's influence was not inoperative. Two of the Wanley part-books[50] contain 10 communion services dating from the last decade of Henry VIII's reign, but only two attempted to follow Cranmer's directions. The chief monument to Cranmer's ideals of music is, of course, Merbecke's *The Book of Common Praier Noted* (1550).

This work helped to resolve some of the difficulties of Anglican musicians. It was the work of a composer of decided Protestant leanings, who, in fact, in Henry's reign had only been reprieved from the stake at the last minute for publishing Protestant writings and for preparing a concordance of the Bible. By the same token, he was *persona gratissima* to Northumberland and Cranmer in Edward VI's reign; his book was published by Richard Grafton, printer to the King's Majesty. From the year of his escape from death in 1543 until his death in 1585, he remained a member of the musical staff at St. George's, Windsor. Like the Prayer-Book, Merbecke's music was intended to be used by congregations. Its best quality was that the music for the Mass and the Divine Office was "singable and never difficult."[51] A typical English compromise, it was reminiscent of plainsong, but of a gambolling Gregorian, with all the rhythmic mannerisms of the period.[52]

The duration of the prayer-book was brief, and not only because of difficulty in singing it.[53] It was unfortunate that Merbecke had prepared his music for what turned out to be only an interim rite. Two years later the text of the second prayer-book was modified

48 Cited E. H. Fellowes, *English Cathedral Music*, p. 25.
49 *Ibid.*, p. 18.
50 In the Bodleian Library, Oxford (MS Mus. Sch. E 420-422).
51 C. Henry Phillips, *The Singing Church*, p. 61.
52 For example, if a long note is surrounded by shorter notes, it requires the accent, and a high note is accented if placed in the midst of low notes and vice versa.
53 Phillips, *Singing Church*, p. 62, says: "a modern performance shows it can easily be sung by the uninitiated."

too greatly to permit Merbecke's music to be retained. The swift onset of Mary's reign a year after the second Book of Common Prayer left the composer with insufficient time or spirit for revision, even if he had intended such. The appropriate eulogy was provided by C. Henry Phillips: "The failure of the book to be retained as a singers' handbook was a blow to congregational worship and so also to the intention of the reformers to create a congregational service. From now on only the service without the music can be described as congregational."[54]

If the prospect for the great choral tradition in ecclesiastical music was bleak in Edward's reign, was it better in Elizabeth's? While the Elizabethan prayer-book was practically a reprinting of the austere book of 1552, its ceremonial was a return to the 1549 book with its more 'Catholic' ornaments. The removal of the "black rubric" offered no consolation for either extreme Protestants or incipient Puritans. Indeed, the criticisms of the returning Protestant exiles were anticipated in the following Royal Injunction of 1559 concerning music:

49. *Item*, because in divers Collegiate and also some parish Churches, heretofore there have been livings appointed for the maintenance of men and children to use singing in the church, by means whereof the laudable science of music has been had in estimation, and preserved in knowledge: the Queen's Majesty, neither meaning in any wise the decay of anything that might conveniently tend to the use and continuance of the said science, neither to have the same so abused in the church, that thereby the common prayer should be the worse understood of the hearers, willeth and commandeth, first that no alterations be made of such assignments of living, as heretofore hath been appointed to the use of singing or music in the Church, but that the same so remain. And that there be a modest distinct song, so used in all parts of the common prayers in the Church, that the same may be as plainly understood, as if it were read without singing, and yet nevertheless, for the comforting of such that delight in music, it may be permitted that in the beginning, or in the end of common prayers, either at morning or evening, there may be sung an Hymn, or such like song, to the praise of Almighty God, in the best sort of melody and music that may be con-

[54] *Ibid.*

veniently devised, having respect that the sentence of the
Hymn may be understood and perceived. . . .[55]

It should be noted that the emphasis on hearing and understanding the words met the Puritan concern, and that the proponents
of metrical psalmody interpreted the latter as falling within the
definition of a "hymn." Otherwise, the entire, widespread practice
of singing metrical psalms was unauthorised. Certainly Puritanism may have been partly responsible for the depreciation of
choral music, but hardly for the despising of church music. Bishop
Robert Horne of Winchester (a Marian exile) prepared an injunction in 1571 that smacked of Puritanism:

> 6. *Item*, that in the choir no more shall be used in song that
> shall drown any word or syllable, or draw out in length or
> shorten any word or syllable, otherwise than by the nature of
> the word it is pronounced in common speech, whereby the
> sentence cannot be well perceived by the hearers. And also the
> often reports or repeating of notes with words or sentences,
> whereby the sense may be hindered in the hearer, shall not be
> used.[56]

As a result of the research of Percy A. Scholes, it is now recognised
that the Puritans have been greatly maligned as haters of music.[57]
They believed in congregational music but not in elaborate and
costly polyphonic music sung by choirs filching the rights of the
congregation.

Whatever the reasons—musical or financial poverty, inflation,
or a narrow Biblicism that restricted popular praise to psalmody—
the result in the early years of Elizabeth's reign has been described
as a "decline in musical standards,"[58] or as "drabness."[59] The
marvel is that the situation should have improved so markedly
later in the reign.

Some idea of the standards of music in the Royal Chapels may
be guessed from the following account of St. George's Chapel,
Windsor in 1575, when the queen was in residence, by the secretary of the visiting Duke of Württemberg: "The castle stands
upon a knoll or hill. In the outer court is a very beautiful and
spacious church, with a low flat roof covered with lead, as is

---

[55] Frere and Kennedy, *Visitation Articles*, III, 22-23.
[56] *Ibid.*, III, 319.
[57] In his *The Puritans and Music in England and New England*; and in his
edition of *The Oxford Companion to Music*, pp. 756-68.
[58] Le Huray, *Music and the Reformation*, p. 39.
[59] See note 18 above.

common with all churches in England. In this church his high-ness listened for more than an hour to the beautiful music, the usual ceremonies, and the English sermon. The music, and espe-cially the organ, was exquisite. At times could be heard cornets, then flutes, then recorders, and other instruments. And there was a little boy who sang so sweetly, and lent such charm to the music with his little tongue, that it was really wonderful to listen to him. Their ceremonies indeed are very similar to those of the papists, with singing and so on. After the music, which lasted a long time, a minister or preacher ascended the pulpit for the sermon, and soon afterwards, it being noon, his highness went to dinner. . . ."[60] Further evidence for the improved attitude to music as the reign developed is to be found in Richard Hooker's paean to music. Music is apt, he averred, for all occasions—"as season-able in grief as in joy; . . . as decent, being added unto actions of the greatest weight and solemnity, as being used when men most sequester themselves from action . . . a thing which all Christian Churches in the world have received, a thing which so many ages have held, . . . a thing which always heretofore the best men and wisest governors of God's people did think they never could commend enough. . . ." Hooker questioned whether "there is more cause to fear lest the want thereof be a maim, than the use a blemish, to the service of God."[61]

The queen would have agreed with Hooker that music was never to be commended enough by the "wisest governors of God's people," among whom she counted herself. Her love of music was intelligent and informed. John Buxton was right in thinking that, "in her judgment the music which was composed for her Chapel Royal or for private enjoyment was a greater glory to her realm than the poetry written for her court, or the plays written for the London theatres."[62] The queen had 60 musicians in her service, both singers in the Chapel Royal and instrumentalists. The Chapel Royal counted among its musicians such distinguished composers as Tallis and Tye (the founders of English cathedral music), Mundy, Byrd, Morley, Bull, Tomkins and Orlando Gibbons. As a mark of her favour, Queen Elizabeth gave the exclusive right to print musical scores jointly to Tallis and Byrd, and later to Morley. Similarly King Henry VIII had arranged at the dissolu-

[60] Le Huray, *Music and the Reformation*, pp. 80-81, using as his source, W. B. Rye, *England as seen by Foreigners*.
[61] *Ecclesiastical Polity*, Book v, secs. 38-39.
[62] Buxton, *Elizabethan Taste*, p. 173.

tion of Waltham Abbey in 1540 for Tallis to be taken into the Chapel Royal, as he had recognised the ability of Christopher Tye. Royal musical patronage showed a considerable insight into musicians and interest in church music.

6. *The Leading Composers*

Literature is a national, music an international, art. It is a pity that the fame of poets and dramatists such as Shakespeare and Marlowe, Ben Jonson and Webster, has eclipsed the international fame of such composers as William Byrd and Thomas Morley. For example, in 1583 the madrigalist Filippo di Monte sent Byrd a setting for eight voices for the opening of the 137th Psalm, *Super flumina*, to which compliment Byrd replied by sending his own setting of *Quomodo cantabimus* from the same psalm.[63] A few years later Morley's work was published in Germany soon after it had been issued in England, and Dowland was for several years lutenist to King Christian IV of Denmark.

One of the interesting facts about the greatest Elizabethan composers is that several of them were Catholics, including Tallis,[64] Byrd, and Morley. Yet they held office in the Chapel Royal of a Protestant queen. As Tallis was Byrd's teacher, so Byrd was Morley's teacher. As long as they fulfilled their musical duties with satisfaction the queen allowed them to think and compose as they pleased. They were able to fulfill their duties with comparative ease. In the Anglican communion service they had to provide music for the *Kyrie* and *Creed* only. For morning and evening prayer the canticles to be set to music were the *Venite*, the *Te Deum*, *Benedictus* (or *Jubilate*), the *Magnificat* and *Nunc Dimittis*. There was ample time, if the spirit was willing, in which to compose hymns or to adapt Latin motets to English texts.

Thomas Tallis (c. 1505-1585), the teacher of Byrd, was second only to his pupil in fame in sixteenth-century England. Today he is remembered chiefly as the composer of the simple harmonisations of the responses in the prayer-books of the Anglican church. Little is known about his life. He was organist of Waltham Abbey when Henry VIII decided to transfer him to the Royal Chapel, and his music reflected its pre-Reformation inspiration in many Latin motets, its post-Reformation utility in the form

---

[63] *Ibid.*, p. 171.
[64] Tallis is generally thought of as a Catholic (e.g. Stevens, *Tudor Church Music*, p. 22), but I owe to Dr. A. L. Rowse the suggestion that the form of Tallis's will indicates that he died a Protestant.

of English anthems. Tallis's five-part *Salvator mundi* and short anthem, "Hear the voice and prayer," have been revived with great acceptance in our own day. The former manages to be nobly serene and moving at the same time. His ingenious and varied output of high quality included five English anthems published in Daye's *Certain Notes* (1560-1563), a few Latin Services, about 50 motets, some of which were published in a joint work with Byrd in 1575, entitled, *Cantiones Sacrae*. Perhaps his most amazing composition was a motet in 40 parts, which ended with the words, *respice humilitatem nostram*, and was an astonishing musical contradiction of the theme of humility. It has been described by Sir Henry Hadow as "an ocean of moving and voluminous sound."[65]

William Byrd (1543-1623) was the most renowned composer of his day. He was, by common consent, the most distinguished musician from the Reformation to the Restoration. Born probably in Lincolnshire, Byrd was organist of Lincoln cathedral at the age of 20 and joint organist with Tallis of the Royal Chapel after 1569. He was the founder of the rich English madrigal tradition; as a liturgical composer he was thought to be the equal of Palestrina and Lassus.[66] An excellent teacher, his pupils included Thomas Morley, Orlando Gibbons, and John Bull.

Byrd was the first famous English composer to have his works published to any extent in his own time, though those published were a fraction of the whole. The joint work with Tallis was the *Cantiones quae ab argumento Sacrae vocantur*, which was dedicated to the queen. But in 1589 and 1591 Byrd published two volumes of his *Cantiones Sacrae* alone. In 1588 the *Psalmes, Sonets, and Songs of Sadnes and Pietie* appeared, with a dedication to Sir Christopher Hatton. Byrd's *Songs of Sundrie Natures* was printed in 1589, and the *Liber Primus Sacrarum Cantionum quinque vocum*, followed by its successor of sacred songs for five voices, in 1591. A volume including three undated masses appeared about 1603. Two books of *Gradualia* came out in 1607, and in 1611 there appeared the *Psalms, Songs, and Sonnets*.

In addition to church music this fertile and profound composer wrote string music, keyboard music, and secular choral

---

[65] *English Music*, pp. 3f.

[66] See the evaluations of Byrd recorded by Percy A. Scholes in *The Oxford Companion to Music*, p. 128b. Morrison Comegys Boyd, in *Elizabethan Music and Musical Criticism*, p. 80, claims that "Byrd's church music rivals that of Palestrina in contrapuntal skill and beauty, and surpasses it in expressiveness."

music, almost all of high quality. His religious music, like that of his mentor Tallis, has great nobility and sometimes attains sublimity. Also, like Tallis, Byrd was able to produce with equal facility services for the Latin and English rites. His technique is brilliant and warm, especially when fitting music to vernacular texts with action in them. In this his work contrasts with the cold perfection of Tallis's Latin works or the simpler earnestness of Tye.[67] It was a style combining dignity with pathos, plumbing the spiritual depths of the sacred texts he set to music. "He believed," says C. Henry Phillips, "in his heart as well as in his mind the dogmas of his faith . . . and could in addition carry over into his music his strong emotional conviction."[68]

Byrd was also an important experimenter. He was able to achieve beauty in the Short Services even when required to compose according to the Cranmerian restriction of a syllable to a note, so pure were his verbal rhythms. In his Second Evening Service he developed a novel idea which would revolutionise English sacred music. He composed certain verses for solo voices with an independent organ accompaniment. For about 50 years afterwards Short Services, alternating solo or duet passages with the music of the whole choir, were composed with great musical and devotional satisfaction.[69]

Byrd's most outstanding religious works are his three masses,[70] his *Gradualia* or Latin motets for the Catholic Daily Offices, and his music for Anglican services. Less consistently good are the early motets of 1575 and the English anthems, though some are outstanding. Byrd and his pupil Gibbons were pioneers in the composition of verse anthems. They often scored the music for viols to accompany the solo passages.[71] His greatest Anglican anthems include "How long shall mine enemies triumph," "O Lord, make thy servant Elizabeth," "Prevent us, O Lord," and "Arise, O Lord!" He also composed special psalms in an elaborate chant style of which the most notable is "O clap your hands." Byrd's work, whether in the form of motet or madrigal, was the glory of Tudor musical accomplishment. E. H. Fellowes believed

[67] See Phillips, *Singing Church*, p. 67.
[68] *Ibid.*, p. 93.
[69] Winfred Douglas, *Church Music in History and Practice*, rev. Leonard Ellinwood, pp. 117-18.
[70] Young, *British Music*, p. 153, writes of Byrd's development of a "sense of the interdependence of liturgical thought, verbal meanings, and rhythm and overtone, and musical techniques" as being "evident throughout the Masses."
[71] Woodfill, *Musicians in English Society*, p. 150.

that few short choruses for four boys' or women's voices could be found in all music to compare with the brilliance of "Rejoice, rejoice" and the chorus of "From the Virgin's womb."[72]

Thomas Morley (1557-1604?) became organist of St. Paul's cathedral in 1591. In 1592 he was appointed a Gentleman of the Royal Chapel. He wrote a famous primer in the form of a dialogue which was being studied 200 years after its initial publication—*A Plaine and Easie Introduction to Practicall Musicke* (1597).[73] The composer of at least two of the settings to Shakespeare's songs, Morley was an expert writer of madrigals when the genre was at its peak. Possibly at no other period was there such harmony between poets and musicians or such willingness to make music the servant of the word. A typical excerpt from Morley's primer makes this clear: "Now having discoursed unto you the composition of three, four, five and six parts with these few ways of canons and catches, it followeth to show you how to dispose your music according to the nature of the words which you are therein to express, as whatsoever matter it be which you have in hand, such a music must you frame to it. You must therefore, if you have a grave matter, apply a grave kind of music to it; if a merry subject, you must make your music also merry, for it will be a great absurdity to use a sad harmony to a merry matter, or a merry harmony to a sad, lamentable, or tragical ditty."[74] The greatest achievements of the age were in vocal music.

Morley's great reputation as a madrigalist may well have over-shadowed his achievement as a church musician. The fact that little of his church music was published also helps to account for its inadequate recognition. In the undated *Whole Booke of Psalmes*, printed by W. Barley, "the assigne of T. Morley," there are four settings of tunes by Morley, two of which, together with a previously unpublished third, appeared in Ravenscroft's Psalter of 1621. Barnard also published a Morning and Evening Service of four and five parts, respectively; an Evening Service of five parts; and a verse anthem, "Out of the Deep." His renowned Burial Service was published by Boyce. There are, however, 10 unpublished motets and several anthems, preces and psalms, responses, and other services yet unpublished.[75]

72 *English Cathedral Music from Edward VI to Edward VII*, p. 81.
73 Morley's *Plaine and Easie Introduction to Practicall Musicke*, ed. R. A. Harman.
74 *Ibid.*, p. 290.
75 From *Grove's Dictionary of Music and Musicians*, ed. H. C. Colles, III, 519a-521a.

Morley's achievement is all the greater, considering that he died at the age of 45. Few of his English anthems have survived. One full anthem begins with the Latin words *Nolo mortem peccatoris*, then breaks into English, "Father, I am thine only Son." This penitential composition was the most suitable for Good Friday. "Out of the deep," with its alternate verses for solo and chorus is also poignant. In the first of his four Services Morley demonstrated an experimental advance even on Byrd, for in addition to solo song with organ accompaniment, he used passages for duet, trio, and quartet of solo voices. His was the earliest setting of the Burial Service in the English prayer-book; it was widely used in the seventeenth and eighteenth centuries. It is a complete setting, including "I am the Resurrection and the Life," "Man that is born of woman," "Thou knowest, Lord," and "I heard a voice from heaven." It is sensitive and beautiful. Morley's prestige in his own time was second only to that of his master, Byrd.[76] In 1614, a decade after Morley's death, Thomas Ravenscroft said of him: "he did shine as the Sun in the Firmament of our Art."[77]

These planets in the firmament of English music, were, it should be remembered, surrounded by a galaxy of stars. Their contemporaries included Tye, cofounder with Tallis of English cathedral music, Whyte, Weelkes, Tompkins, Dowland, Bull, Wilbye, and Orlando Gibbons, although the last came to full maturity in the Jacobean age. This was a golden age for English composers, for in the closing years of Elizabeth's reign, Gloriana had 30 or 40 composers worthy of remembrance, of whom only the most outstanding have been mentioned. They represented "the fine flower of the polyphonic style of composition."[78]

7. Gains and Losses

In considering the impact of the Reformation on English music there are to be seen clear gains and clear losses. On the negative side, there was the loss of much of the sonority and richness, subtlety and intricacy of medieval Catholic polyphonic music. The number and size of choirs in churches were reduced. The significance of the liturgical composer was lessened in Anglicanism, since the Mass required more settings (not forgetting the major Daily Offices) than the Anglican services. He thereupon found

[76] From Fellowes, *English Cathedral Music*, pp. 83-87.
[77] In Ravenscroft's *Brief Discourse* (1614).
[78] Fellowes, *English Cathedral Music*, p. 68.

other outlets for his inventive powers, as in madrigals, as he was in the next century to do in opera. The quality of the music of parish churches was poorer, as metrical psalms were substituted for motets. In brief, as the movement away from purely liturgical music gained momentum, music became increasingly secular in an age of expanding humanism.[79] But before the Tudor period was over Byrd had produced matchless Masses, which were not only the best of his work, but outstanding in the entire range of English music.

There were, however, gains in the influence of the Reformation on music, as on worship. The emphasis on the vernacular led not only to a far greater understanding, but inevitably to an increased popular participation in worship. One immediate advantage was the gain in the simplicity and directness of musical techniques, and the consigning of musical extravagances, such as allowing 15 bars for one syllable, to utter oblivion. The argument for the need to reduce overcomplex and overflorid music to simpler dimensions was so overwhelmingly cogent that it produced not only the Calvinist metrical psalm, the Lutheran chorale, and the Anglican Service, anthem, and psalm tune, but even the model *Missa brevis* of Palestrina. All these were creative responses to new conditions and new demands.

The greatest of the contributions to the popular praise of England and Scotland was metrical psalmody, which dominated the hearts of the people and therefore the parish churches and their homes. These were set to memorable tunes, in which admittedly the French combination of Marot's words to Bourgeois' tunes was never reached in English. Where England did, however, make a significant and original contribution to church music was in the anthem, especially as composed by Byrd and Gibbons.[80] This was a development Purcell would exploit almost a century later.

This transformation of the motet into the anthem, and its development in a responsive direction, occurred almost concurrently with the development of the madrigal. It provided a great opportunity for expression in music. Indeed, it may well be that the primary emphasis on the "Word" in Protestantism (supremely exemplified in the importance of preaching) and the insistence on

[79] Young, *British Music*, p. 151.

[80] William Byrd wrote 17 verse anthems and about 60 full anthems, while Orlando Gibbons wrote 24 verse anthems and 10 full anthems. The most prolific composer in this genre was Thomas Tomkins, another of Byrd's pupils, who composed 50 verse anthems and about 41 full anthems. Le Huray, *Music and Reformation*, p. 224.

the vernacular, led to the great concentration on making choral music the expression or representation of the words of the text in verse anthem and madrigal.[81] When settings were made for verse anthems, which allowed for verses to be sung partly in solo or duet and partly in chorus, Byrd had invented a fertile new vehicle of praise, with the accompaniment of cornets and sackbuts, recorders and viols, or of organs. Richard Farrant and William Mundy also favoured verse anthems. The new style became immensely popular, and immediately so, because there were good, practical reasons for it. The task of performance was placed on the ablest singers (now increasingly difficult to get), and the solo voices stood out against the musical accompaniment.

How colourful and enchanting the new style of anthems was can be appreciated from the impressions of a seventeenth-century listener, who summed up his delight when hearing a solemn anthem, "Wherein a sweet melodious treble or countertenor singeth single and a full choir answereth, much more when two such single voices interchangeably reply to one another, and, at the last, close all together. . . ."[82] This expression of joy in church music is excelled only by John Milton, the son of an Elizabethan and Jacobean composer of motets and madrigals, whose desire was:

> There let the pealing organ blow,
> To the full-voiced quire below,
> In service high and anthems clear,
> As may, with sweetness, through mine ear,
> Dissolve me into ecstasies,
> And bring all heaven before mine eyes.[83]   20

[81] This is well argued in chap. 5 of Stevens, *Music and Poetry in the early Tudor Court*, and partly rebutted by Le Huray in *Music and Reformation*, pp. 135-36. Le Huray points out that at this time the "simultaneous," or horizontal (as contrasted with the "successive" or vertical), technique of writing music, by which all the parts were developed together, also aided expressiveness in music, as shown, for example, in the famous motets of Josquin des Prez, *Absolon fili mi* and *Planxit autem David*. The issue, then, is whether the change in technique followed or preceded the search for expressiveness.

[82] Charles Butler, *The Principles of Musick* (1636), p. 41, cited in Le Huray, *Music and Reformation*, p. 225.

[83] *Il Penseroso*, ll. 153-58.

CHAPTER XII

SPIRITUALITY: CATHOLIC, ANGLICAN, PURITAN

CHRISTIAN public worship was undergirded and supplemented by spirituality. The external observance of the liturgy was merely a mechanical ritualism, a lifeless puppet articulated by rubrical strings, apart from the culture and discipline of the spiritual life. Thus spirit-inspired devotion was the life and power of the liturgy.

It was the spirituality of an age, reflected in Roman Catholic, Anglican, and Puritan primers and books of devotion, that pointed to the highest ideals and deepest motivations of conduct in Tudor England. C. J. Stranks has rightly argued that a study of the spiritual training reflected in prayer books can provide "if not a history, then a pointer to the history of those ideas which determined the day to day conduct of each man with his neighbour."[1]

In addition, while the liturgy was the medium of official, clerically-conceived piety, popular books of private and family devotions, especially when the contents were selected by laymen, were better indications of lay predilections and concerns in prayer. Books of personal and familial prayers were reinforcements of the Christian life corporately mediated by the liturgy. They also supplied new concerns which seemed to be lacking or were not emphasised in the public forms of worship. These new concerns were in the forefront in Protestant books of prayer, because the Reformation gave a great impetus to new forms suitable for family prayers, since these were naturally of greater interest to the married Protestant clergyman than to the ascetical priest or monk of the old faith.

There were, however, unexpected continuities between medieval and renaissance themes and types of prayer, thus linking Catholic and Protestant devotions. In several cases Catholic prayers, slightly or severely doctored, were incorporated into Protestant books of prayer. Unsuspecting fathers of Protestant households were approaching God in the accents of the great Catholic humanist, Erasmus, or of the Spaniard Vives, or even of the English Jesuit, Robert Persons. The most notable case of

---

[1] C. J. Stranks, *Anglican Devotion: Studies in the Spiritual Life of the Church of England between the Reformation and the Oxford Movement*, p. 9.

Catholic devotional influence on a Protestant was that of Persons via Bunny on Baxter. The Jesuit's *First Booke of the Christian Exercise, appertayning to resolution* appeared in 1582 with the avowed aim of convincing the unconverted by stressing those considerations of his own sinfulness and its reward, and God's grace and generosity, with the utmost vividness. This Ignatian type of treatise was bowdlerized by Edmund Bunny, the Puritan minister of Bolton Percy in 1584. He impertinently introduced his purified Persons thus: "I perceived that the booke insuing was willingly read by divers, for the persuasion that it hath to godliness of life, which notwithstanding in manie points was corruptly set downe: I thought good in the end to get the same published againe in better manner. . . ."[2] Persons, improbably purged in the "Bunny Club" edition, went through nine issues in 16 years. Richard Baxter, one of the most attractive Puritan saints of the seventeenth century, was converted to Christianity by reading Bunny's version of Persons. Baxter records that: "a poor Day-Labourer in the Town . . . had an old torn book which he lent my Father, which was called Bunny's Resolution (being written by Parson's the Jesuit, and corrected by Edm. Bunny. . . . And in the reading of this Book (when I was about Fifteen years of Age) it pleased God to awaken my Soul, and shew me the folly of Sinning and the misery of the Wicked, and the unexpressible weight of things Eternal, and the necessity of resolving on a Holy Life, more than I was ever acquainted with before. The same things which I knew before came now in another manner, with Light, and Sense, and Seriousness to my Heart."[3]

There was another reason for studying private and family devotions. It was that here if anywhere sincere religion was likely to be found, despite the notable dissent of Samuel Butler in his novel satirising Victorian hypocrisy, *The Way of All Flesh*. At least, it was likelier to be found in the silence of solitude or in small groups. Certainly, it was only the households prepared at home in prayer that were able genuinely to share in the gatherings of public worship. Mere spectators at public worship were formalists; they had already travelled halfway to hypocrisy. If private and familial piety was strong, so would public worship be vigorous.

2 Bunny's version (1584), sig. 2, as cited in A. C. Southern, *Elizabethan Recusant Prose, 1559-1582*, p. 185. Augustus Jessopp, in his *One Generation of a Norfolk House, a Contribution to Elizabethan History*, p. 90, drew attention to the influence of Persons on Baxter in 1879.
3 *Reliquiae Baxterianae* (1696), p. 3.

If personal and family prayer was rare or sporadic, then worship would be perfunctorily celebrated and sparsely attended. Happily, in the Elizabethan age the printing (and presumably the use) of supplementary books of prayer, combining instruction in the art of prayer with examples of prayer for different persons and situations, was prolific.

1. *The Popularity of Devotional Books*

H. S. Bennett has described the demand for religious books in sixteenth-century England as "insatiable."[4] Louis B. Wright points out that 60 percent of the breviaries, books of hours (or primers), and manuals printed in the first half of the century came from overseas, despite the intermittent restrictions on the importing of books.[5] Two examples of many that might be chosen to illustrate the voracity of the public appetite for books of devotions were those issued by F. G. (especially *A Manual of Prayers*, which went into 15 editions between 1583 and 1613) and Edward Dering, a famous Puritan minister, whose *Godly Private Prayers* was issued in over 20 editions between 1576 and 1626.[6] It is known that more than 80 different collections of private devotions were printed during the reign of Elizabeth I.[7]

Why was the demand so great? Even after allowing for the excitement over the relatively recent invention of printing, and the widespread Renaissance conviction of the efficacy of book learning, in both of which religion shared, the predominance of this particular genre of religious literature has still to be explained. The truth seems to be, as Helen White has suggested, that the greatest controversy in sixteenth-century England was over the issue of religious allegiance. The effect of the controversy was to place a heavier responsibility on the individual for the conduct of his own private religious life.[8] The Protestant principle of the priesthood of all believers was taken in its fullest implications by the Puritan merchants of the middle classes. These, as Wright has shown, except in the most affluent households, lacked the chaplains considered indispensable in aristocratic households. Laymen

---

4 *English Books and Readers, 1475-1557*, p. 65.

5 *Middle-Class Culture in Elizabethan England*, p. 228. On p. 24 Wright claims: "Of all the works that poured from the press in the sixteenth and seventeenth centuries, these [devotional books] had the widest circulation and remained popular longest."

6 See Faye L. Kelly, *Prayer in Seventeenth-Century England*.

7 *Christian Prayers and Meditations*, preface, p. 14.

8 Helen C. White, *Tudor Books of Private Devotion*, pp. 149, 150, 174.

had to act as their own curates, as priests in their own households. The moderate Puritans who were the authors of the majority of English manuals of devotion, "unaware of what they were doing . . . were undermining the future authority of the Church."[9] They taught the London citizen to claim the promises of God in Holy Scripture, to model his prayers on the simple and profound petitions of the Psalmist and on the comprehensive concerns expressed in the Lord's Prayer; if he lacked the gift of extemporaneous utterance, they provided samples and directions for prayer. The heads of middle class households assumed the responsibility for teaching their children, apprentices, and maidservants, the essentials of religious belief, duty, and devotion. The less learned and dedicated his neighbouring clergyman was, the more concerned was the Puritan citizen to be independent of him.

Nor should it be overlooked that the manuals of devotion inculcated important ethical ideals directly or indirectly. They instructed all the members of the household in their social duties and obligations, laying before them the virtues appropriate to their callings in the household, the town, and the state, affirming the duty to be generous toward the poor and handicapped, to overcome abrasive anger with the lenitives of patience and forgiveness, and to cultivate the prudential virtues of honesty, industry, and thrift. They confirmed the oncoming generation in respect for the constituted authority of the queen and her magistrates, if not always for her bishops. The most popular books of devotion were, however, completely uncontroversial.

An example of the compassion taught in the manuals of devotion is selected from one of the most popular manuals of the age, *A Booke of Christian Prayers, collected out of the aunciēt writers and best learned in our tyme, worthy to be read with an earnest mynde of all Christians, in these daungerous and troublesome dayes, that God for Christes sake will yet still be mercyfull unto us.* Issued by the printer John Daye, this volume contains wide marginal illustrations on each page. For example, "A prayer for them that be in povertie" is illustrated with woodcuts of "Charity" feeding the hungry, giving drink to the thirsty, harbouring strangers, clothing the naked, and visiting the sick and prisoners, with the appropriate Biblical references to these Dominical commendations. The prayer itself, though wordy and confusing in combining a meditation and a series of petitions, is moving:

[9] Wright, *Middle-Class Culture*, p. 240.

They that are snarled and intangled in ye extreme penury of things needfull for the body, cannot set theyr myndes upō thee O Lord, as they ought to doe: but when they be disapoynted of the thynges which they doe so mightely desire, their hartes are cast downe, and quayle for excesse of grief. Have pitie upon them therfore O mercyfull Father, and releeve their misery through thine incredible riches, that by thy remouvyng of their urgent necessitie, they may rise up unto thee in mynde. Thou O Lord providest inough for all men with thy most liberall and bountyfull hand: but whereas thy giftes are, in respect of thy goodnesse and free favour, made common to all men, we (through our naughtynesse, nigardshyp, and distrust,) doe make them private and peculiar. Correct thou the thyng whiche our iniquitie hath put out of order: let thy goodnesse supply that which our niggardlynesse hath plucked away. Geve thou meate to the hungry, and drinke to the thirsty: Comfort thou the sorrowfull: Cheare thou up the dismayde: Strengthen thou the weake: Deliver thou them that are prisoners: and geve thou hope and courage to them that out of hart.

O Father of all mercy have compassion of so great misery. O Fountaine of all good thynges, and of all blessednesse, washe thou away these so sundry, so manifold and so great miseries of ours, with one drop of the water of thy mercy, for thyne onely Sonne our Lord & Saviour Jesus Christes sake. Amen.[10]

2. Catholic and Royal Primers

English Catholicism had inherited a great tradition of spirituality,[11] which had blossomed richly in the fourteenth-century mystics, including Richard Rolle of Hampole, the Benedictine Monk of Farne, the anonymous author of *The Cloud of Unknowing*, Walter Hilton, the Lady Julian of Norwich, and Margery Kempe, as well as the unknown visionary who wrote *Piers Plowman*. In addition to nourishing these exponents of high mysticism (some of them monks, some hermits, and some anchoresses), the Catholic

---

[10] Edition of 1581, pp. 53-54.

[11] For an introduction to this tradition see *Pre-Reformation English Spirituality*, ed. James Walsh, which contains biographical essays on English writers on spirituality from Bede to Augustine Baker, with a notable chapter on Thomas More. See also *English Spiritual Writers*, ed. Charles Davis; and David Knowles, *The English Mystical Tradition*.

Church also made provision for the "lower" mysticism of those who wished to love God and serve men in the world. This spiritual food continued to be supplied in the turbulent sixteenth century.

The Catholic devotional books for the laity were usually known as "primers."[12] They had been the most popular medieval manuals of devotion, consisting of brief handbooks of prayer and basic religious instruction. The earliest contained the Psalter and the Litany, and often the Vigils of the Dead were added. By the end of the thirteenth century, while the Litany and the Vigils of the Dead were retained, the Psalter had been reduced to the Seven Penitential Psalms and the important addition was made of the Hours of the Blessed Virgin. These had started as a monastic devotion in Lent and Advent, supplementary to the Day Hours, but they proved so attractive that the lay folk desired to have them at all seasons, especially since special prayers were added which carried indulgences. Much legendary material, some dubious and much unedifying, was included in these prayers. Yet the material provided a tender Marian devotion through which it was hoped that adoration of the infant Saviour might grow, but which also might be stifled by sentimentality.[13]

The standard, or model, primer developed the following features, according to Helen C. White, at the end of the fifteenth century: the Hours of the Blessed Virgin, the Seven Penitential Psalms, the Fifteen Gradual Psalms, the Litany, the Office for the Dead and the Commendations or prayers following that Office.[14] Most primers of the period also included prayers for special occasions and needs; some provided an extensive selection of prayers for private or corporate use, covering many applications. It was in these additions that Protestant editors of prayers were to find attractive possibilities for their own adaptation and invention. Monarchs with an interest in modifying or reforming the country's religion also found primers to be potent instruments of change, until, with the provision of a liturgy in English (as with the Book of Common Prayer of 1549), there was no longer any need for a person to carry a primer to church in order to take an intelligent

12 For two texts see *The Prymer or Lay Folk's Prayer Book*, ed. Henry Littlehales; and *Horae B.V.M. or, Sarum and York Primers, with Kindred Books of the Reformed Roman Use*, ed. E. Hoskyns; see also C. C. Butterworth, *The English Primers, 1529-1545*.

13 Stranks, *Anglican Devotion*, p. 15.

14 "Sixteenth Century English Devotional Literature," in Joseph Quincy Adams Memorial Studies, ed. J. G. McManaway, G. E. Dawson, and E. E. Willoughby, pp. 439-46.

part in the worship, which up to that year had been largely in Latin. The first Anglican prayer-book was made available for use at home as well as for public worship. Thus the disuse of English primers was directly related to the disuse of Latin in the liturgy.[15] The primer lived on usefully in the educational field, for prayers were composed and included in the major pedagogical handbooks of the period, even in the instructions issued by military commanders. The queen herself was reputed to compose private prayers.

Even though the provision of an official vernacular prayer-book for use in churches and homes rendered primers eventually unnecessary, yet while religion was in the process of change, royal primers performed a useful state function and served as intriguing reflectors of royal religious policy.[16] Henry VIII's primer, or "boke of ordinarie prayers," appeared in 1545, six years after the unofficial but important experimental primer of Bishop Hilsey of Rochester. The royal book kept the traditional title and the opening Office of *Dirige*, though the latter was reduced by two-thirds. Psalms of thanksgiving and praise replaced the previous psalms of penance and mercy. The Prayers of the Passion were altered so as to emphasise the practical lessons to be drawn from contemplating the sufferings of Christ, rather than providing a medium for imaginative and emotional unity with the Crucified. The hymns in the Little Hours were changed from the praise of the Virgin Mary to the praise of Christ. The Litany was rewritten so as to lessen the honouring of the saints.

The Primer of Edward VI (1553) was a combination of the Book of Common Prayer and a primer. Those parts of it which provided private devotions showed a transition from the liturgical and universal to the practical and occasional. Here first appeared in an official publication in England the start of the change, "from the general spiritual provision for all sorts of men in one situation to the special prayers for particular classes and conditions of men," which was to dominate the prayers of the later sixteenth and early seventeenth centuries.[17] The chief Protestant writers of prayers

15 See Kelly, *Prayer*.

16 Hilsey's Primer (1539) was used as a means of religious change chiefly through its unqualified Erastianism and its assertion of the eucharistic doctrine of the Real Presence against the sacramentarians.

17 See White, *Tudor Books of Private Devotion*, chaps. 6, 7. Chap. 6 deals chiefly with the reign of Henry VIII, and chap. 7 with the reign of Elizabeth. The convenience of the Primer as an instrument of religious change was that its editors could provide alternative selections of Scripture to mark theological changes, and

during this reign (excluding Archbishop Thomas Cranmer) were Thomas Becon (Cranmer's chaplain) and Bishop Hilsey.

It is interesting that in the rapidly improvised primer of Queen Mary's reign (1555), the Marian prayers and the prayers invoking the saints were only partly restored. The Henrician Primer's borrowings from Vives, and even the compositions of Becon, were rather surprisingly retained.

The Elizabethan Primer of 1559 owed more to the Henrician book of 1545 than to the Edwardian book of 1553. The Marian elements were further reduced, however, and there were changes in the liturgical sections, but the book was substantially that of Elizabeth's father rather than that of her brother.

Since the primers supplied the needs of the lay Catholics there was little incentive to produce supplementary prayers for Catholics until the Holy Communion service in English supplanted the Latin Mass. Apart from one significant exception, almost all the important Catholic works on prayer in English appeared during the reign of Queen Elizabeth, and at a time when it became clear that England's faith and worship were likely to remain Protestant for decades to come.

The exception was the remarkable collection of *Certain Devout and Godly Prayers made in Latin by the Reverend Father in God Cuthbert Tunstall* (1558), which was translated and printed by Thomas Paynell in 1558. A Catholic in doctrine and religious practice, holding a high sacramental doctrine, he was a reluctant Erastian under Henry VIII who had him appointed to the sees successively of London and Durham.[18] He was imprisoned in Edward VI's reign for his opposition to the official religious policy. In 1547 he had voted against the abolition of the chantries, and in 1549 against the Act of Uniformity and the legislative enactments that permitted priests to marry. In 1552 Tunstall was tried for high treason and deprived of his bishopric. Queen Mary reinstated him, but he refused the Oath of Supremacy on the accession of Queen Elizabeth. Here, when it was dangerous, was a man of conscience. His refusal to countenance persecution of Protestants in Queen Mary's reign shows that he was a man of compassion. These qualities are evident in his book of prayers. The most notable

---

alternative special prayers for the same purpose, while retaining the title and format of the volume, so as to give the impression that nothing of any consequence was changed.

[18] For a biography see Charles Sturge, *Cuthbert Tunstall, Churchman, Scholar, Statesman, Administrator.*

expression of them was his "Prayer unto God for the Dead which have no man that prayeth for them," which recalled the suppression of the chantries. It deserves citation in full:

> Have mercy, we beseche the O Lorde god have mercy, for the precyous death sake of thy onelye sonne our Lorde Jesu Chryst, of those soules ye which have no intercessors that remember thē, or that doth put the in remembraunce of them, nor no consolation, nor hope in there [sic] torments, but onelye that they are created and made lyke unto thy similitude and Image, and markyd wt the signacle of thy faith, the whiche other thrugh the negligence of those whiche are alyve, or of the slydynge course of tyme are cleane forgotten of theyr frendes, and posteritie. Spare them o Lord, and defende thy workmanship in them, nor despise not the work of thy handes, but putte forthe thy hande unto them, and beynge delyvered frō the tormente of paynes, brynge them through thy great mercies, the whych are celebratyd and estemyd above all thy workes, to the felowshyp of the hevenly citezins, which livest and reigneste God, through all worldes. So be it.[19]

A. C. Southern divides the recusant books of spirituality produced during Elizabeth's reign into four main categories: general spiritual directories, books on penance, books on the Rosary, and Psalters and Hours. Since the three last categories were traditional, it is in the first category that originality is to be sought. Of the general directories of spirituality, six seem to Southern to merit mention, and only three of these display creativity. They are the previously mentioned Persons, *The First Booke of the Christian Exercise, appertayning to Resolution* (1582), Hide's *Consolatorie Epistle to the afflicted Catholikes* (1580), and the anonymous *Certayne devout Meditations* (1576).

Of these the most impressive is the work of Persons. His mystical aspiration has nothing in common with the desire to be alone with the Alone. At its most lyrical and inspired it is still an ecclesial consummation that is sought—a communion with God in the communion of the saints—as the soul ascends:

> Imagine, besides all this, what a ioye it shall be vnto thy soule at that daye, to meet with all her godlie freendes in heauen, with father, with mother, with brothers, with sisters:

19 *Certain Devout*, sig. **D iii** v.-**D iv.**

with wyfe, with husband, with maister, with scholares: with neyghboures, with familiares, with kynred, with acquayntance: the welcomes, the myrthe, the sweet embracementes that shall be there, the ioye whereof (as noteth S. Cyprian) shalbe vnspeakable. Add to this, the dalye feasting and inestimable triumphe, whiche shalbe there, at the arriuall of new bretheren and sisters coming thither from time to time with the spoyles of their enemies, conquered and vanquyshed in this world. O what a cōfortable sight will it be to see those seates of Angells fallen, filled vpp agayne with men & wemen from day to day? to see the crownes of glorie sett vpon their heades, and that in varietie, according to the varietie of their cōquestes. One for martyrdome, or confession against the persecutor: an other for virginitie or chastitie against the flesh. . . .[20]

Within the same mystical tradition, but derivative as even a good translation must still be, was Richard Hopkins' *Of Prayer and Meditation*, which made available in English the admirably lucid and simple introduction to the spiritual life and the 14 meditations for the mornings and evenings of one week on the major mysteries of the Catholic faith composed by Fray Luis de Granada, the Spanish Dominican who became Provincial of his Order in Portugal.[21] This important book, the *Libro de la Oración*, of 1553, stressed the significance of the prepared heart for meditation. It had a strong influence on St. Francis de Sales in the writing of his *Introduction to the Devout Life*.

In 1575 *Certaine devout and Godly petitions, commonly called Iesus Psalter* appeared. The book consists of 150 petitions, which are divided into 15 decades for the 15 chief petitions, are further divided into three series according to the three stages of the mystical life—purgation, illumination, and union. Attributed to Richard Whytford, a member of the Brigittine Order, they maintain the ascetical tradition of spirituality, even if the arrangement is contrived.

Important as was the work of Fr. Persons in mediating the Ignatian tradition of the sons of Loyola and the work of Hopkins in making available the work of the Spanish mystic, Luis de Granada, they were only a trickle of the flood of Counter-Reforma-

[20] *Ibid.*, sig. 14.
[21] The only English account of Fray Luis de Granada's mysticism is in E. Allison Peers, *Studies in the Spanish Mystics*, vol. 1, pp. 33-76.

tion spirituality that irrigated a desiccated Europe. For the most part, English Catholics in a newly Protestant country were insulated from the new spirituality. For most of the laity the old, worn, dogeared primers had to suffice as guides to spirituality unless they had access to the clandestine Jesuits or to the chaplains of aristocratic families who could be their spiritual directors.

3. *Protestant Manuals of Devotion*

The Protestant books of devotion can be characterised as a supplement to public worship consisting of directions on the values and methods of prayer, with an anthology of prayers suitable for private and household use.

Both Catholic humanists and Protestant reformers encouraged the study of the Scriptures. The new scripturalism found expression in the publication of prayer manuals consisting largely of selections of Scriptures, with indications of how they were helpful in particular situations, such as a Psalm to encourage one in despair to renewed hope in God. The Catholic primers had made good use of the Psalms, and now so did the Protestant manuals. New translations of them appeared, annotated in such a manner as to claim their confirmation of Reformed doctrines; even the "cursing" psalms seemed fitting expression for maledictions of those who clung to superstition and idolatry instead of embracing the true faith. The partisanship of the Psalmist appealed to the partisanship of the Protestant. Thomas Sternhold provided a popular versified English translation of the Psalms, the first edition of which, with 19 Psalms, appeared in 1547, and the second, with 37, in 1549, the year of his death. In 1557 a third edition was issued which included seven more Psalms, the work of John Hopkins. The complete edition of 1562, issued by John Daye, the Marian exile and printer, shaped the public as well as the private piety of Elizabethan England, for it was often included with the Book of Common Prayer. Sternhold was unwilling that the Devil should have the catchiest metres and the most cheerful tunes, so he wrote nine of his psalms in the ballad metre of Chevy Chase. Their appeal, even if the verse was often jogging, was nationwide, so much so that a Roman Catholic author grudgingly paid them this tribute: "There is nothing that hath drawne multitudes to be of their Sects so much, as the singing of their psalmes, in such variable and delightful tunes. These the souldier singeth in warre, the artizans at their worke, wenches spinning and serving, apprentices in their shoppes, and wayfaring men in their travaile, little knowing (God wotte) what a serpent lieth under these sweet flowers."[22]

Clement Marot anticipated Sternhold in 1533 in France by using metres and tunes associated with the boudoir and the ballade. He won the Dauphin's admiration for *ces sanctes chansonettes*.[23] The age produced more elegant English versions of the Psalms, such as those of Philip Sidney or Edmund Spenser,[24] but none so suitable for common praises as the mediocre verse of Sternhold and Hopkins.

Yet another consequence of the Biblicism of the time was the production of anthologies of prayers and spiritual directions, which consisted of Scripture prayers from the Old and New Testaments, or recommendations to prayer drawn from holy writ. One popular book, among many of the same type, was Richard Grafton's *Praiers of the Holi Fathers*, issued about 1540.

It also became popular to include in collections of prayers those which had been composed or used by Protestant reformers or martyrs, or even brief summaries of Christian doctrine composed by such men. For example, Henry Bull's *Christian Praiers and Holie Meditations*[25] in the 1578 edition contains three works of John Bradford, the Protestant martyr, two on prayer and the third a summary of doctrine as well as his last prayer at Smithfield. Furthermore, John Daye's *A Booke of Christian Prayers*,[26] appearing in the same year, has two prayers by Calvin, and one each by Bradford, Knox, and John Foxe.

In due time manuals of devotion of a more specific character were prepared, which had in mind the needs of persons in particular callings, or suffering particular kinds of distress. Such were collections of prayers for the poor or for those about to die. In 1585 William Perkins issued his *A Salve for a Sicke Man, or, A Treatise containing the nature, differences, and kindes of death; as also the right manner of dying well*. The rest of the subtitle indicates how useful Perkins intended his prayers to be: *And It may serve for spirituall instruction to 1. Mariners when they goe*

---

[22] Thomas Harrap, *Tessaradelphus*, cited in White, *Tudor Books of Private Devotion*, p. 44; and in Stranks, *Anglican Devotion*, pp. 14-15.

[23] R. E. Prothero, *The Psalms in Human Life*, p. 51.

[24] Spenser's *Seven Penitential Psalms* are excellent poetry.

[25] All references are to the Henry E. Huntington Library copy of the edition of 1578, although the less comprehensive, first edition is of 1566.

[26] When Bull's and Daye's collections of prayer are examined in detail they will be found to have preserved many Catholic prayers also, both traditional and contemporary.

to sea. 2. Souldiers when they goe to battell. 3. Women when they travell of child. Particularly at the end of the century, with the raging of the plague, uncertainty over Elizabeth's successor, and general disillusionment, the demand for palliatives for the fear of death became widespread; it is noteworthy that the book went through six editions by 1632. Christopher Sutton's *Disce Mori. Learne to Die* (1600) met the same *fin de siècle* need, which is mirrored so hauntingly in the plays of Webster, particularly in the macabre parts of *The Duchess of Malfi.* To come to a good end was almost as important as living a good life; in this concern Renaissance bravado conspired with Christian fortitude, as one realises by reading accounts of the death of a martyr or the end of a magnifico. In each case, "ripeness is all."

Prayers for the poorer people were the special concern of Thomas Tymme and Thomas Dekker, the poet-playwright. Tymme produced in 1598 a handbook of prayers for the lower classes, entitled *The Poor Mans Pater noster, with a preparative to praier: Wherto are annexed divers godly Psalmes and Meditations.* Dekker's *Foure Birds of Noahs Ark* (1609) provides prayers for schoolboys, apprentices, servingmen, servinggirls, seamen, colliers, and even galley slaves. As remarkable as its compassion is the book's application of the Protestant principle of the priesthood of all believers, the conviction that God can be served faithfully and fully outside a monastery while discharging one's calling and duty in the secular world. The most popular of all the books aiming at the common people was Arthur Dent's *The Plain Mans Path-way to Heaven*, which was arranged "Dialogue wise for the better understanding of the simple." Issued first in 1601, it reached 25 editions by 1640.[27]

Part of the Protestant contribution to devotion was the composition of books on the culture and practice of the spiritual life, some of which were very successful. One memorable guide to prayer was Abraham Fleming's *The Diamond of Devotion*, which appeared in 1581. Another was John Norden's *The Pensive Mans Practise* (1584), which went to 40 editions. It contains "very devout and

---

[27] Some indications of the practical Puritan ethos may be divined from the nine signs of damnation as delineated by Dent. They are pride, whoredom, covetousness, contempt of the Gospel, swearing, lying, drunkenness, idleness, and oppression. Four of the seven deadly sins of the medieval Church have been retained, but the extra five have been recruited from the Old Testament, from mercantile prudence, and one—contempt of the Gospel—sums up the rejection of the New Covenant, and under it may be subsumed the other eight vices.

417

necessary prayers for sundry godly purposes, with requisite per-
swasions before every Prayer."

However, the most characteristic and influential Protestant
devotional manuals combined examples of prayer with directions
on the methods, motives, and ends of prayer. They were thus both
directories of prayers and collections of prayers. New as they
seemed, they were, in a sense, merely the Protestant continuation
of the functions of the Catholic and Royal primers. Their novelty
was in their strong Biblicism and their concern for the different
callings and conditions of men. The pioneer English Protestant
work in collecting prayers for many occasions and for various callings was Thomas Becon's *A Pomaunder*[28] *of Prayer* of 1558.[29] It
did not, however, contain directions on prayer. The quality of the
distinctive Protestant directory and anthology combined can be
discerned most readily by a survey of the contents of three most
influential books of this type.

The first of them is *A Godly Meditation*, which was issued first
in 1559 by the printer William Copland and which reappeared
successively under the imprints of Rowland Hall (1562), William
Seres (1564), John Allde (1578), and Elizabeth Allde in 1604,
1607, 1614, and 1632. To this volume were added John Bradford's *Private Prayers and Meditations with other Exercises*. Many
of the devotions were in fact Bradford's translations of the *Excita-
tiones Animi in Deum* of the Spanish Catholic humanist, Ludovicus
Vives (1492-1540). A brief meditation is followed by a short
prayer appropriate to the events of an ordinary day. Waking is to
recall that God intends to awaken the soul from the sleep of sin
and death, and the act of rising is to be a memento of Adam's fall
and "the great benefit of Christ by whose help we do daily arise
from our falling." Meals are to be occasions on which our dependence on God is recalled and also when it is remembered that the
food of the soul is Christ's body broken and his blood shed. Most
similar reflections were the thoughts of Vives, but there were characteristic additions or modifications of John Bradford, the Protestant martyr. For example, while with Vives the safe return home
at night from a journey led to thoughts of gratitude for the safety
and quiet which a house affords and the joys of our eternal home,

[28] A pomander is a sweet-smelling ball of spices carried as a preservative against
infection, and hence metaphorically a handbook of fragrant and protective prayers.
See D. S. Bailey, *Thomas Becon and the Reformation of the Church in England.*
[29] Becon also wrote *The Sick Man's Salve* in 1561, which ran to 17 editions by
1632.

for Bradford imprisoned for the faith, the occasion suggested the darkness of the religious situation under Queen Mary's reign: "So long as the sun is up, wild beasts keep their dens, foxes their burrows, owls their holes, etc., but when the sun is down then come they abroad: so wicked men and hypocrites keep their dens in the Gospel; but, it being taken away, then swarm they out of their holes like bees, as this day doth teach."[30] The last thought for the day is a reflection on sleep: "Think that, as this troublesome day is now past, and night come, and so rest, bed, and pleasant sleep, which maketh most excellent princes and most poor peasants alike; even so after the tumults, troubles, temptations, and tempests of this life, they that believe in Christ have prepared for them an haven and rest most pleasant and joyful. As you are not afraid to enter into your bed, and to dispose yourself to sleep, so be not afraid to die, but rather prepare yourself to it; think that now you are nearer your end by one day's journey, than you are in the morning."[31]

The scheme, whereby the events of each day suggested thoughts which directed the soul to God and the meaning of life, was a good one, even if the reflections themselves tended to be dutiful rather than delightful and often commonplace rather than striking or profound. The judgment of Stranks on this book is sound: "The habit of moralising common incidents and common things, which grew wearisome when carried to excess by later writers, helped to bring the highest thoughts into day to day affairs and emphasises that pathways to God lie all around us."[32] Its most happy fruitage in Anglicanism would be in the poems and devotional works of George Herbert in the following century.

The second volume of great popularity and influence (which also shows the impact of Vives and Bradford) was that published by Henrie Bull, *Christian Praiers and Holie Meditations*, about 1566. No copy of the first edition is known to exist, but the editors of the Parker Society edition issued in 1842 reprinted an edition earlier than 1570. The original compiler of the book was a Fellow of Magdalen College, Oxford, during the reign of Edward VI, who sought refuge on the Continent during Queen Mary's reign and returned during Elizabeth's. His book has the Vives-Bradford structure, but he also drew on the works of reformers other than Bradford and from the primers of Edward VI and Henry VIII.

[30] Ed. A. Townshend, *The Writings of John Bradford*, 2 vols., vol. 2, p. 239.
[31] *Ibid.*, p. 241.
[32] *Anglican Devotion*, p. 25.

The 1578 edition of the same work begins with a lengthy introduction on the benefits, preparation for, and the parts of prayer. "By the benifite of prayer," Bull wrote, "therefore we attaine to those riches which God hath laide up in store for us: for thereby we have familiar accesse to God, and boldlie entering into the sanctuarie of heaven, we put him in mind of his promises: so that now by experience we feele and finde that to be true in deede, which by the worde we did before, but onely beleeve: now we injoy those treasures by prayer, which by faith wee did before but onely beholde in the Gospell of our Lorde Iesus."[33] Then follow two of John Bradford's meditations, the first on prayer in general, and the second on the Lord's Prayer. These provide instruction on how to pray, to whom to pray, and what to pray for, and stress the Divine invitations in Holy Scripture, through which God "allureth us with manie sweet promises to call upon Him."[34] Prayer is of two kinds, petition and thanksgiving. In petitions, requests for spiritual benefits were to take precedence over pleas for corporal benefits. Three reasons of Bradford's are given for asking for corporal benefits, which are almost a theology of Protestant prayer: to acknowledge God as author and giver; to understand that the Divine providence is bountiful; and "that our faith of reconciliation and forgiveness of sins should be exercised through asking of these corporal things."[35] The introduction and Bradford meditations comprise a third of the compilation.

The main part of the book consists of prayers. First there are 14 private prayers of the Vives kind related to the events of the day.[36] Then follows a group of 50 prayers[37] and meditations suitable for use either in families or in public assemblies. These include confessions of sin and many meditations on such themes as the presence of God, the Divine providence, the power, goodness, beauty and other attributes of God, and meditations on the Passion, Resurrection and Glorification of Christ. Some of the prayers are for improvement in the Christian life, as in petitions for true repentance, for increase of faith, for "the true sense and feeling of God's favour and mercy in Christ," and for present help in temptation. Yet other prayers are for particular classes of people: for those persecuted for their faith, for the sick, for those about to die, and for women in childbirth. There are still other prayers to be

---

[33] Bull, *Christian Praiers*, using the 1578 edition.
[34] *Ibid.*, p. 35.   [35] *Ibid.*, p. 59.
[36] *Ibid.*, pp. 132-34.   [37] *Ibid.*, pp. 165-346.

said before and after the preaching of God's Word, and before and after receiving Holy Communion.

The third part[38] consists of psalms and graces before and after a meal. Then follows an amorphous collection of prayers "commonly called Lidley's,"[39] which include petitions for forgiveness, for understanding God's Word, for the leading of a godly life, and the prayer "which Master John Bradford said a little before his death in Smithfeelde." The Litany ensues, including 15 prayers. The book concludes with "A Godly Instruction, conteining the Summe of all the Divinitie necessary for a Christian Conscience," also the work of the Protestant martyr, Bradford.

There is little that is original in this volume. But what is impressive is its comprehensiveness, its possibly unconscious transconfessional borrowings, and the practicality of the instruction of the laity in the spiritual life. These prayers and meditations have a moving simplicity and brevity which verge on the concentrated concision of Roman collects. Here, for example, is the private prayer to be offered before going to bed: "O Lord Jesus Christ, my watchman and keeper, take me into thy protection. Graunt that my body sleeping, my minde may watch for thee, and be made merrie by some sight of that celestiall and heavenly life, wherein thou art the king and prince, together with the father and the holie Ghost, where the angels and holy soules be most happy citizens. Oh purifie my soule, keepe cleane my bodie, that in both I may please thee, sleeping and waking forever. Amen."[40] Echoes of Compline still sound through these words, like a Catholic angelus heard in a Protestant harvest field.

The third volume of wide popularity and influence was John Daye's *A Booke of Christian Prayers and Meditations in English, French, Italian, Spanish, Greek, and Latin*, issued in 1569. The edition of 1581,[41] less cosmopolitan and scholarly and more national and popular in appeal, was entitled *A Booke of Christian Prayers, collected out of the aunciēt writers, and best learned in our tyme, worthy to be read with an earnest mynde of all Christians, in this daungerous and troublesome dayes, that God for Christes sake will yet still be mercyfull unto us.* The title is an interesting product of piety and mercantile prudence. The page opposite the title page illustrates the same spirit with a portrait of Queen Elizabeth at prayer. An attractive feature of this edition is

---

[38] *Ibid.*, pp. 347-62.   [39] *Ibid.*, pp. 362, 408.   [40] *Ibid.*, p. 164.
[41] All succeeding references are to this edition.

that it has marginal illustrations of the Christian virtues and of the various ranks in the social hierarchy, from emperors to peasants, suggesting that their secular callings are potentially sacred.

The work consists of 122 prayers. Their subjects and the order in which they are arranged are similar to the work of Henry Bull examined earlier. It is therefore only worth commenting on its distinctive contribution. It lacks the treatises on spirituality and doctrine that Bull's work has, providing a prefatory note of only a single page on private prayer. It does, however, have three prayers for the universal Church, and three for the queen, as well as 15 for the remission of sins, 14 against despair, four prayers to be said before the receiving of Communion, and four prayers to be said at the visitation of the sick. The Vives-Bradford influence is strong in the opening part of the work, and four prayers are acknowledged to be after St. Augustine.[42] The most pronounced characteristic of the work, apart from the marginal illustrations, is its strongly pragmatic bent, as seen in the number of petitionary prayers, seeking cleanness of heart, faith, love towards Christ, true mortification, "continuance in seeking after Christ,"[43] and "for Christ's direction and success in all our doings."[44] The intercessory element is admirable, as already seen in its fine prayer, "for them that be in povertie," but which is also found in three prayers for love towards our neighbour,[45] a prayer of children for parents,[46] and prayers for the persecuted,[47] for such as are in adversity,[48] and two prayers for our "evilwillers."[49]

4. Catholic and Protestant Spirituality Compared

The most striking difference between Catholic spirituality and Protestant devotions is in the locus and ethos of their spiritual training. For the Catholic it was the monastery or the convent in which the life of perfection was sought. For the sixteenth-century Protestant it was in the family that Christian nurture begins and

[42] Sister Marian Leona, a scholar working at the Huntington Library in San Marino, has been kind enough to let me read her manuscript, as yet unpublished: "Renaissance Prayers: A Sixteenth Century Ecumenical Encounter." She carefully documents the borrowings of the prayers of Juan Luis Vives, or Lodovicus, by Protestant compilers of prayers. His first prayers were published in Bruges in 1535 and appeared in England included in *Preces Privatae* of 1539, and next in the Royal Primer of 1545, which prints seven of them. They appeared in English Protestant manuals of devotion as late as the Elizabeth Allde edition of *A Godly Meditation* of 1633.

[43] Leona, "Renaissance Prayers," p. 103.
[44] *Ibid.*, p. 114.   [45] *Ibid.*, p. 50.   [46] *Ibid.*, p. 49.
[47] *Ibid.*, p. 51.   [48] *Ibid.*, p. 52.   [49] *Ibid.*, p. 54.

matures. For both, of course, the Christian life was nourished in the larger public gatherings for worship, whether at Mass or the elevation of the Word. But the primary foundation of training was different.

The ethos of Catholic and Protestant spirituality was also different. For Catholicism the end of the spiritual life was unity with God, along the arduous path of mysticism. It was, to adapt St. Bonaventura's daring book title, an *itinerarium cordis ac mentis in Deum per ardua*. It involved great sacrifices of earthly love for the Divine love who is also lightning. It meant a perpetual striving for purity, for the will ductile to God's every leading, and an unregretted abandonment of the world. It meant listening to the Dominical counsels of perfection with attentive ear and heart. All of this was easier for Protestants to criticise than to comprehend, for it was a lifelong habit of discipline. For the contemplative man's true end is the vision of God, glimpsed mystically here and now as the mists divide about the towers of felicity's home, but hereafter Jerusalem with all its lights ablaze and God face to face. This was a concept several high Anglicans had cherished; but it did not attract many Protestants. It would rarely occur to a Protestant to define prayer as the means to unity with God as St. John Fisher did so eloquently: "Prayer is like a certeine golden rope or chaine lett downe from heaven, by which we endeavour to draw God to us whereas we are more truly drawne to him. . . . This rope or golden Chaine holy S. Dionisius calleth prayer, which truly is let downe to us from heaven and by God himselfe fastened to our hearts. . . . Lastly, what other thing doth it, but elevates the mind above all things created, soe that att last it is made one spirit with God, fast bound unto him with the Golden line or Chaine of Prayer."[50]

For the much less contemplative and more active Protestant the true end of man was to do the will of God. His end was ethically and, to a lesser extent, mystically conceived. His love for God was proved by obedience. He believed that the true test of Christian discipleship was not to use the adoring titles in approaching Christ, but to do his will. It is significant that the Protestant used the word "service" equally to refer to divine worship and to giving human help. He believed that he should love the brother he had seen rather than be content with adoring the invisible God whose ikon

---

[50] *A Treatise of Prayer and the Fruits and Manner of Prayer* (1640), pp. 19, 20, 22, cited by Kelly, *Prayer in Sixteenth-century England*, p. 29.

was Christ. Such love for Christ as he felt must issue in action for the brethren for whom Christ died. It is, however, to press the contrast between Catholic and Protestant concepts of prayer too far to claim, as Faye Kelly does by implication, that while Fisher never used prayer as a means of obtaining personal favours, "never a means of improving one's self except in the sense of drawing closer to God,"[51] Protestants have no such scruples, since the bulk of prayers for them took the petitionary form. One has only to refer to William Perkins, the greatest of the sixteenth-century Puritan theologians, with his insistence that all prayer is to be grounded on God's word and a seeking of His will,[52] to refute the imputation that Protestant prayers were personal whims. Protestant pragmatism never ceased to be amazed by the grace of God offered to the undeserving, but it was not "lost in wonder, love, and praise" of God; it sought to get on with God's work. No one familiar with the works of charity and mercy undertaken for so many centuries by the Catholic church will lightly accuse that great Church of ethical escapism, but he might consider that some forms of spirituality cultivated by the more exalted monks and hermits, such as the Carthusians and the Calmaldolese, were obsessively and exclusively other-worldly. While never denying the primacy of the love of God as the motivation for all human concern, the Protestants believed that God was best loved in his children, especially the neediest.

Another difference between Catholic and Protestant pieties was dependent on the time when salvation was believed to be complete. For the Protestant, salvation was the accomplished work of Christ.[53] The elect soul was saved when it was called, being predestined to salvation. Acts of sanctification by that soul were then merely the proofs of the "calling," or election. For the Catholic, salvation was complete only when the purified soul emerges into heaven from purgatorial cleansing. Prayers for the dead were therefore a significant part of Catholic piety. They play no part at all

[51] *Ibid.* With this should be compared the view of Gordon Wakefield, *Puritan Devotion*, p. 33: "The Puritans have as much to say about Mystical Union, as about Divine Election in their doctrine of the Church." See also T. F. Torrance, *Kingdom and Church*, pp. 100ff.

[52] Perkins, *Cases of Conscience*, as republished in the posthumous folio edition of the *Works* (1611), vol. 2, pp. 63-64.

[53] One example is John Owen's Eucharistic doctrine of the finished work of Christ for men: "He [Christ] does not tender himself as one that *can* do these things . . . nor does he tender himself as one that *will* do these things upon any conditions as shall be prescribed unto us; but he tenders himself unto our faith as one that *hath* done these things. *Works*, ed. Goold, vol. 9, p. 565.

in Protestant piety. Furthermore, Catholics naturally invoked the prayers of the saints, particularly of the Blessed Virgin, for themselves and for their beloved dead. The intercession of the Mother of Christ or of the saints played no role in Protestant piety. The books of personal and familial prayers mirror these important differences of belief and practice.

Catholics were sufficiently realistic about human nature to know that only a few elect souls have the gifts and the opportunity to become saints and preeminent pointers to God. Protestants, to use C. S. Lewis's metaphor, dropped the honours degree in spirituality for the passing degree a much larger number of students were able to attain. The result was that Protestants did not seek a "religious" vocation. They desired, instead, to be religious in their family life and secular vocations. This is clearly indicated in the multitude of prayers for persons in different callings in the Protestant manuals of devotion. Thomas Becon, for example, in *The flower of godlye praiers* (1561), specified the relationship of the different groups of vocations to their masters, to their work, and to God. Landlords pray, for instance, to God that they may "remember themselves as tennauntes and not racke and strech our rentes" and recalled that they were "straungers and pilgrimes in this world having here no dwelling place, but seking one to come."[54] Becon proposes that merchants pray that they may "so occupye theyr marchaundise without fraude, gile, or deceite, that the common weale may prospere and floryshe with the abundance of worldly things through theyr godly and righteous travailes, unto the glory of thy name."[55]

Catholics and Protestants appear to have intended different purposes for a manual of devotion. To a Catholic a book of devotion meant an account of his duties as a member of the Church, which would spur him to accept its doctrine with firmer faith, provide him with officially approved devotions, as well as kindle his love for Christ and the saints. It might even encourage him to seek mystical experience. For the Protestant, it was thought of as a work in elementary theology, including clear instruction on the nature of prayer, which would enable him to form his own petitions in his own words. For this reason the manual would also include specimen prayers and directions for his public and private duties.

[54] Folio 39 verso.
[55] *Ibid.*, folio 40. See also Thomas Dekker, *Foure Birds of Noahs Arke*, ed. F. P. Wilson, pp. 36-39, for an excellent prayer intended for the use of merchants.

The Protestant believed that "there was no need of ecstasies, trances, visions, in order to lay hold of God. The act of understanding was of itself an act of union."[56] Thus it is not unfair to say that Catholic devotions were churchly in character while Protestant devotions were inclined to encourage independence of ecclesiastical authority and greater dependence on the Bible and the thoughtfulness of the individual.

It may well be true that Catholic and Protestant devotions, while Christ-centred, concentrated on different aspects of the Christian's role. So much of Catholic devotion seems to have been Incarnation-centred, so much of Protestant devotion Cross-centred. Or to put it in another way, the Catholic lovingly concentrated on the human aspects of God the Son: his relationship to his Mother, and his example in defeating temptations in the wilderness, in teaching, healing, praying, and loving; while the Protestant concentrated on Christ's redemptive function, as Mediator of forgiveness, as pioneer and perfector of faith, as well as giver of eternal life to the faithful. It seems as if the Catholic concentrated on the benefactor and the Protestant on the benefits; that would be one reading of the pragmatism of Protestant lay prayers and the preponderance of petitions over thanksgiving and even over intercession. Another interpretation would, as indicated before, stress the contemplative aspect of Catholic and the practical aspects of Protestant piety.

On the whole, it seems true that the Catholics emphasized the human aspects of Christ, while the Protestants concerned themselves with the divine power and gifts of the Saviour and Redeemer who delivered men from sin, suffering, and death. Yet even when this difference has been stressed, it cannot be made absolute. The ever-present crucifixes in Catholic churches, quite apart from the annual renewal of the climactic events of Holy Week, were a reminder that the image of the Virgin Mother with the Holy Infant in her arms, whether hieratically depicted as the throne of wisdom in Romanesque art or as the sublimest evocation of maternal tenderness, as in Gothic art, was, while dominant, far from being the only image of Christ in historic Catholicism. One recalls that certain humanistic and rationalistic forms of Protestantism have seen the role of Christ less as saviour than as teacher and exemplar. The difference is rather a difference of interest, or in the words of

[56] The citation comes from Stranks, *Anglican Devotion*, pp. 246-47; the entire paragraph is dependent on this source of information.

Helen C. White, the expert on Tudor devotional life, "this shift in the attitude towards Christ's life on earth is probably the key to the most significant difference between English devotion in the fifteenth century and the seventeenth."[57]

There were other, minor, differences. One notes, for example, the decidedly national character of Protestant prayers, as contrasted with the typically international Catholic devotions. This was often more than the fulfillment of the Biblical injunction to pray for all in authority; it frequently led to "political" prayers practically indistinguishable from royalist propaganda. Through the various royal primers and the required prayers in the Book of Common Prayer, as well as in the approved Protestant manuals (which required episcopal approval before being given a license for publication), there was a sense of the sacredness of the sovereign, as if he or she were the eye of God. The handbooks implied in their prayers for the queen and the magistrates that the maintenance of religion, the present civil authority, and the social order itself were interdependent. As Faye Kelly so justly remarks: "It is not always easy to be quite sure whether religion is being used to defend the power of the state or whether the power of the state is being marshalled to defend the existing religious order."[58]

Another difference was the Protestant preference for Biblical phraseology, as well as doctrine, wherever and whenever possible, as if it were a talisman against superstition and corrupt tradition. The intention was expressed by Thomas Becon in *The flower of godlye praiers*: "I have travailed to the uttermost of my power to use in these prayers as few words of my own as I could, and to glean out of the fruitful field of the sacred scriptures whatsoever I found meet for every prayer that I made, that when it is prayed, not man, but the Holy Ghost may seem to speak."[59]

The most surprising characteristic in an acrimonious and controversial age was the phenomenon I referred to earlier—the retention by Protestants of the prayers of Catholics, such as those of St. Augustine or the contemporary Erasmus and Juan Luis Vives. St. Augustine's *Confessions* had been a classic of spirituality for a thousand years, and their author seems today to be the foster par-

[57] *Tudor Books of Private Devotion*, pp. 246-47.
[58] Kelley, *Prayer in Sixteenth Century England*, p. 63.
[59] *Works*, vol. 3, pp. 12-13. A further consequence of the Biblicism for Protestants and especially for Puritans is that the latter substituted the patriarchs and the prophets for the Catholic saints as exemplars and companions of the godly way. See Wakefield, *Puritan Devotion*, p. 27.

ent of both Catholic and Protestant Christianity, so that his inclusion is less surprising than that of contemporaries. Erasmus had remarkably put his finger on the pulse of the times; his streamlined Catholicism proved very attractive to Luther and other first-generation Protestants. To cite the prayer of his most commonly used by Protestants is to understand why he spoke to their condition, as in this "Prayer for Peace in the Church" in an age of turbulence and confusion: "Thou seest (O good shepherd) what sundry sortes of wolves have broken into thy shepecotes, of whom every one crieth, Here is Christ, here is Christ, so that if it were possible the very perfecte persons should be brought into errour. Thou seest with what wyndes, with what waves thy sely shyp is tossed, thy ship wherein thi litle flock is in peril to be drouned. And what is nowe lefte, but that it utterly synke and we al perish."[60] What, however, was the secret of the attraction for Vives? It was more than the fact that he had been brought over to England from Bruges to help tutor Princess Mary, the elder daughter of Henry VIII and Catherine of Aragon, and that he had been a lecturer at Oxford, residing at Corpus Christi College. The fact that he was imprisoned by Henry VIII for taking Catherine's side in the "King's Matter" might have endeared him to later Protestants, were it not for the fact that he recanted to gain his freedom. His attraction lay in the informality and naturalness of the prayers, their intensely personal character, and their practicality being so directly related to the everyday experience of every man. Finally, it was a simple matter to adapt his prayers for Protestant use. Whatever the reasons, the phenomenon of prayers that crossed the vast religious divide of the sixteenth century is an intriguing and important one, and a promise and presage for the future.

5. Puritan Spirituality

At the outset Puritan spirituality would seem to be no more than a profoundly Biblical form of devotion lived in deep intensity and seriousness. Apart from such depth of commitment it was undistinguishable from Tudor Anglican spirituality. This is a true, but not wholly true, view. Anglican spirituality approximated that of Puritanism, but it also had a continuing link with Catholicism in Lancelot Andrewes and his successors. Protestantism had its

[60] *The Primer in Englishe and Latin set foorth by the Kinges Maiestie and his Clergie*, sig. R5-R5verso.

Lutheran, Calvinist, and Free church traditions; Puritanism's affinity was not with Lutheranism, but it *was* an English form of Calvinism with great variations which have later been developed into what is now known as the Free church tradition.

The first characteristic of Puritan spirituality was that it flourished best in the life of the family. The Puritan household was a little church, with the father as its father-in-God. His duty was to conduct the weekday prayers both morning and evening for his wife, the children, the servants, and possibly any apprentices. On Sunday, after returning from worship, it was his task to rehearse the children in their catechism and to check whether they had properly memorised and understood the main points of the sermon, and to arrange for reading aloud from the Bible or some other godly books, whether from the sermons of an approved minister or from the Protestant classic, Foxe's *Book of Martyrs*.[61] In these duties he was aided by the Bible, along with the manuals of devotion and prayers, which even provided him with the words of appropriate graces before and after meals. Serious, stringent, inflexibly disciplined as it may seem to later generations, this spirituality was mingled with the joy of psalm-singing[62] to catchy tunes and motivated by the sense that this is exactly how God's elect must live in a world of snares and traps laid by the ungodly, until they gain the reward of joining the everlasting community of the friends of Christ. This was the life of sanctification; this was how "saints" were matured. But they were saints who rejoiced in their marriages and in their children. Theirs was an intramundane asceticism for an extramundane reward. The monastery was the training ground of Catholic spirituality; the family was the seed-plot of Puritan Christian life.

The location of Puritan spirituality in the family was aided by two powerful factors. The first was the fact that religion was not associated with a sacred building for them, as it was for Catholics or for High Church Anglicans. When they eventually separated from the Church of England and erected their own buildings for worship, they called them "meeting-houses," not "churches." For them the word "church" referred exclusively to the company of

61 Lady Margaret Hoby records in her diary that she listened one Lord's day to someone reading from the Book of Martyrs, and on a weekday heard her minister read from the sermons of Richard Greenham. *The Diary of Lady Margaret Hoby*, ed. Dorothy M. Meads, pp. 75, 86.

62 Knappen, in *Tudor Puritanism*, pp. 388-89, calls such "rouse-ments," that is, "emotional trimmings to reach the affections as well as the mind."

God's faithful people, the *communio electorum*, gathered to hear the reading and exposition of the oracles of God. Thus they could by the easiest of transitions conceive of the "church in the house" on the analogy of the gatherings of persecuted Christians of the primitive church, or use the even earlier precedent of the house meetings of the Apostolic church.

The second factor was the new significance given family life by the Reformers. God's covenant, invariably read at Baptismal services, was "to you and your children." This recognition helped weld family life into an enduring solidarity in Christ. The head of the family, who had promised at the Baptism of his children (Puritans did not approve of sponsors taking the vows instead of parents) to supervise their Christian nurture, was duty bound to teach his children the Scriptures and to make sure they understood the main points of Christian doctrine, behaviour, and worship. His promises at Baptism, his commitment in the covenant of church membership, the examples of the godly in the Scriptures which he read as God's marching orders, and the responsibility for which he must account to God at the Great Assize on the Day of Judgment, all confirmed him in the duty of being a prophet and priest to his own household. As J. Mayne wrote in *The English Explained*: "The parents and masters of families are in God's stead to their children and servants . . . every chief householder hath . . . the charge of the souls of his family."[63] The Puritan head of the household was truly fulfilling the role of the priesthood of all believers while conducting family prayers.

All forms of Christian spirituality claim to be Biblical, but each is dependent on a different selection from the Scripture; the proof passages are often taken from the same book of the Bible. Catholic sacramentalism made considerable use of the Fourth Gospel, and especially of the Prologue affirming that Christ, the Word became flesh. This was the foundation of the sacramental principle. Yet the Society of Friends used the same Gospel to affirm the priority and sovereign freedom of the Holy Spirit which, like the wind, blows where it pleases, and to claim that the inner light is the illumination of every man by Christ—assertions which challenge institutional ecclesiasticism to the full. Similarly, the most iconoclastic of the sacramentarian Protestants, the followers of Zwingli the Reformer of Zurich, turned to the sixth chapter of the same

[63] Cited by Hart, *Country Clergy*, p. 200.

Gospel[64] for support for their Memorialist doctrine of the Lord's Supper. Puritan theology and spirituality was predominantly Pauline, rather than Johannine, although the Puritans made good use of Augustine and Calvin as his interpreters.

With Paul, they saw the Christian life in terms of a historic recapitulation which is reenacted in the soul of every man. The second Adam, Christ, undoes the evil effected by the first Adam, and leads the bewildered and frightened soul from paradise lost to paradise regained. He is, in the military metaphor so appropriate for these spiritual men of steel, the captain of their salvation. The Scripture, in which the external Word of God is attested to by the *testimonium internum Spiritus Sancti*, is their iron ration: they would have no truck with tradition. Neither Popes nor bishops, but the Holy Spirit, would be their teacher, with occasional assistance from the Fathers or their own contemporary theologians, all of whom claimed to be following the guidance of the Holy Spirit. They took seriously the words of the Word of God when tempted in the wilderness: "Man shall not live by bread alone, but by every word which proceeds from the mouth of God." Their weapons were, as recorded in Ephesians,[65] the sword of the spirit which is the Word of God, and the shield of faith which extinguishes the flaming darts of the Evil One. Their armour included the helmet of salvation, the breastplate of righteousness. Their protection was integrity and their sandals were the preparation of the pacifying gospel. Their lifelong concern was to fight the good fight of faith; to this end they prayed incessantly, watched the progress of their souls in diaries which recorded their fears, failures, and solaces, sang psalms, and held private sessions of fasting when under the judgments of God, and of thanksgiving at some family occasion for joy. Godly reading, private and family prayers, making and renewing churchly and personal covenants, and diligence in their daily calling and active sharing in worship, together with earnest and active prosecution of the duties of charity, were their way of living the Christian life.

The Puritans had two immense sources of comfort and encouragement. One was their belief on predestination. Its premise—universal depravity—"by levelling all superiority not of the Spirit, enormously enhanced the self-respect of the ordinary man."[66] The

---

64 See John 6:67 as a support of Zwinglian doctrine, while 6:49-51 supports Catholic sacramentalism.
65 Ephesians 6:10-17.
66 William Haller, *The Rise of Puritanism*, pp. 90f.

only true aristocracy was created by God, not by blood or inheritance. It was an aristocracy of the spirit and character. The triumph of God's elect was foreordained: "If God be for us, who can be against us?"[67] The work of salvation, being wholly God's, could not fail. The second source was that Paul had provided a chart of the stages in the progress of the Christian soul on pilgrimage. It was to move from election, through vocation, justification, and sanctification, to glorification. The Puritans *knew* this was the sure path through the vale of tears to eternity. Their ministers encouraged the laity to develop introspection and thus enable them to travel hopefully. As I have mentioned, one popular guide of this type was the work of the moderate Puritan, Arthur Dent, appropriately titled *The Plain Mans Pathway to Heaven*. If the military metaphor be preferred, and the struggle for spirituality was seen as an interior civil war of the spirit against the flesh, then the equivalent book, also of great popularity, was John Downame's *The Christian Warfare*.[68] Then there was always Foxe's *Book of Martyrs* to stir up their latent heroism.

The Puritan spiritual life has been criticised as too intense and rigid. Its introspection has been considered as spiritual valetudinarianism. From a modern standpoint it has seemed not only too inhibited, but also a dull routine. This judgment, however, is caused by a failure to recognise that even if the Puritan was rarely jocular, he had deep springs of joy. Richard Rogers wrote in his diary: "But we must finde that our harty imbracing of the doctrine of god and love of it and labouring after a good consc[ience] to find ioy in Christes redeeming us is that which maketh our lives ioiful, for this cannot by any malice of man nor devil, be taken from us. I had exper[ience] of this of lat."[69] Moreover, on another occasion, he indicated how singing psalms and other activities had cheered him: "my mind hath veary heavenly been exercised in considering, both by med[itation], singing ps. 119: 14, 15, 16, and confer[ence], of godes goodnes in cheering our hartes with the bottomless and unexpressable treasur of his word, and feling of his favour, and inioying of his benef[its]. . . ."[70]

The Puritan spiritual life was earnest, intense, and well-disci-

[67] Romans 8:32.

[68] This book, originally a group of sermons, is described and evaluated in detail in Chap. VII. It is worth recalling that *The Christian Warfare* (pp. 235-47) enumerates 10 "signes and infallible notes of our election."

[69] *Two Elizabethan Puritan Diaries by Richard Rogers and Samuel Ward*, ed. M. M. Knappen, p. 73.

[70] *Ibid.*, p. 99.

plined. It seems rigid, inflexible, and sour to its critics. To Shakespeare, satirising it in the character of Malvolio in *Twelfth Night*, it smacked too much of preciousness and complacency. To Ben Jonson it was a dreadfully boring regimen. In *The Silent Woman* Truewit tells Morose that if he marry one who is "precise" ("Puritan" and "Precisian" were interchangeable terms), he will have to listen to "long-winded exercises, singings, and catechisings." Jonson is even more scathing of Puritan prolixity, pretentiousness, and arguments over predestination in *Bartholomew Fair*, as he describes a rich old Puritan widow seen through the eyes of Quarlous, who says to his friend:

> Dost thou ever think to bring thine ears or stomach to the patience of a dry grace as long as thy table-cloth; and droned out by thy son here . . . till all the meat on thy board has forgot it was that day in the kitchen? or to brook the noise made in a question of predestination, by the good labourers and painful eaters assembled together, put to them by the matron your spouse; who moderates with a cup of wine, ever and anon, and a sentence out of Knox between? Or the perpetual spitting before and after a sober-drawn exhortation of six hours, whose better part was the hum-ha-hum? or to hear prayers groaned out over thy iron chests, as if they were charms to break them? And all this for the hope of two apostle-spoons, to suffer! and a cup to eat a caudle in! for that will be thy legacy. She'll have conveyed her state safe enough from thee, and she be a right widow.

6. *The Products of Piety*

We have seen how different the Catholic and Protestant forms of devotion were. It will be interesting to observe that there were few marked differences between the products of these pieties. For that reason I will compare a Catholic and a Puritan noblewoman, a Catholic and a Protestant martyr, an Anglican bishop and a Puritan minister. Also, I will consider the spirituality of an Anglican squire and an Anglican minister of state.

Lady Magdalen Montague was a notable chatelaine and a Catholic.[71] Hers was an unobtrusive spirituality, so her chaplain informs us, even when she was a young maid of honour at Queen Mary's Court:

[71] See Chap. IV for further information on the Lady Magdalen, Viscountess Montague and her spiritual regimen.

she accustomed to rise from her bed very early, and attiring herself with all possible speed hastened to the chapel, where, kneeling against a wall and the other part of her face covered with her head attire, she accustomed to spend certain hours in devout prayer and to shed abundance of tears before Almighty God. And yet withal would she not be any time from any office of piety prescribed to her and her companions. Neither was she content in this sort to spend the day, but arose from her bed in the night, and prostrate on the ground applied herself to prayer a good part of the night. Which when the Lady Magdalen [her mother] had once perceived, she finding her devotion to be discovered, no otherwise than if she had been apprehended in some notorious lewd fault, falling on her knees, with many tears, she besought her for the honour of God that she would not bewray her secret exercise of piety to any creature while she lived.[72]

As a widow she was deeply concerned for her husband's soul, "for whom she twice every week caused mass to be said, and herself said the office of the dead."[73] Every year she commemorated the anniversary of his death, and thanked God at almost every meal for the bounty that came to her through her husband. At "Little Rome," in one of her country homes, though a Recusant she caused impressive celebrations of the Mass to take place on all the most solemn festivals; each week a sermon was preached in her chapel. Hers was almost a monastic piety, for she heard three Masses each day and read three Offices. Most impressive was her large liberality to the eighty persons or more whom she supported financially to help to maintain their Catholicism in days of persecution:

> She maintained three priests in her house, and gave entertainment to all that repaired to her, and very seldom dismissed any without the gift of an angel. She redeemed two out of prison at her own cost, and attempted the like for others, and gave money to other Catholics both in common and particular. Her alms, distributed every second day at her gates to the poor, were plentiful and such as some of the richer Protestants did calumniate that they augmented the number of beggars and nourished their idleness. When she desisted from her prayers, she accustomed to spend much time in sewing shirts

[72] *An Elizabethan Recusant House, comprising the Life of the Lady Magdalen, Viscountess Montague (1538-1608)*, ed. A. C. Southern, pp. 12-13.
[73] *Ibid.*, p. 23.

or smocks for poor men and women, in which exercise she seemed to take much pleasure; sometimes, also, when she had leisure, she visited the poor in their own houses and sent them either medicines or meat or wood or money, as she perceived their need. . . .[74]

She was no cardboard saint, too good to be true. She was attractive enough to appeal to the wandering eyes of a prince, and brave enough to slap down his intruding hand. Though spiritual, she was also a woman of spirit.

Her Puritan counterpart was Lady Margaret Hoby, who lived at Hackness in Yorkshire and seems to have been almost equally devout, using all opportunities for worship and instruction in the Christian life with great assiduity. She was a notable benefactor of the local Anglican parish church, but her resources (though perhaps not her concern for her neighbours) were considerably less than those of Viscountess Montague, so that we read little about her charities. A typical weekday was spent thus: "After privat praiers I did read of the bible, then I went to publeck [worship?], after to work, then to breakfast: and so about the house: after, to dinner: after dinner I did see Lightes made allmost all the after none, and then I took a Lector, then to supper: after, I hard Mr. Rhodes Read of Grenhame [Greenham the Puritan divine], and then I praied and so went to bed. . . ."[75] The red letter day of the Puritan calendar was the Lord's Day. This is how Lady Hoby spent one: "After privat praier I went to church wher I hard the word preached, and received the sacramentes to my Comfort: after I had given thankes and dined, I walked a whill, and then went to church, whence, after I had hard Catcizising and sarmon, I returned home and wret notes in my bible, and talked of the sarmon and good thinges with Mrs. Ormston: then went to praier, after to supper, then to repetecion of the wholl daies exersice and praiers, hard one of the men reade of the book of marters, and so went to bed."[76]

Both women attended public worship, both took part in private and household prayers; both were obviously helped by the chief sacrament, and enjoyed sermons. The difference between them was that the devotions of Lady Montague are more objective and structured, while those of Lady Hoby are as varied as sermons, extemporaneous prayers, and Biblical note-taking can be, with the

[74] *Ibid.*, p. 40.   [75] *Diary of Lady Margaret Hoby*, p. 86.
[76] *Ibid.*

exception of the catechising, which must have had a formal structure. One doubts whether there was much difference in the time they spent at their devotions or their secular duties, or in the dedication with which they were performed.

Next we compare a Catholic and a Protestant martyr, Thomas More (1478-1535) and John Bradford (c. 1510-1555), each of whom paid the supreme tribute of loyalty unto death. Thomas More was, of course, a famous humanist and friend of Erasmus and Holbein. He was the author of *Utopia* and a legal luminary, who was the first lay Lord Chancellor of England under Henry VIII; he was a man of wit, wisdom, and courage. When he was about 21 he thought seriously about adopting the ascetic life. Taking a lodging at the Charterhouse, he submitted himself to the discipline of a Carthusian monk. From 1499 to 1503 he wore a hairshirt next to his skin, scourged himself every Friday and other fast day, allowed himself only four or five hours sleep each night, lying on the bare ground and using a log for a pillow. Though he abandoned the idea of relinquishing the world, he continued for the rest of his life to be scrupulous in fulfilling his religious duties. Twice married and an exemplary husband and father, he was particularly close to his learned daughter Margaret, whose husband, William Roper, wrote the earliest biography, entitled *The Mirror of Vertue in Worldly Greatness; or, the Life of Syr Thomas More*, during Mary's reign in manuscript form, which was printed in Paris in 1626. His own testament of devotion, intended chiefly for his family, was printed by William Rastell in 1533, as *A Dyaloge of Comfort against Tribulacion*. It was here in prison, like his friend, Bishop John Fisher, that his deep faith and spiritual life were really put to the test as he faced the imminence of death. That harsh, horrific, yet holy event, took place on July 7, 1535. Four centuries later More was canonized by Pope Pius XI.

As a layman More knew that the greatest danger to the spiritual life is an inability to detach oneself from the entanglements of the world. It is for this reason, he believed, that God sends tribulation, a subject which More treated more exhaustively than anyone else before or since, mainly in the three books of his *Dialogue*. Our temptations are our severest test as Christians. Here we do battle with a formidable adversary, the devil, who is both stronger and more brilliant than we are. Should he win, our fate is death, which does not only separate body from soul, but "the whol entire man . . . from the fruition of the very fountain of life, Almighty Glorious

God."[77] Yet there is no reason to be terrified by Satan, for though he can defeat us, God is more than a match for him. Our shield is God's truth, and we cling to God by faith, for which we must continually pray, saying with the apostles, "Lord increase our faith," or with the father of the dumb child, "I believe, good Lord, but help thou the lack of my belief!"[78]

As Germain Marc'hadour pointed out,[79] the Christendom into which More had been born had lost its militant spirit and the decay of chivalry was symptomatic of the softening of the soul's fibre. Erasmus, influenced by Colet as was More, was critical of the passive, compliant, sentimental and fugitive spirituality then dominant. He published his *Enchiridion*, a "handbook" or "hand-dagger" for the Christian knight's daily warfare. More embodied this new type of Christian man, informed and energetic, ready for attack as well as defence. Although he believed that the roles of leadership in the Church were reserved for the priests and bishops who were the salt of the earth spoken of in the Sermon on the Mount, he could not agree that the Church was to be equated with the clergy. He believed that "God hath given everyman cure and charge of his neighbour."[80] He implied that others should share with him the duty of confuting Protestant pamphleteers, but to do that properly meant that they must better understand the faith. There might be a danger in intellectual pride, but the greater danger lay in intellectual laziness. His mind was fully in accord with the principle of Anselm, *fides quaerens intellectum*. More was convinced that less time should be devoted to logic and more to the divine revelation in Holy Scripture. In More's *Life of Pico* there is an appendix in which is found Pico's advice to his nephew, with which More cordially agreed. It reads, in part: "Thou mayst do nothing more pleasant to God, nothing more profitable to thyself, then if thine hand cease not day nor night to turn and read the volume of Holy Scripture. There lieth privily in them a certain heavily strength, quick and effectual, which with a marvellous power transformeth and changeth the reader's mind into the love of God."[81]

Since this marvellous spiritual bread of God's Word is meant

[77] *The Worke of Sir Thomas More Knyght* (1557), ed. William Rastell, p. 1,286. This is the first English edition of the *Works*.
[78] *Ibid.*, p. 1,143.
[79] The essay on the spirituality of Sir Thomas More by Germain Marc'hadour, in *Pre-Reformation English Spirituality*, pp. 224-40, has proved invaluable both for its thoughts and citations in the ensuing five paragraphs.
[80] Rastell, ed., *Worke of Knyght*, pp. 143, 279.
[81] Cited by Marc 'hadour in *English Spirituality*, p. 228, p. 13.

for all men, More made it clear that he fully approved the production of an English translation.[82] The truth of the Scriptures may be misinterpreted by heretics, because they ignore or defy the Catholic church, which was the "perpetual apostle" of Christ.[83] More's spirituality was Christo-centric and Church-centred at the same time. He believed that the fullest life is to be found in Christ; after His glorification, Christ is present chiefly through the word of the Scriptures and through the Blessed Sacrament. Perhaps it was this sacramental element in the piety which distinguished More from the Protestant martyr most clearly. For More the essence of the Eucharist was the communion with God and men that it created and sustained. He wrote: "The *thing* of the Blessed Sacrament is the unity and society of all good folk in the mystical body of Christ. . . . Our Saviour is the worker of this communion: in giving his own very body into the body of every Christian man, he doth in a certain manner incorporate all Christian folk and his own body in one corporation mystical."[84] More was fully convinced of the importance of human solidarity and of the profounder solidarity of souls in the Church, of the Church militant on earth united with the Church triumphant in heaven, and the souls being purified in purgatory. Heaven, indeed, was for More the jubilant society where God's folk are merry together with "the Trinity in his high marvellous majesty, our Savior in his glorious manhood sitting on his throne with his immaculate Mother and all that glorious company."[85]

Marc'hadour suggests that the secret of the exhilaration in More's spirituality was not jocosity, but its ability to teach us "the art of transmuting everything—the drab, the tedious, the tragic—into gladness, through 'clearness of conscience' and the conviction that the Almighty is also the All-loving."[86] This is the meaning of More's own *riddle*: "Though I might have pain, I could not have harm. . . . In such a case a man may lose his head and suffer no harm."[87] The malice of others cannot defeat God. With the Psalmist, More was convinced that underneath are the everlasting arms of God and that death is a rebirth to eternal life.

There were two spurs to send him up to the high country of spirituality if life seemed dull and stale. One was the spur of fear, fear of his own death; the other was the spur of love, of recalling

[82] Rastell, ed., *Worke of Knyght*, pp. 240-47.
[83] *Ibid.*, p. 458.   [84] *Ibid.*, pp. 1,336, 1,347.
[85] *Ibid.*, p. 1,261.   [86] *English Spirituality*, p. 234.
[87] See More's *Letters*, nos. 210, 216.

the death of the Son of God for him. Let them, said More, "spur forth thine horse through the short way of this momentary life to the reward of eternal felicity."[88] More wrote two unfinished treatises on the Passion. He clearly held the truth of the Carthusian motto: *Stat crux dum volvitur orbis.* Christ's Passion was the unmoving, still centre of divine love in the world's whirligig. For More the world was always a prison, but a happy prison if one knows that God is the chief gaoler, and that at death he throws the prison gates wide open for the liberated soul. That was the faith of a great Catholic martyr.

The Protestant martyr, John Bradford, was a much less colourful figure. It seems that he was a native of Manchester and served for some years as secretary to Sir John Harrington, the paymaster of the English forces in France. He then decided on a legal career and is said to have been converted by a sermon of Latimer's so that he was moved to restore monies he had embezzled. He read theology at St. Catharine's Hall, Cambridge, where he took a master's degree and was elected into a fellowship at Pembroke College in the same university. He was appointed chaplain to the sturdy Protestant Bishop Ridley of London. On the accession of Queen Mary in 1553 he was imprisoned for alleged sedition. For a year and a half, while incarcerated, he wrote letters of spiritual advice to many correspondents, also the polemical work, *The Hurt of Hearing Mass.* His prayers, as we have seen, became the current coin of Protestant devotions, because of their sincerity, moving eloquence, and profound convictions, and because his Christian witness was sealed with the blood of martyrdom. He died at Smithfield "where he was led with a great company of weaponed men to conduct him thither,"[89] so that the huge crowd of sympathisers would not attempt to rescue him.

Such a martyrdom had been prepared for by a life of deep consecration. It was reported of him while at St. Catharine's Hall that: "He used in the morning to go to the common prayer in the college where he was, and after that he used to make some prayer with his pupils in his chambers; but not content with this he then repaired to his own secret prayer and exercise in prayer by himself, as one that had not yet prayed to his own mind; for he was wont to say to his familiars, 'I have prayed with my pupils, but I have not yet prayed with myself.' "[90]

88 *Worke of Knyght,* p. 13.
89 *Writings of Bradford,* ed. Townshend, vol. 1, p. xi.
90 *Ibid.,* p. xix.

The life of the Tudor laity was so encircled by prayer that it is no surprise to learn of William Cecil, Lord Burghley, Queen Elizabeth's chief minister of state and the busiest man in England, that he was not too busy to attend prayers. According to the report printed in Francis Peck's *Desiderata Curiosa*,[91] Burghley always attended morning and evening prayer, and never missed a sermon on Sunday or failed to attend Holy Communion on the first Sunday of the month. Late in his life, particularly, in addition to all his other alms-givings, he gave his chaplain 20 shillings for the poor when at a sermon.

An equally devout Anglican layman was Squire Bruen of Bruen Stapleford in Cheshire, who succeeded to the family estates in 1587. His special care was to conduct the family prayers, that he considered "the very goads and spurs unto godliness, the life and sinews of grace and religion." Rising at 3 or 4 a.m. in summer and 5 a.m. in winter, he prayed in private, read and meditated on some portion of Scripture, and composed a fair copy of "some part of such sermons as he had by a running hand taken from the mouth of the preacher." Then he summoned his household to prayers, which comprised an introductory prayer imploring divine blessing on their exercises of devotion, the singing of psalms, the reading and exposition of Scripture, ending with a prayer of thanksgiving. Similar prayers were offered in the evening. Each Sunday the squire walked to Tarvin church, a distance of about a mile, "calling all his family about him, leaving neither cook nor butler behind him, nor any of his servants, but two or three to make the doores and tend the horse untill their returne." On his journey, he collected as many tenants and neighbours as possible, marching at the head of his Christian company with a joyous heart, and leading them in the singing of psalms.

Such was his reputation for holiness that a number of the gentry in the neighbourhood entrusted their children to his care and direction. Besides his own family and servants, he once had as many as 21 boarders in his home. He was a man of liberal bounty. Every week the poor people of Chester and surroundings, in addition to his own parishioners, flocked to the gates of Stapleford House to receive of his generosity.[92] The picture is idyllic, and

[91] The work was published in London in 1732, over a century after Burghley's death. The reference is to p. 45, the information from which is summarised by Stranks, *Anglican Devotion*, p. 26.

[92] The information on Squire Bruen is derived from Hart, *Man in the Pew, 1558-1660*, pp. 193-94.

perhaps far from typical, but it is a vignette of Anglican piety to set beside those of George Herbert as Rector of Bemerton or the Ferrar family at Little Gidding in the next, turbulent, century.

My final comparison is of a dedicated Puritan minister, Richard Rogers, and a gifted Anglican Bishop, Nicholas Ridley of London. Rogers is already familiar as the writer of practical theological treatises on the Christian life, which were an impressive psychology of prayer in themselves.[93] It is fortunate that his private diary[94] has survived and that it is so informative on the subject of his spiritual methods for maintaining the vigour of the Christian life. It appears that Rogers had a programme of meditation on such themes as humility, the forgiveness of sins, the preparation for the seeking of God, and peace with God. In addition, however, he set up regular times for study of the Bible, for private prayer and for family prayers.[95] Other forms of disciplining the spiritual life included keeping the diary itself as both an outlining of the goal, and a record of failure and of determination to overcome failure and the preparation of sermons for weekly occasions. An important Puritan custom that he used was the making of a covenant or solemn vow. As Haller rightly insists, "instead of promising to go on a pilgrimage or to make gifts to a shrine, the Puritan in time of stress, would vow to keep a better course, or even to cause others to do so. . . ."[96] Other forms of spiritual stimulus were psalm-singing, conferences with fellow Christians on spiritual matters, and attendance at public worship and the Lord's Supper. Not least in the Puritan's calendar were occasions of fasting. Of one such occasion, Rogers wrote interestingly that it provoked him to write his own *Seven Treatises* which ran to five editions before being abridged. Here is his account: "The 6 of this month we fasted betwixt our selves min[isters] to the stirring upp of ourselves to greater godlines. Veary good thinges we gathered to this purpose, ef[esians] 1:1-2, and then we determined to bring into writinge a direction for our lives, which might be both for ourselves and others."[97] It is not by the originality of his spiritual discipline that the Puritan is known, but rather the assiduity, endurance, and intensity of his spirituality which made the minimal concessions to the natural man. Like Jesuit spirituality, it was intended for vertebrate Christians.

[93] See Chap. VII.
[94] See M. M. Knappen, *Two Elizabethan Puritan Diaries by Richard Rogers and Samuel Ward.*
[95] *Ibid.*, pp. 65, 96.   [96] *Ibid.*, p. 8.   [97] *Ibid.*, p. 69.

I conclude with an account of the prayer and study schedule of an exemplary Edwardian Bishop, Nicholas Ridley of London. This is how he maintained the culture of the spiritual life in Fulham Palace, as recorded by Foxe the martyrologist: "Duly every morning, as soon as his apparel was done upon him, he went forthwith to his bedchamber, and there upon his knees prayed the space of half an hour; which being done, immediately he went to his study (if there came no other business to interrupt him) where he continued till ten of the clock, and then came to the common prayer, daily used in his house. The prayers being done, he went to dinner where he used little talk, except otherwise occasion by some had been ministered, and then it was sober, discreet and wise, and sometimes merry, as cause required."[98]

Piety and humour have not always been conjoined and it is intriguing that the Catholic martyr More and the Protestant martyr Ridley combined holiness with hilarity. Like so many good men in this century of divided religious allegiances, they were forced to become Catholic or Protestant instead of remaining merely Christian. Though tragically divided in life, in their deaths they were undivided, and share, so one may hope, an equal eschatology.

From the vertiginous standpoint of the present, the differences of spirituality between Catholic, Anglican, and Puritan in the Tudor age seem small. In fact, the difference between Anglican and Puritan spirituality of equal fervour was negligible, so much so that the label "Protestant" adequately covers both before the advent of the Caroline divines. All three forms of Christian spirituality rely on the inspiration of the purifying and empowering Holy Spirit. All three acknowledge that they are constrained and motivated by the love of Christ for the undeserving. All three produce the same effects: the security of faith, the serenity of hope, the peace and joy of believing, and the compassionate charity that seeks the good of others. Finally, all three pieties are the iron rations of pilgrims seeking the same goal—the shining citadels of the communion (and community) of saints.

[98] *The Acts and Monuments of John Foxe*, ed. S. R. Cattley and G. Townsend, 8 vols., VII, 408.

BIBLIOGRAPHY

1. Liturgical Texts
2. Periodicals
3. Sources in English Literature
4. Books

1. SELECTED LITURGICAL TEXTS

(Arranged Chronologically by Denomination)

ANGLICAN

Cranmer's Liturgical Projects. Ed. J. Wickham Legg. Henry Bradshaw Society, vol. 50, 1915.

The Order of Communion, 1548. Ed. H. A. Wilson. Henry Bradshaw Society, vol. 34, 1908.

First Prayer Book of Edward VI, 1549. Ed. Vernon Staley. 1903.

The First and Second Prayer Books of King Edward VI. Ed. E.C.S. Gibson. Everyman edition, 1910.

The Liturgies of 1549 and 1552 with other Documents set forth in the Reign of Edward VI. Ed. J. Ketley. Cambridge: Parker Society, 1844.

John Marbeck. *The Book of Common Praier Noted*. 1550.

Liturgies and Occasional Forms of Prayer set forth in the Reign of Queen Elizabeth. Ed. W. K. Clay. Cambridge: Parker Society, 1851.

Private Prayers put forth by authority during the Reign of Queen Elizabeth. Ed. W. K. Clay. Cambridge: Parker Society, 1851.

Thomas Sternhold and John Hopkins. *The Whole Book of Psalmes collected into Englysh metre*. 1562.

[Homilies]. *Certain Sermons appointed by the Queen's Majesty* . . . 1574. Reprinted and ed. G. E. Corrie. Cambridge, 1850.

PURITAN

John Knox's Genevan Service Book, 1556. Ed. William D. Maxwell. Edinburgh, 1931.

A Booke of the Forme of Common Prayers, Administration of the Sacraments, &c. agreable to Gods Worde, and the Use of the Reformed Churches. Published in 1584 and 1586, respectively, as the Waldegrave and Middelburg Puritan Liturgies; reissued, ed. Peter Hall. *Fragmenta Liturgica*, vol. 1, Bath, 1848, *Reliquiae Liturgicae*, vol. 1, Bath, 1847.

REFORMED

Richter, A. L., ed. *Die evangelischen Kirchenordnungen d. sechszehnten Jahrhunderts*. 2 vols. Weimar, 1846.

Smend, Julius. *Die evangelischen deutschen Messen bis zu Luthers Deutscher Messe.* Göttingen, 1896.
Terry, Richard R. *Calvin's First Psalter* (1539). Facsimile edition, 1932.
Thompson, Bard, ed. and intro. *Liturgies of the Western Church.* Cleveland and New York, 1962. This is a wide selection of Catholic, Lutheran, and a variety of Reformed Rites.

ROMAN CATHOLIC

The Prymer or Lay Folk's Prayer Book. Ed. Henry Littlehales. Early English Text Society, 1897.
The Sarum Missal in English. Tr. F. E. Warren. 2 vols. 1911.
The Use of Sarum. Ed. W. H. Frere. 2 vols. Cambridge, 1898-1901.
The Sarum Missal. Ed. J. Wickham Legg. Oxford, 1916.
Breviarium Romanum Quignonianum. Ed. J. Wickham Legg. Cambridge, 1888.
The Second Recension of the Quignon Breviary. Ed. J. Wickham Legg. 2 vols. Henry Bradshaw Society, 1908-1912.

2. PERIODICALS

DENOMINATIONAL

ANGLICAN
Theology

BAPTIST
Transactions of the Baptist Historical Society

CONGREGATIONALIST
Transactions of the Congregational Historical Society

ROMAN CATHOLIC
The Downside Review, Benedictine Abbey of Downside, near Bath, England.
Worship, St. John's Benedictine Abbey, Collegeville, Minnesota.

INTERDENOMINATIONAL
Church History, Chicago.
Journal of Ecclesiastical History, London.
Journal of Theological Studies, Oxford.
Studia Liturgica, Rotterdam, Holland.

3. SOURCES IN ENGLISH LITERATURE

Incidental use was made of Sir Thomas Wyatt and Edmund Spenser as writers of psalms more suitable for private meditation than for public praise. Passing mention was also made of Thomas Dekker's remarkable prayers for artisans.

It was noted that the three major competing religious traditions, Catholic, Anglican, and Puritan, had their own poets. In Ben Jonson the Catholics had a major dramatist and poet, and a superb lyricist in the Jesuit, Robert Southwell.

The Anglicans could boast of John Donne, writer of sensual love poems and profoundly spiritual poems, in whom erudition and wit were commingled, and who—as Dean of St. Paul's cathedral, London —was a great pulpit luminary.

The Puritans counted the gentle allegorist, Edmund Spenser in their number, partly because of his championship of Archbishop Grindal (as "Algrind" in *The Shepherd's Calendar*), who refused to put down the Puritan "prophesyings" even at royal command. But their greatest poet, Milton, was not born until the next century.

Both Anglicans and Catholics were intensely critical of the Puritans, for their intensity of religion outraged the moderation and easy-going ways of Erastian Anglicanism. Their Biblicism posed a threat to Catholicism's appeal to tradition. Thus it was that Shakespeare and Ben Jonson helped to create the stereotype of the Puritan as one who was a hypocrite in faith and a blue-nosed busybody who hated others to be happy. Shakespeare's caricature is Malvolio in *Twelfth Night* and Jonson's is Mr. Zeal-of-the-land-Busy in *Bartholomew Fair*.

Both Webster and Donne are referred to for their expression of a Jacobean *fin de siècle* preoccupation with death and disintegration, which finds expression alike in the divine and amatory poems and sermons of Donne and in *The Duchess of Malfi* of Webster. The latter's *White Devil* was used for an interesting quotation which satirized the growing Jacobean use of tomb hatchments in church as status symbols.

4. BOOKS

(Except where otherwise indicated, all books were published in London)

Abraham, Gerald, ed. *The Age of Humanism, 1540-1630*. The New Oxford History of Music, vol. 4. 1968.
Addison, W. *The English Country Parson*. 1948.
———. *Worthy Dr. Fuller*. 1951.
Addleshaw, G.W.O. and F. Etchells. *The Architectural Setting of Anglican Worship*. 1948.
Allen, William. *An Apologie and true declaration of the institution and endevours of the two English Colleges*. 1581.
Anderson, Anthony. *The Sheild of our Safetie set foorth*. 1581.
Anderson, M. D. *The Imagery of British Churches*. 1955.
Andrewes, Bartimaeus. *Certaine verie godlie and profitable Sermons upon the fifth chapiter of the Songe of Solomon*. 1595.
Andrewes, Lancelot. *Ninety-Six Sermons*. 5 vols. Oxford, 1850-1856.
[Anglican homilies]. *Certain Sermons appointed by the Queen's Majesty.* . . . [1574]. Reprinted and ed. by G. E. Corrie. Cambridge, 1850.
[Anglican homilies]. *The Two Books of Homilies*. Ed. J. Griffiths. Oxford, 1859.
Arber, Edward, ed. *A Brief Discussion of the Troubles begun at*

Frankfort in the year 1554 about the Book of Common Prayer and Ceremonies. 1908.

Babbage, Stuart B. *Puritanism and Richard Bancroft.* 1962.

Baillie, Robert. *A Dissuasive from the Errours of the Time.* . . . 1645.

Bancroft, Richard. *Dangerous Positions and Proceedings . . . under pretence of Reformation and for the Presbiteriall Discipline.* 1593.

———. *A Sermon preached at Paules Crosse . . . Anno 1588.*

Barnes, Robert. *Works.* Ed. Foxe, 1573.

Barrow, Henry. *The Writings of Henry Barrow, 1587-1590.* Ed. Leland H. Carlson. 1962.

———. *The Writings of Henry Barrow, 1590-1591.* Ed. Leland H. Carlson. 1964.

Baskerville, G. *English Monks and the Suppression of the Monasteries.* 1937.

Baum, J. G. *Première Liturgie des églises reformées de France de l'an 1533.* Strassburg, 1859.

[Baxter, Richard]. *Reliquiae Baxterianae.* 1696.

Bayne, Peter, ed. *Puritan Documents.* 1862.

Becon, Thomas. *The Catechism of Thomas Becon.* Ed. J. Ayre. Cambridge: Parker Society, 1844.

———. *The Jewel of Joy.* 1560.

Betjeman, John, ed. *Collins Guide to English Parish Churches.* Rev. edn. 1959.

Bennett, H. S. *English Books and Readers, 1475-1557.* Cambridge, 1952.

Benz, Ernst. *The Eastern Orthodox Church, Its Thought and Life.* Tr. by Richard and Clara Winston from *Geist und Leben der Ostkirche.* Hamburg, 1957. New York, 1963.

Birt, H. N. *The Elizabethan Religious Settlement.* 1907.

Bishop, Edmund. *Liturgia Historica, Papers on the Liturgy and Religious Life of the Western Church.* Oxford, 1918.

Black, J. B. *The Reign of Elizabeth, 1558-1603.* 2nd edn., 1959.

Blench, J. W. *Preaching in England in the Late Fifteenth and Sixteenth Centuries.* Oxford, 1964.

Blomfield, Reginald. *A History of Renaissance Architecture in England, 1500-1800.* 2 vols. 1897.

Bolton, Robert. *Works.* 4 vols. 1631-1641.

Bond, F. *An Introduction to English Church Architecture from the 11th to the 16th Century.* 2 vols. 1913.

Bonnet, Jules, ed. *Calvin's Letters.* London and Edinburgh, 2 vols., 1855-1857.

Booty, John E., ed. *An Apology of the Church of England by John Jewel.* Ithaca, N.Y., 1963.

———. *John Jewel as Apologist of the Church of England.* 1963.

Bouyer, Louis. *Life and Liturgy.* South Bend, Ind., 1956.

Boyd, Morrison Comegys. *Elizabethan Music and Musical Criticism.* 2nd edn. Philadelphia, 1962.

Bradford, John. *Works.* Ed. A. Townshend. 2 vols. Cambridge: Parker Society, 1848, 1853.

Bradshaw, William. *English Puritanisme, containing the maine opin-*

ions of the rigidest sort of those that are called Puritanes in the realm of England. 1605.

Bridgett, T. E. and T. F. Knox. *The True Story of the Catholic Hierarchy deposed by Queen Elizabeth with fuller memoirs of its last two survivors.* 1889.

Brightman, F. E. *The English Rite.* 2 vols. 1915.

Brinkworth, E. R., ed. *The Archdeacon's Court, 1584.* 2 vols. Oxfordshire Record Society, 1942, 1946.

Brilioth, Yngve. *Eucharistic Faith and Practice, Evangelical and Catholic.* 1930.

Bromiley, G. W. *Thomas Cranmer, Theologian.* 1956.

Brook, Benjamin. *The Lives of the Puritans.* 3 vols. 1813.

Brook, Stella. *The Language of the Book of Common Prayer.* 1965.

Brook, V.J.K. *A Life of Archbishop Cranmer.* 1962.

Brooks, Peter. *Thomas Cranmer's Doctrine of the Eucharist.* 1965.

Browne, Robert. *A Booke which sheweth the life and manners of all true Christians.* Middelburg, Holland, 1582.

————. *A Treatise of Reformation without tarrying for anie.* Middelburg, 1582.

————. *A True and Short Declaration. . . .* Middelburg, Holland, 1582.

————. *The Writings of Robert Harrison and Robert Browne.* Ed. A. Peel and L. H. Carlson, 1953.

Buchberger, M., ed. *Lexikon für Theologie.* 10 vols. 1930-1938.

Bukofzer, Manfred E. *Music in the Baroque Era from Monteverdi to Bach.* New York, 1947.

Bull, Henry. *Christian Praiers and Holie Meditations.* 1578.

Burrage, Champlin. *The Early English Dissenters in the light of recent research.* 2 vols. Cambridge.

Burton, William. *Davids Evidence.* 1596.

Butler, Charles. *The Principles of Musick.* 1636.

Butterworth, C. C. *The English Primers, 1529-1545.* Philadelphia, 1953.

Buxton, John. *Elizabethan Taste.* 1963.

Calendar of State Papers, Domestic Series, of the Reigns of Edward VI, Mary and Elizabeth. Ed. R. Lemon, 1856.

Calendar of State Papers, Domestic Series, of the Reign of Elizabeth. Ed. M.A.E. Green, 1872.

Calvin, John. *Commentary on I Corinthians. Corpus Reformatorum.* 101 vols. vol. 49.

————. *Letters.* Ed. Jules Bonnet. 2 vols. London and Edinburgh, 1855-1857.

————. *Institutes of the Christian Religion.* Tr. Thomas Norton, 1599.

Cambridge History of English Literature. Ed. A. W. Ward and A. R. Waller. 15 vols. Cambridge, 1907-1932.

[Campion, Edmund]. Richard Simpson, *Life of Edmund Campion.* 1896.

Caraman, Philip, ed. *The Other Face, Catholic Life under Elizabeth I.* 1950.

447

Cardwell, Edward. *Documentary Annals of the Reformed Church of England.* 2 vols. Oxford, 1844.

———. *History of Conferences and other Proceedings connected with the Revision of the Book of Common Prayer, 1558-1690.* Oxford, 1840.

———. *Synodalia.* 2 vols. Oxford, 1842.

Carleton, George. *Life of Bernard Gilpin.* 1629.

Carlson, Leland H., ed. *The Writings of Henry Barrow, 1587-1590.* 1962.

———. *The Writings of Henry Barrow, 1590-1591.* 1964.

———. *The Writings of John Greenwood, 1587-1590.* 1962.

Carpenter, S. C. *The Church in England, 1597-1688.* 1954.

Cartwright, Thomas. *A Commentary on the Epistle of St. Paul written to the Colossians.* 1612.

———. *A Treatise of the Christian Religion.* 1616.

[Catholic Record Society]. *Miscellanea.* vol. 3, no. 3, Catholic Record Society, 1906.

[Catholic Record Society]. *Miscellanea. Bedingfield Papers, etc.* no. 7, 1909.

Chester, Allan G. *Hugh Latimer, Apostle to the English.* Philadelphia, 1954.

Clapham, Henock. *A Chronological Discourse.* 1609.

Clark, Francis. *Eucharistic Sacrifice and the Reformation.* 1960.

Clark, W. K. Lowther, ed. *Liturgy and Worship.* 1932.

Clarke, Basil and John Betjeman. *English Churches.* 1964.

Clay, W. K., ed. *Liturgies and Occasional Forms of Prayer set forth in the Reign of Queen Elizabeth.* Cambridge: Parker Society, 1847.

———. *Private Prayers put forth by authority during the Reign of Queen Elizabeth.* Cambridge: Parker Society, 1851.

Clebsch, William. *England's Earliest Protestants, 1520-1535.* New Haven, 1964.

Coleridge, S. T. *Notes on English Divines. Works,* vol. 4. Ed. H. M. Coleridge, 1853.

Collins, A. Jefferies, ed. *Manuale ad usum percelebris ecclesie Sarisburiensis.* vol. 91. Henry Bradshaw Society, 1960.

Collinson, Patrick. *The Elizabethan Puritan Movement.* Berkeley, Calif., 1967.

Concilium Tridentinum Diariorum, Actorum, Epistolarum, Tractatuum. Nova Collectio edidit Societas Goerresiana. Fribourg and Barcelona, 1919.

Cooper, Thomas. *An Admonition to the People of England.* 1589.

Coulton, G. G. *Art and the Reformation.* Oxford, 1928, 2nd edn., 1953.

Coverdale, Myles. *The Goostly psalmes and spirituall songs drawen out of the holy scripture. . . .* c. 1539.

———. *A faythful and most godly treatise concernynge the sacrament.* Trans. of Calvin's *De Coena Domini.* c. 1549.

———. *Writings and Translations of Myles Coverdale, Bishop of Exeter.* Ed. G. Pearson, Cambridge: Parker Society, 1844.

Cox, J. C. *English Church Fittings, Furniture and Accessories.* 1923.

————— and C. B. Ford. *The Parish Churches of England.* 7th edn., 1954.

Cranmer, Thomas. *Cranmer's Liturgical Projects.* Ed. H. Wickham Legg, vol. 1. Henry Bradshaw Society, 1915.

—————. *Miscellaneous Writings and Letters.* Ed. J. E. Cox, Cambridge: Parker Society, 1846.

—————. *The Remains of Thomas Cranmer.* Ed. H. Jenkyns. 4 vols. Oxford, 1833.

—————. *Writings and Disputations . . . Relative to the Sacrament of the Lord's Supper.* Ed. J. E. Cox, Cambridge: Parker Society, 1844.

Cross, F. L., ed. *The Oxford Dictionary of Church History.* Corrected impression of 1963.

Curtis, Mark H. *Oxford and Cambridge in Transition, 1558-1642.* Oxford, 1959.

Darby, Harold S. *Hugh Latimer.* 1953.

Davies, Horton. *The Worship of the English Puritans.* 1948.

—————. *From Watts and Wesley to Maurice, 1690-1850, From Newman to Martineau, 1850-1900, The Ecumenical Century, 1900-1965. Worship and Theology in England,* vols. 3-5. Princeton, 1961-1965.

Davis, Charles, ed. *English Spiritual Writers.* 1961.

Dawley, Powell M. *John Whitgift and the English Reformation.* New York, 1954.

Daye, John, ed. *A Book of Christian Prayers, collected out of the auncient writers and best learned in our tyme. . . .* 1578.

Dearmer, Percy. *Songs of Praise Discussed.* 1933.

de Félice, Paul. *Les Protestants D'Autrefois.* 4 vols., 2nd edn. Paris, 1897-1902.

Dekker, Thomas. *Foure Birds of Noahs Ark.* 1609.

Dent, Arthur. *The Plain Mans Path-way to Heaven.* 1601.

Dering, Edward. *A Sermon preached before the Quenes maiestie . . . the 25 day of February, Anno 1569.*

—————. *Godly Private Prayers.* 1576.

Devlin, Christopher. *The Life of Robert Southwell, Poet and Martyr.* 1956.

D'Ewes, Simonds, ed. *Journals of all the Parliaments during the reign of Queen Elizabeth.* 1862.

Dickens, A. G. *The English Reformation.* 1964.

—————. *Lollards and Protestants in the Diocese of York, 1509-1558.* 1959.

Dickinson, F. H., ed. *Missale ad usum insignis et praeclarae ecclesiae Sarum.* Burntisland, 1861-1883.

Dictionary of National Biography. Ed. Leslie Stephen and Sidney Lee. 63 vols. 1885-1900.

Dix, Gregory. *The Shape of the Liturgy.* 1945.

Dixon, R. W. *History of the Church of England.* 6 vols. 1891-1902.

Dod, John and Robert Cleaver. *Ten Sermons . . . of the Lords Supper.* 1609.

Donaldson, Gordon. *The Making of the Scottish Prayer Book of 1637.* Edinburgh, 1954.

Donne, John. *Essays in Divinity*. Ed. E. M. Simpson. Oxford, 1952.

Douen, O. *Clement Marot et le psautier huguenot*. 2 vols. Paris, 1878-1879.

Douglas, Winfred. *Church Music in History and Practice*. Rev. and ed. L. Ellinwood, 1963.

Doumergue, E. *Jean Calvin*. 7 vols. Lausanne, 1899-1927.

Dowden, John. *Further Studies in the Prayer Book*. 1908.

————. *The Workmanship of the Prayer Book in its Literary and Liturgical Aspects*. 1899.

Downame, John. *The Christian Warfare*. 1604.

Drysdale, A. H. *History of the Presbyterians in England, their Rise, Decline and Revival*. 1889.

Dugmore, C. W. *Eucharistic Doctrine in England from Hooker to Waterland*. 1942.

————. *The Mass and the English Reformers*. 1958.

Dumoutet, E. *Le désir de voir l'Hostie*. Paris, 1926.

Egli, E. and G. Finster, eds. *Huldreich Zwinglis Sämtliche Werke*. Corpus Reformatorum, 13 vols. Leipzig, 1905-1935.

Elton, G. R. "The Reformation in England." In The New Cambridge Modern History, vol. 2. 1958.

————. *England Under the Tudors*. 1955.

————. *The Tudor Revolution in Government*. 1953.

Erichson, A. *Die Calvinische und die Altstrassburgische Gottesdienst ordnung*. Strassburg, 1894.

Eusden, John D., ed. *The Marrow of Theology, William Ames, 1576-1663*. Boston and Philadelphia, 1968.

Farner, Oskar. *Hulrych Zwingli*. 3 vols. Zurich, 1943-1954.

Fellowes, Edmund H. *English Cathedral Music from Edward VI to Edward VII*. 3rd edn., 1946.

Fendt, Leonhard. *Die Lutherische Gottesdienst des 16. Jahrhunderts*. Munich, 1921.

Field, J. E. *English Liturgies of 1549 and 1661*. 1930.

Fisher, John. *The English Works of John Fisher*. Ed. J.E.B. Mayor. 1935.

————. *A Treatise of Prayer and the Fruits and Manner of Prayer*. Paris, 1640.

Fleming, Abraham. *The Diamond of Devotion*. 1581.

Fitzgerald, W., tr. and ed. *A Disputation on Holy Scripture against the Papists, especially Bellarmine and Stapledon*. Cambridge: Parker Society, 1849.

Foxe, John. *Acts and Monuments*. Ed. S. R. Cattley and G. Townsend. 8 vols. 1837-1841.

————. *A Sermon of Christ crucified preached at Paules Crosse the Friday before Easter, commonly called, Goodfryday*. 1570.

Frere, Walter Howard. *A Collection of His Papers on Liturgical and Historical Subjects*. Eds. J. H. Arnold and E.G.P. Wyatt. Alcuin Club Collections, no. 35, 1940.

————. *The English Church in the Reigns of Elizabeth and James*. 1904.

————. *The Marian Reaction in its relation to the English clergy.* 1896.

————. *The Use of Sarum.* 2 vols. Cambridge, 1898-1901.

———— and C. E. Douglas. *Puritan Manifestos.* 1954. 2nd edn. N. Sykes.

———— and W. M. Kennedy. *Visitation Articles and Injunctions.* 3 vols. 1910.

Frith, John. *A booke made by Iohn Frith prisoner in the tower of London.* 1533.

Fuller, Thomas. *The Church History of Britain.* Ed. J. S. Brewer. 6 vols. Oxford, 1845.

————. *The Holy State.* Edn. of 1698.

————. *The Worthies of England.* Ed. J. Freeman, 1952.

G. F. *A Manual of Prayers.* 1583.

Gairdner, James. *The English Church in the Sixteenth Century.* 1902.

Garrett, C. H. *The Marian Exiles.* Cambridge, 1938.

Garside, Charles. *Zwingli and the Arts.* New Haven, Conn., 1966.

Gasquet, F. A. and E. Bishop. *Edward VI and the Book of Common Prayer.* 1891, 3rd edn., 1928.

Gee, Henry. *The Elizabethan Clergy and the Settlement of Religion, 1558-1564.* Oxford, 1898.

————. *The Elizabethan Prayer-Book and Ornaments.* 1920.

———— and W. J. Hardy. *Documents illustrative of English Church History.* 1896 and 1921.

George, Katherine and Charles. *The Protestant Mind of the English Reformation.* Princeton, N.J., 1961.

Gerard, John. *John Gerard. The Autobiography of an Elizabethan.* Tr. from the Latin by Philip Caraman, intro. by Graham Greene. 1951.

Geree, John. *The Character of an old English Puritane or Nonconformist.* 1646.

Gibson, E.C.S., ed. *The First and Second Prayer Books of King Edward VI.* Everyman edn., 1910.

————. *The Thirty-Nine Articles of the Church of England.* 1898.

Gilpin, Bernard. *A Sermon preached in the Court at Greenwitch before King Edward the Sixth, the first Sonday after the Epiphany, Anno Domini 1552.* Reissued 1630.

Glass, Norman. *The Early History of the Independent Church at Rothwell.* 1871.

Greenham, Richard. *Godly instructions for the due examinations and directions of all men.* 1599.

Greenslade, S. L. *The Work of William Tyndale.* 1938.

Greenwood, John. *The Writings of John Greenwood, 1587-1590.* Ed. Leland H. Carlson. 1962.

[Grey Friars]. *Grey Friars of London Chronicle.* Ed. J. G. Nichols. vol. 53. Camden Society, 1852.

Griffiths, John, ed. *The Two Books of Homilies Appointed to be Read in Churches.* Oxford, 1859.

Griffiths, Olive M. *Religion and Learning: A Study of English Presbyterian Thought.* 1935.

Grindal, Edmund. *The Remains of Archbishop Grindal.* Ed. W. Nicholson. Cambridge: Parker Society, 1843.

―――. *The History of the Life and Acts of Edmund Grindal.* Ed. J. Strype. Oxford, 1821.

Grove's Dictionary of Music and Musicians. Ed. H. C. Colles. 3rd edn. New York, 1927.

Hadow, Sir Henry. *English Music.* 1931.

Hall, Peter, ed. *Fragmenta Liturgica.* 7 vols. Bath, 1848.

―――. *Reliquiae Liturgicae.* 5 vols. Bath, 1847.

Haller, William. *Foxe's Book of Martyrs and the Elect Nation.* New York, 1963.

―――. *Liberty and Reformation in the Puritan Revolution.* New York, 1955.

―――. *The Rise of Puritanism, 1570-1643.* New York, 1938.

Harbison, E. Harris. *The Christian Scholar in the Age of the Reformation.* New York, 1956.

Harding, Thomas. *An Answer to Maister Iuelles Chalenge.* Antwerp, 1565.

―――. *A Confutation of a Booke intituled An Apologie of the Church of England.* Antwerp, 1565.

Harrison. *Harrison's Description of England in Shakspere's Youth.* Ed. F. J. Furnivall. 2 vols. 1877.

Harrison, Robert and Robert Browne. *The Writings of Robert Harrison and Robert Browne.* Ed. A. Peel and Leland H. Carlson. 1953.

Hart, A. Tindal. *The Country Clergy in Elizabethan and Stuart Times, 1558-1660.* 1958.

―――. *The Man in the Pew, 1558-1660.* 1966.

Harvey, A. E. *Martin Bucer in England.* Marburg, 1906.

Harvey, John. *Tudor Architecture.* 1949.

Hautecour, Louis. *Histoire de l'architecture classique en France.* vol. 1. Paris, 1943.

Hay, George. *The Architecture of Scottish Post-Reformation Churches, 1560-1843.* Oxford, 1947.

Herr, Alan Fager. *The Elizabethan Sermon, A Survey and Bibliography.* Philadelphia, 1940.

Heylyn, P. *Ecclesia Restaurata, the History of the Reformation of the Church of England.* 1661.

Higham, Florence. *Catholic and Reformed: A study of the Anglican Church, 1559-1662.* 1962.

Hill, Christopher. *Puritanism and Revolution.* 1958.

―――. *Society and Puritanism.* 1964.

Hillerdal, Gunnar. *Reason and Revelation in Richard Hooker.* Lund, 1962.

Hoby, Lady Margaret. *The Diary of Lady Margaret Hoby.* Ed. Dorothy M. Meads. 1933.

Holl, Karl. *The Cultural Significance of the Reformation.* Trans. of *Die Kultur bedeuting der Reformation*, 1911, revised 1948, Tübingen. New York, 1959.

Hooker, Richard. *The Ecclesiastical Polity and other Works of Richard Hooker.* Ed. Benjamin Hanbury. 3 vols. 1830.

Hooper, John. *The Early Writings of John Hooper.* Ed. S. Carr. Cambridge: Parker Society, 1843.
———. *The Later Writings of Bishop Hooper.* Ed. C. Nevinson. Cambridge: Parker Society, 1852.
Hopf, C. *Martin Bucer and the English Reformation.* Oxford, 1946.
Hoskyns, E. *Horae B. V. M. or Sarum and York Primers, with Kindred Books of the Reformed Roman Use.* 1901.
Howell, Wilbur Samuel. *Logic and Rhetoric in England, 1500-1700.* Princeton, N.J., 1956.
Hudson, W. S. *John Ponet, Advocate of Limited Monarchy.* Chicago, 1942.
Hughes, Philip. *The Reformation in England.* 3 vols. 1950-1954.
———. *Rome and the Counter-Reformation in England.* 1944.
Hughes, Philip Edgcumbe. *The Theology of the English Reformers.* 1965.
Hutchinson, F. E. *Cranmer and the English Reformation.* Reissued 1951.
Jackson, S. M. *Huldreich Zwingli.* New York, 1901.
Jacobs, H. E. *The Lutheran Movement in England.* 1892.
Jefferson, H.A.L. *Hymns in Christian Worship.* 1950.
Jenkyns, Henry, ed. *The Remains of Thomas Cranmer.* 4 vols. Oxford, 1833.
Jessopp, Augustus. *One Generation of a Norfolk House, A Contribution to Elizabethan History.* 1879.
Jewel, John. *An Apology of the Church of England.* Ed. J. E. Booty. Ithaca, N.Y., 1963.
———. *The Works of John Jewel, Bishop of Salisbury.* Ed. J. Ayre. 4 vols. Cambridge: Parker Society, 1845-1850.
Jones, Rufus M. *Spiritual Reformers in the Sixteenth and Seventeenth Centuries.* 1914.
Jungmann, Josef A. *Missarum Sollemnia; eine genetische Erklarung der römischen Messe.* 2 vols. Wien, 1948-1949. Trans. into English by Francis A. Brunner as *The Mass of the Roman Rite; Its Origins and Development (Missarum Sollemnia).* 2 vols. 2nd edn. of 1949. New York, 1951-1955.
Kelly, Faye L. *Prayer in Seventeenth-century England.* Gainesville, Fla., 1966.
Kennedy, W. M. *Elizabethan Episcopal Administration.* 3 vols. Alcuin Club Collections, 1924.
Kennedy, W.P.M. *Parish Life under Queen Elizabeth.* 1914.
Kidd, B. J., ed. *Documents illustrative of the Continental Reformation.* Oxford, 1911.
["King's Book," i.e. Henry VIII's]. *A Necessary Doctrine and erudition for any chrysten man, set furth by the Kynges maiesty of Englande &c.* 1534.
Kirk, Kenneth E. *The Study of Theology.* 1939.
Knappen, Marshall M. *Tudor Puritanism; A Chapter in the History of Idealism.* Chicago, 1939.
———. ed. *Two Elizabethan Puritan Diaries by Richard Rogers and Samuel Ward.* 1930.

Knewstub, John. *The Lectures of John Knewstub on the Twentieth chapter of Exodus and certaine other places of Scripture.* 1579.

Knowles, David. *The English Mystical Tradition.* 1961.

Knox, D. B. *The Doctrine of Faith in the Reign of Henry VIII.* 1961.

Knox, John. *Works.* Ed. David Laing. 6 vols. Edinburgh, 1895.

Knox, S. J. *Walter Travers: Paragon of Elizabethan Puritanism.* 1962.

Kramm, H. H. *The Theology of Martin Luther.* 1947.

Lasco, John à. *Works.* Ed. A. Kuyper. 2 vols. Amsterdam, 1866.

Latimer, Hugh. *Sermons and Remains.* Ed. G. E. Corrie. 2 vols. Cambridge: Parker Society, 1844-1845.

Leatherbarrow, J. S. *The Lancashire Elizabethan Recusants.* N.s., vol. 110. Manchester: Chetham Society, 1947.

Legg, J. Wickham, ed. *Breviarium Romanum Quignonianum.* Cambridge, 1888.

———. *The Sarum Missal.* 1916.

———. *The Second Recension of the Quignon Breviary.* 2 vols. Henry Bradshaw Society, 1908-1912.

———. *Some Principles and Services of the Prayer Book historically considered.* 1899.

Le Huray, Peter. *Music and the Reformation in England, 1549-1660.* London and New York, 1967.

Leys, M.D.R. *The Catholics in England: 1559-1829, A Social History.* 1961.

Lister, Joseph. *The Autobiography of Joseph Lister.* Ed. T. Wright, 1842.

Littlehales, Henry, ed. *The Prymer or Lay Folk's Prayer Book.* Early English Text Society, 1897.

Lloyd, Charles. *Formularies of Faith.* Oxford, 1825.

Luther, Martin. *Werke.* 58 vols. Weimar, 1883-1963.

———. *Works* in English translation. Ed. A. Spaeth and H. E. Jacobs. 6 vols. Philadelphia, 1915-1932.

———. *Works* in English translation. Ed. J. Pelikan and H. T. Lehmann. 54 vols. St. Louis and Philadelphia, 1955-1968.

Mackerness, E. D. *A Social History of English Music.* 1964.

Mackie, J. D. *The Early Tudors, 1485-1558.* 1952.

Maclure, Millar. *The St. Paul's Cross Sermons, 1534-1642.* Toronto, 1958.

Magee, Brian. *The English Recusants: A Study of the Post-Reformation Catholic Survival and the Operation of the Recusancy Laws.* 1938.

Mahrenholz, Chrishard. *Das evangelische Kirchengesangbuch. Ein Bericht über seine Vorgeschichte.* Kassel and Basle, 1950.

———. *Luther und die Kirchenmusik.* Kassel, 1937.

Manningham, John. *The Diary of John Manningham.* Ed. John Bruce. Camden Society, 1868.

Marbeck, John. *The Book of Common Praier Noted.* 1550.

Marchant, Ronald A. *The Puritans and the Church Courts in the Diocese of York, 1560-1642.* 1960.

Marsden, J. B. *History of the Early Puritans to 1642.* 1850.

Marshall, John S. *Hooker and Anglican Tradition.* Sewanee, Tenn., 1963.
Martz, Louis L. *The Poetry of Meditation.* 1954.
Mather, Cotton. *Magnalia Christi Americana.* 1702.
Mathew, David. *Catholicism in England.* 1936.
Maxwell, William D. *The Book of Common Prayer and the Worship of Non-Anglican Churches.* 1950.
————. *John Knox's Genevan Service Book, 1556.* Edinburgh, 1931.
Mayor, J.E.B., ed. *The English Works of John Fisher.* 2nd edn. Early English Text Society, extra series xxvii, 1935.
McDonnell, Kilian. *John Calvin, the Church, and the Eucharist.* Princeton, N.J., 1967.
McGinn, D. J. *The Admonitions Controversy.* New Brunswick, N.J., 1949.
McGrath, Patrick. *Papists and Puritans Under Elizabeth I.* 1967.
McLelland, J. C. *The Visible Words of God; An Exposition of the Sacramental Theology of Peter Martyr Vermigli, A.D. 1500-1562.* Grand Rapids, Mich., 1957.
McManaway, J. G., G. E. Dawson, and E. E. Willoughby, eds. *Joseph Quincy Adams Memorial Studies.* Washington, D.C., 1948.
McMillan, William. *The Worship of the Scottish Reformed Church, 1550-1638.* 1931.
Mercer, Eric. *English Art, 1553-1625.* Oxford History of English Art, vol. 7. Oxford, 1962.
Meyer, Arnold Oskar. *England and the Catholic Church under Queen Elizabeth.* Trans. from the German by J. R. McKee. 1916.
Meyer, Carl S., ed. *Cranmer's Selected Writings.* 1961.
————. *Elizabeth I and the Religious Settlement of 1559.* 1960.
Michell, G. A. *Landmarks in Liturgy.* 1961.
Micklem, Nathaniel, ed. *Christian Worship.* Oxford, 1936.
Miller, Perry. *The New England Mind: The Seventeenth Century.* New York, 1939.
Miller, Perry and T. H. Johnson. *The Puritans.* New York, 1938.
Moorman, J.R.H. *History of the Church in England.* 1953.
More, Sir Thomas. *Letters.* Ed. E. E. Rogers. Princeton, N.J., 1947.
————. *The Worke of Sir Thomas More Knyght.* Ed. William Rastell, 1557; also a modern facsimile ed. W. E. Campbell, 1931.
Morison, Stanley. *English Prayer Books, An Introduction to the Literature of Christian Public Worship.* Cambridge, 1943.
Morley, Thomas. *A Plaine and Easie Introduction to Practicall Musicke* [1597]. Ed. R. A. Harman. 1952.
Morris, Christopher. *Political Thought in England: Tyndale to Hooker.* 1953.
Muller, J. A. *Stephen Gardiner and the Tudor Reaction.* New York, 1926.
Neal, Daniel. *The History of the Puritans.* 5 vols. 1822.
Neale, J. E. *Elizabeth I and Her Parliaments, 1559-1581.* 2 vols. 1953.
————. *Elizabeth I and Her Parliaments, 1581-1601.* 1958.
————. *The Elizabethan House of Commons.* 1950.

Neale, J. E. *England's Elizabeth*. 1958.

Nettl, Paul. *Luther and Music*. Philadelphia, 1948.

New, John H. A. *Anglican and Puritan, the Basis of their Opposition, 1558-1640*. Palo Alto, Calif., 1964.

New Catholic Encyclopedia. 15 vols. New York, 1967.

New Schaff-Herzog Encyclopedia of Religious Knowledge. 2 supplementary vols. Ed. Lefferts A. Loetscher. Grand Rapids, Mich., 1955.

Newman, John Henry. *Parochial and Plain Sermons*. 8 vols. 1868.

————. *Sermons preached on Various Occasions*. 1857.

Nichols, J. G., ed. *Chronicle of the Grey Friars of London*. 1852.

Nichols, James Hastings. *Corporate Worship in the Reformed Tradition*. Philadelphia, 1968.

Nicholson, William, ed. *The Remains of Edmund Grindal*. Cambridge: Parker Society, 1843.

Norden, John. *The Pensive Mans Practise*. 1584.

Nowell, A. *A True report of the Disputation or rather private conference etc.* [concerning Edmund Campion]. 1583.

Nuttall, Geoffrey F. *The Holy Spirit in Puritan Faith and Experience*. Oxford, 1946.

————. *Visible Saints: The Congregational Way, 1640-1660*. Oxford, 1957.

Nuttall, Geoffrey F. and Owen Chadwick. *From Uniformity to Unity, 1662-1962*. 1962.

Oman, Charles. *English Church Plate, 1597-1830*. 1957.

Original Letters relative to the English Reformation. Ed. Hastings Robinson. 2 vols. Cambridge: Parker Society, 1846-1847.

Owen, John. *Works*. Ed. W. H. Goold. 16 vols. Edinburgh, 1850-1853.

Owst, G. R. *Literature and Pulpit in Mediaeval England*. 2nd edn. Oxford, 1961.

————. *Preaching in Mediaeval England*. Cambridge, 1926.

Oxley, J. E. *The Reformation in Essex to the Death of Mary*. Manchester, 1965.

Ozinga, Murk D. *De Protestantische Kerkengebouw in Nederland*. Amsterdam, 1929.

Paget, Eusebius. *A godly Sermon preached at Detford*. 1586.

Parker, Matthew. *The Life and Acts of Matthew Parker*. Ed. J. Strype. 3 vols. Oxford, 1821.

————. *Correspondence from 1535 to 1575*. Ed. J. Bruce and T. T. Perowne. Cambridge: Parker Society, 1853.

Parker, T.H.L., ed. *The English Reformers*. 1966.

————. *The Oracles of God: An Introduction to the Preaching of John Calvin*. 1947.

Parker, T. M. *The English Reformation to 1558*. 1950.

Patrick, Millar. *Four Centuries of Scottish Psalmody*. London and New York, 1949.

Pauck, Wilhelm. *Das Reich Gottes Auf Erden*. Berlin and Leipzig, 1928.

BIBLIOGRAPHY

Peacham, Henry. *The Complete Gentleman, The Truth of our Times, and The Art of Living in London.* Ed. Virgil B. Heltzel. Ithaca, N.Y., 1962.

Pearson, A. F. Scott. *Church and State: Political Aspects of Sixteenth Century Puritanism.* Cambridge, 1928.

―――――. *Thomas Cartwright and Elizabethan Puritanism.* Cambridge, 1925.

Peel, Albert. *The First Congregational Churches.* Cambridge, 1920.

―――――, ed. *The Notebook of John Penry.* Camden Society, 1944.

―――――. *The Seconde Parte of a Register.* vol. 1. Cambridge, 1915.

―――――and Leland H. Carlson, eds. *Cartwrightiana.* 1951.

―――――. *The Writings of Robert Harrison and Robert Browne.* 1953.

Peers, E. Allison. *Spanish Mysticism.* 1924.

―――――. *Studies in the Spanish Mystics.* 2 vols. 1927.

Pelikan, Jaroslav J. "Luther and the Liturgy." In *More About Luther.* Martin Luther Lectures. vol. 2. Decorah, Iowa, 1958.

Perkins, William. *The Combat between Christ and the Devil.* 2nd edn. 1606.

―――――. *The Foundation of the Christian Religion gathered into six principles.* 1595.

―――――. *A Golden Chaine. . . .* 1600.

―――――. *A Salve for a Sicke Man.* 1595.

―――――. *The whole treatise of the cases of conscience.* 1604.

―――――. *The Workes of that Famous and Worthy Minister of Christ in the University of Cambridge, M. William Perkins.* 3 vols. Cambridge, 1613.

Persons, Robert. *A brief discourse containing certain reasons Why Catholikes refuse to goe to Church.* Douai, 1601.

―――――. *Letters and Memorials of Father Robert Persons, S.J.* Ed. L. Hicks. no. 37, Catholic Record Society, 1942.

Pettit, Norman. *The Heart Prepared: Grace and Conversion in Puritan Spiritual Life.* New Haven, Conn. and London, 1966.

Pevsner, N. *The Englishness of English Art.* 1956.

Pidoux, Pierre. *Le Psautier huguenot du 16ᵉ siècle.* 2 vols. Kassel and Basle, 1962.

Phillips, C. Henry. *The Singing Church.* 1945.

Phillips, C. S. *Hymnody Past and Present.* London and New York, 1937.

Pierce, William. *An Historical Introduction to the Marprelate Tracts.* 1908.

Pierce, William, ed. *The Marprelate Tracts, 1588, 1589.* 1911.

Playfere, Thomas. *The Whole Sermons of that Eloquent Divine of Famous Memory: Thomas Playfere, Doctor of Divinitie.* 1625.

Pocock, Nicholas, ed. *Troubles connected with the Prayer Book of 1549.* Camden Society, N.S., xxxvii, 1884.

Pollard, A. F. *The History of England from the Accession of Edward VI to the Death of Elizabeth (1547-1603).* 1910.

―――――. *Thomas Cranmer and the English Reformation, 1489-1556.* 1904.

Pollard, Arthur. *English Sermons.* 1963.

457

Pollen, J. H. *The English Catholics in the Reign of Queen Elizabeth.* 1920.

Porter, H. C. *Reformation and Reaction in Tudor Cambridge.* Cambridge, 1958.

Potter, G. R. and E. M. Simpson, eds. *The Sermons of John Donne.* 10 vols. Berkeley, Calif., 1953-1962.

Powicke, F. M. *The Reformation in England.* 1941.

Pratt, Waldo S. *The Music of the French Psalter of 1562: A Historical Survey and Analysis.* New York, 1939.

Primus, John H. *The Vestments Controversy.* Kampen, 1960.

Proctor, F. and W. H. Frere. *A New History of the Book of Common Prayer.* 1920.

Prothero, R. E. *The Psalms in Human Life.* 1903.

Pruett, Gordon E. "Thomas Cranmer and the Eucharistic Controversy in the Reformation." Unpublished Ph.D. thesis. Princeton University, 1968.

Purvis, J. S., ed. *Tudor Parish Documents of the Diocese of York.* Cambridge, 1948.

Pützer, Fritz. *Prediger des englischen Barok.* Bonn, 1929.

[Quinones, Francisco]. *The Second Recension of the Quignon Breviary.* Ed. J. W. Legg. 2 vols. Henry Bradshaw Society, 1908-1912.

Ratcliff, Edward C. *The booke of common prayer of the Churche of England: its making and revisions MDXLIX—MDCLIX set forth in eighty illustrations with Introduction and Notes.* 1949.

Ratramnus of Corbie. *Opera.* In *Patrologia Latina.* Ed. J. P. Migne. Paris, 1852. vol. 121.

Ravenscroft, Thomas. *A Brief Discourse. . . .* 1614.

——. *The Whole Book of Psalms.* 1621.

Read, Conyers. *The Tudors.* 1936.

Reed, Luther D. *The Lutheran Liturgy.* Philadelphia, 1947.

Richardson, Cyril C. *Zwingli and Cranmer on the Eucharist: Cranmer dixit et contradixit.* Evanston, Ill., 1949.

Richter, A. L. *Die evangelischen Kirchenordnungen d. sechszehnten Jahrhunderts.* 2 vols. Weimar, 1846.

Ridley, Jasper. *Nicholas Ridley: A Biography.* 1957.

——. *Thomas Cranmer.* 1962.

Ridley, Nicholas. *A Brief Declaration of the Lord's Supper.* Ed. H.G.C. Moule, 1895.

——. *Works.* Ed. H. Christmas. Cambridge: Parker Society, 1841.

Robinson, H., ed. *Original Letters Relative to the English Reformation.* 2 vols. Cambridge: Parker Society, 1846-1847.

——. *Zurich Letters.* 2 vols. Cambridge: Parker Society, 1842-1845.

Robinson, John. *A Iustification of Separation from the Church of England.* 1610.

——. *A Treatise of the Lawfulness of Hearing of Ministers.* 1634.

——. *Works.* Ed. Robert Ashton. 3 vols. 1852.

Rogers, Richard. *Seven Treatises . . . leading and guiding to true happiness both in this life and in the life to come.* 1603.

Rowse, A. L. *The England of Elizabeth.* 1953.

——. *Tudor Cornwall.* 1941.

Rupp, E. Gordon. *Six Makers of English Religion, 1500-1700.* 1957.
————. *Studies in the Making of the English Protestant Tradition.* Cambridge, 1949.
Rye, W. B. *England as seen by Foreigners.* 1865.
Sanders, Nicholas. *The Rise and Growth of the Anglican Schism* [1585]. Ed. D. Lewis. 1877.
Sandys, Edwin. *The Sermons of Edwin Sandys, D.D.* Ed. John Ayre. 2 vols. Cambridge: Parker Society, 1841-1842.
Schmidt, Hermanus A. P. *Introductio in Liturgiam Occidentalem.* Rome, Freiburg, Barcelona, 1960.
Schmidt-Clausing, F. *Zwingli als Liturgiker.* Göttingen, 1952.
Schnapper, E. B. *The British Union-Catalogue of Early Music printed before the year 1801.* 2 vols. 1957.
Schnorrenberg, John M. "Early Anglican Architecture, 1558-1662: Its Theological Implications and Its Relation to the Continental Background." Unpublished Ph.D. thesis, Princeton University, 1964.
Scholes, Percy A., ed. *The Oxford Companion to Music.* 1938.
————. *The Puritans and Music in England and New England.* London, 1934, 2nd edn., New York, 1966.
Schuler, M. and J. Schulthess, eds. *Zwinglii Opera.* vol. 4. Zurich, 1841.
Seebohm, F. *The Oxford Reformers, Colet, Erasmus, and More.* 2nd edn., 1869.
Shepherd, Massey H., Jr. *The Oxford American Prayer Book Commentary.* New York, 1951.
Shirley, J. *Richard Hooker and Contemporary Political Ideas.* 1949.
Simpson, Alan. *Puritanism in Old and New England.* Chicago, 1955.
Simpson, E. M., ed. John Donne. *Essays in Divinity.* Oxford, 1952.
Sisson, C. J. *The Judicious Marriage of Mr Hooker and the Birth of the Laws of Ecclesiastical Polity.* 1940.
Smallman, Basil. *The Background of Passion Music.* 1957.
Smend, Julius. *Die evangelischen deutschen Messen bis zu Luthers Deutscher Messe.* Göttingen, 1896.
Smith, Henry. *The Sermons of Maister Henrie Smith gathered into one Volume.* 1593.
Smith, H. Maynard. *Henry VIII and the Reformation.* 1948.
Smith, L. C. *Tudor Prelates and Politics, 1536-1558.* Princeton, N.J., 1953.
Smithen, F. J. *Continental Protestantism and the English Reformation.* 1928.
Smyth, C. H. *Cranmer and the Reformation under Edward VI.* Cambridge, 1926.
Smyth, John. *The Differences of the Churches of the Separation.* 1608.
————. *The Works of John Smyth.* Ed. W. T. Whitley. Cambridge, 1915.
Southern, A. C., ed. *An Elizabethan Recusant House, comprising the Life of the Lady Magdalen, Viscountess Montague (1538-1608).* 1950.
————. *Elizabethan Recusant Prose, a Historical and Critical Account*

of the Books of the Catholic Refugees printed and published abroad and at secret presses in England. 1955.

Southgate, W. M. *John Jewel and the Problem of Doctrinal Authority.* Cambridge, Mass., 1962.

Southwell, Robert. Christopher Devlin. *The Life of Robert Southwell Poet and Martyr.* 1956.

————. *The Book of Robert Southwell.* Ed. Isabel M. Hood. Oxford, 1926.

Staley, Vernon, ed. *First Prayer Book of Edward VI, 1549.* 1903.

Stanford, Charles V. and Cecil Forsyth. *The History of Music.* 1916.

Stephens, W.R.W. and W. Hunt. *History of the English Church.* 9 vols. 1899-1912.

Sternhold, Thomas and John Hopkins. *The Whole Booke of Psalmes collected into Englysh metre.* 1562.

Stevens, Denis W. *Tudor Church Music.* New York, 1955, London, 1961.

Stevens, J. *Music and Poetry in the Early Tudor Court.* 1961.

Stockwood, John. *A Sermon Preached at Paules Crosse on Barthelmew Day.* 1578.

Stow, John. *Survey of London.* Ed. C. L. Kingsford. 2 vols. Oxford, 1908.

Stranks, C. J. *Anglican Devotion: Studies in the Spiritual Life of the Church of England between the Reformation and the Oxford Movement.* 1961.

Strype, J. *Annals of the Reformation.* 4 vols. Oxford, 1824.

————. *Ecclesiastical Memorials.* 3 vols. Oxford, 1824.

————. *The History of the Life and the Acts of Edmund Grindal.* Oxford, 1821.

————. *The Life and Acts of John Whitgift.* 3 vols. Oxford, 1827.

————. *The Life and Acts of Matthew Parker.* 3 vols. Oxford, 1821.

————. *Memorials of Thomas Cranmer.* 3 vols. in 4. Oxford, 1848-1854.

Sturge, Charles. *Cuthbert Tunstall, Churchman, Scholar, Statesman, Administrator.* 1938.

Summerson, John. *Architecture in Britain, 1530 to 1830.* The Pelican History of Art. Harmondsworth and Baltimore, 1954.

Surtz, Edward. *The Works and Days of John Fisher.* Cambridge, Mass., 1967.

Sutton, Christopher. *Disce Mori. Learne to Die.* 1600.

Swete, H. B. *Church Services and Service-Books before the Reformation.* Rev. edn. by A. J. MacLean. 1930.

Sykes, Norman. *The Church of England and Non-Episcopal Churches in the Sixteenth and Seventeenth Centuries.* 1948.

Terry, Richard R. *Calvin's First Psalter* [1539]. Facsimile edn. 1932.

Thompson, Bard, ed. and selection. *Liturgies of the Western Church.* Cleveland and New York, 1962.

Theisen, Reinhold. *Mass Liturgy and the Council of Trent.* Collegeville, Minn., 1965.

T'Hooft, W. A. Visser. *Rembrandt and the Gospel.* New York and Cleveland, 1960.

Thornton, L. S. *Richard Hooker, A Study of His Theology.* 1924.

Tillyard, E.M.W. *The Elizabethan World Picture.* 1945.

Tomlinson, J. T. *The Prayer Book, Articles, and Homilies.* 1897.

Topsell, Edward. *Times Lamentation, or an exposition of the prophet Joel in sundry sermons or meditations.* 1599.

————. *The Reward of Religion. Delivered in sundry Lectures upon the Booke of Ruth, wherein the godly may see their dayly both inward and outward trials, with the presence of God to assist them, and his mercies to recompense them.* 1597.

Torbet, R. G. *The History of the Baptists.* Philadelphia, 1950.

Torrance, Thomas F. *Kingdom and Church.* Edinburgh, 1956.

Travers, Walter. *A full and plaine declaration of Ecclesiastical Discipline.* Zurich, 1574.

Trevelyan, George M. *England under the Stuarts.* 1933.

Trimble, W. R. *The Catholic Laity in Elizabethan England.* Cambridge, Mass., 1954.

Tunstall, Cuthbert. *De Corporis et Sanguinis Domini in Eucharistia.* Paris, 1554.

Tyndale, William. *Doctrinal Treatises.* Ed. H. Walter. Cambridge: Parker Society, 1848.

————. *The Whole Workes of W. Tyndall, John Frith, and Doct. Barnes, three worthy Martyrs, and principall teachers of the Churche of England.* . . . 1572.

Tyrer, Ralph. *Five godlie sermons.* . . . 1602.

Underhill, E. B., ed. *Fenstanton, Warboys and Hexham Records.* 1847.

Usher, R. G. *The Reconstruction of the English Church.* 2 vols. New York, 1910.

————. *The Presbyterian Movement . . . illustrated by the minute book of the Dedham Classis.* Camden Society, 1905.

Vajta, Vilmos. *Luther on Worship.* Philadelphia, 1958.

Van de Poll, G. J. *Martin Bucer's Liturgical Ideas.* Assen, 1954.

Vaux, Laurence. *A Catechisme or Christian Doctrine.* Manchester: Chetham Society, 1885.

Verstegan, Richard. *The Letters and Despatches of Richard Verstegan.* Ed. A. G. Petti. no. 52. Catholic Record Society, 1959.

Wakefield, Gordon A. *Puritan Devotion, Its place in the Development of Christian Piety.* 1957.

Walker, Williston. *The Creeds and Platforms of Congregationalism.* New York, 1893.

Wallace, Ronald S. *Calvin's Doctrine of the Word and Sacraments.* Edinburgh, 1953.

Walsh, James, ed. *Pre-Reformation English Spirituality.* New York, 1965.

Walton, Izaak. *Life of Hooker.* In Hooker, *Works.* Ed. J. Keble. Oxford, 1836.

Waterhouse, Ellis. *Painting in Britain, 1530-1790.* Pelican History of Art. London and Baltimore, 1953.

Welsby, P. A. *George Abbott: The unwanted Archbishop.* 1962.

Wendel, François. *Calvin.* 1963.

West, W.M.S. "John Hooper and the Origins of Puritanism." A sum-

461

mary of an unpublished doctoral dissertation submitted to the Faculty of Theology of the University of Zurich; printed for private circulation in 1955; which appeared in *The Baptist Quarterly*, London, Oct. 1954 to April 1955.

White, Helen C. *The Tudor Books of Private Devotion.* Madison, Wis., 1951.

———. *Tudor Books of Saints and Martyrs.* Madison, Wis., 1963.

White, Thomas. *A Sermon preached at Pawles Crosse on Sunday the thirde of November, 1577.* 1577.

White, W. D. "John Henry Newman, Anglican Preacher." Unpublished Ph.D. thesis, Princeton University, 1968.

Whitgift, John. *Works.* Ed. J. Ayre. 3 vols. Cambridge: Parker Society, 1851-1853.

———. *The Life and Acts of John Whitgift.* Ed. J. Strype. 3 vols. Oxford, 1822.

Wilkins, D. *Concilia Magnae Britanniae. . . .* 4 vols. 1737.

Williams, George Hunston. *The Radical Reformation.* Philadelphia, 1962.

Wilson, H. A., ed. *The Order of Communion, 1548.* Henry Bradshaw Society. vol. 34. 1908.

Woodfill, Walter. *Musicians in English Society from Elizabeth to Charles I.* Princeton, N.J., 1953.

Woodforde, Christopher. *English Stained and Painted Glass.* Oxford, 1954.

Woodhouse, H. F. *The Doctrine of the Church in Anglican Theology, 1547-1603.* 1945.

Wright, Louis B. *Middle-Class Culture in Elizabethan England.* Chapel Hill, N.C., 1935.

Wriothesley, Charles. *A Chronicle of England.* Ed. W. D. Hamilton. 2 vols. N.S., XI, parts i and ii. 1875-1877.

Young, Percy. *A History of British Music.* 1967.

[Zurich]. *The Zurich Letters.* Ed. H. Robinson. Cambridge: Parker Society, 1846.

———. *The Zurich Letters, 1558-1579*, ed. H. Robinson, Parker Society, Cambridge, 1842.

INDEX

I. Index of Persons

Abbott, Abp. George, 243, 309
Abraham, Gerald, 378n., 383n.
Ab Ulmis, John, 199
Aburne, Robert, 332
Adams, J. Q., 410n.
Addleshaw, G.W.O., 363n., 364, 365, 368n.
Agatha, St., 21, 372
Agnes, St., 21
A Lasco, John, 79n., 106f., 110-112n., 191, 192n., 201, 204f., 299n., 350
Alexander, 107
Allde, Elizabeth, 418
Allde, Richard, 418
Allen, Cardinal, 152, 156
Allison, 390
Ames, William, 52, 58n., 63n., 257, 305n., 326
Amos, 239
Anderson, Anthony, 311, 316n., 317
Andrewes, Bartimaeus, 310, 317
Andrewes, Bp. Lancelot, xix, 67, 218, 227, 232f., 236, 242, 253, 294, 304, 307, 310, 314, 324, 368n., 428
Annas, 146
Anselm, St., 437
Anthony, St., 21
Apolline, St., 21
Aquinas, St. Thomas, 65n.
Arber, Edward, 40n., 211n.
Arnold, J. H., 168n., 207n.
Ashton, Robert, 269n.
Athanasius, St., 147
Augustine, St., of Canterbury, 173
Augustine, St., of Hippo, 25, 55, 91, 96, 102, 114, 244, 253, 422, 427, 431
Aylmer, Bp. John, 46, 290n., 300, 328
Ayre, J., 102n., 121n., 233n., 235n., 236n., 382n.

Babbage, Stuart B., 295n.
Bailey, D. S., 418n.
Baillie, Robert, 63n., 432n.
Baker, Augustine, 409n.
Bancroft, Abp., 40, 51n., 60n., 232, 241, 277f., 295
Barbara, St., 21
Barnard, 401
Barley, W., 401
Barnes, Robert, 19n., 80n., 237
Baro, Peter, 59n.
Barrett, William, 59n.
Barrow, Henry, 60, 257f., 327-329, 333f.

Baum, J. G., 79n., 209n.
Baxter, Richard, 275n., 406
Bayne, Peter, 54n.
Baynes, Paul, 257
Beaumont, Robert, 239, 241n.
Becon, Thomas, 17, 20, 21n., 102f., 229, 234f., 243, 381f., 412, 418, 425, 427
Becket, Abp. Thomas à, 134
Bede, the Venerable, 46, 409n.
Belloc, Hilaire, 6
Bembo, Cardinal, 137
Bennett, H. S., 407
Bentham, Bp. Thomas, 47n.
Benz, Ernst, 374n.
Berengarius, 91
Bernini, 147
Betjeman, John, 359, 368n., 369n.
Beza, Theodore, 244, 380
Bibliander, 83n.
Bilney, T., 237, 248
Bishop, Edmund, 147, 168n., 169n., 187n., 188n., 191, 202n.
Blake, William, 388
Blaurer, Thomas, 195
Blench, J. W., 235, 237, 239n., 250, 304n., 305n., 309, 318
Blunt, 341
Boleyn, Queen Anne, 151
Bolton, Robert, 65
Bonaventura, St., 423
Bonhoeffer, Dietrich, 18n.
Bonner, Bp. Edmund, 103, 134f., 187, 199f., 229, 350
Bonnet, Jules, 55n.
Booty, John E., 10n., 21n., 29n.
Boreman, Christopher, 153n.
Bourgeois, Louis, 380, 385, 403
Bouyer, Louis, 149n.
Boyce, 401
Boyd, Morrison C., 384n., 399n.
Bradford, John, xix, 34n., 228, 243, 303, 416, 418, 419, 420-422, 436, 439
Bradshaw, William, 70, 295
Brady, Nicholas, 387
Brewster, William, 341
Bridges, John, 33n., 240n.
Bridgett, T. E., 138n.
Brightman, F. E., 167n., 168n., 170n., 179n., 180n., 194n., 208n., 381n.
Brigit, St., 132
Brilioth, Abp. Yngve, 79n.
Bromiley, G. W., 117n.
Brook, Benjamin, 290n.

II. Index of Places (and Churches)

(Churches are assumed to be Anglican except when otherwise indicated)

III. Index of Topics

476

INDEX